Remote Sensing Data Compression

Remote Sensing Data Compression

Editors

Vladimir Lukin
Benoit Vozel
Joan Serra-Sagristà

MDPI • Basel • Beijing • Wuhan • Barcelona • Belgrade • Manchester • Tokyo • Cluj • Tianjin

Editors
Vladimir Lukin
National Aerospace University
Ukraine

Benoit Vozel
University of Rennes 1
France

Joan Serra-Sagristà
Universitat Autonoma Barcelona
Catalonia

Editorial Office
MDPI
St. Alban-Anlage 66
4052 Basel, Switzerland

This is a reprint of articles from the Special Issue published online in the open access journal *Remote Sensing* (ISSN 2072-4292) (available at: http://www.mdpi.com).

For citation purposes, cite each article independently as indicated on the article page online and as indicated below:

LastName, A.A.; LastName, B.B.; LastName, C.C. Article Title. *Journal Name* **Year**, *Volume Number*, Page Range.

ISBN 978-3-0365-2303-3 (Hbk)
ISBN 978-3-0365-2304-0 (PDF)

Contents

About the Editors

Vladimir V. Lukin received a master's degree (Hons.) in radio engineering from the Kharkov Aviation Institute (now National Aerospace University), Kharkiv, Ukraine, in 1983. He presented the thesis of Candidate of Technical Science in 1988 and Doctor of Technical Science in 2002 in digital signal processing (DSP) for remote sensing. He is currently the Chairman of and a Professor with the Department of Information–Communication Technologies at the National Aerospace University. Since 1995, he has worked in cooperation with colleagues at Tampere University of Technology, Tampere, Finland. His research interests include digital signal/image processing, remote sensing data processing, image filtering, and compression.

Benoit Vozel holds a State Engineering degree, an M.Sc. degree in control and computer science, and a Ph.D. degree in applied sciences from Ecole Centrale de Nantes, Nantes, France, which were awarded in 1991 and 1994, respectively. Since 1995, he has been with the Engineering School of Applied Sciences and Technology, University of Rennes 1, Lannion, France, where he is a member of the SAR and Hyperspectral Multimodal Imaging and Signal Processing, Electromagnetic Modeling (SHINE) Team at the Institute of Electronics and Telecommunications of Rennes (IETR UMR CNRS 6164). His research interests include the blind estimation of noise characteristics, blind image restoration, image registration, image compression, adaptive multichannel signal and image processing, machine and deep learning, and remote sensing data processing.

Joan Serra-Sagristà received a Ph.D. degree in computer science from the Universitat Autònoma de Barcelona (UAB), Barcelona, Spain, in 1999. He is currently a Full Professor with the Department of Information and Communications Engineering, UAB. From September 1997 to December 1998, he worked at the University of Bonn, Germany, funded by the DAAD. He has coauthored more than 150 publications. Dr. Serra-Sagristà was a recipient of the Spanish Intensification Young Investigator Award in 2006. He serves or has served as an Associate Editor for the *IEEE Transactions on Geoscience and Remote Sensing* and the *IEEE Transactions on Image Processing,* and as Committee Chair for the Data Compression Conference.

remote sensing

Editorial

Editorial to Special Issue "Remote Sensing Data Compression"

Benoit Vozel [1], Vladimir Lukin [2,*] and Joan Serra-Sagristà [3]

[1] Engineering School of Applied Sciences and Technology, University of Rennes 1, 22305 Lannion, France; benoit.vozel@univ-rennes1.fr
[2] Department of Information and Communication Technologies, National Aerospace University, 61070 Kharkov, Ukraine
[3] Department of Information and Communications Engineering, Universitat Autònoma de Barcelona, Cerdanyola del Valles, 08290 Barcelona, Catalonia, Spain; joan.serra@uab.cat
*   Correspondence: v.lukin@khai.edu

Abstract: A huge amount of remote sensing data is acquired each day, which is transferred to image processing centers and/or to customers. Due to different limitations, compression has to be applied on-board and/or on-the-ground. This Special Issue collects 15 papers dealing with remote sensing data compression, introducing solutions for both lossless and lossy compression, analyzing the impact of compression on different processes, investigating the suitability of neural networks for compression, and researching on low complexity hardware and software approaches to deliver competitive coding performance.

Keywords: remote sensing data compression; lossless compression; lossy compression; compression impact; neural networks; computational complexity

Citation: Vozel, B.; Lukin, V.; Serra-Sagristà, J. Editorial to Special Issue "Remote Sensing Data Compression". *Remote Sens.* **2021**, *13*, 3727. https://doi.org/10.3390/rs13183727

Received: 28 August 2021
Accepted: 16 September 2021
Published: 17 September 2021

Publisher's Note: MDPI stays neutral with regard to jurisdictional claims in published maps and institutional affiliations.

1. Overview of the Issue: Remote Sensing Data Compression

Announcing this Special Issue, the following was considered. First, a huge amount of data is acquired each day by different remote sensing systems and these data must be transferred to image processing centers, stored, and delivered to customers. Due to various restrictions, data compression is strongly desired or necessary. Second, there is a wide diversity of methods that can be used, requirements to compression and their priority, types, and properties of images to be processed, practical implementation aspects, etc. Our intention was to collect papers focused on advances in lossless and lossy compression, multi- and hyperspectral image compression, radar image compression, applications of remote sensing data compression; compression standards, practical implementation of image compression techniques, data compression hardware and software, impact of data compression on solving classification and identifications tasks.

As a result of our work as guest editors, 21 submissions have been received from which 6 have been rejected. The accepted publications cover a wide variety of questions.

Five papers relate to lossless and near lossless methods with application to multi- and hyperspectral data.

The paper "Analysis of Variable-Length Codes for Integer Encoding in Hyperspectral Data Compression with the k^2-Raster Compact Data Structure" by Chow, K., Tzamarias, D.E.O., Hernández-Cabronero, M., Blanes, I., Serra-Sagristà, J. [1], examines various variable-length encoders that provide integer encoding to hyperspectral scene data within a k^2-raster compact data structure. This structure leads to a compression ratio similar to that produced by some classical compression techniques while also providing direct access for query to its data elements without requiring any decompression. The selection of the integer encoder is critical for a competitive performance (compression ratio and access time). Different integer encoders, such as Rice, Simple9, Simple16, PForDelta codes, and DACs are investigated. Further, a method to determine an appropriate k value for building a competitive k^2-raster compact data structure is discussed.

The paper "Using Predictive and Differential Methods with K²-Raster Compact Data Structure for Hyperspectral Image Lossless Compression" by Chow, K., Tzamarias, D.E.O., Blanes, I., Serra-Sagristà, J. [2] extends the previous paper by proposing a lossless coder for real-time processing and compression of hyperspectral images. After applying either a predictor or a differential encoder by exploiting the close similarity between neighboring bands, it uses the k^2-raster compact data structure to further reduce the bit rate. Experiments show that using k^2-raster alone already achieves much lower rates (up to 55% reduction), and with preprocessing, the rates are further reduced (up to 64%). Finally, experimental results show that prediction produces higher rates reduction than differential encoding.

The paper "Compression of Hyperspectral Scenes through Integer-to-Integer Spectral Graph Transforms" by Tzamarias, D.E.O., Chow, K., Blanes, I., Serra-Sagristà, J. [3] exploits the redundancies found between consecutive spectral components and within components themselves through the use of spectral graph filterbanks, such as the GraphBior transform. Such graph based filterbank transforms do not yield integer coefficients, making them appropriate only for lossy image compression schemes. In the paper, two integer-to-integer transforms are introduced for the purpose of the lossless compression, and its performance as a spatial transform is assessed.

The paper "Performance Impact of Parameter Tuning on the CCSDS-123.0-B-2 Low-Complexity Lossless and Near-Lossless Multispectral and Hyperspectral Image Compression Standard" by Blanes, I., Kiely, A., Hernández-Cabronero, M., Serra-Sagristà, J. [4] studies the performance impact related to different parameter choices for the new CCSDS-123.0-B-2 Low-Complexity Lossless and Near-Lossless Multispectral and Hyperspectral Image Compression standard. This standard supersedes CCSDS-123.0-B-1 and extends it by incorporating a new near-lossless compression capability, as well as other new features. Experimental results include data from 16 different instruments with varying detector types, image dimensions, number of spectral bands, bit depth, level of noise, level of calibration, and other image characteristics. Guidelines are provided on how to adjust the parameters in relation to their coding performance impact.

The paper "High-Performance Lossless Compression of Hyperspectral Remote Sensing Scenes Based on Spectral Decorrelation" by Hernández-Cabronero, M., Portell, J., Blanes, I., Serra-Sagristà, J. [5] investigates the most advantageous compression–complexity trade-off in hyperspectral image (HSI) compression. Compression performance and execution time results are obtained for a set of 47 HSI scenes produced by 14 different sensors in real remote sensing missions. Assuming only a limited amount of energy is available, obtained data suggest that the FAPEC algorithm yields the best trade-off. When compared to the CCSDS 123.0-B-2 standard, FAPEC is 5.0 times faster and its compressed data rates are on average within 16% of the CCSDS standard. In scenarios where energy constraints can be relaxed, CCSDS 123.0-B-2 yields the best average compression results.

There are three papers that deal with compression impact on classification and segmentation.

The paper "Lossy Compression of Multichannel Remote Sensing Images with Quality Control" by Lukin, V., Vasilyeva, I., Krivenko, S., Li F., Abramov, S., Rubel, O., Vozel, B., Chehdi, K., and Egiazarian, K. [6] studies a dependence between classification accuracy of maximum likelihood and neural network classifiers that have been applied to three-channel images and visual quality of compressed images. It is demonstrated that the classification accuracy starts to decrease faster when image quality due to increasing compression ratio reaches a distortion visibility threshold. In addition, classification accuracy depends essentially on the training methodology: training carried out for lossy compressed data seems preferable over training on undistorted data.

The paper "Lossy Compression of Multispectral Satellite Images with Application to Crop Thematic Mapping: A HEVC Comparative Study" by Miloš Radosavljević, Branko Brkljač, Predrag Lugonja, Vladimir Crnojević, Željen Trpovski, Zixiang Xiong, and Dejan Vukobratović [7] provides a comprehensive analysis of the HEVC still-image intra coding while applied to multispectral satellite images acquired by the Landsat-8's OLI and Sentinel-2's multispectral instrument. In the specific context of a crop classification application, HEVC's intra coding is shown to maintain approximately the same classification accuracy of a random forest pixel-based classifier for CR up to 150:1 while it is only up to 70:1 with JPEG 2000. It also achieves a better trade-off between compression gain and image quality, both visually and in terms of PSNR values, as compared to standard JPEG 2000.

The paper "Spectral Imagery Tensor Decomposition for Semantic Segmentation of Remote Sensing Data through Fully Convolutional Networks" by Josué López, Deni Torres, Stewart Santos, and Clement Atzberger [8] suggests a whole framework, called HOOI-FCN, to perform compression of an input RS-image followed by semantic classification. A Tucker decomposition-based mapping with preservation of the features of the classes of interest transforms the input third-order tensor into a core tensor with the same spatial resolution but a lower number of bands. This is done by means of the higher order orthogonal iteration (HOOI) algorithm. A fully convolutional network (FCN) is next considered to classify the core tensor at the pixel level. HOOI-FCN is shown to achieve high performance metrics competitive with some RS-multispectral images semantic segmentation state-of-the-art methods on Sentinel-2 images while significantly reducing computational complexity and processing time.

Lossy compression of data for unmanned aerial vehicle (UAV) is also attracting interest, as witnessed by two papers.

The paper "Real-Time Hyperspectral Data Transmission for UAV-Based Acquisition Platforms" by Melián, J.M., Jiménez, A., Díaz, M., Morales, A., Horstrand, P., Guerra, R., López, S., and López, J.F. [9] focuses on rapid compression of hyperspectral data prior to their transmission using two different NVIDIA boards—the Jetson Xavier NX and the Jetson Nano. The obtained results show the possibility of achieving real-time performance if the Jetson Xavier NX is used for all the configurations that could be applied in real missions.

The paper "FPGA-Based On-Board Hyperspectral Imaging Compression: Benchmarking Performance and Energy Efficiency against GPU Implementations" by Julián Caba, María Díaz, Jesús Barba, Raúl Guerra, Jose A. de la Torre, and Sebastián López [10] proposes a highly optimized implementation using integer arithmetic of the lossy compression algorithm for hyperspectral image systems. The purpose is to comply with the high-frame requirement imposed by a UAV-based sensing platform. The single-core version of the FPGA-based solution onto a heterogeneous Zynq-7000 SoC chip allows setting the baseline scenario of compressed hyperspectral image blocks at 200 FPS, using a small number of FPGA resources and low power consumption. Moreover, it is shown that a multi-core FPGA-based version can reach the same level of performance as the most efficient embedded GPU-based implementations.

Two papers concern neural network use in image compression.

The paper "Reduced-Complexity End-to-End Variational Autoencoder for on Board Satellite Image Compression" by de Oliveira, V.A., Chabert, M., Oberlin, T., Poulliat, C., Bruno, M., Latry, C., Carlavan, M., Henrot, S., Falzon, F., and Camarero, R. [11] concentrates on design of a complexity-reduced variational autoencoder with attempt to meet the constraints dealing with board satellite compression, time, and memory complexities. A simplified entropy model that preserves the adaptability to the input image is proposed. It is shown that the designed complexity-reduced autoencoder provides a better rate-distortion trade-off compared to the Consultative Committee for Space Data Systems standard CCSDS 122.0-B.

The paper "Spectral–Spatial Feature Partitioned Extraction Based on CNN for Multispectral Image Compression" by Kong, F., Hu, K., Li, Y., Li, D., and Zhao, S. [12] puts forward a multispectral image compression framework that is fully based on a convolutional neural network (CNN). The novelty concerns the feature extraction module, divided into spectral and spatial parallel parts. The testing is carried out for datasets acquired by Landsat-8 and WorldView-3 satellites. A better performance is shown in comparison to JPEG 2000, 3D-SPIHT and ResConv, another CNN-based algorithm.

Compression acceleration is also addressed in the paper "An FPGA Accelerator for Real-Time Lossy Compression of Hyperspectral Images" by Daniel Báscones, Carlos González, and Daniel Mozos [13], which derives a custom FPGA implementation of the costliest part (tier 1 coder within JPEG2000) of the JYPEC algorithm, a lossy hyperspectral compression algorithm that combines PCA and JPEG2000. The main goal is to accelerate it significantly to bring the full algorithm execution time down as much as possible and even below the real-time constraint. An average acceleration of 3.6 is verified when the FPGA accelerated algorithm is applied to six hyperspectral images, four from the Spectrir library and two from the CCSDS 123 dataset.

Finally, two papers are devoted to compressive sensing.

The paper "Compressive Underwater Sonar Imaging with Synthetic Aperture Processing" by Choi, H., Yang, H., and Seong, W. [14] deals with synthetic aperture sonars (SAS) in underwater imaging. SAS imaging algorithms that employ compressive sensing are considered and verified through simulation and experimental data. A better resolution compared to the ω-k algorithms with minimal performance degradation by side lobes are demonstrated in simulations. Experimental data show the method's robustness with respect to sensor loss.

The paper "A Task-Driven Invertible Projection Matrix Learning Algorithm for Hyperspectral Compressed Sensing" by Dai, S., Liu, W., Wang, Z., and Li, K. [15] proposes a hyperspectral compressed sensing algorithm with low complexity and strong real-time performance. It is based on a task-driven invertible projection matrix learning algorithm aiming at solving the problems of long time-consuming and low reconstruction accuracy of compressed sensing-based reconstruction algorithms. Experiments performed on Indian Pine AVIRIS hyperspectral dataset show that, compared with the traditional compressed sensing algorithm, the proposed compressed sensing algorithm has higher reconstruction accuracy and improved real-time performance by more than a hundred times, thus leading to great application prospects in the field of hyperspectral image compression.

2. Conclusions

From the fifteen papers published in this Special Issue, we can state that compression of remote sensing data is today an active research area with new direction appearing and attracting attention of scientists engaged in the design of methods to meet customers' expectations. We hope that the readers will enjoy this Special Issue.

Funding: This research was partially funded by the Spanish Ministry of Economy and Competitiveness and the European Regional Development Fund under Grant RTI2018-095287-B-I00 and by the Catalan Government under Grant 2017SGR-463.

Institutional Review Board Statement: Not applicable.

Informed Consent Statement: Not applicable.

Acknowledgments: We are grateful for having the opportunity to lead this Special Issue. We would like to thank the journal editorial team and reviewers for conducting the review process and all the authors for contributing their work.

Conflicts of Interest: The authors declare no conflict of interest.

References

1. Chow, K.; Tzamarias, D.E.O.; Hernández-Cabronero, M.; Blanes, I.; Serra-Sagristà, J. Analysis of Variable-Length Codes for Integer Encoding in Hyperspectral Data Compression with the k^2-Raster Compact Data Structure. *Remote Sens.* **2020**, *12*, 1983. [CrossRef]
2. Chow, K.; Tzamarias, D.E.O.; Blanes, I.; Serra-Sagristà, J. Using Predictive and Differential Methods with K^2-Raster Compact Data Structure for Hyperspectral Image Lossless Compression. *Remote Sens.* **2019**, *11*, 2461. [CrossRef]
3. Tzamarias, D.E.O.; Chow, K.; Blanes, I.; Serra-Sagristà, J. Compression of Hyperspectral Scenes through Integer-to-Integer Spectral Graph Transforms. *Remote Sens.* **2019**, *11*, 2290. [CrossRef]

4. Blanes, I.; Kiely, A.; Hernández-Cabronero, M.; Serra-Sagristà, J. Performance Impact of Parameter Tuning on the CCSDS-123.0-B-2 Low-Complexity Lossless and Near-Lossless Multispectral and Hyperspectral Image Compression Standard. *Remote Sens.* **2019**, *11*, 1390. [CrossRef]
5. Hernández-Cabronero, M.; Portell, J.; Blanes, I.; Serra-Sagristà, J. High-Performance Lossless Compression of Hyperspectral Remote Sensing Scenes Based on Spectral Decorrelation. *Remote Sens.* **2020**, *12*, 2955. [CrossRef]
6. Lukin, V.; Vasilyeva, I.; Krivenko, S.; Li, F.; Abramov, S.; Rubel, O.; Vozel, B.; Chehdi, K.; Egiazarian, K. Lossy Compression of Multichannel Remote Sensing Images with Quality Control. *Remote Sens.* **2020**, *12*, 3840. [CrossRef]
7. Radosavljević, M.; Brkljač, B.; Lugonja, P.; Crnojević, V.; Trpovski, Ž.; Xiong, Z.; Vukobratović, D. Lossy Compression of Multispectral Satellite Images with Application to Crop Thematic Mapping: A HEVC Comparative Study. *Remote Sens.* **2020**, *12*, 1590. [CrossRef]
8. López, J.; Torres, D.; Santos, S.; Atzberger, C. Spectral Imagery Tensor Decomposition for Semantic Segmentation of Remote Sensing Data through Fully Convolutional Networks. *Remote Sens.* **2020**, *12*, 517. [CrossRef]
9. Melián, J.M.; Jiménez, A.; Díaz, M.; Morales, A.; Horstrand, P.; Guerra, R.; López, S.; López, J.F. Real-Time Hyperspectral Data Transmission for UAV-Based Acquisition Platforms. *Remote Sens.* **2021**, *13*, 850. [CrossRef]
10. Caba, J.; Díaz, M.; Barba, J.; Guerra, R.; López, J.A. FPGA-Based On-Board Hyperspectral Imaging Compression: Benchmarking Performance and Energy Efficiency against GPU Implementations. *Remote Sens.* **2020**, *12*, 3741. [CrossRef]
11. Alves de Oliveira, V.; Chabert, M.; Oberlin, T.; Poulliat, C.; Bruno, M.; Latry, C.; Carlavan, M.; Henrot, S.; Falzon, F.; Camarero, R. Reduced-Complexity End-to-End Variational Autoencoder for on Board Satellite Image Compression. *Remote Sens.* **2021**, *13*, 447. [CrossRef]
12. Kong, F.; Hu, K.; Li, Y.; Li, D.; Zhao, S. Spectral–Spatial Feature Partitioned Extraction Based on CNN for Multispectral Image Compression. *Remote Sens.* **2021**, *13*, 9. [CrossRef]
13. Báscones, D.; González, C.; Mozos, D. An FPGA Accelerator for Real-Time Lossy Compression of Hyperspectral Images. *Remote Sens.* **2020**, *12*, 2563. [CrossRef]
14. Choi, H.-M.; Yang, H.-S.; Seong, W.-J. Compressive Underwater Sonar Imaging with Synthetic Aperture Processing. *Remote Sens.* **2021**, *13*, 1924. [CrossRef]
15. Dai, S.; Liu, W.; Wang, Z.; Li, K. A Task-Driven Invertible Projection Matrix Learning Algorithm for Hyperspectral Compressed Sensing. *Remote Sens.* **2021**, *13*, 295. [CrossRef]

remote sensing

MDPI

Article

Compressive Underwater Sonar Imaging with Synthetic Aperture Processing

Ha-min Choi [1], Hae-sang Yang [1] and Woo-jae Seong [1,2,*]

[1] Department of Naval Architecture and Ocean Engineering, Seoul National University, Seoul 08826, Korea; 0595gkals@snu.ac.kr (H.-m.C.); coupon3@snu.ac.kr (H.-s.Y.)
[2] Research Institute of Marine Systems Engineering, Seoul National University, Seoul 08826, Korea
* Correspondence: wseong@snu.ac.kr; Tel.: +82-2-880-7332

Abstract: Synthetic aperture sonar (SAS) is a technique that acquires an underwater image by synthesizing the signal received by the sonar as it moves. By forming a synthetic aperture, the sonar overcomes physical limitations and shows superior resolution when compared with use of a side-scan sonar, which is another technique for obtaining underwater images. Conventional SAS algorithms require a high concentration of sampling in the time and space domains according to Nyquist theory. Because conventional SAS algorithms go through matched filtering, side lobes are generated, resulting in deterioration of imaging performance. To overcome the shortcomings of conventional SAS algorithms, such as the low imaging performance and the requirement for high-level sampling, this paper proposes SAS algorithms applying compressive sensing (CS). SAS imaging algorithms applying CS were formulated for a single sensor and uniform line array and were verified through simulation and experimental data. The simulation showed better resolution than the ω-k algorithms, one of the representative conventional SAS algorithms, with minimal performance degradation by side lobes. The experimental data confirmed that the proposed method is superior and robust with respect to sensor loss.

Keywords: compressive sensing; synthetic aperture sonar; underwater sonar imaging

Citation: Choi, H.-m.; Yang, H.-s.; Seong, W.-j. Compressive Underwater Sonar Imaging with Synthetic Aperture Processing. *Remote Sens.* **2021**, *13*, 1924. https://doi.org/10.3390/rs13101924

Academic Editor: Vladimir Lukin

Received: 31 March 2021
Accepted: 12 May 2021
Published: 14 May 2021

Publisher's Note: MDPI stays neutral with regard to jurisdictional claims in published maps and institutional affiliations.

1. Introduction

Synthetic aperture sonar (SAS) is a technique that repeatedly transmits and receives pulses while the sonar is moving and coherently synthesizes the received signals to obtain a high-resolution image [1–3]. By synthesizing multiple pings, it is possible to achieve the effect of a sonar operating with an aperture larger than the actual sonar aperture, therefore called a "synthetic aperture" sonar. Compared to other techniques for obtaining underwater images, such as side-scan sonar, SAS obtains images with a high resolution [4] and is used in various fields such as crude oil exploration, geological exploration, and for military purposes such as in mine detection [5,6].

Conventional SAS methods reconstruct the image by performing Fourier transform and matched filtering in the slant-range or in the azimuth domain. Conventional SAS methods are classified into back-projection in the spatial–temporal domain [1], correlation in the spatial–temporal domain [7], range-Doppler in the range-Doppler domain [8], wavenumber in the wavenumber domain [9,10], and chirp-scaling in the wavenumber domain [11], contingent on whether Fourier transform is performed in the slant-range or in the azimuth domain. To form a synthetic aperture requires sampling following Nyquist theory in the time domain according to the traditional signal processing technique, and dense sampling in the spatial domain alongside the sonar movement is also required. Because conventional SAS signal processing techniques pass through a matched filter, side lobes are generated, resulting in the deterioration of image reconstruction performance [12,13].

This paper proposes SAS imaging algorithms that apply the compressive sensing (CS) framework to compensate for disadvantages associated with conventional SAS signal

7

processing techniques. CS is a technique used to restore a sparse signal from a small number of measurements [14]. Under suitable conditions, CS obtains a better resolution than conventional signal processing and suppresses side lobes. In addition, admittedly under suitable conditions, CS obtains exact solutions even at a low sampling level, which violates Nyquist theory. In recent years, studies related to the CS framework have been conducted in various fields, such as medical imaging fields—including MRI and ultrasound imaging—and sensor networks [15–20]. In the estimation of the direction-of-arrival (DOA), which is a classical source-localizing method, CS is applied to increase the number of employed sensors or observations, thereby enhancing localization performance [21,22]. Additionally, studies have been conducted on the application of sparse reconstruction to synthetic aperture radar (SAR) [23–28] and SAS [29,30]. Many studies have applied CS to SAR, but as far as could be determined, few studies have been conducted on SAS, especially underwater.

In [29], a method that applies CS to SAS imaging is presented, which estimates the reflectivity function in the area of interest using all the given data. When some of the data were excluded, results were good, showing that large data reduction is possible. However, this study does not show results for actual underwater acoustic conditions; it only shows results for the ultrasonic synthetic aperture laboratory system using assumed point targets. The fact that the laboratory results have not been verified against actual underwater experimental data has significant consequences. Targets in real underwater environments are generally not point targets but targets with continuous characteristics. Therefore, if the reflectivity function is estimated instantly, as in the method proposed in [29], the shape of the target will not be properly revealed and only segments with high reflectivity will be obtained. In [30], CS was applied to SAS to obtain a parsimonious representation to utilize aspect- or frequency-specific information. By way of simulation and employing real underwater experimental data, it was verified that the strategy using aspect- or frequency-specific information was effective. However, the method proposed in [30], which uses an iterative method called the alternating direction method of multipliers (ADMM), has a limitation in that it is unstable because convergence is highly dependent on the regularization parameter. Therefore, we propose a stable method that does not use an iterative method and that expresses the characteristics of a real underwater target by dividing data and repeatedly estimating the reflectivity function of the area of interest.

This study offers three main contributions: First, the proposed method (called the CS-SAS algorithm for simplicity) that shows better reconstruction performance compared to a conventional SAS algorithm. The proposed algorithms are SAS algorithms formulated from the perspective of the CS framework and in accordance with the CS characteristics. Less aliasing occurs and high-resolution results are obtained. Section 3 explains that the proposed method outperforms one of the conventional SAS algorithms, the ω-k algorithm. Second, the proposed algorithms are more robust in the absence of sensor data. Because conventional SAS algorithms require sampling frequency according to Nyquist theory in the time and spatial domains, conventional SAS algorithms are not resistant to sensor failure or data loss in the sonar system. Conversely, the proposed algorithms are robust, as indicated later on in this paper. Third, few studies apply CS to SAS underwater and, therefore, this study is meaningful in that it applies simulation data and actual underwater experimental data.

The remainder of this paper is organized into four sections. In Section 2, the geometry of the SAS system and the ω-k algorithm—which is a representative conventional SAS algorithm—are described. In Section 3, the basic theory of CS is described, and SAS algorithms using CS are proposed. In Section 4, the performance of the proposed method is verified by comparing the results of applying the CS-SAS and the ω-k algorithms to the simulation and experimental data. Finally, conclusions are presented in Section 5.

In the following, vectors are represented by bold lowercase letters, and matrices are represented by bold capital letters. The l_p-norm of a vector $\mathbf{x} \in \mathbb{C}^N$ is defined as

$\|\mathbf{x}\|_p = \left(\sum_{i=1}^{N} |x_i|^p \right)^{1/p}$. The imaginary unit $\sqrt{-1}$ is denoted as j. The operators $^T, ^*$ denote the transpose and conjugate operators, respectively.

2. Basic SAS Imaging

2.1. SAS Geometry

The general geometry of SAS is depicted in Figure 1. The direction of y along which the sonar moves is defined as azimuth or the cross-range axis, the direction x perpendicular to y is the slant-range, the size of the sonar is D, and the synthetic aperture and the distance the sonar travels is $2L$. The basic concept of SAS is that the sonar moves from $-L$ to L along the cross-range axis, and then transmits and receives signals to synthesize the received signals scattered back from the targets to obtain underwater images.

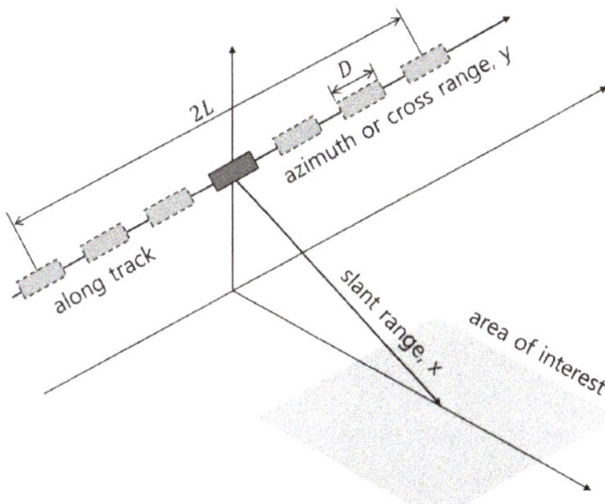

Figure 1. SAS geometry.

One of the standards for evaluating the level of performance of an SAS system is the resolution of the reconstructed images. The slant-range resolution Δ_x of the SAS was determined by the matched filtering process as follows:

$$\Delta_x = \frac{c\pi}{\omega_{bd}} = \frac{c}{2f_{bd}},\tag{1}$$

where ω_{bd} is the bandwidth of the transmitted signal, and c is the sound speed.

The cross-range resolution Δ_y can be derived simply through the following development. Consider a single frequency signal as the simplest signal model. Then, the -3 dB main lobe width of an l-length uniform line array is λ/l, the main lobe width θ_{SAS} of the synthetic aperture is $\lambda/2L$. For the slant-range of the target R, $L \simeq R\theta$ and $\theta = \lambda/D$. Therefore, the cross-range resolution Δ_y can be expressed as

$$\Delta_y = RD/(2R) = D/2.\tag{2}$$

For convenience, a single frequency signal was assumed, but it is known that cross-range resolution $\Delta_y = D/2$ even if a signal with bandwidth like LFM signal is used [28,31]. From Equation (2), it is clear that the cross-range resolution of the synthetic aperture sonar is independent of range and frequency. This independence makes it possible to reconstruct high-resolution images over a long range [32,33].

2.2. Wavenumber Domain Algorithm (ω k Algorithm)

The wavenumber domain algorithm, a representative conventional SAS algorithm, was used as a baseline method. The wavenumber domain algorithm is a method that obtains an image using 2-D Fourier transform of recorded signals and is also called the ω-k algorithm because signal processing is performed in the frequency domain [1,8]. The signal of duration T_p transmitted from the sonar, is denoted by $p(t)$, and the signal received by the sonar at position u is denoted by $s(t,u)$. The signal $s(t,u)$ can be expressed as the sum of the signals scattered by the N-targets as follows:

$$s(t,u) = \sum_{n=1}^{N} \sigma_n p\left(t - \frac{2\sqrt{x_n^2 + (y_n - u)^2}}{c}\right)\left(0 < t - \frac{2\sqrt{x_n^2 + (y_n - u)^2}}{c} < T_p\right) \quad (3)$$

where σ_n is the target strength of the n-th target, and x_n and y_n are the slant-range and cross-range of the n-th target, respectively. The 2-D Fourier transform of Equation (3) using the stationary-phase principle [34] gives

$$S(\omega, k_u) = \sum_{n=1}^{N} P(\omega)\sigma_n \exp\left(-j\sqrt{4k^2 - k_u^2}\, x_n - jk_u y_n\right). \quad (4)$$

where ω is the angular frequency, k is the wavenumber, and k_u is the azimuth wavenumber. By changing the coordinates as shown in Equation (5), it can be arranged as in Equation (6).

$$k_x = \sqrt{4k^2 - k_u^2}, \ k_y = k_u. \quad (5)$$

$$S(\omega, k_u) = \sum_{n=1}^{N} P(\omega)\sigma_n \exp\left(-jk_x x_n - jk_y y_n\right). \quad (6)$$

The function of the distribution of the targets is expressed in Equation (7), and its Fourier transform is expressed as Equation (8).

$$f_0 = \sum_{n=1}^{N} \sigma_n \delta(x - x_n, y - y_n). \quad (7)$$

$$F_0(k_x, k_y) = \sum_{n=1}^{N} \sigma_n \exp\left(-jk_x x_n - jk_y y_n\right). \quad (8)$$

Equation (9) can be obtained by combining Equations (6) and (8).

$$S(\omega, k_u) = P(\omega) F_0. \quad (9)$$

Therefore, the distribution of the targets can be estimated through the following relationship:

$$F(k_x, k_y) = P^*(\omega) S(\omega, k_u) = |P(\omega)|^2 \sum_{n=1}^{N} \sigma_n \exp\left(-jk_x x_n - jk_y y_n\right). \quad (10)$$

The mapping—Equation (5) from (ω, k_u) to (k_x, k_y), called Stolt mapping [35]—involves interpolation from the data. The interpolation process can be made more rational through the following spatial shift formulation:

$$F_b(k_x, k_y) = F(k_x, k_y) \exp\left(jk_x X_c + jk_y Y_c\right) \quad (11)$$

where subscript b indicates baseband conversion, and X_c, Y_c are the centers of the area of interest in the slant-range and cross-range, respectively. In Equation (11), the exponential term performs a spatial shift function, which performs a function similar to the carrier removal process in spectrum demodulation. It enables interpolation in a slowly varying region while moving the entire swath down to the origin of the spatial coordinates.

Because the received signal is scattered by a target with a target strength of 1 in the center of the area of interest, its Fourier transform can be expressed as Equations (12) and (13), respectively, and Equation (11) can be summarized as follows:

$$s_0(t, u) = p\left(t - \frac{2\sqrt{X_c^2 + (Y_c - u)^2}}{c}\right),$$ (12)

$$S_0(\omega, k_u) = P(\omega)\exp(-jk_x X_c - jk_y Y_c),$$ (13)

$$F_b(k_x, k_y) = S(\omega, k_u)S_0^*(\omega, k_u).$$ (14)

The flow chart of the ω-k algorithm is shown in Figure 2 [1].

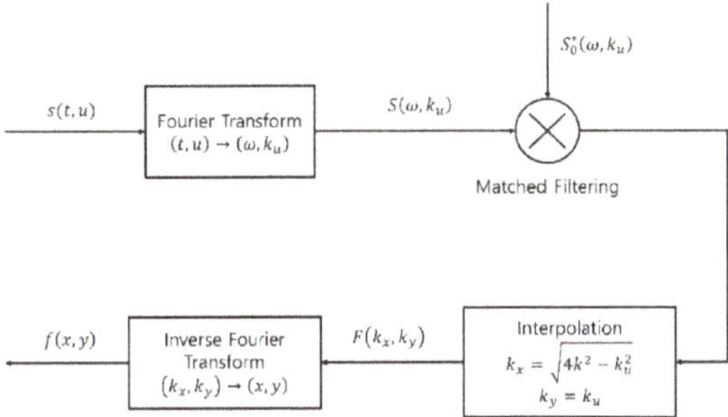

Figure 2. Flow chart of the ω-k algorithm.

3. SAS Algorithm with CS Framework (CS-SAS)

3.1. Compressive Sensing

Compressive sensing is a method or framework for solving linear problems, such as $y = Ax$ for sparse signal x [36]. $x \in \mathbb{C}^N$ is an unknown signal vector that we want to reconstruct. The unknown signal vector x is a *k-sparse* vector, where x is *k-sparse*, meaning that $\|x\|_0 = k$, that is, x has only k non-zero elements. $y \in \mathbb{C}^M$ is a measurement vector consisting of measured values. In many realistic problems, $A \in \mathbb{C}^{M \times N}$—called a sensing matrix—is introduced to represent the problem as a linear relationship, such as $y = Ax$. When the dimension of the measurement vector y is smaller than the dimension of x, that is, $M \ll N$, the $y = Ax$ problem becomes an underdetermined problem and has numerous solutions, making it impossible to specify x. Using the sparse property of x, it is possible to specify a unique and exact solution among countless feasible solutions of the underdetermined problem. The sparsity is imposed by the sparsity constraint l_0-norm. The l_0-norm minimization problem is formulated as follows:

$$\min_{x \in \mathbb{C}^N} \|x\|_0 \text{ subject to } y = Ax.$$ (15)

However, the l_0-norm minimization problem, Equation (15), is an NP-hard problem that is computationally intractable. To deal with the NP-hard problem, various methods have been developed such as l_1-norm relaxation or greedy algorithms represented by orthogonal match-pursuit.

One of the most representative methods for solving the compressive sensing problem is l_1-norm relaxation, which solves the problem by replacing l_0-norm with l_1-norm. The l_0-norm minimization problem, Equation (15), can be relaxed by reformulating as

$$\min_{x \in \mathbb{C}^N} \|x\|_1 \text{ subject to } y = Ax.$$ (16)

In the presence of noise, a sparse solution $\mathbf{x}$ can be obtained by the following equation:

$$\min_{\mathbf{x} \in \mathbb{C}^N} \|\mathbf{x}\|_1 \text{ subject to } \|\mathbf{y} - \mathbf{Ax}\|_2 < \epsilon. \tag{17}$$

Equations (16) and (17) are called basis pursuit (BP) and basis pursuit denoising (BPDN) problems, respectively. The larger the hyperparameter ϵ, the sparser the optimized solution $\mathbf{x}$. Oppositely, the smaller the ϵ, the more optimized solution $\mathbf{x}$ fits the data. Therefore, it is important to assume a suitable hyperparameter. However, finding a suitable hyperparameter is complex and deemed to be outside the scope of this study.

In this study, the SAS image was obtained by solving the BPDN problem using the tool provided by CVX [37].

3.2. CS-SAS Algorithm for Single Sensor

To handle SAS imaging problems from the perspective of compressive sensing, the problem must first be well defined as the $\mathbf{y} = \mathbf{Ax}$ problem. To formulate a compressive algorithm in which a single sensor is in linear motion, the signal reflected by the targets and returned to the single sensor needs to be considered first. When a signal $p(t)$ is transmitted from a single sensor located in $\mathbf{r}_u = [x_0, \, y_0]$ and reflected by N-targets, the received signal $s(t, \mathbf{r}_u)$ can be written as

$$s(t, \mathbf{r}_u) = \sum_{k=1}^{N} \sigma_k p(t - \tau(\mathbf{r}_k, \mathbf{r}_u)), \tag{18}$$

$$\tau(\mathbf{r}_k, \, \mathbf{r}_u) = 2\frac{|\mathbf{r}_k - \mathbf{r}_u|}{c} = 2\frac{\sqrt{(x_k - x_0)^2 + (y_k - y_0)^2}}{c}, \tag{19}$$

where σ_k, $k = 1, \ldots, N$ is the target strength of the k-th target, $\tau(\mathbf{r}_k, \, \mathbf{r}_u)$ is a function representing travel time, x_k is the slant-range of the k-th target, and y_k is the cross-range of the k-th target. When $p(t)$ is a continuous wave (CW) signal with a pulse duration of T_p and carrier frequency of f_c, Equation (19) can be rewritten as Equation (21).

$$p(t) = e^{j\omega t} = e^{j2\pi f_c t} \left(0 \leq t \leq T_p\right), \tag{20}$$

$$s(t, \mathbf{r}_u) = \sum_{k=1}^{N} \sigma_k exp(j\omega(t - \tau(\mathbf{r}_k, \mathbf{r}_u)))\left(0 \leq t - \tau(\mathbf{r}_k, \mathbf{r}_u) \leq T_p\right). \tag{21}$$

The above is a formulation of the signal received at one location, $\mathbf{r}_u$. This can be expanded to the expression for a single sensor.

The operation of a single sensor sonar can be expressed as shown in Figure 3. The total number of pings is N_p, the single sensor—transmitter and receiver—corresponding to the m-th ping, is u_m, and the position of u_m is r_{u_m}, $m = 1, \ldots, N_p$. The center point of the area of interest is (X_c, Y_c), the half-size of the area of interest in the range is X_0, and the half-size of the area of interest in the cross-range is Y_0. By dividing the area of interest into $N_x \times N_y$ grids, assuming that there is a virtual target σ_k at each grid point, the signal $s(t, \mathbf{r}_{u_m})$ received at the m-th ping can be written as follows:

$$s(t, \mathbf{r}_{u_m}) = \sum_{k=1}^{N_x N_y} \sigma_k exp(j\omega(t - \tau(\mathbf{r}_k, \mathbf{r}_u)))\left(0 \leq t - \tau(\mathbf{r}_k, \mathbf{r}_{u_m}) \leq T_p\right). \tag{22}$$

The signal vector $\mathbf{s}_i$, which consists of signals received at a specific time t_i at each position of the sonar, can be written as

$$\mathbf{s}_i = \left[s(t_i, \mathbf{r}_{u_1}), s(t_i, \mathbf{r}_{u_2}), \ldots, s\left(t_i, \mathbf{r}_{u_{N_p}}\right)\right]^T. \tag{23}$$

Combining Equations (22) and (23), s_i can be expressed as the product of the target strength vector σ of the virtual targets and the sensing matrix A_i:

$$s_i = A_i \sigma, \tag{24}$$

$$\sigma = \left[\sigma_1, \ldots, \sigma_{N_x N_y}\right]^T, \tag{25}$$

$$A_i(m,k) = exp(j\omega(t_i - \tau(\mathbf{r}_k, \mathbf{r}_{u_m}))) \quad (0 \leq t_i - \tau(\mathbf{r}_k, \mathbf{r}_{u_m}) \leq T_p), \tag{26}$$

where $A_i(m,k)$ denotes an element corresponding to the m-th row and k-th column of A_i. In the CS system, where the length of $\mathbf{y}$ is M and the length of $\mathbf{x}$ is N and k-*sparse*, $\mathbf{x}$ is successfully recovered when $M \geq O(k \log(N/k))$ measurements are used [36,38]. In the proposed algorithm, if the length N_p of the received signal s_i is too small compared to the length $N_x N_y$ of σ, an accurate solution cannot be obtained. Therefore, in the case where the length of the signal s_i is too small, signals for a total of N_t times are collected to form a long signal vector $\mathbf{s}$ and, similarly, corresponding matrices are collected to form a long sensing matrix A:

$$\mathbf{s} = A\sigma, \tag{27}$$

$$\mathbf{s} = \left[s_i{}^T, s_{i+1}{}^T, \ldots, s_{i+N_t-1}{}^T\right]^T, \tag{28}$$

$$A = \left[A_i{}^T, A_{i+1}{}^T, \ldots, A_{i+N_t-1}{}^T\right]^T, \tag{29}$$

where $\mathbf{s} \in \mathbb{C}^{N_t N_u}$, $A \in \mathbb{C}^{N_t N_u \times N_x N_y}$, and $\sigma \in \mathbb{C}^{N_x N_y}$. σ is estimated by solving the following BP or BPDN problems:

$$\min_{\sigma \in \mathbb{C}^{N_x N_y}} \|\sigma\|_1 \text{ subject to } \mathbf{s} = A\sigma, \tag{30}$$

$$\min_{\sigma \in \mathbb{C}^{N_x N_y}} \|\sigma\|_1 \text{ subject to } \|\mathbf{s} - A\sigma\|_2 < \epsilon. \tag{31}$$

The σ value, determined by applying Equation (30) or (31), is a target strength vector obtained from using only the signals received in consecutive N_t time snapshots from t_i. Therefore, to obtain the target strength vector for all the area of interest, the l_1-norm minimization process must be repeated for all time snapshots corresponding to the area of interest. The final image was compiled by adding all the σ values obtained in each process.

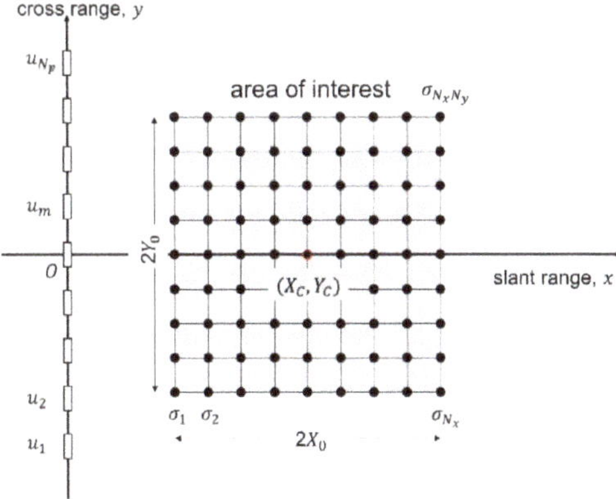

Figure 3. Description for a single sensor sonar operation: u_m is the single sensor corresponding to the m-th ping, N_p is the total number of pings, σ_k is a virtual target, N_x is the number of the grid in the x direction, N_y is the number of the grid in the y direction, (X_c, Y_c) is the center point of the area of interest, X_0 is the half-size of the area of interest in the x direction, Y_0 is the half-size of the area of interest in the y direction.

When obtaining solutions for BP or BPDN, the elements of $\boldsymbol{\sigma}$ with a large l_1-norm of the corresponding sensing matrix column vector tend to have a non-zero value. To eliminate this bias, each column vector a_i of the sensing matrix $\mathbf{A}$ and the received signal $\mathbf{s}$ is normalized by their l_2-norms [14].

$$\mathbf{s}_{\text{normal}} = \frac{\mathbf{s}}{\|\mathbf{s}\|_2}, \tag{32}$$

$$\mathbf{A}_{\text{normal}} = \left[\frac{a_1}{\|a_1\|_2}, \frac{a_2}{\|a_2\|_2}, \dots, \frac{a_{N_x N_y}}{\|a_{N_x N_y}\|_2} \right]^T. \tag{33}$$

Therefore, Equations (30) and (31) are modified, and the final solution is obtained by compensating the l_2-norms to the $\boldsymbol{\sigma}$ estimated from Equation (34) or Equation (35) as follows:

$$\min_{\boldsymbol{\sigma} \in \mathbb{C}^{N_x N_y}} \|\boldsymbol{\sigma}\|_1 \text{ subject to } \mathbf{s}_{\text{normal}} = \mathbf{A}_{\text{normal}} \boldsymbol{\sigma}, \tag{34}$$

$$\min_{\boldsymbol{\sigma} \in \mathbb{C}^{N_x N_y}} \|\boldsymbol{\sigma}\|_1 \text{ subject to } \|\mathbf{s}_{\text{normal}} - \mathbf{A}_{\text{normal}} \boldsymbol{\sigma}\|_2 < \lambda. \tag{35}$$

3.3. CS-SAS Algorithm for Uniform Line Array

The algorithm proposed in Section 3.2 is a method used for a single sensor. However, in many cases, a uniform array sonar consisting of one transmitter and multiple receivers is used, and it needs to be extended to the algorithm for a uniform line array. The algorithm for the uniform linear array introduced in this section is the same as the algorithm for a single sensor, except for the travel time function. The transmitter in the m-th ping is u_{tm}, $m = 1, \dots, N_p$, and its position vector is $\mathbf{r}_{u_{tm}}$. The number of receivers in the physical array is N_u, and the n-th receiver in the physical array is u_n, $n = 1, \dots, N_u$. Therefore, Equations (22) and (23) can be reformulated as follows:

$$s(t, \mathbf{r}_{u_n}, \mathbf{r}_{u_{tm}}) = \sum_{k=1}^{N_x N_y} \sigma_k exp(j\omega(t - \tau(\mathbf{r}_k, \mathbf{r}_{u_n}, \mathbf{r}_{u_{tm}})))(0 \le t - \tau(\mathbf{r}_k, \mathbf{r}_{u_n}, \mathbf{r}_{u_{tm}}) \le T_p), \tag{36}$$

$$\tau(\mathbf{r}_k, \mathbf{r}_{u_n}, \mathbf{r}_{u_{tm}}) = \frac{|\mathbf{r}_k - \mathbf{r}_{u_n}| + |\mathbf{r}_k - \mathbf{r}_{u_{tm}}|}{c}. \tag{37}$$

The signal vector composed of measurements received at a specific time t_i of each receiver in the j-th ping is denoted as $\mathbf{s}_{i,j}$. Therefore, the signal vectors $\mathbf{s}_{i,j}$ can be arranged as

$$\mathbf{s}_{i,j} = \mathbf{A}_{i,j} \boldsymbol{\sigma}, \tag{38}$$

$$\mathbf{s}_{i,j} = \left[s\left(t_i, \mathbf{r}_{u_1}, \mathbf{r}_{u_{tj}}\right), s\left(t_i, \mathbf{r}_{u_2}, \mathbf{r}_{u_{tj}}\right), \dots, s\left(t_i, \mathbf{r}_{u_{N_u}}, \mathbf{r}_{u_{tj}}\right) \right]^T, \tag{39}$$

$$\mathbf{A}_{i,j}(m,k) = exp\left(j\omega\left(t_i - \tau\left(\mathbf{r}_k, \mathbf{r}_{u_m}, \mathbf{r}_{u_{tj}}\right)\right)\right) \quad \left(0 \le t_i - \tau\left(\mathbf{r}_k, \mathbf{r}_{u_m}, \mathbf{r}_{u_{tj}}\right) \le T_p\right). \tag{40}$$

Similar to the previous section, the following formulas are obtained:

$$\mathbf{s} = \mathbf{A}\boldsymbol{\sigma}, \tag{41}$$

$$\mathbf{s} = \left[\mathbf{s}_{i,1}^T, \mathbf{s}_{i,2}^T, \dots, \mathbf{s}_{i,N_p}^T, \mathbf{s}_{i+1,1}^T, \mathbf{s}_{i+1,2}^T, \dots, \mathbf{s}_{i+N_t-1,N_p}^T \right]^T, \tag{42}$$

$$\mathbf{A} = \left[\mathbf{A}_{i,1}^T, \mathbf{A}_{i,2}^T, \dots, \mathbf{A}_{i,N_p}^T, \mathbf{A}_{i+1,1}^T, \mathbf{A}_{i+1,2}^T, \dots, \mathbf{A}_{i+N_t-1,N_p}^T \right]^T. \tag{43}$$

The CS framework can be applied by formulation as above.

4. Results

In this section, the performance of the proposed algorithms is demonstrated by comparing the results of applying the ω-k algorithm and the CS-SAS algorithms to both the simulation and experimental data. In the simulation and in the experiment, the carrier frequencies of the CW signal were 400 and 455 kHz, respectively, whereas the sampling frequencies were 25 or 50 kHz and, therefore, the ω-k algorithm included the baseband process.

The following shows that the CS-SAS algorithms exhibit superior performance in terms of resolution and noise robustness and indicates how to become robust when combating conditions where sensors are not working or data are lost.

4.1. Simulation Results

4.1.1. Simulation Results for Single Sensor

The ω-k and CS-SAS algorithms were compared for various cases. For the single-sensor SAS, five cases were simulated. The basic simulation environment was a single-sensor sonar operated at 0.02 m intervals from -5 to 5 in the cross-range axis, that is, $N_p = 501$. The signal $p(t)$ was a CW signal with carrier frequency $f_c = 400$ kHz and pulse duration $T_p = 0.1$ ms. The center point of the area of interest was $[X_c, Y_c] = [250, 0]$, the half-size of the area of interest in slant-range was $X_0 = 0.5$ m, the half-size of the area of interest in cross-range was $Y_0 = 1$ m, the sampling frequency was $f_s = 50$ kHz, and the area of interest consisted of $N_X = 101$ and $N_Y = 101$ uniform grid points. The sound speed was $c = 1500$ m/s. Twelve-point targets were placed, as described in Figure 4.

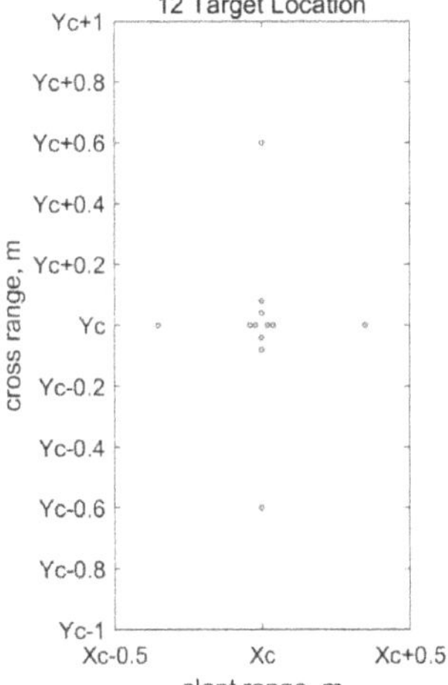

Target Location $[x, y]$
$[X_c + 0.2,\ Y_c]$
$[X_c - 0.2,\ Y_c]$
$[X_c + 0.04,\ Y_c]$
$[X_c - 0.04,\ Y_c]$
$[X_c + 0.35,\ Y_c]$
$[X_c - 0.35,\ Y_c]$
$[X_c,\ Y_c + 0.04]$
$[X_c,\ Y_c - 0.04]$
$[X_c,\ Y_c + 0.08]$
$[X_c,\ Y_c - 0.08]$
$[X_c,\ Y_c + 0.6]$
$[X_c,\ Y_c - 0.6]$

Figure 4. Locations of the 12 targets.

All simulation environments for single-sensor sonar were fundamentally the same as the basic simulation environment described above and were performed by changing the noise level, X_c, f_s, and sonar interval, as shown in Table 1.

Table 1. Simulation Case for Single Sensor.

Case	Noise Existence	X_c (m)	f_s (kHz)	Sonar Interval (m)
1	X	250	50	0.02
2	X	250	25	0.02
3	X	150	50	0.2
4	X	150	50	0.4
5	O	250	50	0.02

- Case 1

As shown in Figure 5, it was confirmed that the images obtained through the CS-SAS algorithm accurately distinguished 12 targets and had a good azimuth resolution. In addition, it was confirmed that there was no performance degradation caused by the side lobes. However, the result for the ω-k algorithm is unable to distinguish between targets adjacent to each other in the center of the area of interest. In addition, much aliasing occurs, especially in the azimuth direction. On account of the influence of matched filtering and Fourier transform, the exact position of the point targets cannot be obtained, resulting in a blurred result.

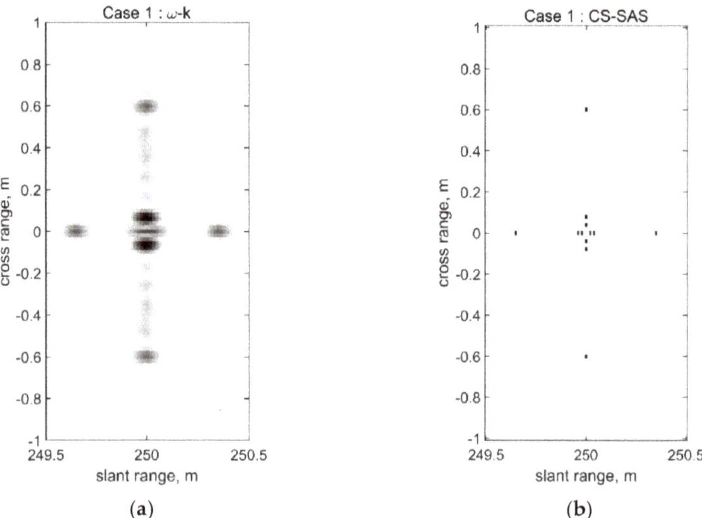

Figure 5. Simulation results for Case 1 for single sensor: (**a**) result for the ω-k algorithm; (**b**) result for the CS-SAS algorithm for single sensor.

- Case 2

The results of the simulation are shown in Figure 6. In the case of the ω-k algorithm, because the sampling frequency is reduced by half compared to Case 1, the resolution in the slant-range direction is reduced. The aliasing at the center of the area of interest has a larger value than the values at the four target locations, $[250 \pm 0.02, 0]$ and $[250 \pm 0.04, 0]$. Nonetheless, the image obtained using the CS-SAS algorithm yielded an accurate target image.

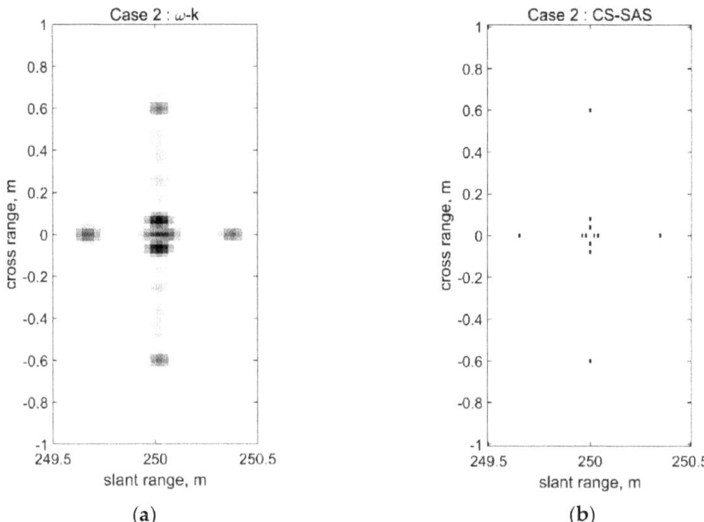

(a)   (b)

Figure 6. Simulation results for Case 2 for single sensor: (**a**) result for the ω-k algorithm; (**b**) result for the CS-SAS algorithm for single sensor.

- Case 3

The results are shown in Figure 7. The CS-SAS algorithm accurately fetches the images of 12 targets. However, the ω-k algorithm requires a Fourier transform in the space domain, which violates Nyquist theory because the sampling level in the space domain is reduced to 1/10. Therefore, aliasing occurred, and the image of the target could not be properly obtained. The eight target points in the center were not distinguishable, and the remaining four points were difficult to determine.

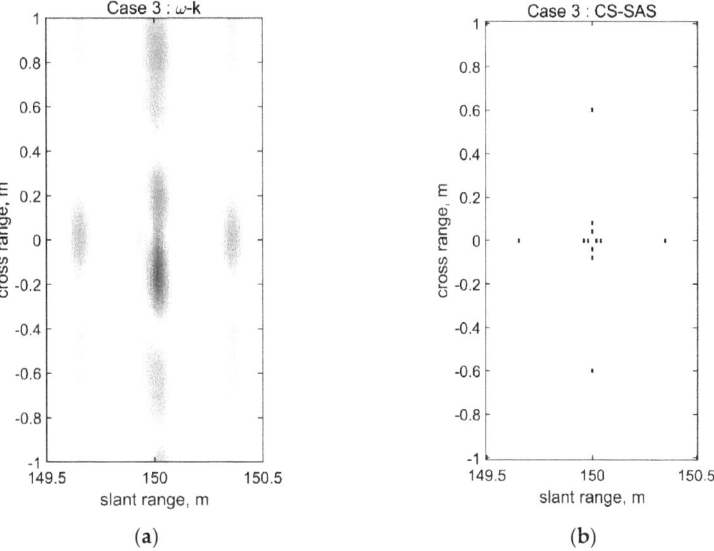

(a)   (b)

Figure 7. Simulation results for Case 3 for single sensor: (**a**) result for the ω-k algorithm; (**b**) result for the CS-SAS algorithm for single sensor.

- Case 4

  The results are depicted in Figure 8. Even when the spatial sampling is reduced to a level of 1/20 and when some side lobes occur, the CS-SAS algorithm still fetches the image of 12 targets, whereas the ω-k algorithm fails to depict the proper image of the targets.

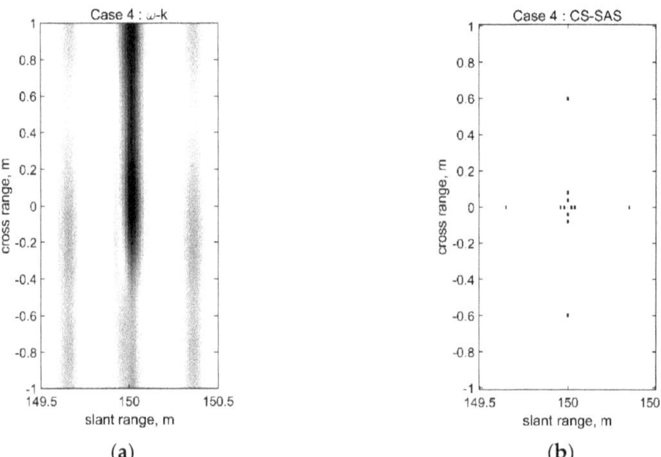

Figure 8. Simulation results for Case 4 for single sensor: (**a**) result for the ω-k algorithm; (**b**) result for the CS-SAS algorithm for single sensor.

- Case 5

  Noisy conditions with signal-to-noise ratios (SNRs) of 20, 10, and 5 dB were simulated. The results in Figure 9 indicate that the CS-SAS algorithm applied to an environment with SNRs of 20 dB and 10 dB, obtained an almost accurate image of the targets. However, when the SNR was 5 dB, the values at the grid point between [250, ±0.04] and [250, ±0.08] were greater than the values at [250, ±0.04] and [250, ±0.08], and an accurate image could not be obtained. As the noise became louder, a degree of degradation occurred. Nevertheless, the CS-SAS algorithm was still superior to the ω-k algorithm in terms of resolution and sidelobe suppression.

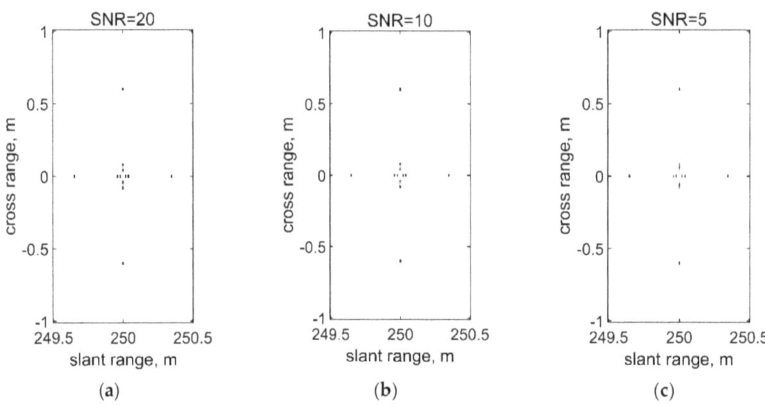

Figure 9. Simulation results of CS-SAS algorithm for Case 5 for single sensor: (**a**) SNR = 20 dB; (**b**) SNR = 10 dB; (**c**) SNR = 5 dB.

4.1.2. Simulation Results for Uniform Line Array

The CS-SAS and ω-k algorithms for a uniform line array were simulated in two cases. The first simulation case is as follows: The simulation environments of sampling frequency f_s and sound velocity c, excluding sonar configuration and X_c, are the same as those of simulation Case 1 for a single sensor. The array has 20 receivers, with a 0.04 m spacing between receivers, as shown in Figure 10a. By moving 0.4 m between pings, a total of 25 pings were shot. The source position of the uniform line array is 0.1 m away from the first sensor in the cross-range direction. Conditions for other simulations are the same as for Case 1, except that the sensor spacing of the array is different. The array has two receivers, with 0.4 m spacing between receivers, as shown in Figure 10b. By moving 0.4 m between pings, a total of 25 pings were shot. The source position of the uniform line array is 0.1 m away from the first sensor in the cross-range direction.

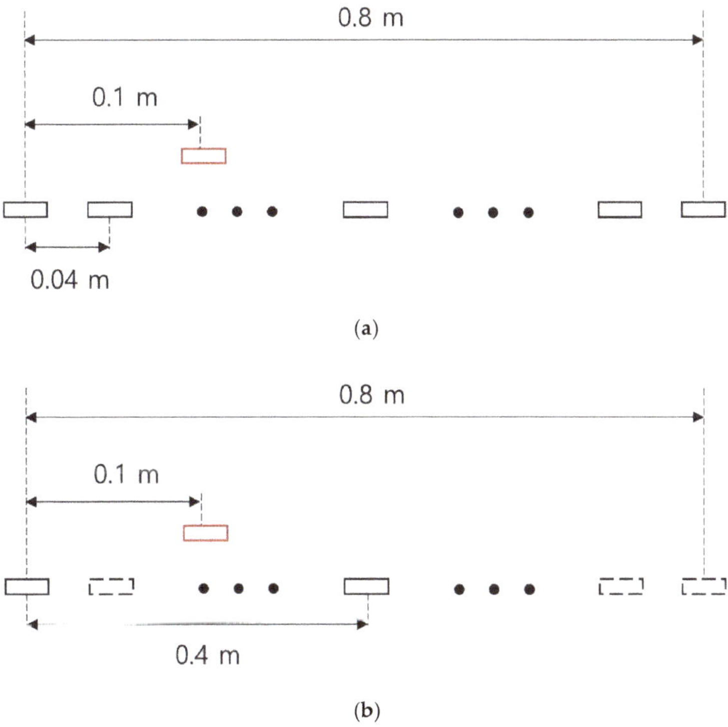

Figure 10. Description of line array in: (**a**) the first case of simulation; (**b**) the second case of simulation; the red box is a transmitter, and the black boxes are receivers. The black-dash boxes in (**b**) are actually empty but marked for comparison with (**a**).

- Case 1

The results for Case 1 are shown in Figure 11. The CS-SAS algorithm accurately fetched images of the 12 targets. Contrarily, the result for the ω-k algorithm showed aliasing. In particular, the total number of sensors used were $500 = 20 \times 25$, which compares favorably to the simulation environment of a single sensor; however, the interval between the sensors doubled to 0.04, and the spatial sampling interval also doubled, resulting in aliasing near [250, ±0.07].

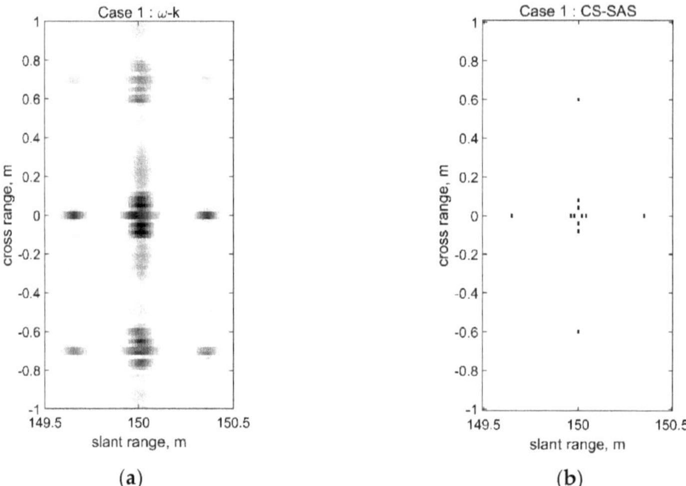

(a)   (b)

Figure 11. Simulation results for Case 1 for uniform line array: (**a**) result for the ω-k algorithm; (**b**) result for the CS-SAS algorithm for uniform line array.

- Case 2

The results for Case 2 are shown in Figure 12. The synthetic aperture is the same as in Case 1, but the sensor spacing is increased 10 times. Severe aliasing occurred in the ω-k results as well as the inability to properly identify the targets. However, the CS-SAS algorithm was significantly better distinguished.

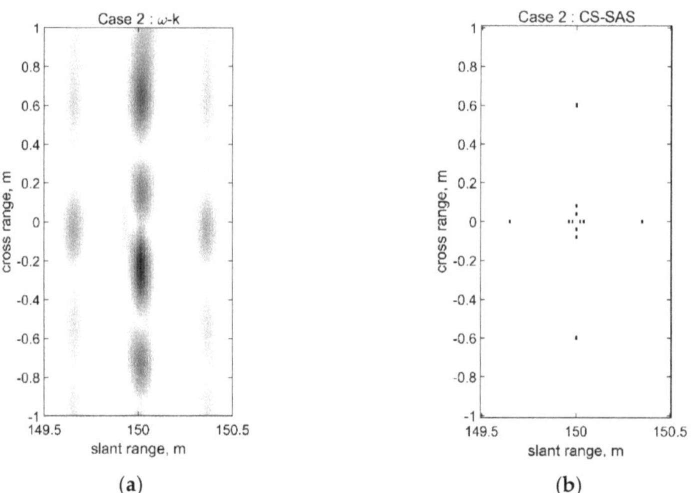

(a)   (b)

Figure 12. Simulation results for Case 2 for uniform line array: (**a**) result for the ω-k algorithm; (**b**) result for the CS-SAS algorithm for a uniform line array.

The performances of the CS-SAS and ω-k algorithms were compared using simulation results for a single sensor and uniform line array. In the case of the ω-k algorithm, even under the most naïve simulation conditions, adjacent targets could not be distinguished and aliasing occurred, whereas in the case of the proposed algorithm, because CS was applied and sidelobes were rather suppressed, high-resolution results were obtained. In effect, CS-SAS has clearly distinguished targets under harsher conditions by increasing spatial sampling or reducing f_s, and has obtained accurate locations and shown robustness in noisy situations. This is possible because the measured signal can be expressed in a sparse representation for a certain domain, and CS can significantly lower the sampling rate and has robust resistance to noise [36,39].

4.2. Experimental Data Results

This was a water tank experiment conducted by SonaTech Inc. (Santa Barbara, CA, USA). As shown in Figure 13a, an experiment was performed to obtain images of two rings in a water tank. As shown in Figure 13b, the sonar has one transmitter and 32 receivers. After transmitting and receiving the signal once, the transmitter moves 616.5 mm and then sends and receives the next signal. This process was repeated seven times to receive signals from 224 locations. The ping signal $p(t)$ is a CW signal with carrier frequency f_c = 455 kHz and pulse duration T_p = 0.3 ms, the sampling frequency is f_s = 50 kHz, and the sound speed c = 1480 m/s. There are two ring-shaped targets of approximate length of major axis 1.5 m each in the slant-range of 7 to 10 m and a cross-range of -2.4385 to 2.4385 m.

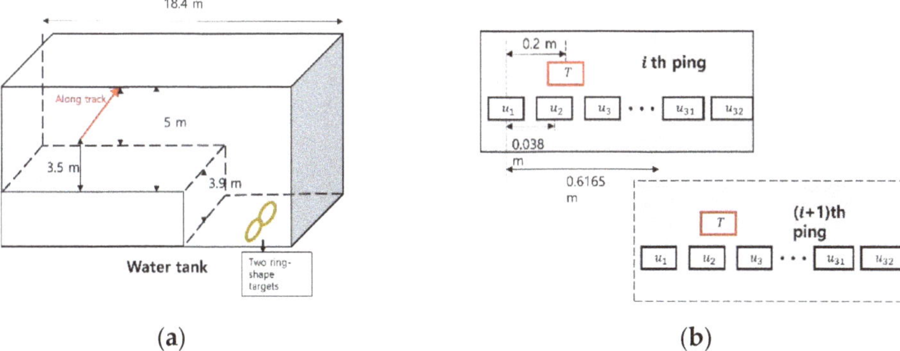

Figure 13. Water tank experiment overview: (**a**) water tank structure; (**b**) synthetic aperture line-array sonar.

The raw data recorded in the slant-range are shown in Figure 14a. The CS-SAS result was derived by dividing the area of interest into a uniform grid of N_X = 101 and N_Y = 101. The results of the ω-k and CS-SAS algorithms are shown in Figure 14b,c, respectively.

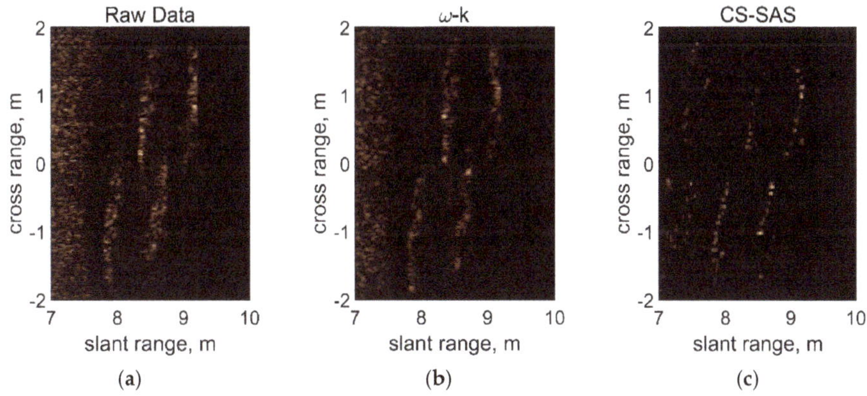

Figure 14. (**a**) Raw data; (**b**) result for the ω-k algorithm; (**c**) result for the CS-SAS algorithm.

From the raw data, the targets can be seen in the form of rings, but the shape appears thick, and it is difficult to accurately determine the location of the targets. In the results of the ω-k algorithm, the shapes are slightly thinner, but aliasing is severe in the azimuth direction, and it appears that there are several rings. The results of the CS-SAS algorithm construct tworing-shaped targets. Because the CS-SAS algorithm attempts to bring an image with as little target distribution as possible, the side lobes are suppressed to obtain thin ring-shaped targets.

To examine whether the CS-SAS algorithm is robust under conditions where some of the sensors are broken, the result was derived by assuming a situation in which data from some sensors were lost. The experiment was divided into two cases: one case where the sensor failed uniformly (Sensor Loss: Uniform, SLU) and another case where the sensor failed randomly (Sensor Loss: Random, SLR). The percentages of sensors that did not malfunction and operated normally are also indicated in the results. The results are displayed in Figure 15.

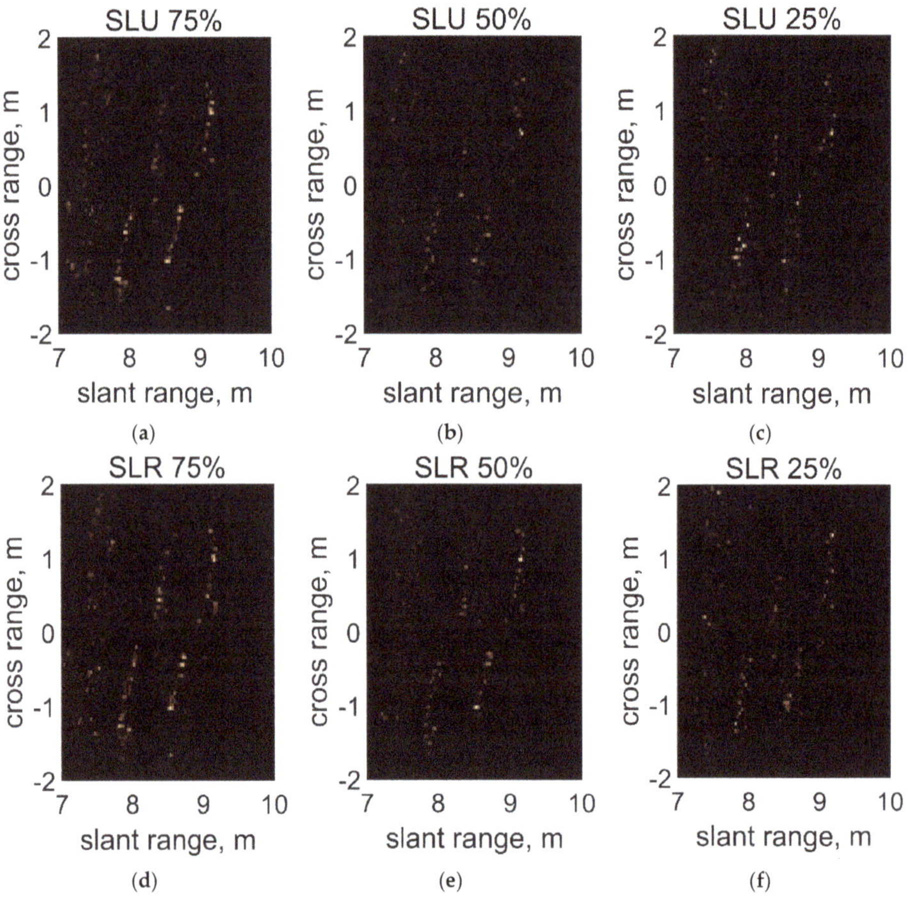

Figure 15. Results for some sensor loss conditions. Uniformly broken case (SLU) where the percentage of the remaining sensors are: (**a**) 75%; (**b**) 50%; (**c**) 25%. Randomly broken case (SRU) where the percentage of the remaining sensors are (**d**) 75%; (**e**) 50%; (**f**) 25%.

In addition, peak signal-to-noise ratio (PSNR) and structural similarity index measure (SSIM), which are representative image quality measurement indicators [40,41], were calculated for quantitative comparison. PSNR is an index that evaluates loss information for the quality of the generated or compressed images and is expressed by the peak signal/mean square error (MSE) term. It has a unit of dB, and a higher value indicates less loss, i.e., higher image quality. SSIM is an index designed to evaluate differences in human visual quality rather than numerical errors. SSIM quality is evaluated in three aspects: luminance, contrast, and structural. However, since the correct answer image is not known, PSNR and SSIM were calculated for all sensor loss situations using Figure 14c as a reference image. The calculation results are shown in Table 2.

Table 2. PSNR and SSIM.

	SLU 75%	**SLU 50%**	**SLU 25%**
PSNR (dB)	33.7691	29.2229	28.1871
SSIM	0.8850	0.6947	0.6029
	SLR 75%	**SLR 50%**	**SLR 25%**
PSNR (dB)	32.9409	30.6011	28.4189
SSIM	0.8609	0.7324	0.6178

In the case of some conventional methods such as ω-k, a Fourier transform in space is performed. Therefore, it is difficult to obtain results freely in the form of an array because linear sampling is not possible in space when some sensors in the array fail. However, the CS-SAS algorithm does not perform Fourier transform in space and has a formulation that is free in the form of an array and, therefore, it is easy to obtain a result in a sensor loss situation. In addition, it is difficult to detect significant performance degradation of up to 75% for both SLU and SLR, and 50% of the SLR show particularly good results; note that CS has the best performance for random array or random down sampling [42,43]. Table 2 also shows that the random array results are generally better. In Table 2, it can be seen that the image quality of SLR is high for both 50% and 25%. At 75%, the indicators of SLR are worse than at 75% of SLU. Because there is little deterioration in image quality in 75% of cases, it can be seen that the PSNR and SSIM simply show how similar the reconstruction results are to Figure 14c, rather than showing the results of image quality deterioration. When it reaches approximately 25%, both SLU and SLR seem to blur to some extent, but in terms of side lobe suppression, it still shows no inferiority over the ω-k algorithm.

Using the CS-SAS algorithm in this study made it possible to obtain a higher resolution image than when using the conventional synthetic aperture sonar algorithm—the ω-k algorithm—and made it possible to reduce the problem of aliasing which also occurs in the conventional method. In addition, even with less spatial sampling, better results were obtained than compared to the conventional algorithm, and it was confirmed to be robust even when some sensors failed. Good results can be expected even if the number of sensors are reduced during actual sonar operation, and as a consequence, cost reduction is possible. Moreover, it is durable because it presents robust characteristics in failure situations. Results of actual experimental data were also observed, and it is expected that satisfying results will be obtained in the event that the CS-SAS algorithm is applied to a natural underwater environment.

5. Conclusions

In this paper, we proposed an algorithm that applies compressive sensing (CS) to a synthetic aperture sonar (SAS) under the assumption that the target distribution in water is sparse. Through simulation, it was confirmed that the proposed algorithm produces images with better resolution than the conventional SAS algorithm, the ω-k algorithm. In addition, because images obtained by the proposed method present very few and small side lobes, no deterioration of imaging performance occurs. Furthermore, even in the case of sampling at a low level that violates Nyquist theory in the time and space domain, a higher quality target image was obtained than with the ω-k algorithm.

Real environment applicability was revealed for the proposed method when comparing the results with actual experimental data. The results confirm that aliasing is reduced and side lobes are suppressed when applying the compressive sensing method. Contrarily, the ω-k algorithm does not obtain accurate target images due to aliasing. Importantly, it was confirmed that the proposed method is robust in the event of some sensors of the sonar system failing or when some data are lost.

Author Contributions: Conceptualization, H.-m.C. and H.-s.Y.; methodology, H.-m.C.; software, H.-m.C.; validation, H.-s.Y. and W.-j.S.; formal analysis, H.-m.C.; investigation, H.-m.C.; resources, H.-s.Y. and W.-j.S.; data curation, H.-m.C.; writing—original draft preparation, H.-m.C.; writing—review and editing, H.-s.Y. and W.-j.S.; visualization, H.-m.C.; supervision, W.-j.S.; project administration, H.-s.Y.; funding acquisition, W.-j.S. All authors have read and agreed to the published version of the manuscript.

Funding: This research received no external funding.

Data Availability Statement: The data presented in this study are not publicly available because these data belong to SonaTech Inc.

Acknowledgments: This research was supported by the Agency for Defense Development in Korea under Contract No. UD190005DD and the SonaTech Inc under Contract No. C180005.

Conflicts of Interest: The authors declare no conflict of interest.

References

1. Soumekh, M. *Synthetic Aperture Radar Signal. Processing with MATLAB Algorithms*, 1st ed.; Wiley-Interscience: New York, NY, USA, 1999.
2. Michael, P.H.; Peter, T.G. Synthetic Aperture Sonar: A Review of Current Status. *IEEE J. Ocean. Eng.* **2009**, *34*, 207–224.
3. Hyun, A. A Study on Robust Synthetic Aperture Sonar Signal Processing for Multipath Environment using SAGE Algorithm. Ph.D. Thesis, Seoul National University, Seoul, Korea, 2016.
4. Gustafson, E.; Jalving, B. HUGIN 1000 Arctic Class AUV. In Proceedings of the OTC Arctic Technology Conference, Houston, TX, USA, 7–9 February 2011.
5. Hansen, R.E. Introduction to Synthetic Aperture Sonar. In *Sonar Systems*; Kolev, N.Z., Ed.; IntechOpen: Rijeka, Croatia, 2011; Chapter 1; pp. 3–28.
6. Gough, P.T.; Hawkins, D.W. Unified Framework for Modern Synthetic Aperture Imaging Algorithms. *Int. J. Imaging Syst. Technol.* **1997**, *8*, 343–358. [CrossRef]
7. Barber, B.C. Theory of Digital Imaging from Orbital Synthetic-Aperture Radar. *Int. J. Remote Sens.* **1985**, *6*, 1009–1057. [CrossRef]
8. Bamler, R. A Comparison of Range-Doppler and Wavenumber Domain SAR Focusing Algorithms. *IEEE Trans. Geosci. Remote Sens.* **1992**, *30*, 706–713. [CrossRef]
9. Cafforio, C.; Prati, C.; Rocca, F. SAR Data Focusing using Seismic Migration Techniques. *IEEE Trans. Aerosp. Electron. Syst.* **1991**, *27*, 194–207. [CrossRef]
10. Stolt, R.H. Migration by Fourier Transform. *Geophysics* **1978**, *43*, 23–48. [CrossRef]
11. Cumming, I.; Wong, F.; Raney, K. A SAR Processing Algorithm with No Interpolation. In Proceedings of the Geoscience and Remote Sensing Symposium, Houston, TX, USA, 26–29 May 1992; pp. 376–379.
12. Marston, T.M.; Marston, P.L.; Williams, K.L. Scattering Resonances, Filtering with Reversible SAS Processing, and Applications of Quantitative Ray Theory. In Proceedings of the OCEANS MTS/IEEE, Seattle, WA, USA, 20–23 September 2010; pp. 1–9.
13. Synnes, S.; Hunter, A.J.; Hansen, R.E.; Sæbø, T.O.; Callow, H.J.; Vossen, R.V.; Austeng, A. Wideband Synthetic Aperture Sonar Backprojection with Maximization of Wavenumber Domain Support. *IEEE J. Oceanic Eng.* **2017**, *42*, 880–891. [CrossRef]
14. Elad, M. *Sparse and Redundant Representations: From Theory to Applications in Signal. and Image Processing*, 2nd ed.; Springer: New York, NY, USA, 2010; pp. 1–359.
15. Krueger, K.R.; McClellanm, J.H.; Scott, W.R. Efficient Algorithm Design for GPR Imaging of Landmines. *IEEE Trans. Geosci. Remote Sens.* **2015**, *53*, 4010–4021. [CrossRef]
16. Aberman, K.; Eldar, Y.C. Sub-Nyquist SAR via Fourier Domain Range Doppler Processing. *IEEE Trans. Geosci. Remote Sens.* **2017**, *55*, 6228–6244. [CrossRef]
17. Lustig, M.; Donoho, D.L.; Pauly, J.M. Sparse MRI: The Application of Compressed Sensing for Rapid MR Imaging. *Magn. Reson. Med.* **2007**, *58*, 1182–1195. [CrossRef]
18. Lustig, M.; Donoho, D.L.; Santos, J.M.; Pauly, J.M. Compressed Sensing MRI. *IEEE Signal. Process. Mag.* **2008**, *25*, 72–82. [CrossRef]
19. Haupt, J.; Bajwa, W.U.; Raz, G.; Nowak, R. Toeplitz Compressed Sensing Matrices with Applications to Sparse Channel Estimation. *IEEE Trans. Inf. Theory* **2010**, *56*, 5862–5875. [CrossRef]

20. Tur, R.; Eldar, Y.C.; Friedman, Z. Innovation Rate Sampling of Pulse Streams with Application to Ultrasound Imaging. *IEEE Trans. Signal. Process.* **2011**, *59*, 1827–1842. [CrossRef]
21. Xenaki, A.; Gerstoft, P. Compressive Beamforming. *J. Acoust. Soc. Am.* **2014**, *136*, 260–271. [CrossRef]
22. Gerstoft, P.; Xenaki, A.; Mecklenbräuker, C.F. Multiple and Single Snapshot Compressive Beamforming. *J. Acoust. Soc. Am.* **2015**, *138*, 2003–2014. [CrossRef]
23. Güven, H.E.; Güngör, A.; Çetin, M. An Augmented Lagrangian Method for Complex-Valued Compressed SAR Imaging. *IEEE Trans. Comput. Imag.* **2016**, *2*, 235–250. [CrossRef]
24. Güngör, A.; Çetin, M.; Güven, H.E. Autofocused Compressive SAR Imaging based on the Alternating Direction Method of Multipliers. In Proceedings of the 2017 IEEE Radar Conference, Seattle, WA, USA, 8–12 May 2017; pp. 1573–1576.
25. Zhao, B.; Huang, L.; Zhang, J. Single Channel SAR Deception Jamming Suppression via Dynamic Aperture Processing. *IEEE Sens. J.* **2017**, *17*, 4225–4230. [CrossRef]
26. Çetin, M.; Stojanović, I.; Önhon, Ö.; Varshney, K.; Samadi, S.; Karl, W.C.; Willsky, A.S. Sparsity-Driven Synthetic Aperture Radar Imaging: Reconstruction, Autofocusing, Moving targets, and Compressed Sensing. *IEEE Signal. Process. Mag.* **2014**, *31*, 27–40. [CrossRef]
27. Amin, M. *Compressive Sensing for Urban. Radar*, 1st ed.; CRC Press: Boca Raton, FL, USA, 2014.
28. Wei, S.J.; Zhang, X.L.; Shi, J.; Xiang, G. Sparse Reconstruction for SAR Imaging Based on Compressed Sensing. *Prog. Electromagn. Res.* **2010**, *109*, 63–81. [CrossRef]
29. Leier, S.; Zoubir, A.M. Aperture Undersampling using Compressive Sensing for Synthetic Aperture Stripmap Imaging. *EURASIP J. Adv. Signal. Process.* **2014**, *156*. [CrossRef]
30. Xenaki, A.; Pailhas, Y. Compressive Synthetic Aperture Sonar Imaging with Distributed Optimization. *J. Acoust. Soc. Am.* **2019**, *146*, 1839–1850. [CrossRef] [PubMed]
31. Ji, X.; Zhou, L.; Cong, W. Effect of Incorrect Sound Velocity on Synthetic Aperture Sonar Resolution. In Proceedings of the 2019 MATEC Web Conf. 2019, Sibiu, Romania, 5–7 June 2019. [CrossRef]
32. Cutrona, L.J. Comparison of Sonar System Performance Achievable using Synthetic-Aperture Techniques with the Performance Achievable by More Conventional Means. *J. Acoust. Soc. Am.* **1975**, *58*, 336–348. [CrossRef]
33. Cutrona, L.J. Additional Characteristics of Synthetic-Aperture Sonar Systems and a Further Comparison with Nonsynthetic-Aperture Sonar Systems. *J. Acoust. Soc. Am.* **1977**, *61*, 1213–1217. [CrossRef]
34. Soumekh, M. *Fourier Array Imaging*; Prentice-Hall: Hoboken, NJ, USA, 1994.
35. Gough, P.T.; Hawkins, D.W. Imaging Algorithms for a Strip-Map Synthetic Aperture Sonar: Minimizing the Effects of Aperture Errors and Aperture Undersampling. *IEEE J. Ocean. Eng.* **1997**, *22*, 27–39. [CrossRef]
36. Donoho, D.L. Compressed Sensing. *IEEE Trans. Inf. Theory.* **2006**, *52*, 1289–1306. [CrossRef]
37. Grant, M.; Boyd, S.; Ye, Y. CVX: Matlab Software for Disciplined Convex Programming. Available online: http://cvxr.com/cvx (accessed on 1 December 2020).
38. Baraniuk, R.G.; Cevher, V.; Duarte, M.; Hegde, C. Model-Based Compressive Sensing. *IEEE Trans. Inf. Theory* **2010**, *56*, 1982–2001. [CrossRef]
39. Candes, E.J.; Romberg, J.; Tao, T. Robust Uncertainty Principles: Exact Signal Reconstruction from Highly Incomplete Frequency Information. *IEEE Trans. Inf. Theory* **2006**, *52*, 489–509. [CrossRef]
40. Horè, A.; Ziou, D. Image Quality Metrics: PSNR vs. SSIM. In Proceedings of the 2010 20th International Conference on Pattern Recognition, Istanbul, Turkey, 23–26 August 2010; pp. 2366–2369.
41. Rehman, A.; Wang, Z. Reduced–Reference SSIM Estimation. In Proceedings of the 2010 IEEE 17th International Conference on Image Processing, Hong Kong, China, 26–29 September 2010; pp. 289–292.
42. Baraniuk, R.G. Compressive Sensing [Lecture Notes]. *IEEE Signal. Process. Mag.* **2007**, *24*, 118–121. [CrossRef]
43. Baraniuk, R.; Davenport, M.; DeVore, R.; Wakin, M. A Simple Proof of the Restricted Isometry Property for Random Matrices. *Constr. Approx.* **2008**, *28*, 253–263. [CrossRef]

remote sensing

MDPI

Article

Real-Time Hyperspectral Data Transmission for UAV-Based Acquisition Platforms

José M. Melián *, Adán Jiménez, María Díaz, Alejandro Morales, Pablo Horstrand, Raúl Guerra, Sebastián López and José F. López

Institute for Applied Microelectronics (IUMA), University of Las Palmas de Gran Canaria (ULPGC), 35001 Las Palmas, Spain; ajimenez@iuma.ulpgc.es (A.J.); mdmartin@iuma.ulpgc.es (M.D.); amorales@iuma.ulpgc.es (A.M.); phorstrand@iuma.ulpgc.es (P.H.); rguerra@iuma.ulpgc.es (R.G.); seblopez@iuma.ulpgc.es (S.L.); lopez@iuma.ulpgc.es (J.F.L.)
* Correspondence: jmelian@iuma.ulpgc.es

Abstract: Hyperspectral sensors that are mounted in unmanned aerial vehicles (UAVs) offer many benefits for different remote sensing applications by combining the capacity of acquiring a high amount of information that allows for distinguishing or identifying different materials, and the flexibility of the UAVs for planning different kind of flying missions. However, further developments are still needed to take advantage of the combination of these technologies for applications that require a supervised or semi-supervised process, such as defense, surveillance, or search and rescue missions. The main reason is that, in these scenarios, the acquired data typically need to be rapidly transferred to a ground station where it can be processed and/or visualized in real-time by an operator for taking decisions on the fly. This is a very challenging task due to the high acquisition data rate of the hyperspectral sensors and the limited transmission bandwidth. This research focuses on providing a working solution to the described problem by rapidly compressing the acquired hyperspectral data prior to its transmission to the ground station. It has been tested using two different NVIDIA boards as on-board computers, the Jetson Xavier NX and the Jetson Nano. The *Lossy Compression Algorithm for Hyperspectral Image Systems* (HyperLCA) has been used for compressing the acquired data. The entire process, including the data compression and transmission, has been optimized and parallelized at different levels, while also using the Low Power Graphics Processing Units (LPGPUs) embedded in the Jetson boards. Finally, several tests have been carried out to evaluate the overall performance of the proposed design. The obtained results demonstrate the achievement of real-time performance when using the Jetson Xavier NX for all the configurations that could potentially be used during a real mission. However, when using the Jetson Nano, real-time performance has only been achieved when using the less restrictive configurations, which leaves room for further improvements and optimizations in order to reduce the computational burden of the overall design and increase its efficiency.

Keywords: real-time compression; on-board compression; real-time transmission; hyperspectral images; UAVs

Citation: Melián, J.M.; Jiménez, A.; Díaz, M.; Morales, A.; Horstrand, P.; Guerra, R.; López, S.; López, J.F. Real-Time Hyperspectral Data Transmission for UAV-Based Acquisition Platforms. *Remote Sens.* **2021**, *13*, 850. https://doi.org/10.3390/rs13050850

Academic Editors: Vladimir Lukin, Benoit Vozel and Joan Serra-Sagristà

Received: 22 January 2021
Accepted: 22 February 2021
Published: 25 February 2021

Publisher's Note: MDPI stays neutral with regard to jurisdictional claims in published maps and institutional affiliations.

1. Introduction

During the last years, there has been a great interest in two very different technologies that have demonstrated to be complementary, unmanned aerial vehicles (UAVs) and hyperspectral imaging. In this way, the combination of both allows to extend the range of applications in which they have been independently used up to now. In one hand, UAVs have been employed in a great variety of uses due to the advantages that these vehicles have over traditional platforms, such as satellites or manned aircrafts. Among these advantages are the high spatial and spectral resolutions that can be reached in images obtained from UAVs, the revisit time and a relative low economical cost. The monitoring of road-traffic [1–4], searching and rescuing operations [5–7], or security and surveillance [8,9] are some examples of civil applications that have adopted UAVs-based technologies. On the other hand, hyperspectral imaging technology has also gained an important popularity increment due to the high amount of information that this kind of images is able to produce, which is specially valuable for applications that need to detect or distinguish different kind of materials that may look like very similar in a traditional digital image. Accordingly, it can be inferred that the combination of these two technologies by mounting hyperspectral cameras onto drones would bring important advances in fields like real-time target detection and tracking [10].

Despite the benefits that can be obtained combining UAVs and hyperspectral technology, it is necessary to take into account that both also have important limitations. Referring to UAVs, they have a flight autonomy that is limited by the capacity of their batteries, a restriction that gets worse as the weight carried by the drone increases with additional elements, such as cameras, gimbal, on-board PC, or sensors. In reference to the hyperspectral technology, the analysis of the large quantity of data provided by this kind of cameras to extract conclusions for the targeted application implies a high computational cost and makes it difficult to achieve real-time performance. So far, two main approaches have been followed in the state-of-the-art of hyperspectral UAVs-based applications:

1. **Storing the acquired data into a non-volatile memory and off-board processing it once back in the laboratory.** The main advantage of this approach is its simplicity and the possibility of processing the data without any limitation in terms of computational resources. However, it is not viable for applications that require a real-time response, and it does not allow the user to visualize the acquired data during the mission. For instance, following this strategy, the drone could fly over a region, scan it with the hyperspectral camera, and store all of the acquired data into a solid state drive (SSD), so, once the drone lands, it is powered off and the disk is removed for extracting the data and processing it [11–13]. However, if, by any reason, the acquisition parameters were not correctly configured or the data was not captured as expected, this will not be detected until finishing the mission and extracting the data for its analysis.

2. **On-board processing the acquired hyperspectral data.** This is a more challenging approach due to the high data acquisition rate of the hyperspectral sensors as well as the limited number of computational resources available on-board. Its main advantage is that the acquired data could be potentially analysed in real-time or near real-time and, so, decisions could be taken on the fly according to the obtained results. However, the data analysis methods that can be on-board executed are very limited [10,14,15]. Additionally, since, while following this approach, the data are fully processed on-board and never transmitted to the ground station during the mission, it cannot be used for any supervised or semi-supervised application in which an operator is required for visualizing the acquired data or its results and taking decisions according to it. Some examples of this kind of applications are surveillance and search and rescue missions.

This work focuses on providing an alternative solution for a UAV-based hyperspectral acquisition platform to achieve greater flexibility and adaptability to a larger number of applications. Concretely, the aforementioned platform consists of a visible and near-infrared (VNIR) camera, the Specim FX10 pushbroom scanner, [16] mounted on a DJI Matrice 600 [17], which also carries a NVIDIA Jetson board as on-board computer. A detailed description of this platform can be found in [10]. Two different Jetson boards have been used in the experiments and it can be set in the platform as on-board computer, the Jetson Nano [18] and the Jetson Xavier NX [19]. They are both pretty similar in shape and weight, although the Jetson Xavier NX has a higher computational capability as well as a higher price.

The main goal of this work is to provide a working solution that allows us to rapidly download the acquired hyperspectral data to the ground station. Once the data are in the ground station, they could be processed in a low latency manner and both the hyperspectral data or the results obtained after the analysis could be visualized by an operator in real-time. This is advantageous for many applications that involve a supervised or semi-supervised procedure, such as defense, surveillance, or search and rescue missions. Currently, these kind of applications are mostly carried out by using RGB first person view cameras [1,2], but they do not take advantage of the possibilities that are offerred by the extra information provided by the hyperspectral sensors. Figure 1 graphically describes the desired targeted solution. To achieve this goal, it is assumed that a Wireless Local Area Network (WLAN) based on the 802.11 standard, which was established by the Institute of Electrical and Electronics Engineers (IEEE), will be available during the flight, and that both the on-board computer and the ground station will be connected to it to use it as data downlink. It will be considered to be real-time transmissions those situations in which the acquisition frame rate and the transmission frame rate reach the same value, although there may be some latency between the capturing process and the transmission one.

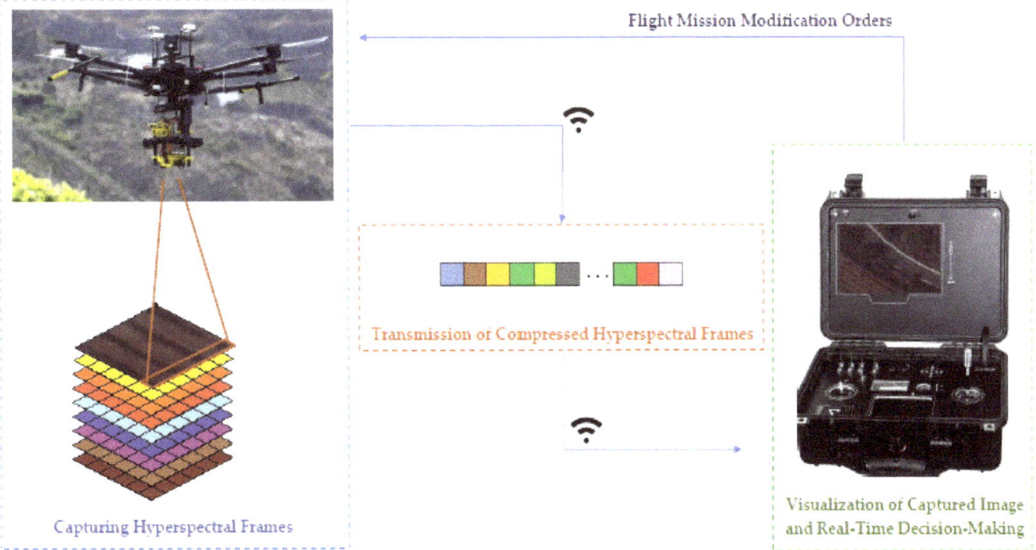

Figure 1. Overall description of the targeted design purpose.

According to our experience in the flight missions carried up to now, the frame rate of the Specim FX10 camera mounted in our flying platform is usually set to a value between 100 and 200 frames per second (FPS) and, so, these are the targeted scenarios to be tested in this work. The Specim FX10 camera produces hyperspectral frames of 1024 pixels with up to 224 spectral bands each, storing them using 12 or 16 bits per pixel per band (bpppb). This means that the acquisition data rate goes from 32.8 MB/s (100 FPS and 12 bpppb) to 87.5 MB/s (200 FPS and 16 bpppb). These values suggest the necessity of carrying out the hyperspectral data compression before its transmission to the ground station in order to achieve real-time performance. For that purpose, the *Lossy Compression Algorithm for Hyperspectral Image Systems* (HyperLCA) has been chosen. This specific compressor has been selected due to the fact that it offers several advantages with respect to other state-of-the art solutions. One one side, it presents a low computational complexity and high level of parallelism in comparison with other transform-based compression approaches. On the other side, the HyperLCA compressor is able to achieve high compression ratios while preserving the relevant information for the ulterior hyperspectral applications. Additionally, the HyperLCA compressor independently compresses blocks of hyperspectral pixels without any spatial alignment required, which makes it especially advantageous for compressing hyperspectral data that were captured by pushbroom or whiskbroom sensors and providing a natural error-resilience behaviour. All of these advantages were deeply analyzed in [20].

The data compression process reduces the amount of data to be downloaded to the ground station, which makes the achievement of a real-time data transmission viable. However, it is also necessary to achieve a real-time performance in the compression process. For doing so, the HyperLCA compressor has been parallelized taking advantage of the available LPGPU included in the Jetson boards as it was done in [21]. However, multiple additional optimizations have been developed in this work with respect to the one presented in [21] to make the entire system work as expected, encompassing the capturing process, the on-board data storing, the hyperspectral data compression and the compressed data transmission.

The document is structured, as follows. Section 2 contains the description of the algorithm used in this work for the compression of the hyperspectral data. Section 3 displays the information of the on-board computers that have been tested in this work. The proposed design for carrying out the real-time compression and transmission of the acquired hyperspectral data is fully described in Section 4.1. Section 4.2 contains the data and resources that are used for the experiments and Section 5 shows the description of the experiments and obtained results. Finally, the obtained conclusions are summarized in Section 7.

2. Description of the HyperLCA Algorithm

The HyperLCA compressor is used in this work to on-board compress the hyperspectral data, reducing its volume before being transmitted to the ground station, as it was previously specified. This compressor is a lossy transform-based algorithm that presents several advantages that satisfy the necessities of this particular application very well. The most relevant are [20]:

- It is able to achieve high compression ratios with reasonable good rate-distortion ratios and keeping the relevant information for the ulterior hyperspectral images analysis. This allows considerably reducing the data volume to be transmitted without decreasing the quality of the results obtained by analyzing the decompressed data in the ground station.
- It has a reasonable low computational cost and it is highly parallelizable. This allows for taking advantage of the LPGPU available on-board to speed up the compression process for achieving a real-time compression.
- It permits independently compressing each hyperspectral frame as it is captured. This makes possible to compress and transmit the acquired frames in a pipelined way, enabling the entire system to continuously operate in an streaming fashion, with just some latency between the acquired frame and the frame received in the ground station.
- It permits setting the minimum desired compression ratio in advance, making it possible to fix the maximum data rate to be transmitted prior to the mission, thus ensuring that it will not saturate the donwlink.

The compression process within the HyperLCA algorithm consists of three main compression stages, which are a spectral transform, a preprocessing stage, and the entropy coding stage. In this work, each compression stage independently processes one single hyperspectral frame at a time. The spectral transform sequentially selects the most different pixels of the hyperspectral frame using orthogonal projection techniques. The set of selected pixels is then used for projecting this frame, obtaining a spectral decorrelated and compressed version of the data. The preprocessing stage is executed after the spectral transform for adapting the output data for being entropy coded in a more efficient way. Finally, the entropy coding stage manages the codification of the extracted vectors using a Golomb–Rice coding strategy. In addition to these three compression stages, there is one extra initialization stage, which carries out the operations that are used for initializing the compression process according to the introduced parameters. Figure 2 graphically shows these compression stages, as well as the data that the HyperLCA compressor share between them.

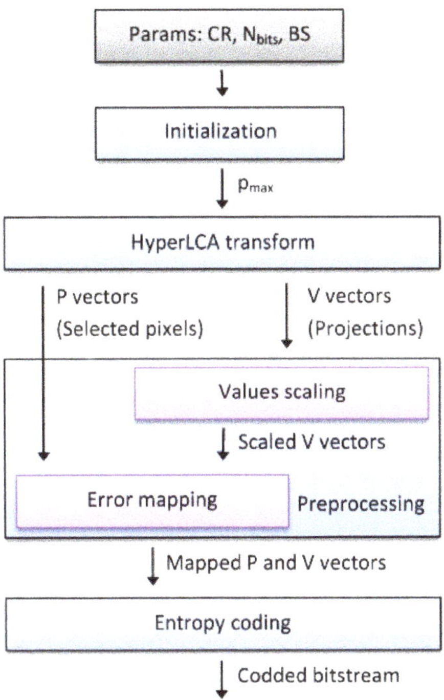

Figure 2. Flowchart of the HyperLCA compressor.

2.1. HyperLCA Input Parameters

The HyperLCA algorithm needs three input parameters in order to be configured:

1. **Minimum desired compression ratio (CR)**, defined as the relation between the number of bits in the real image and the number of bits of the compressed data.
2. **Block size (BS)**, which indicates the number of hyperspectral pixels in a single hyperspectral frame.
3. **Number of bits used for scaling the projection vectors (N_{bits})**. This value determines the precision and dynamic range to be used for representing the values of the V vectors.

2.2. HyperLCA Initialization

Once the compressor has been correctly configured, its initialization stage is carried out. This initialization stage consists on determining the number of pixel vectors and projection vectors (p_{max}) to be extracted for each hyperspectral frame, as shown in Equation (1), where DR refers to the number of bits per pixel per band of the hyperspectral image to be compressed and N_{bands} refers to the number of spectral bands. The extracted pixel vectors are referred as P and the projection vectors are referred as V in the rest of this manuscript.

$$p_{max} \leq \frac{DR \cdot N_{bands} \cdot BS}{CR \cdot (DR \cdot N_{bands} + N_{bits} \cdot BS)} \tag{1}$$

The p_{max} value is used as an input of the HyperLCA transform, which is the most relevant part of the HyperLCA compression process. The compression that is achieved within this process directly depends on the number of selected pixels, p_{max}. Selecting more pixels provides better decompressed images, but lower compression ratios.

2.3. HyperLCA Transform

For each hyperspectral frame, the HyperLCA Transform calculates the average pixel, also called centroid. Subsequently, the centroid pixel is used by the HyperLCA Transform to select the first pixel as the most different from the average. This is an unmixing like strategy used in the HyperLCA compressor for selecting the most different pixels of the data set to perfectly preserve them through the compression–decompression process. This results in being very useful for many remote sensing applications, like anomaly detection, spectral unmixing, or even hyperspectral and multispectral image fusion, as demonstrated in [22–24].

The HyperLCA Transform provides most of the compression ratio that was obtained by the HyperLCA compressor and also most of its flexibility and advantages. Additionally, it is the only lossy part of the HyperLCA compression process. Algorithm 1 describes, in detail, the process followed by the HyperLCA Transform for a single hyperspectral frame. The pseudocode assumes that the hyperspectral frame is stored as a matrix, M, with the hyperspectral pixels being placed in columns. The first step of the HyperLCA Transform consists on calculating the centroid pixel, c, and subtracting it to each pixel of M, obtaining the centralized frame, M_c. The transform modifies the M_c values in each iteration.

Figure 3 graphically describes the overall process that is followed by the HyperLCA Transform for compressing a single hyperspectral frame. As shown in this figure, the process that is followed by the HyperLCA Transform mainly consists of three steps that are sequentially repeated. First, the brightest pixel in M_c, which is the pixel with more remaining information, p_i, is selected (lines 2 to 7 of Algorithm 1). After doing so, the vector v_i is calculated as the projection of the centralized frame, M_c, in the direction spanned by p_i (lines 8 to 10 of Algorithm 1). Finally, the information of the frame that can be represented with the extracted p_i and v_i vectors is subtracted from M_c, as shown in line 11 of Algorithm 1.

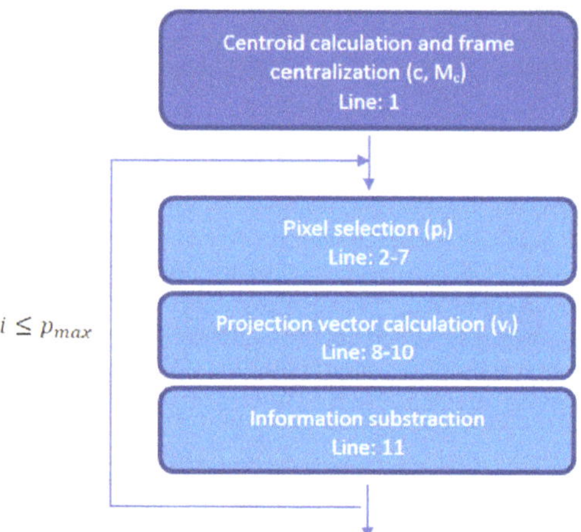

Figure 3. Flowchart of the HyperLCA Transform.

Accordingly, M_c contains the information that is not representable with the already selected pixels, P, and V vectors. Hence, the values of M_c in a particular iteration, i, would be the information lost in the compression–decompression process if no more pixels p_i and v_i vectors are extracted.

Algorithm 1 HyperLCA transform.

Inputs:

$M = [r_1, ..., r_{N_p}]$, $p_{\max}$

Outputs:

c, $P = [p_1, ..., p_{p_{\max}}]$, $V = [v_1, ..., v_{p_{\max}}]$

Declarations:

c; {Centroid pixel.}

$P = [p_1, ..., p_{p_{\max}}]$; {Extracted pixels.}

$V = [v_1, ..., v_{p_{\max}}]$; {Projected image vectors.}

$M_c = [x_1, ..., x_{N_p}]$; {Centralized version of M.}

Algorithm:

1: {Centroid calculation and frame centralization (c, M_c)}

2: **for** $i = 1$ **to** $p_{\max}$ **do**

3:     **for** $j = 1$ **to** N_p **do**

4:         $b_j = x_j^t \cdot x_j$

5:     **end for**

6:     $j_{\max} = \arg\max(b_j)$

7:     $p_i = r_{j_{\max}}$

8:     $q = x_{j_{\max}}$

9:     $u = x_{j_{\max}} / \left((x_{j_{\max}})^t \cdot x_{j_{\max}} \right)$

10:     $v_i = u^t \cdot M_c$

11:     $M_c = M_c - v_i \cdot q$

12: **end for**

2.4. HyperLCA Preprocessing

This stage is crucial in the HyperLCA compression algorithm to adapt the HyperLCA Transform output data for being entropy coded in a more efficient way. This compression stage encompasses two different parts.

2.4.1. Scaling V Vectors

After executing the HyperLCA Transform, the resulting V vectors contain the projection of the frame pixels into the space spanned by the different orthogonal projection vector u_i in each iteration. This results in floating point values of V vector elements between -1 and 1. These values need to be represented using integer data type for their codification. Hence, vectors V can be easily scaled in order to fully exploit the dynamic range available according to the N_{bits} used for representing these vectors and avoid loosing too much precision in the conversion, as shown Equation (2). After doing so, the scaled V vectors are rounded to the closest integer values.

$$v_{j_{\text{scaled}}} = (v_j + 1) \cdot (2^{N_{\text{bits}} - 1} - 1) \tag{2}$$

2.4.2. Error Mapping

The entropy coding stage takes advantage of the redundancies within the data to assign the shortest word length to the most common values in order to achieve higher compression ratios. In order to facilitate the effectiveness of this stage, the output vectors of the HyperLCA Transform, after the preprocessing of V vectors, are independently lossless processed to represent their values using only positive integer values closer to zero than the original ones, using the same dynamic range. To do this, the HyperLCA algorithm makes use of the prediction error mapper described in the Consultative Committee for Space Data Systems (CCSDS) that are recommended standard for lossless multispectral and hyperspectral image compression [25].

2.5. HyperLCA Entropy Coding and Bitstream Generation

The last stage of the HyperLCA compressor corresponds to a lossless entropy coding strategy. The HyperLCA algorithm makes use of the Golomb–Rice algorithm [26] where each single output vector is independently coded.

Finally, the outputs of the aforementioned compression stages are packaged in the order that they are produced, generating the compressed bitstream. The first part of the bitstream consists of a header that includes all of the necessary information to correctly decompress the data.

The detailed description of the coding methodology followed by the HyperLCA compressor as well as the exact structure of the compressed bitstream can be found in [20].

3. On-Board Computers

The on-board computer is one of the key elements of the entire acquisition platform, as described in [10]. It is in charge of managing the overall mission, controlling the UAV actions and flying parameters (direction, speed, altitude, etc.), as well as the data acquisition controlling all of the sensors available on-board, including the Specim FX10 hyperspectral camera. For doing so, two different Nvidia Jetson boards have been separately tested in this work, the Jetson Nano [18] and the Jetson Xavier NX [19]. These two boards present very good computational capabilities as well as many connection interfaces that facilitate the integration of all the necessary devices into the system, while, at the same time, are characterized by a high computational capability in relation to their reduced size, weight and power consumption. Additionally, their physical characteristics are generally better than the ones that are presented by the Jetson TK1, the board originally used in the first version of our UAV capturing platform, described in [10]. This is a remarkable point when taking into account the restrictions in terms of available space and load weight in the drone. Table 1 summarizes the main characteristics of the Jetson TK1, whereas, in Tables 2 and 3, the main characteristics of the two Jetson boards tested in this work are summarized.

In addition to the management of the entire mission and acquisition processes, the on-board computer is also used in this work for carrying out the necessary operations to compress and transmit in real-time the acquired hyperspectral data to the ground station. For doing so, the compression process, which is the most computational demanding one, has been accelerated while taking advantage of the LPGPUs integrated into these Jetson boards.

Table 1. Technical specifications of the NVIDIA Jetson TK1 development board used in the original flight platform.

Board Model	Jetson TK1
LPGPU	GPU NVIDIA Kepler with 192 CUDA cores (uptto 326 GFLOPS)(Model GK20)
CPU	NVIDIA 2.43 GH < ARM quad-core CPU with Cortex A15 battery saving shadow core
Memory	16GB fast eMMC 4.51 (routed to SDMMC4)
Weight	500 g
Dimensions	$127 \times 127 \times 25.4$ mm
Power consumption	1 W minimum

By default, both the Jetson Nano and Jetson Xavier NX are prepared for running the operating system in a Secure Digital card (SD). However, this presents limitations in terms of memory speed as well as in the maximum amount of data that can be stored. To overcome this issue, an external SSD that is connected through an USB3.0 port has been used in the Jetson Nano. The Jetson Xavier NX is already prepared to integrate a Non-Volatile Memory

Express (NVMe) 2.0 SSD keeping the system more compact and efficient. However, this memory was not yet available during this work and, hence, all of the experiments executed in the Jetson Xavier NX were carried out using just the SD card memory.

Additionally, a Wifi antenna is required. to be able to connect to the WLAN to transmit the compressed hyperspectral data to the ground station. While the Jetson Xavier NX already integrates one, the Jetson Nano does not. Hence, an external USB2.0 TP-Link TL-WN722N Wifi antenna [27] has been included in the experiments that were carried out using the Jetson Nano.

Table 2. Technical specifications of the NVIDIA Jetson Nano and NVIDIA Jetson Xavier NX.

Board Model	Jetson Nano
LPGPU	GPU NVIDIA Maxwell with 128 CUDA cores (Model GM200)
CPU	Quad-core ARM A57 @ 1.43 GHz
Memory	4 GB 64-bit LPDDR4 25.6 GB/s
Weight	140 g
Dimensions	99.06 × 78.74 × 27.94 mm
Power consumption	5W minimum
Board model	Jetson Xavier NX
LPGPU	GPU NVIDIA Volta with 384 CUDA cores and 48 Tensor Cores (Model GV100)
CPU	6-core NVIDIA Carmel ARM v8.2 64-bit CPU + 6MB L2 + 4MB L3
Memory	8 GB 128-bit LPDDR4x 51.2 GB/s
Weight	180 g
Dimensions	99.06 × 78.74 × 27.94 mm
Power consumption	10 W minimum

Table 3. Technical specifications of the LPGPU in NVIDIA Jetson Nano and NVIDIA Jetson Xavier NX.

Technical Specifications	GM200	GV100
LPGPU	Jetson Nano	Jetson Xavier NX
Number of CUDA cores	640	384
Number of Tensor Cores	-	48
Number of Streaming multiprocessors (SM)	5	6
Compute Capability	5.2	7.0
Warp Size	32	32
Maximum blocks per SM	32	32
Maximum warps per SM	64	64
Maximum threads per SM	2048	2048
32 bit registers per SM	64 KB	64 KB
Maximum shared memory per SM	96 KB	96 KB

4. Materials and Methods

4.1. Proposed Methodology

As already mentioned, the goal of this work is to provide a working solution that allows for rapidly downloading the acquired hyperspectral data to the ground station, where it could be visualized in real-time. It is assumed that a WLAN will be available during the flight, and that both the on-board computer and the ground station will be connected to it to use it as data downlink. Additionally, it is necessary to independently compress each acquired hyperspectral frame in real-time, so that it can be rapidly transferred to the ground station without saturating the donwlink. A previous work was already published, where the viability of carrying out the real-time compression of the hyperspectral data using the HyperLCA algorithm and some NVIDIA Jetson boards was tested [21]. In that work, it was assumed that the hyperspectral data that were collected by the Specim FX10 camera would be directly placed in Random-access memory (RAM), and that the compression process would directly read from it. Figure 4 graphically shows the workflow that was proposed in that work.

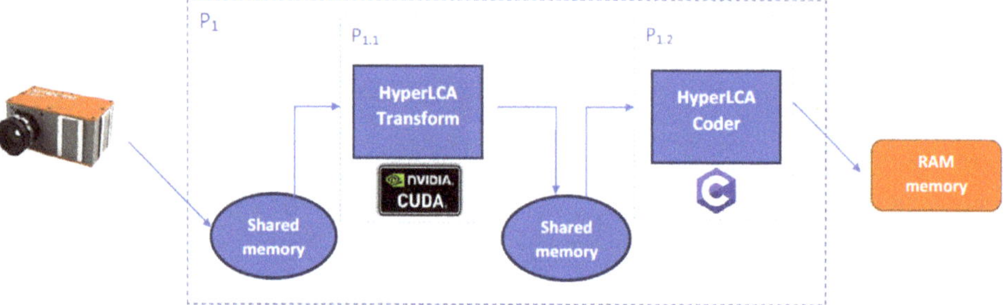

Figure 4. Previous work design for the real-time compression of the hyperspectral data.

Ring buffers are used to prevent the saturation of the RAM memory of the board. These ring buffers allow for a maximum number of frames to be stored at the same time in the memory. Once that the last position of the ring buffer is written, it returns to the first one, overwriting the information that is present in this position. This methodology is able to achieve real-time performance and very high compression speeds, as demonstrated in [21]. For doing so, the HyperLCA Transform, which is the most computational demanding part of the HyperLCA compressor, has been implemented into the LPGPU that is available in the Jetson board using a set of self developed kernels that were programmed using CUDA. However, this methodology presents some weaknesses that need to be overcome for the purpose of this work:

- **Information lost if any part of the process delays**. If anything affects the performance of the compression or transmission process during the mission, so that one of them is delayed, part of the hyperspectral data will be lost. This is due to the fact that the data are never written to a non-volatile memory on-board, and that the ring buffers will be overwritten after a while.
- **Original uncompressed data are not preserved.** The captured hyperspectral frames are just stored in the corresponding ring buffer in RAM memory, from where it is read by the compression process. However, they are never stored in a non-volatile memory and, hence, just the compressed-decompressed data will be available for the analysis. Because the HyperLCA compressor is a lossy transform-based one, the original data cannot be recovered.
- **Restricted input data format.** The overall compression process that is presented in [21] was developed assuming that the captured hyperspectral frames would be placed in the RAM memory in Band Interleaved by Pixel (bip) format using 16 bits unsigned integer values and coded in little-endian. However, many of the hyperspectral pushbroom cameras, such as the Specim FX10, work in Band Interleaved by Line (bil) format and use different precision and endianness.
- **Real-time data transmission not tested.** While, in [21], it was proved that it was possible to achieve real-time compression performance using the HyperLCA compressor in the Jetson boards, the transmission to the ground station was not tested.

A new design is proposed in this work to overcome these weaknesses and achieve the desired performance. Figure 5 shows a graphic description of this design.

The main changes with respect to the design previously proposed in [21] are:

- **Acquired frames and compressed frames are stored in non-volatile memory**. Each acquired hyperspectral frame is independently stored in a single file in the non-volatile memory of the on-board computer. Both the compression and transmission processes read from these files. By doing so, it is guaranteed that all of the frames will be compressed and transmitted to the ground station, even if a delay occurs in any process. Additionally, the original uncompressed hyperspectral data can be extracted from the on-board computer once that the mission finishes and the UAV lands.
- **Flexible input data format.** The compression process has been adapted to be able to process data in different input formats and data types, namely (bip and bil, little-endian, and big-endian), as well as being able to adapt to 12 and 16 bits resolutions. This makes the proposed design more flexible and adaptable to different hyperspectral cameras.
- **Real-time data transmission tested.** The transmission of the compressed hyperspectral data from the on-board computer to the ground station has been tested in this work, thus verifying the possibility of achieving real-time performance.

Each of these changes is explained in detail in the subsequent sections.

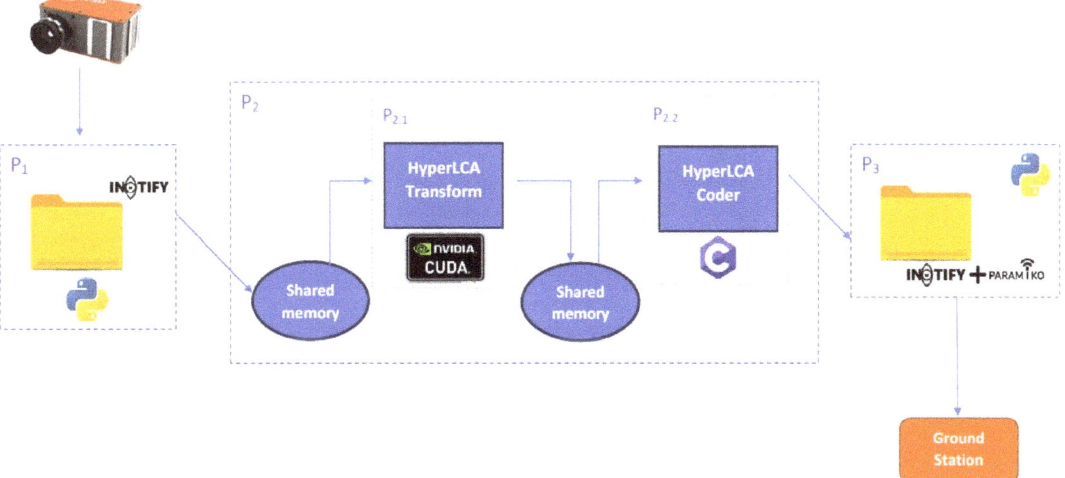

Figure 5. Proposed design for the real-time compression and transmission of the hyperspectral data.

4.1.1. Use of Non Volatile Memory

As previously mentioned, in the design that is described in [21], both the captured frames and the compressed frames are stored in ring buffers in RAM memory. This may result in frames being lost if any of the compression or transmission processes delay during mission. In order to avoid this potential failure, the design proposed in this work stores each frame captured by the Specim FX10 hyperspectral camera as a single file in non-volatile memory. These files are read to feed the compression process, and the compressed bitstreams for each frame are also stored as files to non-volatile memory. By doing so, even if the compression or transmission process delays for a while, all of the frames will be compressed and transmitted to the ground station with a certain latency. Additionally, the original frames keep being stored in the on-board computer and they can be extracted from it once the mission finishes and the UAV lands.

Despite the clear advantages of this new methodology, it also presents some limitations. First of all, the writing and reading speed in the non-volatile memory is usually lower than the corresponding speed of the RAM memory, thus making it more challenging to achieve real-time performance. Additionally, it is necessary to add new steps to the overall process to manage the writing and reading stages of the produced files and the synchronization between the data capturing, compression, and transmission processes.

Different levels of parallelism have been employed in order to carry out this methodology. A graphic representation of the overall process is displayed in Figure 5. First of all, a Python process (P1 in Figure 5) is executed using the Python module, named iNotify, which permits event-driven-programing. This process reacts every time that a file is created in the target directory by the capturing process, reading the stored hyperspectral frame and then loading it into the RAM memory for the compression process (P2 in Figure 5).

Secondly, another process (referred to as P2 in Figure 5) reads the hyperspectral frames to be compressed from the RAM memory, compresses them, and stores the resulting bitstream in a new file. This process involves the execution of the HyperLCA Transform, implemented in the LPGPU included in the Jetson board, and the HyperLCA entropy coder, executed in the CPU. For synchronizing the processes included inside P2, a set of ring buffers and memory pointers is used. Concretely, two shared memory ring buffers are used for transferring the data between the different subprocesses:

- **Input Frame Ring Buffer**. Part of the shared memory where the hyperspectral data to be compressed are placed.
- **Transform Output Data**. Part of the shared memory where the results of the HyperLCA Transform are stored for its codification.

Additionally, four single shared memory pointers are used to synchronize the execution of the different parallel subprocesses:

- **Last Captured Frame Index**. Index indicating the last frame captured by the hyperspectral camera.
- **Last Transformed Frame Index**. Index specifying the last transformed frame.
- **Last Coded Frame Index**. Index indicating the last coded frame.
- **Process Finished**. Boolean value indicating if the overall process should stop.

Finally, another Python process (P3 in Figure 5) is executed using the Python module named iNotify, together with the Python module, named Paramiko. This process reacts every time that a compressed bitstream is written to a file by process P2 in the target directory and transmits it to the ground station via SSH connection.

Altogether, these processes (P1 to P3) result in the next synchronized behavior, which keeps allowing both the real-time compression and the real-time transmission processes while at the same time is able to recover itself from losses of capabilities and permits to analyze the original captured data after the drone lands and the data are extracted from the disk.

1. The Specim FX10 VNIR sensor captures a new frame and stores it in a specific folder placed into a non-volatile memory. In this moment, the Python module named iNotify, which permits event-driven-programing, detects the creation of a new hyperspectral file that is loaded into the *Input Frame Ring Buffer* and the *Last Captured Frame Index* is increased. If the HyperLCA Transform is delayed, it is detected by this process by checking the *Last Tranformed Frame Index*, preventing overwriting the data in the *Input Frame Ring Buffer* until the HyperLCA Transform moves to the next frame.
2. The HyperLCA Transform is continuously checking the *Last Captured Frame Index* to know when a new frame is loaded in the *Input Frame Ring Buffer*. In that moment, the frame is copied from the shared memory to the device memory to be compressed by the HyperLCA Transform. Once this process finishes, the corresponding bitstream is copied to the *Transform Output Data* and the *Last Transformed Frame Index* is increased. If the codification process delays, it is detected by this process by checking the *Last Coded Frame Index*, preventing overwriting the data in the *Transform Output Data Ring Buffer* until the HyperLCA Coder moves to the next frame.
3. The entropy coder is continuously checking the *Last Transformed Frame Index* to know when a new frame has been compressed by the HyperLCA Transform and stored in

the *Transform Output Data*. After that, the frame is coded by the entropy coder and the resultant compressed frame is written into a specific folder within a non-volatile memory.

4. Once this file is written in the specific folder, the iNotify Python module detects this new element and triggers the transmission process of the compressed frame to the ground station via Secure Shell Protocol (SSH) while using the Python module, named Paramiko.

4.1.2. Flexible Input Data Format

As already described, the HyperLCA Transform stage, which was implemented into the LPGPU available in the Jetson board in [21], expects the input data to be formatted as 16 bits unsigned integers, in little-endian and BIP order. In order to make it more flexible and adaptable to any input data format, the first kernel of the HyperLCA Transform has been replaced by a set of six new kernels that support the conversion from different input data formats to the one that is required by the next kernel of the HyperLCA Transform. Each of these kernels first converts the data to 16 bits unsigned integer, in little-endian and BIP order, as expected by the previous version, and then casts these values to floating point arithmetic, as it is required by the subsequent HyperLCA Transform operations. The kernel to be executed in each case is selected from the CPU while taking the input format of the hyperspectral frames to be compressed into account. The six new proposed kernels are:

- **uint16_le_bip_to_float**. This kernel is used when the input image is formatted as unsigned 16 bits integers, in bip format and little-endian.
- **uint16_be_bip_to_float**. This kernel is used when the input image is formatted as unsigned 16 bits integers, in bip format and big-endian.
- **uint16_le_bil_to_float**. This kernel is used when the input image is formatted as unsigned 16 bits integers, in bil format and little-endian.
- **uint16_be_bil_to_float**. This kernel is used when the input image is formatted as unsigned 16 bits integers, in bil format and big-endian.
- **uint12_bip_to_float**. This kernel is used when the input image is formatted as unsigned 12 bits integers, in bip format.
- **uint12_bil_to_float**. This kernel is used when the input image is formatted as unsigned 12 bits integers, in bil format.

4.1.3. Packing Coded Frames

As it has been stated before in Section 4.1.1, the hyperspectral data are transmitted from the UAV to the ground station once the compressed frames are stored in non-volatile-memory. In one hand, this is possible due to the use of the SSH protocol that establishes a secured connection between the on-board computer that was installed on the drone and the ground station. On the other hand, it is the Python module, named Paramiko, which implements that SSH connection generating a client-server (drone-ground station) functionality for the transmission of the hyperspectral information.

As already described, the compression process is individually applied to each frame and, so, each single frame could be written to a single file. Because the transmission process is triggered every time that a file is written in the specified location using the Python iNotify module, the transmission process would be triggered once per frame and a SSH transmission would be executed for each of them. However, this may lead to a situation in which the overhead produced in the SSH connection for initiating the transmission of each data file is closed to the time that is needed to transmit the data. Thereby, in order to overcome this issue, the compression process may accumulate a certain amount of compressed frames to write them all together in a single file, thus reducing the communication overhead at the cost of slightly increasing the transmission latency. Figure 6 illustrates this strategy, where each block corresponds to a compressed hyperspectral frame.

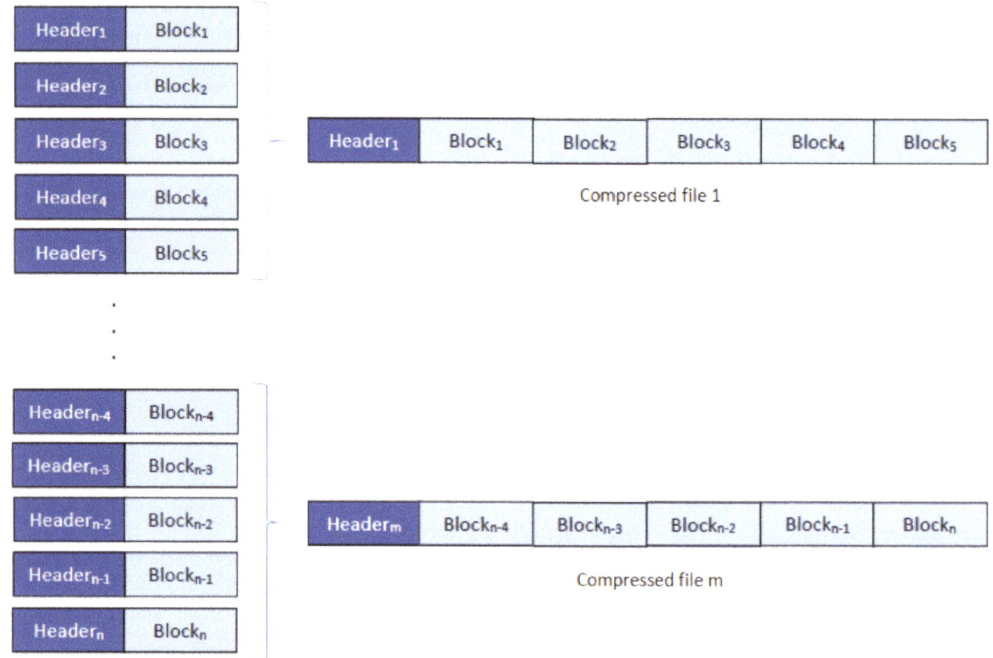

Figure 6. Strategy of packing various compressed frames into a single file.

4.2. Data Set Used in the Experiments

The proposed design for the real-time hyperspectral data compression and transmission has been tested in this work carrying out multiple experiments and using real hyperspectral data that are captured by the aforementioned UAV-based acquisition platform [10]. This design includes a compression stage using the HyperLCA algorithm, whose performance, in terms of compression quality, has been evaluated in detail in previous works [20,21,28]. Accordingly, this work exclusively focuses in the development and validation of a design that can achieve a real-time hyperspectral data compression and transmission in a real scenario.

An available hyperspectral dataset that was obtained from one of the performed missions has been used to simplify the experiments execution and the evaluation process. This image is a subset of the hyperspectral data collected during a flight campaign over a vineyard area in a village called Tejeda located in the center of the island of Gran Canaria. The exact coordinates of the scanned terrain are 27°59'35.6"N 15°36'25.6"W (graphically indicated in Figure 7a. The flight was performed at a height of 45 m over the ground and at a speed of 4.5 m/s with the hyperspectral camera capturing frames at 150 FPS. This resulted in a ground sampling distance in line and across line of approximately 3 cm. This flight mission consisted of 12 waypoints that provided six swathes, but just one of them was used in the experiments that were carried out in this work. The ground area covered by this swath is highlighted in the Google Map picture displayed in Figure 7b. In particular, a smaller portion of 825 subsequent hyperspectral frames of 1024 pixels and 160 spectral bands each was selected for the experiments (as highlighted in Figure 7c. Despite the Specim FX10 camera used for acquiring this data can collect up to 224 bands, the first 20 bands and the last 44 ones were discarded during the capturing process due to a low *Signal-to-Noise-Ratio* (SNR), as described in [29].

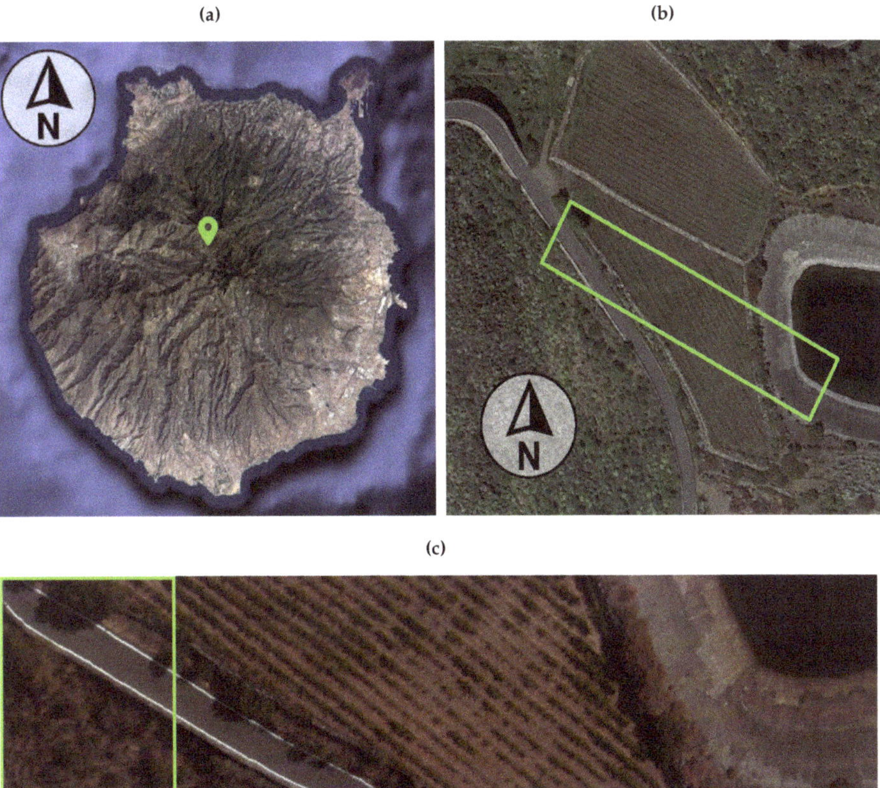

Figure 7. Graphical description of the appearance and location of the terrain corresponding to the image used in this work for the experiments. (**a**) Google Maps pictures indicating the terrain location on the island of Gran Canaria. (**b**) Google Maps pictures indicating the area covered during the selected swath of the flight campaign over the vineyard. (**c**) RGB respresentation of the real hyperspectral data, highlighting the portion of it that was used for the experiments.

Using the described hyperspectral image, an acquisition simulation process has been developed, which independently stores each of the 825 hyperspectral frames as a single file in a target directory and at user defined speed. This process has been used in the experiments to emulate the camera capturing process with the possibility of controlling the simulated capturing speed, while, at the same time, allowing for the execution of all the experiments with the same data and in similar conditions.

The Specim FX10 camera used for collecting the aforementioned images measures the incoming radiation using a resolution of 12 *bpppb*. However, it can be configured in two different modes, 12-*Packed* and 12-*Unpacked* to store each value using 12, or 16 *bpppb*, respectively. In the 12-*Unpacked* mode, four zeros are padded at the beginning of each sample. With independence of the mode used, the hyperspectral frames are always stored in *bil* format. However, in order to test the proposed design with more details, the acquired image has been off-board formatted to store it using different configurations. On one side, it has been stored using the 12-*Packed*, *bil* format, which is the most efficient format that is produced by the sensor. This has been labeled as *uint12-bil* in the results. On the other side, it has been stored as 12-*Unpacked*, *bip* format, which is not a format that can be directly produced by the sensor and requires reordering the data. Additionally, this format requires more memory, which makes it less efficient. However, this format is the one for which the reference HyperLCA compressor and its parallel implementation proposed in [21] were originally designed, and allows for using it without including any of the additional data transformation described in Section 4.1.2. This has been labeled as *uint16-bip* in the results. The results that could be obtained with any other combination of data precision and samples order would be in the range of the results that were obtained for these two combinations.

5. Results

Several experiments have been carried out in this work to evaluate the performance of the proposed designed for real-time hyperspectral data compression and transmission. First, the achievable transmission speed that could be obtained without compressing the acquired hyperspectral frames is analyzed in Section 5.1. The obtained results verify that the data compression is necessary for achieving a real-time transmission. Secondly, the maximum compression speed that can be obtained without parallelizing the compression process using the available GPU has been tested in Section 5.2. The obtained results demonstrate the necessity of parallelizing the process using the available GPU to achieve real-time performance. Finally, the results that can be obtained with the proposed design taking advantage of the different levels of parallelism and the available GPU are tested in Section 5.3, demonstrating its suitability for the real-time hyperspectral data compression and transmission.

In the following Tables 4–7, the simulated capturing speed and the maximum compressing and transmission speeds are referred to as Capt, Comp, and Trans, respectively. In addition, the WiFi signal strength is named Signal in the aforementioned tables of results, a value that is directly obtained from the WLAN interface during the transmission process.

5.1. Maximum Transmission Speed without Compression

First of all, the maximum transmission speed that could be achieved if the acquired frames were not compressed has been tested. For doing so, the targeted capturing speed has been set to 100 FPS, since it is the lowest frame rate typically used in our flight missions. Table 4 displays the obtained results.

Table 4. Transmission speed without compressing the frames.

Input Parameters			Jetson Xavier NX Speed (FPS)			Jetson Nano Speed (FPS)		
Format	FPS	Packing	Capt	Trans	Signal	Capt	Trans	Signal
uint12-bil	100	1	99	26	42/100	98	13	100/100
uint12-bil	100	5	103	36	45/100	101	18	100/100
uint16-bip	100	1	101	21	47/100	93	11	100/100
uint16-bip	100	5	100	30	38/100	105	15	100/100

As it can be observed, when independently transmitting each frame (Packing = 1), the transmission speed is too low in relation to the capturing speed, even capturing at 100 FPS. When transmitting five frames at a time, the transmission speed increases, but it is still very low. It can also be observed that, when using the *uint12-bil* format, the transmission speed is approximately 25 percent faster than when using the *uint16-bip* one. This makes sense, since the amount of data per file when using the *uint12-bil* format is 25 percent lower than the *uint16-bip* one. These results verify the necessity of compressing the acquired hyperspectral data before transmitting it to the ground station.

Additionally, it can also be observed in the results that the Jetson Xavier NX is able to achieve a higher transmission rate than the Jetson Nano, even if the Jetson Nano has a higher WiFi signal strength. This could be due to the fact that the Jetson Nano is using an external TP-Link TL-WN722N WiFi antenna that is connected through USB2.0 interface, while the WiFi antenna used by the Jetson Xavier NX is integrated in the board.

5.2. Maximum Compression Speed without Gpu Implementation

Once the necessity of carrying out the compression of the acquired hyperspectral data has been demonstrated, the necessity of speeding up this process by carrying out a parallel implementation is to be tested. For doing so, the compression process within the HyperLCA algorithm has been serially executed in the CPU integrated in both boards for evaluating the compression speed. The capturing speed has been set to the minimum value used in our missions (100 FPS), as done in Section 5.1. The compression parameters have been set to the less restrictive values typically used for the compression within the HyperLCA algorithm (CR = 20 and N_{bits} = 12). From the possible combinations of input parameters for the HyperLCA compressor, these values should lead to the fastest compression results according to the analysis done in [21], where the HyperLCA compressor was implemented onto a Jetson TK1 and Jetson TX2 developing boards. Additionally, the data format used for this experiment is *uint16-bip*, since this is the data format for which the reference version of the HyperLCA compressor is prepared. By using this data format, the execution of extra data format transformations that would lead to slower results is prevented. Table 5 displays the obtained results.

Table 5. Compression speed without using parallelism.

Input Parameters					Jetson Xavier NX Speed (FPS)		Jetson Nano Speed (FPS)	
Format	CR	N_{bits}	FPS	Packing	Capt	Comp	Capt	Comp
uint16-bip	20	12	100	1	103	23	96	9

The compression speed achieved without speeding up the process is too low and far from a real-time performance, as it can be observed in Table 5. This demonstrates the necessity of a parallel implementation that takes advantage of the available LPGPUs integrated in the Jetson boards for increasing the compression speed. It can also be observed that the Jetson Xavier NX offers a better performance that the Jetson Nano.

5.3. Proposed Design Evaluation

In this last experiment, the performance of the proposed design has been evaluated for both the Jetson Xavier NX and the Jetson Nano boards. The capturing speed has been set to the minimum and maximum values that are typically used in our flying missions, which are 100 FPS and 200 FPS, respectively. Two different combinations of compression parameters have been tested. The first one, CR = 20 and N_{bits} = 12, corresponds to the less restrictive scenario and should lead to the fastest compression results according to [21]. Similarly, the second one, CR = 12 and N_{bits} = 8, corresponds to the most restrictive case and it should lead to the slowest compression results, as demonstrated in [21]. Furthermore, both the *uint16-bip* and *uint12-bil* data formats have been tested. Finally, the compressed frames have been packed in two different ways, individually and in groups of 5. All of the obtained results are displayed in Table 6.

The Jetson Xavier NX is always capable of compressing more than 100 frames per second, regardless of the configuration used, as observed in Table 6. However, it is only capable of achieving 200 FPS in the compression in the less restrictive scenario, which is *uint12-bil*, CR = 20, N_{bits} = 12 and Packing = 5. Additionally, when capturing at 200 FPS, there are other scenarios in which real-time compression is almost achieved, producing compression speeds up to 190 FPS or 188 FPS. On the other hand, the Jetson Nano only presents a poorer compression performance, achieving real-time compression speed in the less restrictive scenario.

Regarding the transmission speed, it can be observed that, when packing the frames in groups of five, the transmission speed is always the same as the compression speed, but in the two fastest compression scenarios in which the compression speed reaches 200 and 188 FPS, the transmission speed stays at 185 and 162 FPS, respectively. This may be to the fact that the WiFi signal strength is not high enough (36/100 and 39/100) in these two tests. Nevertheless, two additional tests have been carried out for these particular situations with the Jetson Xavier NX, whose results are displayed in Table 7. A real-time transmission is achieved when increasing the packing size to 10 frames, as shown in this table. It can be also observed that in these two tests real-time compression has also been achieved. This may be due to the fact that a lower number of files are being written and read by the operating system. This suggests that a faster performance could be obtained in the Jetson Xavier NX using a solid stage disk (SSD) instead of the SD card that has been used so far in these experiments.

Table 6. Speed results for the proposed design.

Input Parameters					Jetson Xavier NX Speed (FPS)				Jetson Nano Speed (FPS)			
Format	CR	N_{bits}	FPS	Packing	Capt	Comp	Trans	Signal	Capt	Comp	Trans	Signal
uint12—bil	20	12	100	1	104	104	75	48/100	93	93	56	100/100
				5	106	106	106	37/100	101	101	101	100/100
			200	1	197	190	77	47/100	185	105	59	100/100
				5	202	200	185	36/100	208	110	110	100/100
	12	8	100	1	104	104	69	48/100	91	44	44	100/100
				5	104	104	104	37/100	94	44	44	100/100
			200	1	205	140	73	47/100	205	44	44	100/100
				5	187	138	138	37/100	187	45	45	100/100
uint16—bip	20	12	100	1	102	102	75	47/100	85	82	39	100/100
				5	103	103	103	38/100	98	83	83	100/100
			200	1	189	166	65	46/100	194	82	52	100/100
				5	193	188	162	39/100	186	84	84	100/100
	12	8	100	1	100	99	68	47/100	99	36	36	100/100
				5	101	101	101	38/100	99	36	36	97/100
			200	1	196	119	49	38/100	188	35	35	100/100
				5	171	110	110	37/100	208	36	36	100/100

Table 7. Speed results for the proposed design when increasing the packing size to 10 frames.

Input Parameters					Jetson Xavier NX Speed (FPS)			
Format	CR	N_{bits}	FPS	Packing	Capt	Comp	Trans	Signal
uint12-bil	20	12	200	10	206	206	206	37/100
uint16-bip	20	12	200	10	203	203	203	48/100

Additionally, the time at which each hyperspectral frame has been captured, compressed, and transmitted in these two last experiments is graphically displayed in Figure 8. Figure 8a shows the results that were obtained for the *uint12-bil* data format, while Figure 8b shows the values that were obtained for the *uint16-bip* one. In these two graphics, it can be observed how, on average, the slope of the line representing the captured frames (in blue color), the line representing the compressed frames (in orange), and the line representing the transmitted frames (in green color) is the same, indicating a real-time performance. It can also be observed the effect of packing the frames in groups of 10 in the transmission line. Finally, Figure 8a also shows the effect of a short reduction in the transmission speed due to a lower WiFi signal quality. As it can be observed in this figure, the transmission process is temporarily delayed and so is the compression one to prevent writing new data to the shared memory before processing the existing one. After a while, both the connection and the overall process are fully recovered, demonstrating the benefits of the proposed methodology previously exposed in Section 4.1.

Finally, all of the obtained results demonstrate that packing the compressed frames reduces the communication overhead and accelerates the transmission process. It can also be observed that using the data in *uint12-bil* format accelerates the compression process, since less data are been transferred through the shared memory, demonstrating the benefits of the developed kernels for allowing the compression process to be adapted to any input data format.

In general, the results that were obtained for the Jetson Xavier NX demonstrate that the proposed design can be used for real-time compressing and transmitting hyperspectral images at most of the acquisition frame rates typically used in our flying missions and using different configuration parameters for the compression. Additionally, it could be potentially faster if a solid stage disk (SSD) memory were used instead of the SD card that was used in the experiments. Regarding the Jetson Nano, a real-time performance was hardly achieved in just the less restrictive scenarios.

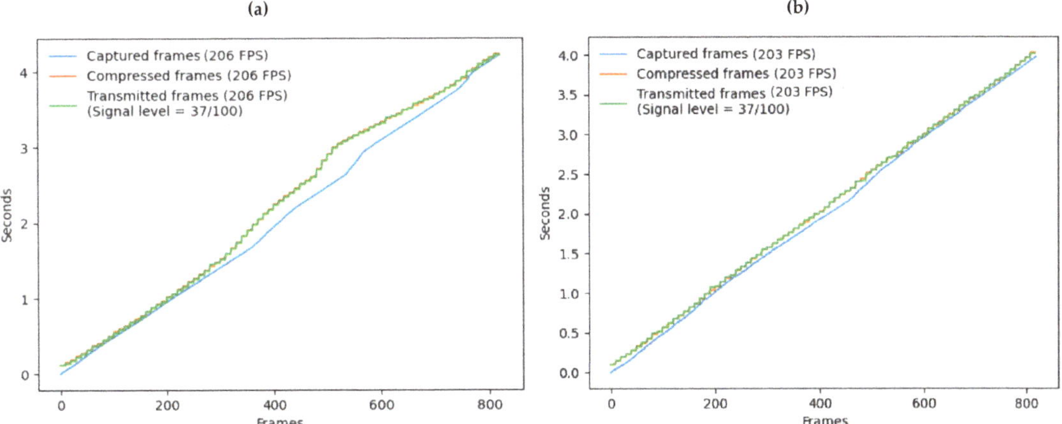

Figure 8. Graphical representation of the time in which each hyperspectral frame is captured, compressed and transmitted for the experiments displayed in Table 7. (**a**) *uint12-bil* version. (**b**) *uint16-bip* version.

6. Discussion

The main goal of this work is to be able to transmit the hyperspectral data captured by an UAV-based acquisition platform to a ground station in such a way that it can be analyzed and/or visualized in real-time to take decisions on the flight. The experiments conducted in Section 5.1 show the necessity of compressing the acquired hyperspectral data in order to achieve a real-time transmission. Nevertheless, the achievement of a real-time compression performance on-board this kind of platforms is not an easy task when considering the high data rate that is produced by the hyperspectral sensors and the limited computational resources available on-board. This fact has been experimentally tested in Section 5.2. While considering this, the compression algorithm to be used must meet several requirements, including a low computational cost, a high level of parallelism that allows for taking advantage of the LPGPUs available on-board to speed up the compression process, being able to guarantee high compression ratios, and integrate an error resilience nature to ensure that the compressed data can be delivered as expected. These requirements are very similar to those found in the space environment, where the acquired hyperspectral data must be rapidly compressed on-board the satellite to save transmission bandwidth and storage space, using limited computational resources and ensuring an error resilience behaviour.

The HyperLCA compressor has been selected in this work for carrying out the compression of the acquired hyperspectral data, since this compressor was originally developed for the space environment and satisfies all the mentioned requirements [20,28]. This compressor has been tested in previous works against those that were proposed by the

Consultative Committee for Space Data Systems (CCSDS) in their *Recomended Standards (Blue Books)* [25]. Concretely, in [20], it was compared with the Karhunen–Loeve transform-based approach described in the *CCSDS 122.1-B-1 standard* (Spectral Preprocessing Transform for Multispectral and Hyperspectral Image Compression) [30] as the best spectral transform for decorrelating hyperspectral data in terms of accuracy. The results shown in [20] indicate that the HyperLCA transform is able to achieve a similar decorrelation performance, but at a much lower computational cost and introducing extra advantages, such as higher levels of parallelism and an error resilience behavior. In [20], the HyperLCA compressor was also tested against the lossless prediction-based approach that was proposed in the *CCSDS 123.0-B-1 standard* (Lossless Multispectral and Hyperspectral Image Compression) [31]. As expected, the compression ratios that can be achieved by a lossless approach are very far from those that are required by the application targeted in this work. The CCSDS has recently published a new version of this prediction-based algorithm, making it able to behave not only as a lossless solution, but also as a near-lossless one to achieve higher compression ratios. This new solution has been published under the *CCSDS 123.0-B-2 standard* (Low-Complexity Lossless and Near-Lossless Multispectral and Hyperspectral Image Compression) [32] and it represents a very interesting alternative to be considered in future works.

On the other hand, although presenting a new methodology to transmit hyperspectral information from a UAV to a ground station in real-time is the main goal of this research work, the ulterior hyperspectral imaging applications have been always taken into account during the development process. This means that, while it is necessary to carry out lossy-compression to meet the compression ratios that are imposed by the acquisition data rates and transmission bandwidth, the quality of the hyperspectral data received in the ground station has to be good enough to preserve the desired performance in the targeted applications. As previously described in this work, the HyperLCA compressor was developed following an unmixing-like strategy in which the most different pixels present in each image block are perfectly preserved through the compression-decompression process. This is traduced in the fact that most of the information that is lost in the compression process corresponds to the image noise, while the relevant information is preserved, as demonstrated in [20,29]. In [20], the impact of the compression-decompression process within the HyperLCA algorithm was tested when using the decompressed images for hyperspectral linear unmixing, classification, and anomaly detection, demonstrating that the use of this compressor does not negatively affect the obtained results. This specific study was carried out while using well known hyperspectral datasets and algorithms, such as the *Pavia University* data set coupled with the *Support Vector Machine* (SVM) classifier, or the *Rochester Institute of Technology* (RIT) and the *World Trade Center* (WTC) images coupled with the *Orthogonal Subspace Projection Reed-Xiaoli* (OSPRX) anomaly detector. The work presented in [29] carries out a similar study, just for anomaly detection, but using the hyperspectral data that were collected by the acquisition platform used in this work and with the exact same configurations, both in the acquisition stage and compression stage. Concretely, the data used in this work, as described in Section 4.2, are a reduced subset of the hyperspectral data used in [29].

Finally, all of this work has been developed while assuming that a Wireless Local Area Network (WLAN), based on the 802.11 standard, will be available during the flight, and that both the on-board computer and ground station will be connected to it to use it as data downlink. Further research works are needed to increase the availability and range of this kind of networks or to be able to integrate the proposed solution with another wireless transmission technologies to make it available to a wider range of remote sensing applications.

7. Conclusions

In this paper, a design for the compression and transmission of hyperspectral data acquired from an UAV to a ground station has been proposed so that it can be analysed and/or visualized by an operator in real-time. This opens the possibility of taking advantage of the spectral information collected by the hyperspectral sensors for supervised or semi-supervised applications, such as defense, survilliance, or search and rescue missions.

The proposed design assumes that a WLAN will be available during the flight, and that both the on-board computer and the ground station will be connected to it to use it as data downlink. This design can work with different input data formats, which allows for using it with most of the hyperspectral sensors present in the market. Additionally, it preserves the original hyperspectral data in a non-volatile memory, producing two additional advantages. On one side, if the connection is lost for a while during the mission, the information is not lost, and the process will go on once the connection is recovered, guaranteeing that all of the acquired data are transmitted to the ground station. On the other side, once the mission finishes and the drone lands, the real hyperspectral data can be extracted from the non-volatile memory without compression if a more detailed analysis is required.

The entire design has been tested using two different boards from NVIDIA that integrate LPGPUs, the Jetson Xavier NX, and the Jetson Nano. The LPGPU has been used for accelerating the compression process, which is required for decreasing the reduction of the data volume for its transmission. The compression has been carried out using the HyperLCA algorithm, which permits achieving high compression ratios with a relatively high rate-distortion relation and at a reduced computational cost.

Multiple experiments have been executed to test the performance of all the stages that build up the proposed design for both the Jetson Xavier NX and Jetson Nano boards. The results obtained for the Jetson Xavier NX demonstrate that the proposed design can be used for real-time compressing and transmitting hyperspectral images at most of the acquisition frame rates typically used in our flying missions and using different configuration parameters for the compression. Additionally, it could be potentially faster if a solid stage disk (SSD) memory was used instead of the SD card that was used in these experiments. On the other hand, when using the Jetson Nano, a real-time performance was achieved in just the less restrictive scenarios.

Future research lines may include the optimization of the proposed design for reducing its computational burden, so that it can achieve a more efficient performance, especially when using boards with more limitations in terms of computational resources as an on-board computer.

Author Contributions: R.G. proposed the design for the real-time compression and transmission of the hyperspectral data. J.M.M. programmed the necessary GPU code, aided by M.D. who created the simulation platform. J.F.L. acquired the hyperspectral image from an UAV; S.L. conceived and designed the simulations; A.M. and A.J. performed the simulations; P.H. supervised the technical work and paper reviews. All authors contributed to the interpretation of the results and the writing of the paper. All authors have read and agreed to the published version of the manuscript.

Funding: This research has been funded by the Ministry of Economy and Competitiveness (MINECO) of the Spanish Government (PLATINO projecto, no. TEC2017-86722-C4-1 R), the European Comission and the European Regional Development Fund (FEDER) under project APOGEO (grant number MAC/1.1.b/226) and the Agencia Canaria de Investigación, Innovación y Sociedad de la Información (ACIISI) of the Conserjería de Economía, Industria, Comercio y Conocimiento of the Gobierno de Canarias, jointly with the European Social Fund (FSE) (POC2014-2020, Eje 3 Tema Prioritario 74 (85%)).

Informed Consent Statement: Not applicable.

Data Availability Statement: The data presented in this study are available on request from the corresponding author.

Conflicts of Interest: The authors declare no conflict of interest.

References

1. Ro, K.; Oh, J.S.; Dong, L. Lessons learned: Application of small uav for urban highway traffic monitoring. In Proceedings of the 45th AIAA Aerospace Sciences Meeting and Exhibit, Reno, NV, USA, 8–11 January 2007; p. 596.
2. Chen, Y.M.; Dong, L.; Oh, J.S. Real-time video relay for uav traffic surveillance systems through available communication networks. In Proceedings of the 2007 IEEE Wireless Communications and Networking Conference, Hong Kong, China, 11–15 March 2007; pp. 2608–2612.
3. Qu, Y.; Jiang, L.; Guo, X. Moving vehicle detection with convolutional networks in UAV videos. In Proceedings of the 2016 2nd International Conference on Control, Automation and Robotics (ICCAR), Hong Kong, China, 28–30 April 2016; pp. 225–229.
4. Zhou, H.; Kong, H.; Wei, L.; Creighton, D.; Nahavandi, S. Efficient road detection and tracking for unmanned aerial vehicle. *IEEE Trans. Intell. Transp. Syst.* **2014**, *16*, 297–309. [CrossRef]
5. Silvagni, M.; Tonoli, A.; Zenerino, E.; Chiaberge, M. Multipurpose UAV for search and rescue operations in mountain avalanche events. *Geomat. Nat. Hazards Risk* **2017**, *8*, 18–33. [CrossRef]
6. Scherer, J.; Yahyanejad, S.; Hayat, S.; Yanmaz, E.; Andre, T.; Khan, A.; Vukadinovic, V.; Bettstetter, C.; Hellwagner, H.; Rinner, B. An autonomous multi-UAV system for search and rescue. In Proceedings of the First Workshop on Micro Aerial Vehicle Networks, Systems, and Applications for Civilian Use, Florence, Italy, 18 May 2015; pp. 33–38.
7. Doherty, P.; Rudol, P. A UAV search and rescue scenario with human body detection and geolocalization. In Proceedings of the Australasian Joint Conference on Artificial Intelligence, Gold Coast, Australia, 2–6 December 2007; pp. 1–13.
8. Li, Z.; Liu, Y.; Hayward, R.; Zhang, J.; Cai, J. Knowledge-based power line detection for UAV surveillance and inspection systems. In Proceedings of the 2008 23rd International Conference Image and Vision Computing New Zealand, Christchurch, New Zealand, 26–28 November 2008; pp. 1–6.
9. Semsch, E.; Jakob, M.; Pavlicek, D.; Pechoucek, M. Autonomous UAV surveillance in complex urban environments. In Proceedings of the 2009 IEEE/WIC/ACM International Joint Conference on Web Intelligence and Intelligent Agent Technology, Milan, Italy, 15–18 September 2009; Volume 2, pp. 82–85.
10. Horstrand, P.; Guerra, R.; Rodríguez, A.; Díaz, M.; López, S.; López, J.F. A UAV platform based on a hyperspectral sensor for image capturing and on-board processing. *IEEE Access* **2019**, *7*, 66919–66938. [CrossRef]
11. Zarco-Tejada, P.J.; Guillén-Climent, M.L.; Hernández-Clemente, R.; Catalina, A.; González, M.; Martín, P. Estimating leaf carotenoid content in vineyards using high resolution hyperspectral imagery acquired from an unmanned aerial vehicle (UAV). *Agric. For. Meteorol.* **2013**, *171*, 281–294. [CrossRef]
12. Vanegas, F.; Bratanov, D.; Powell, K.; Weiss, J.; Gonzalez, F. A novel methodology for improving plant pest surveillance in vineyards and crops using UAV-based hyperspectral and spatial data. *Sensors* **2018**, *18*, 260. [CrossRef] [PubMed]
13. Mitchell, J.J.; Glenn, N.F.; Anderson, M.O.; Hruska, R.C.; Halford, A.; Baun, C.; Nydegger, N. Unmanned aerial vehicle (UAV) hyperspectral remote sensing for dryland vegetation monitoring. In Proceedings of the 2012 4th Workshop on Hyperspectral Image and Signal Processing: Evolution in Remote Sensing (WHISPERS), Shanghai, China, 4–7 June 2012; pp. 1–10.
14. Valentino, R.; Jung, W.S.; Ko, Y.B. A design and simulation of the opportunistic computation offloading with learning-based prediction for unmanned aerial vehicle (uav) clustering networks. *Sensors* **2018**, *18*, 3751. [CrossRef] [PubMed]
15. Freitas, S.; Silva, H.; Almeida, J.; Silva, E. Hyperspectral imaging for real-time unmanned aerial vehicle maritime target detection. *J. Intell. Robot. Syst.* **2018**, *90*, 551–570. [CrossRef]
16. Specim, Specim FX Series Hyperspectral Cameras. Available online: http://www.specim.fi/fx/ (accessed on 4 May 2019).
17. DJI, MATRICE 600 PRO. Available online: https://www.dji.com/bg/matrice600 (accessed on 4 May 2019).
18. NVIDIA, Jetson Nano Developer Kit. Available online: https://developer.nvidia.com/embedded/jetson-nano-developer-kit (accessed on 4 May 2019).
19. NVIDIA, Jetson Xavier NX. Available online: https://www.nvidia.com/es-es/autonomous-machines/embedded-systems/jetson-xavier-nx/ (accessed on 4 May 2019).
20. Guerra, R.; Barrios, Y.; Díaz, M.; Santos, L.; López, S.; Sarmiento, R. A new algorithm for the on-board compression of hyperspectral images. *Remote Sens.* **2018**, *10*, 428. [CrossRef]
21. Díaz, M.; Guerra, R.; Horstrand, P.; Martel, E.; López, S.; López, J.F.; Sarmiento, R. Real-time hyperspectral image compression onto embedded GPUs. *IEEE J. Sel. Top. Appl. Earth Obs. Remote Sens.* **2019**, *12*, 2792–2809. [CrossRef]
22. Díaz, M.; Guerra, R.; López, S.; Sarmiento, R. An Algorithm for an Accurate Detection of Anomalies in Hyperspectral Images With a Low Computational Complexity. *IEEE Trans. Geosci. Remote Sens.* **2018**, *56*, 1159–1176. [CrossRef]
23. Guerra, R.; Santos, L.; López, S.; Sarmiento, R. A new fast algorithm for linearly unmixing hyperspectral images. *IEEE Trans. Geosci. Remote Sens.* **2015**, *53*, 6752–6765. [CrossRef]
24. Guerra, R.; López, S.; Sarmiento, R. A computationally efficient algorithm for fusing multispectral and hyperspectral images. *IEEE Trans. Geosci. Remote Sens.* **2016**, *54*, 5712–5728. [CrossRef]
25. Consultative Committee for Space Data Systems (CCSDS), Blue Books: Recommended Standards. Available online: https://public.ccsds.org/Publications/BlueBooks.aspx (accessed on 16 January 2021).
26. Howard, P.G.; Vitter, J.S. Fast and efficient lossless image compression. In Proceedings of the Data Compression Conference, 1993. DCC'93, Snowbird, UT, USA, 30 March–2 April 1993; pp. 351–360.
27. TP-LINK, TP-LINK-TL-WN722N. Available online: https://www.tp-link.com/es/home-networking/adapter/tl-wn722n/#overview (accessed on 12 November 2020).

28. Guerra, R.; Barrios, Y.; Díaz, M.; Baez, A.; López, S.; Sarmiento, R. A hardware-friendly hyperspectral lossy compressor for next-generation space-grade field programmable gate arrays. *IEEE J. Sel. Top. Appl. Earth Obs. Remote Sens.* **2019**, *12*, 4813–4828. [CrossRef]
29. Díaz, M.; Guerra, R.; Horstrand, P.; López, S.; López, J.F.; Sarmiento, R. Towards the Concurrent Execution of Multiple Hyperspectral Imaging Applications by Means of Computationally Simple Operations. *Remote Sens.* **2020**, *12*, 1343. [CrossRef]
30. Consultative Committee for Space Data Systems (CCSDS), Blue Books: Recommended Standards. Spectral Preprocessing Transform for Multispectral and Hyperspectral Image Compression. Available online: https://public.ccsds.org/Pubs/122x1b1.pdf (accessed on 17 February 2021).
31. Consultative Committee for Space Data Systems (CCSDS), Blue Books: Recommended Standards. Lossless Multispectral and Hyperspectral Image Compression. Available online: https://public.ccsds.org/Pubs/123x0b1ec1s.pdf (accessed on 17 February 2021).
32. Consultative Committee for Space Data Systems (CCSDS), Blue Books: Recommended Standards. Low-Complexity Lossless and Near-Lossless Multispectral and Hyperspectral Image Compression. Available online: https://public.ccsds.org/Pubs/123x0b2c2.pdf (accessed on 17 February 2021).

remote sensing

MDPI

Article

Reduced-Complexity End-to-End Variational Autoencoder for on Board Satellite Image Compression

Vinicius Alves de Oliveira [1,2,*], Marie Chabert [1], Thomas Oberlin [3], Charly Poulliat [1], Mickael Bruno [4], Christophe Latry [4], Mikael Carlavan [5], Simon Henrot [5], Frederic Falzon [5] and Roberto Camarero [6]

[1] IRIT/INP-ENSEEIHT, University of Toulouse, 31071 Toulouse, France; marie.chabert@toulouse-inp.fr (M.C.); charly.poulliat@toulouse-inp.fr (C.P.)
[2] Telecommunications for Space and Aeronautics (TéSA) Laboratory, 31500 Toulouse, France
[3] ISAE-SUPAERO, University of Toulouse, 31055 Toulouse, France; thomas.oberlin@isae-supaero.fr
[4] CNES, 31400 Toulouse, France; Mickael.Bruno@cnes.fr (M.B.); Christophe.Latry@cnes.fr (C.L.)
[5] Thales Alenia Space, 06150 Cannes, France; mikael.carlavan@thalesaleniaspace.com (M.C.); simon.henrot@thalesaleniaspace.com (S.H.); frederic.falzon@thalesaleniaspace.com (F.F.)
[6] ESA, 2201 AZ Noordwijk, The Netherlands; roberto.camarero@esa.int
* Correspondence: Vinicius.Oliveira@irit.fr

Abstract: Recently, convolutional neural networks have been successfully applied to lossy image compression. End-to-end optimized autoencoders, possibly variational, are able to dramatically outperform traditional transform coding schemes in terms of rate-distortion trade-off; however, this is at the cost of a higher computational complexity. An intensive training step on huge databases allows autoencoders to learn jointly the image representation and its probability distribution, possibly using a non-parametric density model or a hyperprior auxiliary autoencoder to eliminate the need for prior knowledge. However, in the context of on board satellite compression, time and memory complexities are submitted to strong constraints. The aim of this paper is to design a complexity-reduced variational autoencoder in order to meet these constraints while maintaining the performance. Apart from a network dimension reduction that systematically targets each parameter of the analysis and synthesis transforms, we propose a simplified entropy model that preserves the adaptability to the input image. Indeed, a statistical analysis performed on satellite images shows that the Laplacian distribution fits most features of their representation. A complex non parametric distribution fitting or a cumbersome hyperprior auxiliary autoencoder can thus be replaced by a simple parametric estimation. The proposed complexity-reduced autoencoder outperforms the Consultative Committee for Space Data Systems standard (CCSDS 122.0-B) while maintaining a competitive performance, in terms of rate-distortion trade-off, in comparison with the state-of-the-art learned image compression schemes.

Keywords: remote sensing; lossy compression; on board compression; transform coding; rate-distortion; JPEG2000; CCSDS; learned compression; neural networks; variational autoencoder; complexity

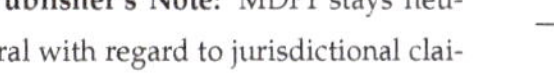

check for **updates**

Citation: Alves de Oliveira, V.; Chabert, M.; Oberlin, T.; Poulliat, C.; Bruno, M.; Latry, C.; Carlavan, M.; Henrot, S.; Falzon, F.; Camarero, R. Reduced-Complexity End-to-End Variational Autoencoder for on Board Satellite Image Compression. *Remote Sens.* **2021**, *13*, 447. https://doi.org/10.3390/rs13030447

Academic Editor: Cicily Chen

Received: 18 December 2020
Accepted: 22 January 2021
Published: 27 January 2021

Publisher's Note: MDPI stays neutral with regard to jurisdictional claims in published maps and institutional affiliations.

1. Introduction

Satellite imaging has many applications in oceanography, agriculture, biodiversity conservation, forestry, landscape monitoring, geology, cartography or military surveillance [1]. The increasing spectral and spatial resolutions of on board sensors allow obtaining ever-better quality products, at the cost of an increased amount of data to be handled. In this context, on board compression plays a key role to save transmission channel bandwidth and to reduce data-transmission time [2]. However, it is subject to strong constraints in terms of complexity. Compression techniques can be divided into two categories: lossless and lossy compression. Lossless compression is a reversible technique that compresses data without loss of information. The entropy measure, which quantifies the information contained in a source, provides a theoretical boundary for lossless compression, e.g., the lowest

51

attainable compression bit-rate. For optical satellite images, the typical lossless compression rate that can be achieved is less than 3:1 [3]. On the other side, lossy compression achieves high compression rates through transform coding [4] and the optimization of a rate-distortion trade-off. Traditional frameworks for lossy image compression operate by linearly transforming the data into an appropriate continuous-valued representation, quantizing its coefficients independently, and then encoding this discrete representation using a lossless entropy coder. To give on-ground examples, JPEG (Joint Photographic Experts Group) uses a discrete cosine transform (DCT) on blocks of pixels followed by a Huffman coder whereas JPEG2000 [5] uses an orthogonal wavelet decomposition followed by an arithmetic coder. In the context of on board compression, the consultative committee for space data systems (CCSDS), drawing on the on-ground JPEG2000 standard, recommends the use of the orthogonal wavelet transform [6]. However, the computational requirements of the CCSDS have been considerably reduced with respect to JPEG2000, taking into account the significant hardware constraints in payload image processing units of satellites. This work follows the same logic; however in the context of learned image compression. The idea is to propose a reduced-complexity learned compression scheme, considering on board limitations regarding computational resources due to hardware and energy consumption constraints.

In recent years, artificial neural networks appeared as powerful data-driven tools to solve problems previously addressed with model-based methods. Their independence from prior knowledge and human efforts can be regarded as a major advantage. In particular, image processing has been widely impacted by convolutional neural networks (CNNs). CNNs have proven to be successful in many computer vision applications [7] such as classification [8], object detection [9], segmentation [10], denoising [11] and feature extraction [12]. Indeed, CNNs are able to capture complex spatial structures in the images through the convolution operation that exploits local information. In CNNs, linear filters are combined with non-linear functions to form deep learning structures doted of a great approximation capability. Recently, end-to-end CNNs have been successfully employed for lossy image compression [13–16]. Such architectures jointly learn a non-linear transform and its statistical distribution to optimize a rate-distortion trade-off. They are able to dramatically outperform traditional compression schemes regarding this trade-off; however at the cost of a high computational complexity.

In this paper, we start from the state-of-the-art CNN image compression schemes [13,16] to design a reduced-complexity framework in order to adapt to satellite image compression. Please note that the second one [16] is a sophistication of the first one [13] that leads to higher performance at the cost of an increased complexity, by better adapting to the input image. More precisely, the variational autoencoder [16] allows reaching state-of-the-art compression performance, close to the one of BPG (Better Portable Graphics) [17] at the expense of a considerable increase in complexity with respect to [13], reflected by a runtime increase between 20% and 50% [16]. Our objective is to find an intermediary solution, with similar performance as [16] and similar or lower complexity as [13]. The first step is an assessment of the complexity of these reference frameworks and a statistical analysis of the transforms they learn. The objective is to simplify both the transform derivation and the entropy model estimation. Indeed, apart from a reduction of the number of parameters required for the transform, we propose a simplified entropy model that still preserves the adaptivity to the input image (as in [16]) and thus maintain compression performance.

The paper is organized as follows. Section 2 presents some background on learned image compression and details two interesting frameworks. Section 3 performs a complexity analysis of these frameworks and a statistical analysis of the transform they learn. Based on these analyses, a complexity-reduced architecture is proposed. After a subjective analysis of the resulting decompressed image quality, Section 4 quantitatively assesses the performance of this architecture on a representative set of satellite images. A comparative complexity study is performed and the impact of the different design options on the compression performance is studied, for various compression rates. A discussion regarding the compatibility of the proposed architecture complexity with the current and future satellite resources is then held. Section 5 concludes the paper. The symbols used in this paper are listed in Appendix A.

2. Background: Autoencoder Based Image Compression

Autoencoders were initially designed for data dimension reduction similar to e.g., Principal Component Analysis (PCA) [7]. In the context of image compression, autoencoders are used to learn a representation with low entropy after quantization. When devoted to compression, the autoencoder is composed of an analysis transform and a synthesis transform connected by a bottleneck that performs quantization and coding. Please note that the dequantization process is integrated in the synthesis transform. An auxiliary autoencoder can also be used to infer the probability distribution of the representation as in [16]. In this paper, we focus on two reference architectures: [13] displayed in Figure 1 (left) and [16] displayed in Figure 1 (right).

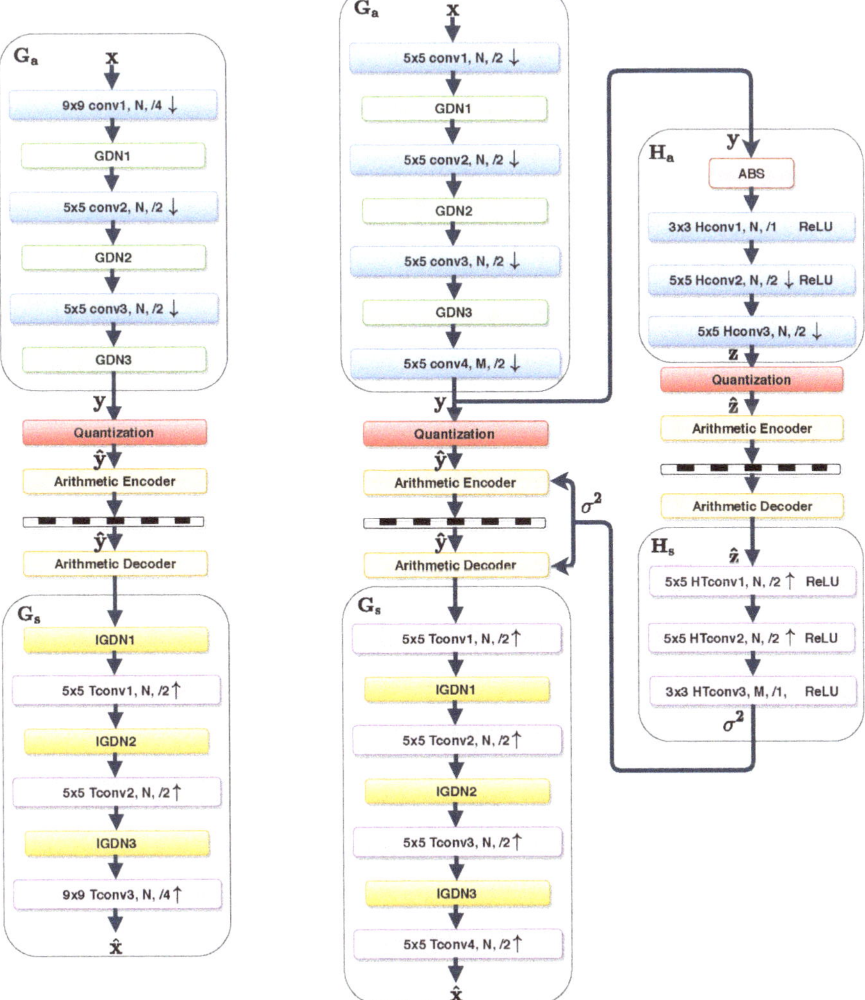

Figure 1. Architecture of the autoencoder [13] (**left**) and of the variational autoencoder [16] (**right**).

The first one [13] is composed of a single autoencoder. The second one [16] is composed of a main autoencoder (slightly different than the one in [13]) and of an auxiliary one which aims to infer the probability distribution of the latent coefficients. Recall that the second architecture is an upgraded version of the first one regarding both the design of the analysis and synthesis transforms and the estimation of the entropy model.

2.1. Analysis and Synthesis Transforms

In the main autoencoder (Figure 1 (left) and left column of Figure 1 (right)), the analysis transform G_a is applied to the input data $\mathbf{x}$ to produce a representation $\mathbf{y} = G_a(\mathbf{x})$. After the bottleneck, the synthesis transform G_s is applied to the quantized representation $\hat{\mathbf{y}}$ to reconstruct the image $\hat{\mathbf{x}} = G_s(\hat{\mathbf{y}})$. These representations are derived through several layers composed of filters each followed by a non-linear activation function. The learned representation is multi-channel (the output of a particular filter is called a channel or a feature) and non-linear. As previously mentioned, the analysis and synthesis transforms proposed in [16] result from improvements (mainly parameter adjustments) of the ones proposed in [13]. Thus, for brevity, the following description focuses on [16]. In [16], the analysis (resp. synthesis) transform G_a (resp. G_s) is derived through 3 convolutional layers each composed of N filters with kernel support $n \times n$ associated with parametric activation functions called generalized divisive normalizations (GDN) (resp. Inverse Generalized Divisive Normalizations (IGDN)) and a downsampling (resp. upsampling) by a factor 2. These three convolutional layers are linked to the input (resp. output) of the bottleneck by a convolution layer composed of $M > N$ (resp. N) filters with the same kernel support but without activation function. Please note that the last layer of the synthesis transform is composed of $M > N$ filters and leads to the so-called wide bottleneck that offers increased compression performance according to [16,18]. Contrarily to usual parameter-free activation functions (e.g., ReLU, sigmoid,...), GDN and IGDN are parametric functions that implement an adaptive normalization. In a given layer, the normalization operates through the different channels independently on each spatial location of the filter outputs. If $v_i(k,l)$ denotes the value indexed by (k,l) of the output of the ith filter, the GDN output is derived as follows:

$$GDN(v_i(k,l)) = \frac{v_i(k,l)}{(\beta_i + \sum_{j=1}^{N} \gamma_{ij} v_j^2(k,l))^{1/2}} \quad \text{for } i = 1,\ldots,N. \tag{1}$$

The IGDN is an approximate inverse of the GDN, derived as follows:

$$IGDN(v_i(k,l)) = v_i(k,l) \left(\beta_i' + \sum_{j=1}^{N} \gamma_{ij}' v_j^2(k,l) \right)^{1/2} \quad \text{for } i = 1,\ldots,N. \tag{2}$$

According to Equation (1) (resp. Equation (2)), the GDN (resp. IGDN) for channel i is defined by $N+1$ parameters denoted by β_i and γ_{ij} for $j = 1,\ldots,N$ (resp. β_i' and γ_{ij}' for $j = 1,\ldots,N$). Finally $N(N+1)$ parameters are required to define the GDN/IGDN in each layer. The learning and the storage of these parameters are required. However, GDN has been shown to reduce statistical dependencies [19,20] and thus it appears particularly appropriate for transform coding. According to [19], the GDN better estimates the optimal transform than conventional activation functions for a wide range of rate-distortion trade-offs. GDN/IGDN, while intrinsically more complex than usual activation functions, are prone to boost the compression performance especially in case of a low number of layers, thus affording a low global complexity for the network.

2.2. Bottleneck

The interface between the analysis transform and the synthesis transform, the so-called bottleneck, is composed of a quantizer that produces the discrete-valued vector $\hat{\mathbf{y}} = Q(\mathbf{y})$, an entropy encoder and its associated decoder. Recall that the dequantization is performed by the synthesis transform Gs (and by Hs in the case of the variational autoencoder). A standard entropy coding method, such as arithmetic, range or Huffman coding [21–23] losslessly compress the quantized data representation by exploiting its statistical distribution. The bottleneck thus requires a statistical model of the quantized learned representation.

2.3. Parameter Learning and Entropy Model Estimation

2.3.1. Loss Function: Rate Distortion Trade-Off

The autoencoder parameters (filter weights, GDN/IGDN parameters and representation distribution model) are jointly learned through the optimization of a loss function involving the rate $R(\hat{\mathbf{y}})$ and the distortion $D(\mathbf{x}, \hat{\mathbf{x}})$ between the original image $\mathbf{x}$ and the reconstructed image $\hat{\mathbf{x}}$. The rate-distortion criterion, denoted as $J(\mathbf{x}, \hat{\mathbf{x}}, \hat{\mathbf{y}})$, writes as the weighted sum:

$$J(\mathbf{x}, \hat{\mathbf{x}}, \hat{\mathbf{y}}) = \lambda D(\mathbf{x}, \hat{\mathbf{x}}) + R(\hat{\mathbf{y}}), \tag{3}$$

where λ is a key parameter that tunes the rate-distortion trade-off.

- The rate R achieved by an entropy coder is lower-bounded by the entropy derived from the actual discrete probability distribution $m(\hat{\mathbf{y}})$ of the quantized vector $\hat{\mathbf{y}}$. The rate increase comes from the mismatch between the probability model $p_{\hat{\mathbf{y}}}(\hat{\mathbf{y}})$ required for the coder design and $m(\hat{\mathbf{y}})$. The bit-rate is given by the Shannon cross entropy between the two distributions:

$$H(\hat{\mathbf{y}}) = \mathbb{E}_{\hat{\mathbf{y}} \sim m}\left[-log_2 p_{\hat{\mathbf{y}}}(\hat{\mathbf{y}})\right], \tag{4}$$

  where means distributed according to. The bit-rate is thus minimized if the distribution model $p_{\hat{\mathbf{y}}}(\hat{\mathbf{y}})$ is equal to the distribution $m(\hat{\mathbf{y}})$ arising from the actual distribution of the input image and from the analysis transform G_a. This highlights the key role of the probability model.
- The distortion measure D is chosen to account for image quality as perceived by a human observer. Due to its many desirable computational properties, the mean square error (MSE) is generally selected. However, a measure of perceptual distortion may also be employed such as the multi-scale structural similarity index (MS-SSIM) [24].

The loss function defined in Equation (3) is minimized through gradient descent with back-propagation [7] on a representative image training set. However, this requires the loss function to be differentiable. In the specific context of compression, a major hurdle is that the derivative of the quantization function is zero everywhere except at integers, where it is undefined. To overcome this difficulty, a quantization relaxation is considered in the backward pass (i.e., when back-propagating the gradient of the error). Ballé et al. (2016) [13] proposed to back-propagate an independent and identically distributed (i.i.d.) uniform noise, while Theis et al. (2017) [14] proposed to replace the derivative of the quantization function with a smooth approximation. In both cases, the quantization is kept as it is in the forward pass (i.e., when processing an input data).

2.3.2. Entropy Model

As stressed above, a key element in the end-to-end learned image compression frameworks is the entropy model defined through the probability model $p_{\hat{\mathbf{y}}}(\hat{\mathbf{y}})$ assigned to the quantized representation for coding.

- Fully factorized model: For simplicity, in [13,14], the approximated quantized representation was assumed independent and identically distributed within each channel and the channels were assumed independent of each other, resulting in a fully factorized distribution:

$$p_{\tilde{\mathbf{y}}|\boldsymbol{\psi}}(\tilde{\mathbf{y}}|\boldsymbol{\psi}) = \prod_i p_{\tilde{y}_i|\psi^{(i)}}(\tilde{y}_i), \tag{5}$$

where index i runs over all elements of the representation, through channels and through spatial locations, $\boldsymbol{\psi}^{(i)}$ is the distribution model parameter vector associated with each element. As mentioned previously, for back-propagation derivation during the training step, the quantization process ($\hat{\mathbf{y}} = Q(\mathbf{y})$) is approximated by the addition of an i.i.d uniform noise $\Delta\mathbf{y}$, whose range is defined by the quantization step. Due to the adaptive local normalization performed by GDN non-linearities, the quantization step can be set to one without loss of generality. Hence the quantized representation $\hat{\mathbf{y}}$, which is a discrete random variable taking values in $\mathbb{Z}$, is modelled by the continuous random vector $\tilde{\mathbf{y}}$ defined by:

$$\tilde{\mathbf{y}} = \mathbf{y} + \Delta\mathbf{y} \tag{6}$$

taking values in $\mathbb{R}$. The addition of the uniform quantization noise leads to the following expression for $p_{\tilde{y}_i|\boldsymbol{\psi}^{(i)}}(\tilde{y}_i)$ defined through a convolution by a uniform distribution on the interval $[-1/2, 1/2]$:

$$p_{\tilde{y}_i|\boldsymbol{\psi}^{(i)}}(\tilde{y}_i) = p_{y_i|\boldsymbol{\psi}^{(i)}}(y_i) * \mathcal{U}(-1/2, 1/2). \tag{7}$$

For generality, in [13], the distribution $p_{y_i|\boldsymbol{\psi}^{(i)}}(y_i)$ is assumed non parametric, namely without predefined shape. In [13,14], the parameter vectors are learned from data during the training phase. This learning, performed once and for all, prohibits adaptivity to the input images during operational phase. Moreover, the simplifying hypothesis of a fully factorized distribution is very strong and not satisfied in practice, elements of $\hat{\mathbf{y}}$ exhibiting strong spatial dependency as observed in [16]. To overcome these limitations and thus to obtain a more realistic and more adaptive entropy model, [16] proposed a hyperprior model, derived through a variational autoencoder, which takes into account possible spatial dependency in each input image.

- Hyperprior model: Auxiliary random variables $\tilde{\mathbf{z}}$, conditioned on which the quantized representation $\tilde{\mathbf{y}}$ elements are independent, are derived from $\mathbf{y}$ by an auxiliary autoencoder, connected in parallel with the bottleneck (right column of Figure 1 (right)). The hierarchical model hyper-parameters are learned for each input image in operational phase. Firstly, the hyperprior transform analysis H_a produces the set of auxiliary random variables $\mathbf{z}$. Secondly, $\mathbf{z}$ is transformed by the hyperprior synthesis transform H_s into a second set of random variables σ. In [16], $\mathbf{z}$ distribution is assumed fully factorized and each representation element $\tilde{y}_i$, knowing $\mathbf{z}$, is modeled by a zero-mean Gaussian distribution with its own standard deviation σ_i. Finally, taking into account the quantization process, the conditional distribution of each quantized representation element is given by:

$$\tilde{y}_i|\tilde{\mathbf{z}} \sim \mathcal{N}\left(0, \sigma_i^2\right) * \mathcal{U}\left(-\frac{1}{2}, \frac{1}{2}\right). \tag{8}$$

The rate computation must take into account the prior distribution of $\tilde{\mathbf{z}}$, which has to be transmitted to the decoder with the compressed data, as side information.

In the following we derive the complexity of the analysis and synthesis transforms, involved both in the main and in the auxiliary autoencoders of [16]. We also perform a statistical analysis of the learned transform on a representative set of satellite images. The objective is to propose an entropy model simpler while adaptive, because more specifically tailored than the two previous ones.

3. Reduced-Complexity Variational Autoencoder

In the literature, the design of learned image compression frameworks hardly takes into account the computational complexity: the objective is merely to obtain the best performance in terms of rate-distortion trade-off. However, in the context of on board compression, a trade-off between performance and complexity has also be considered to take into account the strong computational constraints. Our focus here is to propose a complexity-reduced alternative to the state-of-the-art structure [16] minimizing the impact on the performance. Please note that this complexity reduction must be essentially targeted at the coding part of the framework, most subject to the on board constraints. A meaningful indicator of the complexity reduction is the number of network parameters. Indeed, its reduction has a positive impact not only on the memory complexity, but also on the time complexity and on the training difficulty. Indeed, the optimization problem involved in the training step applies on a smaller number of parameters. This is an advantage even if the training is performed on ground. The convergence is thus obtained with less iterations and thus within a shorter time. Moreover, the complexity reduction has also a positive impact on the ease to upload the final model to the spacecraft (once commissioning is finished and the training with real data completed) as the up-link transmission is severely limited.

3.1. Analysis and Synthesis Transforms

3.1.1. Complexity Assessment

First, let characterize the computational complexity of a convolutional layer composing the analysis and synthesis transforms. Let N_{in} denote the number of features at the considered layer input. In the particular case of the network first layer, N_{in} is the number of channels of the input image ($N_{in} = 1$ for a panchromatic image) else N_{in} is the number of filters of the previous layer. Let N_{out} denote the number of features at this layer output, i.e., the number of filters of this layer. As detailed in Section 2, in [16], $N_{out} = N$ for each layer of the analysis and synthesis transforms except for the last one of the main auto-encoder analysis transform and the last one of the auxiliary auto-encoder synthesis transform composed of M filters with $M > N$ and thus for these layers $N_{out} = M$. As in [13,16], we consider square filters with size $n \times n$. The number of parameters associated with the filtering part of the layer is:

$$\text{Param}^f = (n \times n \times N_{in} + \delta) \times N_{out}. \tag{9}$$

The term δ is equal to 1 when a bias is introduced and is equal to 0 otherwise. Please note that this bias is rarely used in the considered architectures (except in Tconv3, as displayed in Figure 1). The filtering is applied to each input channel after downsampling (respectively upsampling). The downsampled (resp. upsampled) input channel if of size $s_{out} \times s_{out}$ with $s_{out} = s_{in}/D$ (respectively $s_{out} = s_{in} \times D$) where D denotes the downsampling (respectively upsampling) factor and $s_{in} \times s_{in}$ is the size of a feature at the filter input. Floating points operations Operation^f for the filtering operation is thus:

$$\text{Operation}^f = \text{Param}^f \times s_{out} \times s_{out}. \tag{10}$$

GDN/IGDN perform a normalization of a filter output with respect to the other filter outputs. According to Section 2, the number of parameters and the number of operations of each GDN/IGDN are expressed by:

$$\text{Param}^g = (N_{out} + 1) \times N_{out}$$
$$\text{Operation}^g = \text{Param}^g \times s_{out} \times s_{out}. \tag{11}$$

Since the number of layers is already very low for the considered architectures, the reduction of the complexity of the analysis and synthesis transforms may target, according to the previous complexity assessment, the number of filters per layer, the size of these filters and the choice of the activation functions. Our proposal below details our strategy for complexity reduction.

3.1.2. Proposal: Simplified Analysis and Synthesis Transforms

Our approach to reduce the complexity of the analysis and synthesis transforms, while maintaining an acceptable rate-distorsion trade-off, is to fine-tune the parameters of each layer (number of filters and filter sizes) and to consider the replacement of GDN/IGDN by simpler non-parametric activation functions.

In a first step, we propose a fine-tuning of the number of filters composing the convolutional layers of the analysis and synthesis transforms. The state-of-the-art frameworks generally involve a large number of filters with the objective of increasing the network approximation capability [13,16]. However, a high number of filters also implies a high number of parameters and operations in the GDN/IGDN, as it increases the depth of the tensors (equal to the number of filters) at their input. Apart from a harder training, a large number of filters comes with a high number of operations as well as a high memory complexity, which is problematic in the context of on board compression. In [16], the number of filters at the bottleneck (M) and for the other layers (N) are fixed according to the target bit-rate: the higher the target bit-rate, the higher M and N. Indeed, higher bit rates mean lower distortion and thus increased network approximation capabilities to learn accurate analysis and synthesis transforms [16]. This principle also applies to the auxiliary autoencoder implementing the hyperprior illustrated in Figure 1 (right column of the right part). In the present paper, we propose and evaluate a reduction of the number of filters in each layer for different target rate ranges. In particular, we investigate the impact of M on the attainable performance when imposing a drastic reduction of N. The question is whether there is a real need for a high global number of filters (high N and M) or whether a high bottleneck size (low N and high M) is sufficient to achieve good performance at high rates. For that purpose, we impose a low value of N (typically $N = 64$) and we consider increasing values of M defined by $M = 2N, 3N, 4N, 5N$ to determine the minimum value of M that leads to an acceptable performance in a given rate range.

In a second step, we investigate the replacement of the GDN/IGDN by non-parametric activation functions. As previously mentioned, according to [19], GDN/IGDN allow obtaining good performance even with a low number of layers. However, for the sake of completeness, we also test their replacement by ReLU functions. Finally, we propose to evaluate the effect of the filter kernel support. The main autoencoder in [16] is entirely composed of filters with kernel support $n \times n$. The idea then is to test different kernel supports that is $(n - 2) \times (n - 2)$ and $(n + 2) \times (n + 2)$.

3.2. Reduced Complexity Entropy Model

3.2.1. Statistical Analysis of the Learned Transform

This section first performs a statistical analysis of each feature of the learned representation in the particular case of satellite images. A similar statistical analysis has been conducted in the case of natural images in [25] with the objective to properly design the quantization in [13]. The probability density function related to each feature, averaged on a representative set of natural images, was estimated through a normalized histogram. The study showed that most features can be accurately modelled as Laplacian random variables. Interestingly, a similar result has also been analytically demonstrated in [26] for block-DCT coefficients of natural images under the assumption that the variance is constant on each image block and that its values on the different blocks are distributed according to an exponential or a half-normal distribution. We conducted the statistical analysis on the representation obtained by the main autoencoder as defined in [16], with $N = 128$ and $M = 192$, but used alone, as in [13], without auxiliary autoencoder. Indeed, on one side the main autoencoder in [16] benefits from improvements with respect to the one in [13] and on another side, the auxiliary autoencoder is not necessary in this statistical study. This autoencoder is trained on a representative dataset of satellite images and for rates between 2.5 bits/pixel and 3 bits/pixel. First, as an illustration, let consider the satellite image displayed in Figure 2. This image of the city of Cannes (French Riviera) is a 12-bit simulated panchromatic Pléiades image with size 512×512 and resolution 70 cm.

Figure 2. Simulated 12-bit Pléiades image of Cannes with size 512×512 and resolution 70 cm.

Figure 3 shows the 1st 32×32 feature derived from the Cannes image and its normalized histogram with Laplacian fitting.

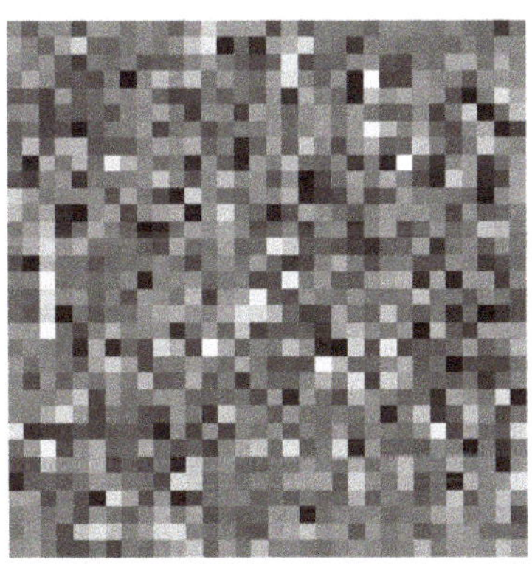

(**a**) 1st feature.

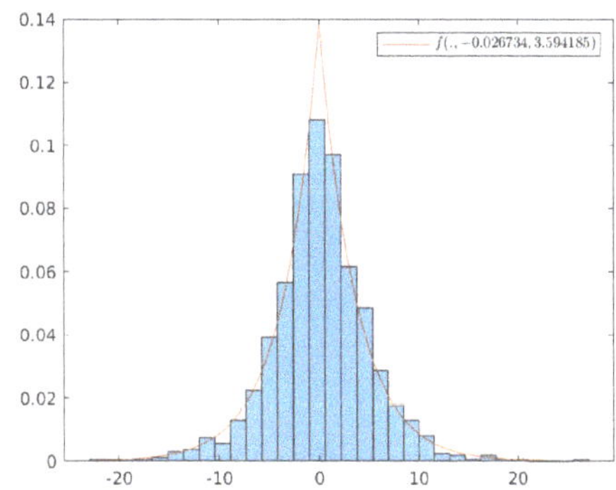

(**b**) Normalized histogram and Laplacian fitting.

Figure 3. First feature of Cannes image representation, its normalized histogram with Laplacian fitting.

On this example, the fitting with an almost centered Laplacian seems appropriate. According to the Kolmogorov-Smirnov goodness-of-fit test [27], 94% of the features derived from this image follow a Laplacian distribution with a significance level $\alpha = 5\%$. Recall that the Laplacian distribution $\text{Laplace}(\mu, b)$ is defined by:

$$f(\zeta, \mu, b) = \frac{1}{2b}\left(-\frac{|\zeta - \mu|}{b}\right) \text{ for } \zeta \in \mathbb{R}, \tag{12}$$

where μ is the mean value and $b > 0$ is a scale parameter related to the variance by $Var(\zeta) = 2b^2$.

To extend this result, the representation (composed of $M = 192$ feature maps) was derived for 16 simulated 512×512 Pléiades images. Figure 4 shows the normalized histograms (derived from 16 observations) of some feature maps (each of size 32×32) and the Laplacian fitting.

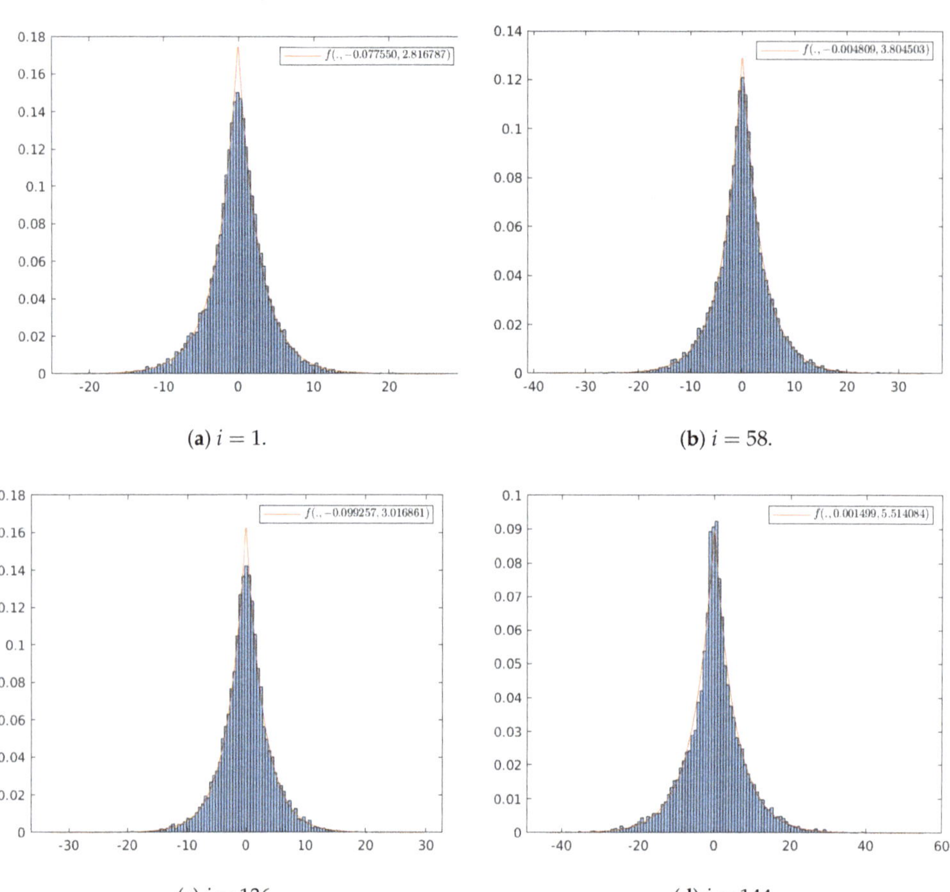

(a) $i = 1$.

(b) $i = 58$.

(c) $i = 136$.

(d) $i = 144$.

Figure 4. Normalized histogram of the ith feature map and Laplacian fitting $f(., \mu, b)$.

Most of the feature maps have a similar normalized histogram. According to Kolmogorov-Smirnov goodness-of-fit test [27], 94% of the features derived from this image follow a Laplacian distribution with a significance level $\alpha = 5\%$. Please note that the Gaussian distribution, $\mathcal{N}(\mu, \sigma^2)$ with a small value of μ, also fits the features albeit to a lesser extent. The remaining non-Laplacian feature maps (6% of the maps for this example) stay close to the Laplacian distribution. Figure 5 displays two representative examples of non-Laplacian feature maps. Please note that the first one (19th feature map) is far from Laplacian, this feature appears as a low-pass approximation of the input image. However, this is a very particular case: the second displayed feature has a typical behavior, not so far from a Laplacian distribution.

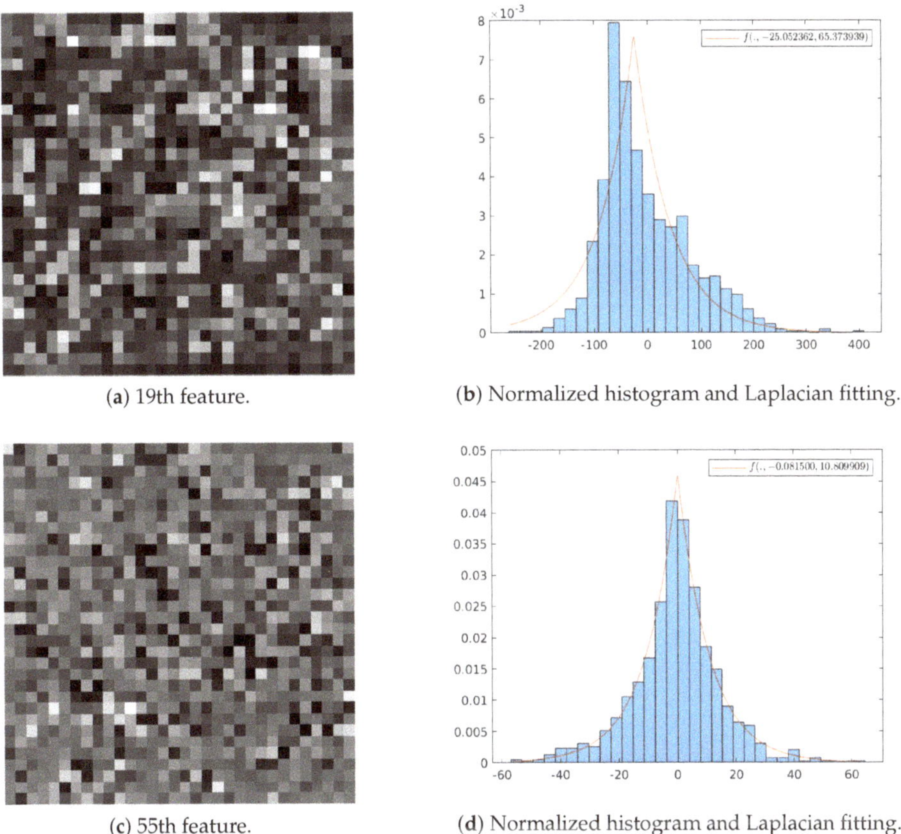

(a) 19th feature.

(b) Normalized histogram and Laplacian fitting.

(c) 55th feature.

(d) Normalized histogram and Laplacian fitting.

Figure 5. Normalized histogram of the *i*th feature map and Laplacian fitting $f(., \mu, b)$.

3.2.2. Proposal: Simplified Entropy Model

The entropy model simplification aims at achieving a compromise between simplicity and performance while preserving the adaptability to the input image. In [13], the representation distribution is assumed fully factorized and the statistical model for each feature is non-parametric to avoid a prior choice of a distribution shape. This model is learned once, during the training. In [16], the strong independence assumption leading to a fully factorized model is avoided by the introduction of the hyperprior distribution, whose parameters are learned for each input image even in the operational phase. Both models are general and thus suitable to a wide variety of images; however the first one implies a strong hypothesis of independence and prohibits adaptivity while the second one is computationally expensive. Based on the previous analysis, we propose the following parametric model for each of the M features. Consider the jth feature elements y_{i_j} for $i_j \in I_j$, where I_j denotes the set of indexes covering this feature:

$$y_{i_j} \sim \text{Laplace}(0, b_j) \text{ (resp. } y_i^j \sim \mathcal{N}(0, \sigma_j^2))$$

$$\text{with: } b_j = \sqrt{Var(y_i^j)/2} \text{ (resp. } \sigma_j^2 = Var(y_i^j)). \tag{13}$$

The problem then boils down to the estimation of a single parameter per feature referred to as the scale b_j (respectively the standard deviation σ_j) in the case of the Laplacian (resp. Gaussian) distribution. Starting from [16], this proposal reduces the complexity at two levels. First, the hyperprior autoencoder, including the analysis H_a and synthesis H_s transforms, is removed. Second, the side information initially composed of the compressed auxiliary random variable set (**z**) of size $8 \times 8 \times M$ now reduces to a $M \times 1$ vector of variances. The auxiliary network simplification is displayed on the right part of Figure 6.

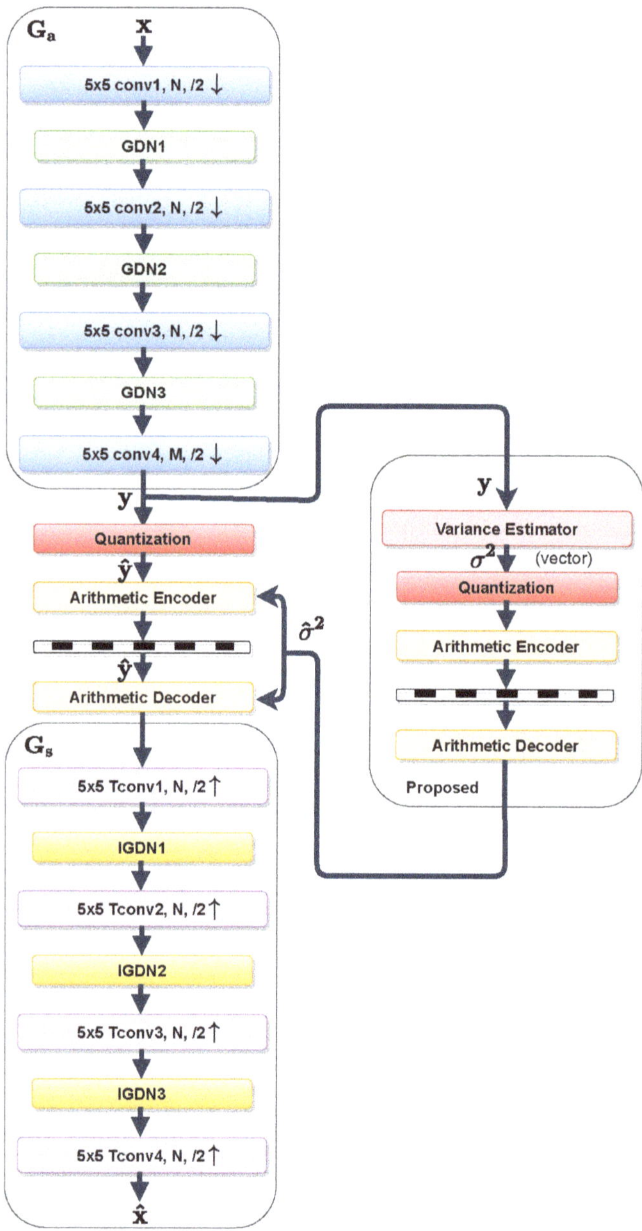

Figure 6. Proposed architecture after entropy model simplification: main autoencoder (**left column**) and simplified auxiliary autoencoder (**right column**).

In the next section, the performance of these proposals and of their combinations are studied for different rate ranges.

4. Performance Analysis

This section assesses the performance, in terms of rate-distortion, of the proposed method in comparison with the CCSDS 122.0-B [6], JPEG2000 [5] and with the reference methods [13,16]. Beforehand, a subjective image quality assessment is proposed. Although informal, it allows comparing the artefacts produced by learned compression and by the CCSDS 122.0-B [6].

4.1. Implementation Setup

To assess the relevance of the proposed complexity reductions, experiments were conducted using TensorFlow. The batch size (i.e., the number of training samples to work through before the parameters are updated) was set to 8 and up to 1M iterations were performed. Both training and validation datasets are composed of simulated 12-bit Pléiades panchromatic images provided by the CNES, covering various landscapes (i.e., desert, water, forest, industrial, cloud, port, rural, urban). The reference learned frameworks for image compression are designed to handle 8-bit RGB natural images and they generally target low rates (typically up to a maximum of 1.5 bits/pixel). In contrast, on board satellite compression handles 12-bit panchromatic images and targets higher rates (from 2bits/pixel to 3.5 bits/pixel). The training dataset is composed of 8M of patches (of size 256×256) randomly cropped from 112 images (of size 585×585). The validation dataset is composed of 16 images (of size 512×512). MSE was considered to be the distortion metric for training. The rate and distortion measurements were averaged across the validation dataset for a given value of λ. Please note that the value of λ has to be set by trial and error for a targeted rate range.

In addition to the MSE, we also evaluate those results in terms of MS-SSIM. Please note that they exhibit a similar behavior even if the models were trained for the MSE only. The proposed framework is compared with the CCSDS 122.0-B [6], JPEG2000 [5] and with the reference methods [13,16] implemented for values of N and M recommended by their authors for particular rate ranges.

- `Ballé(2017)-non-parametric-N` refers to the autoencoder [13] and is implemented for $N = 192$ (respectively $N = 256$) for rates below 2 bits/pixel (respectively above 2bits/pixel).
- `Ballé(2018)-hyperprior-N-M` refers to the variational autoencoder [16] and is implemented for $N = 128$ and $M = 192$ (respectively $N = 192$ and $M = 320$) for rates below 2 bits/pixel (respectively above 2 bits/pixel).

4.2. Subjective Image Quality Assessment

At low rates, JPEG2000 is known to produce quite prominent blurring and ringing artifacts, which is particularly visible in high-frequency textures [5]. This is also the case for the CCSDS 122.0-B [6]. Figure 7a,b shows the original image of the city of Blagnac, which is a 12-bit simulated panchromatic Pléiades image with size 512×512 and resolution 70 cm. The figure also shows the image compressed by CCSDS 122.0 (c) and the image compressed by the reference learned compression architecture `Ballé(2017)-non-parametric-N` (d), for a low compression rate (1.15 bits/pixel). The image obtained through learned compression appears closest to the original one than the image obtained through the CCSDS.

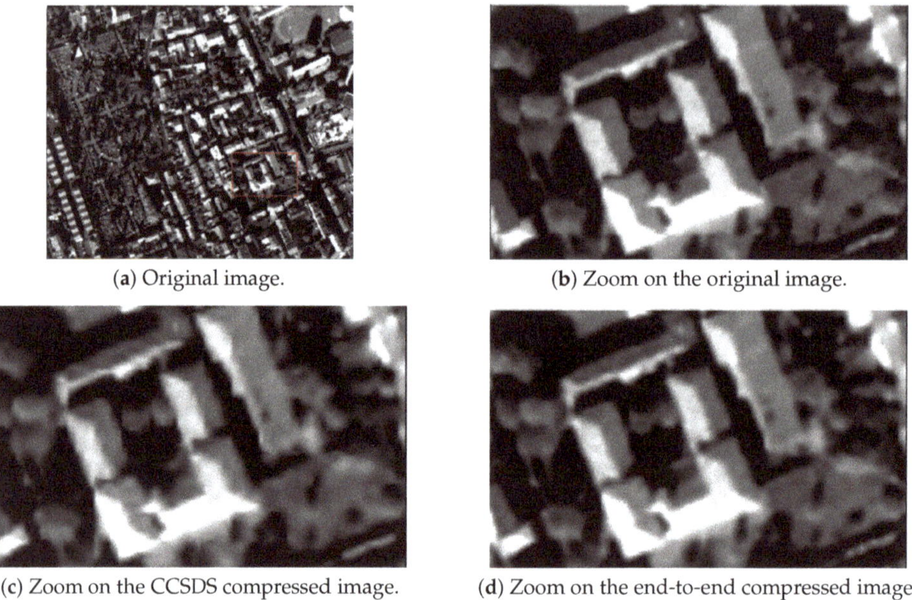

(**a**) Original image.

(**b**) Zoom on the original image.

(**c**) Zoom on the CCSDS compressed image.

(**d**) Zoom on the end-to-end compressed image.

Figure 7. Subjective image quality analysis (R = 1.15 bits/pixel).

For medium luminances, far less artifacts, such as blurring and flattened effects are observed. In particular, the building edges are sharper. The same is true for low luminances, corresponding to shaded areas: the ground markings are sharp and less flatened areas are observed. As shown in Figure 8, for a higher rate of 1.66 bits/pixel, the two reconstructed images are very close. However, the image obtained through learning compression remains closest to the original one, especially in shaded areas.

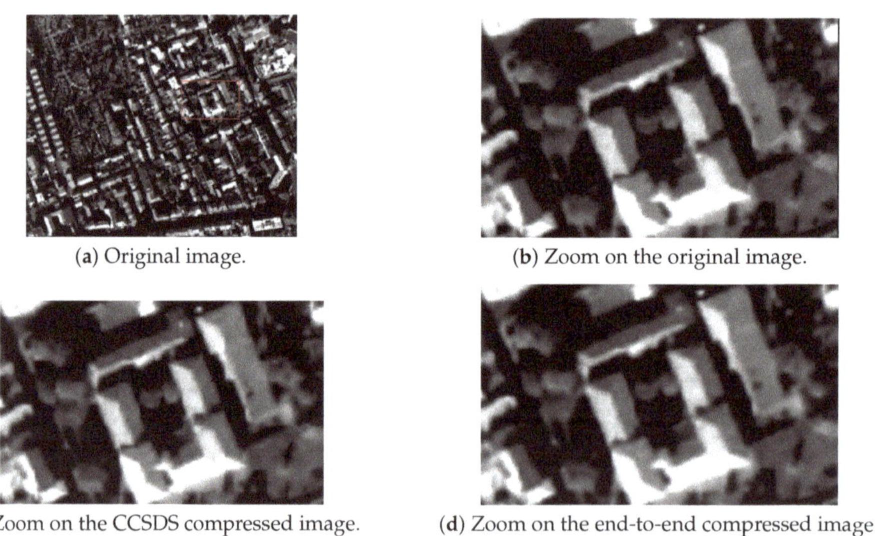

(**a**) Original image.

(**b**) Zoom on the original image.

(**c**) Zoom on the CCSDS compressed image.

(**d**) Zoom on the end-to-end compressed image.

Figure 8. Subjective image quality analysis (R = 1.66 bits/pixel).

Finally, as shown in Figure 9 for an even higher rate of 2.02 bits/pixel, CCSDS 122.0 still produces flattened effects, especially in low variance areas. The learned compression method leads to a reconstructed image that is closest to the original one. Even though the stadium ground markings slightly differs from the original, the image quality is overall preserved.

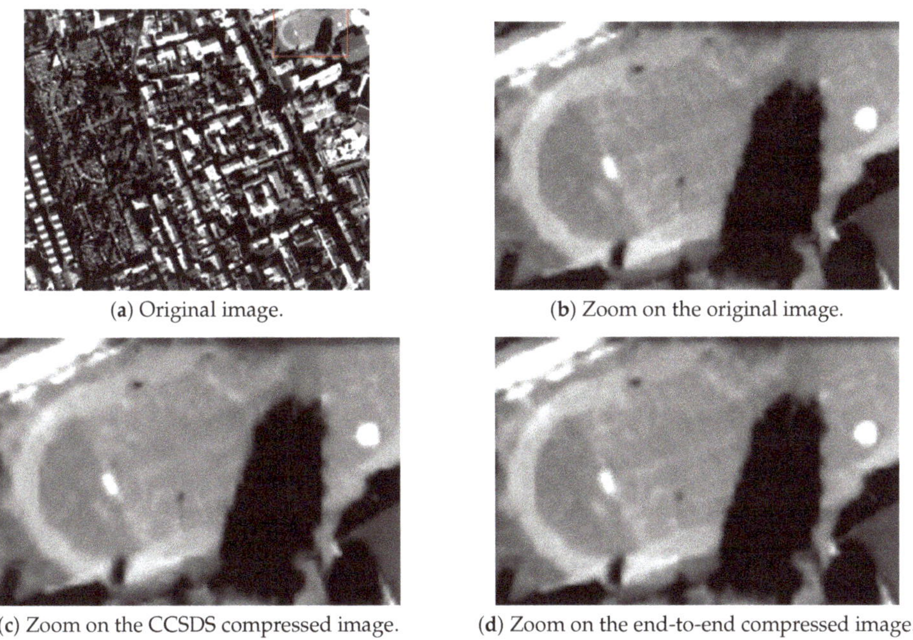

(a) Original image.

(b) Zoom on the original image.

(c) Zoom on the CCSDS compressed image.

(d) Zoom on the end-to-end compressed image.

Figure 9. Subjective image quality analysis—R = 2.02 bits/pixel.

Finally, for various rates, the learned compression method [13] does not suffer from the troublesome artifacts induced by the CCSDS 122.0, leading to a more uniform image quality. The same behavior was observed for the proposed method.

In the following, an objective performance analysis is performed in terms of rate-distortion trade-off for the CCSDS 122.0-B [6], JPEG2000, the reference methods [13,16] and the proposed ones.

4.3. Impact of the Number of Filter Reduction

4.3.1. At Low Rates

We first consider architectures devoted to low rates, say up to 2 bits/pixel. Starting from [16] denoted as `Ballé(2018)-hyperprior-N128-M192`, the number of filters N (for all layers, apart from the one just before the bottleneck) is reduced from $N = 128$ to $N = 64$, keeping $M = 192$ for the layer just before the bottleneck. This reduction is applied jointly to the main autoencoder and to the hyperprior one. The proposed simplified architecture is termed `Ballé(2018)-s-hyperprior-N64-M192`. The complexity of this model is evaluated in terms of number of parameters and of floating point operation per pixel (FLOPp) in Table 1 .

Table 1. Detailed complexity of `Ballé(2018)-s-hyperprior-N64-M192`.

Layer	Filter Size		Channels		Output		Parameters	FLOPp
	n	n	N_{in}	N_{out}	s_{out}	s_{out}		
conv1	5	5	1	64	256	256	1664	4.16×10^2
GDN1							4160	1.04×10^3
conv2	5	5	64	64	128	128	102,464	6.40×10^3
GDN2							4160	2.60×10^2
conv3	5	5	64	64	64	64	102,464	1.60×10^3
GDN3							4160	0.65×10^2
conv4	5	5	64	192	32	32	307,392	1.2×10^3
Hconv1	3	3	192	64	32	32	110,656	4.32×10^2
Hconv2	5	5	64	64	16	16	102,464	1.00×10^2
Hconv3	5	5	64	64	8	8	102,464	0.25×10^2
HTconv1	5	5	64	64	16	16	102,464	1.00×10^2
HTconv2	5	5	64	64	32	32	102,464	4.00×10^2
HTconv3	3	3	64	192	32	32	110,784	4.32×10^2
Tconv1	5	5	192	64	64	64	307,264	4.80×10^3
IGDN1							4160	0.65×10^2
Tconv2	5	5	64	64	128	128	102,464	6.40×10^3
IGDN2							4160	2.60×10^2
Tconv3	5	5	64	64	256	256	102,464	2.56×10^4
IGDN3							4160	1.04×10^3
Tconv4	5	5	64	1	512	512	1601	1.60×10^3
Total							1,683,969	5.2264×10^4

Table 2 compares the complexity of `Ballé(2018)-s-hyperprior-N64-M192` to the reference `Ballé(2018)-hyperprior-N128-M192`.

Table 2. Comparative complexity of the global architectures-Case of target rates up to 2 bits/pixel.

Method	Parameters	FLOPp	Relative
`Ballé(2018)-hyperprior-N128-M192`	5,055,105	1.9115×10^5	1.00
`Ballé(2018)-s-hyperprior-N64-M192`	1,683,969	5.2264×10^4	0.27
`Ballé(2018)-s-laplacian-N64-M192`	1,052,737	5.0774×10^4	0.265

The complexity of the proposed simplified architecture is 73% lower in terms of FLOPp with respect to the reference method. Now let consider the impact on compression performance of the reduction of N. Figure 10 displays the performance, in terms of MSE and MS-SSIM, of our different proposed solutions, of the reference learned methods and of the JPEG2000 [5] and CCSDS 122.0-B [6] standards. The gray curve portions indicate that the values of N and M are not recommended for this rate range (above 2 bits/pixel). In this first experiment, we are mainly concerned by the comparison of the blue lines: the solid one for the reference method `Ballé(2018)-hyperprior-N128-M192` and the dashed one for the proposal `Ballé(2018)-hyperprior-N64-M192`.

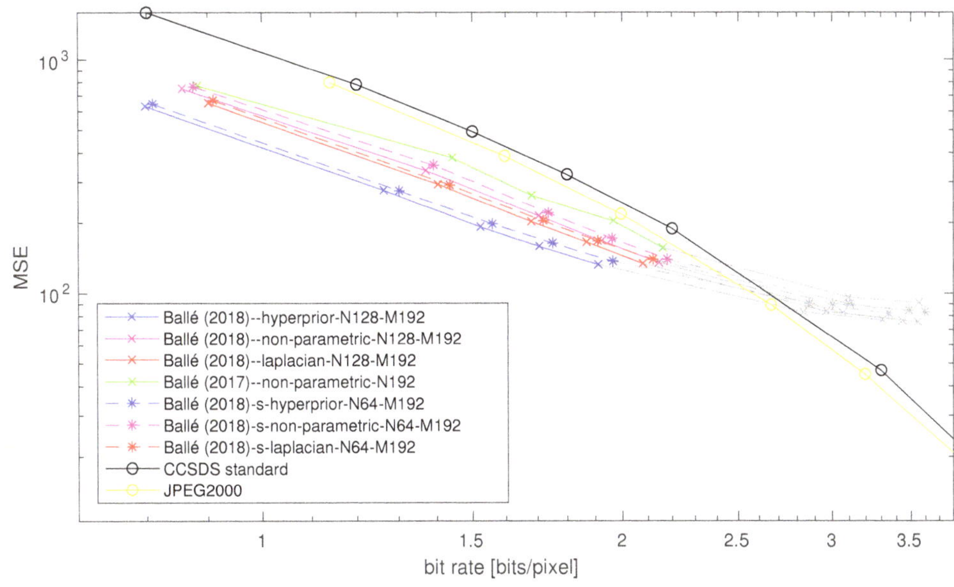

(**a**) Log–log scale. Distortion measure: MSE.

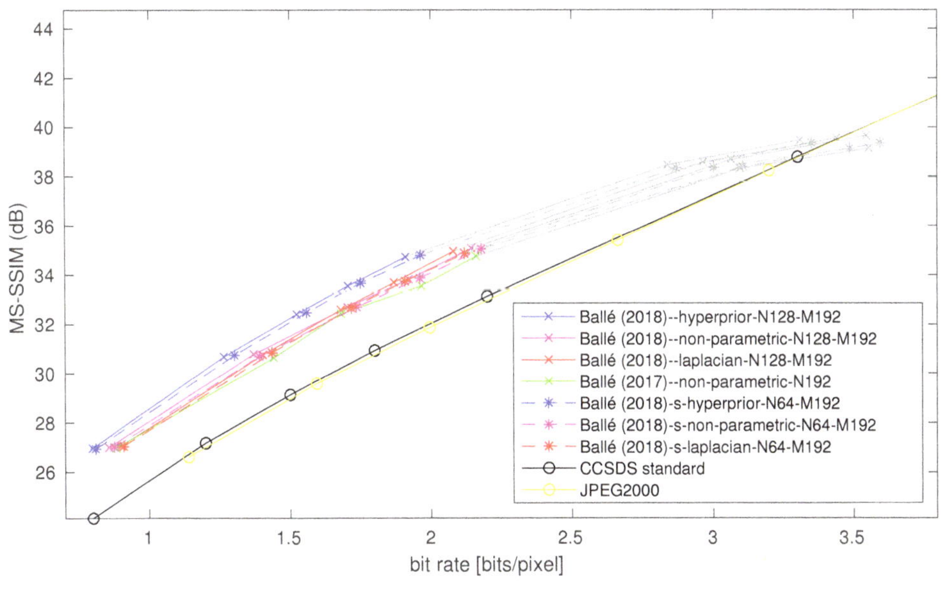

(**b**) Distortion measure: MS-SSIM (dB).

Figure 10. Rate-distortion curves for the considered learned frameworks and for the CCSDS 122.0-B [6] and JPEG2000 [5] standards in terms of MSE and MS-SSIM (dB) (derived as $-10\log_{10}(1-\text{MS-SSIM})$)-Case of rates up to 2 bits/pixel.

As expected, `Ballé(2018)-s-hyperprior-N64-M192` achieves a rate-distortion performance close to the one of `Ballé(2018)-hyperprior-N128-N192` [16], both in terms of MSE and MS-SSIM, for rates up to 2 bits/pixel. We can conclude that the decrease in performance is very small, keeping in mind the huge complexity reduction. Please note that our proposal outperforms by far CCSDS 122.0-B [6], JPEG2000 [5] standards as well as `Ballé(2017)-non-parametric-N192` [13].

4.3.2. At High Rates

Now let consider the architecture devoted to higher rates, say above 2 bits/pixel. For such rates, the reference architectures involve a high number of filters ($N = 256$ in [13], $N = 192$ and $M = 320$ in [16]). Starting from [16], we reduced the number of filters to $N = 64$ in all layers except the one before the bottleneck, keeping $M = 320$. The proposal `Ballé(2018)-s-hyperprior-N64-M320` is compared to the reference `Ballé(2018)-hyperprior-N192-M320` but also to `Ballé(2017)-non-parametric-N256` and to JPEG2000 [5] and CCSDS 122.0-B [6] standards. Table 3 compares the complexity of `Ballé(2018)-s-hyperprior-N64-M320` to the reference `Ballé(2018)-hyperprior-N192-M320`.

Table 3. Comparative complexity of the global architectures-Case of target rates above 2 bits/pixel.

Method	Parameters	FLOPp	Relative
`Ballé(2018)-hyperprior-N192-M320`	11,785,217	4.3039×10^5	1.00
`Ballé(2018)-s-hyperprior-N64-M320`	1,683,969	5.6966×10^4	0.13
`Ballé(2018)-s-laplacian-N64-M320`	1,052,737	5.4774×10^4	0.1273

The complexity of the proposed simplified architecture is 87% lower in terms of FLOPp with respect to the reference method. Now let consider the impact on compression performance of the reduction of N. Figure 11 displays the performance, in terms of MSE only, of our different proposed solutions, of the reference learned methods and of the JPEG2000 and CCSDS standards. The MS-SSIM shows the same behaviour.

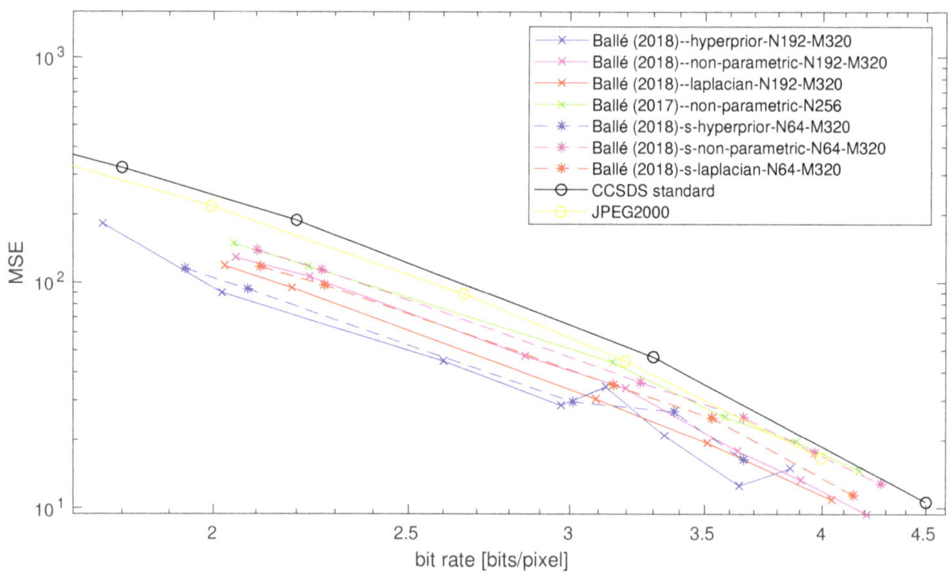

Figure 11. Rate-distortion curves at higher rates for learned frameworks and for the CCSDS 122.0-B [6] and JPEG2000 [5] standards for MSE in log-log scale Case of high rates (above 2 bits/pixel).

Theses curves show that the simplified architectures (e.g., resulting from a decrease of *N*) by far outperform the JPEG2000 and CCSDS standard even at high rates, while showing a low decrease in performance with respect to the reference architectures. Note however that, for both the reference (`Ballé(2018)-hyperprior-N192-M320`) and the simplified (`Ballé(2018)-s-hyperprior-N64-M320`) variational models, a training of 1M iterations seems insufficient for the highest rates. Indeed, due to the auxiliary autoencoder implementing the hyperprior, the training has conceivably to be longer, which can be a disadvantage in practice. This may be an additional argument to propose a simplified entropy model.

4.3.3. Summary

As an intermediary conclusion, for either low or high bit rates, a drastic reduction of *N* starting from the reference architecture [16], does not decrease significantly the performance, both in MSE and in MS-SSIM, while it leads to a complexity decrease of more than 70%. These results are interesting since it was mentioned in [13,16,19] that structures of reduced complexity would not be able to perform well at high rates.

4.4. Impact of the Bottleneck Size

As previously highlighted, the bottleneck size (*M*) plays a key role in the performance of the considered architectures. Thus, we now consider a fixed low value of *N* ($N = 64$) and then we vary the bottleneck size ($M = 128, 192, 256$ and 320). This experiment, performed on the proposed architecture integrating the simplified entropy model `Ballé(2018)-s-laplacian-N64-M`, allows quantifying the impact of *M* on the performance in terms of both MSE and MS-SSIM for increasing values of the target rate. Figure 12 shows the rate-distortion averaged over the validation dataset. According to the literature, high bit rates require a large global number of filters [16].

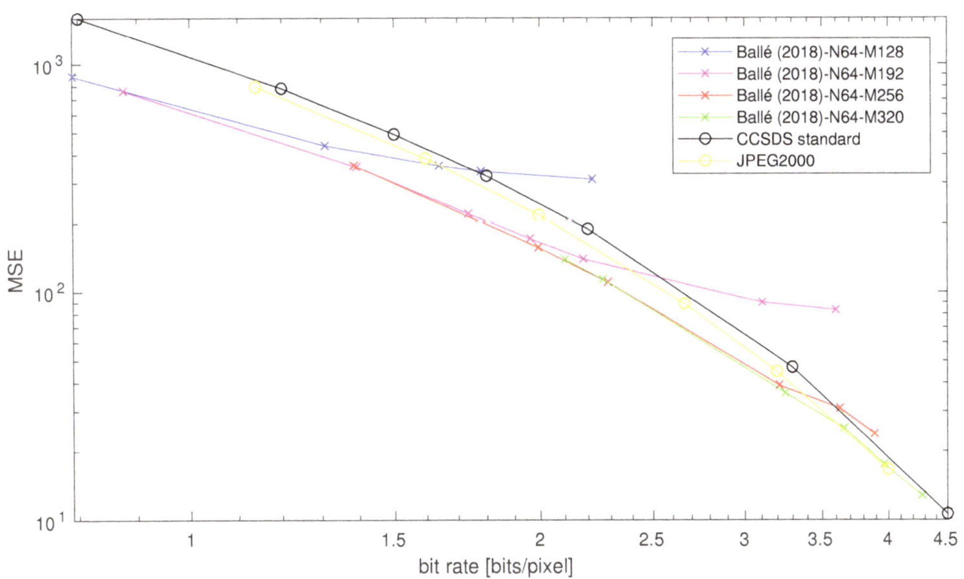

(**a**) Log–log scale. Distortion measure: MSE.

Figure 12. *Cont.*

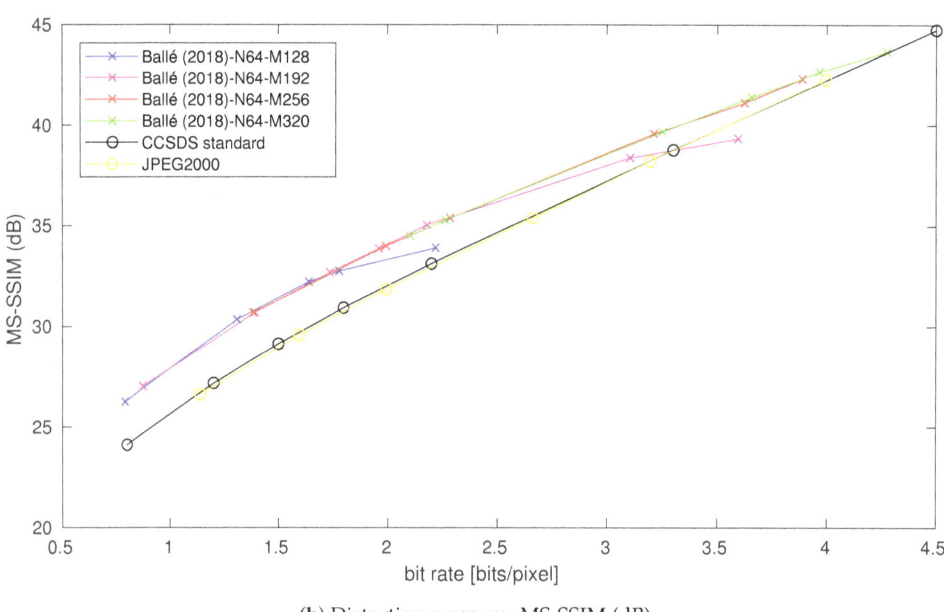

(**b**) Distortion measure: MS-SSIM (dB).

Figure 12. Impact of the bottleneck size in terms of MSE and MS-SSIM (dB) (derived as $-10\log_{10}(1-\text{MS-SSIM})$) .

Figure 12 shows that increasing the bottleneck size M only, while keeping N very small, allows maintaining the performance as the rate increases. As displayed in Figure 12, while the performance reaches a saturation point for a given bottleneck size, it is possible to renew its dynamic by increasing M only. This result is consistent since the number of output channels (M), just before the bottleneck, corresponds to the number of features that must be compressed and transmitted. It therefore makes sense to produce more features at high rates for a better reconstruction of the compressed images. Interestingly, this figure allows establishing in advance the convolution layer dimensions (N and M) for a given rate range, taking into account a complexity concern.

4.5. Impact of the Gdn/Igdn Replacement in the Main Autoencoder

The original architecture `Ballé(2018)-hyperprior-N128-M192` of [16], involving GDN/IGDN non-linearities, is compared with the architecture obtained after a full ReLU replacement, except for the last layer of the decoder part. Indeed, this layer involves a sigmoid activation function for constraining the pixel interval mapping between 0 and 1 before the quantization. Figure 13 shows the rate-distortion averaged over the validation dataset in terms of both MSE and MS-SSIM.

As claimed in [19], GDN/IGDN perform better than ReLU for all rates and especially at high rates. Thus, although GDN/IGDN increase the number of parameters to be learned and stored, as well as the number of FLOPp, on one side this increase represents a small percentage of the overall structure with respect to conventional non-linearities [19]. On the other side, GDN/IGDN lead to a dramatic performance boost. In view of these considerations, the complexity reduction in this paper does not target the GDN/IGDN. However, their replacement by simpler activation functions can be envisioned in future work to take into account on board hardware requirements.

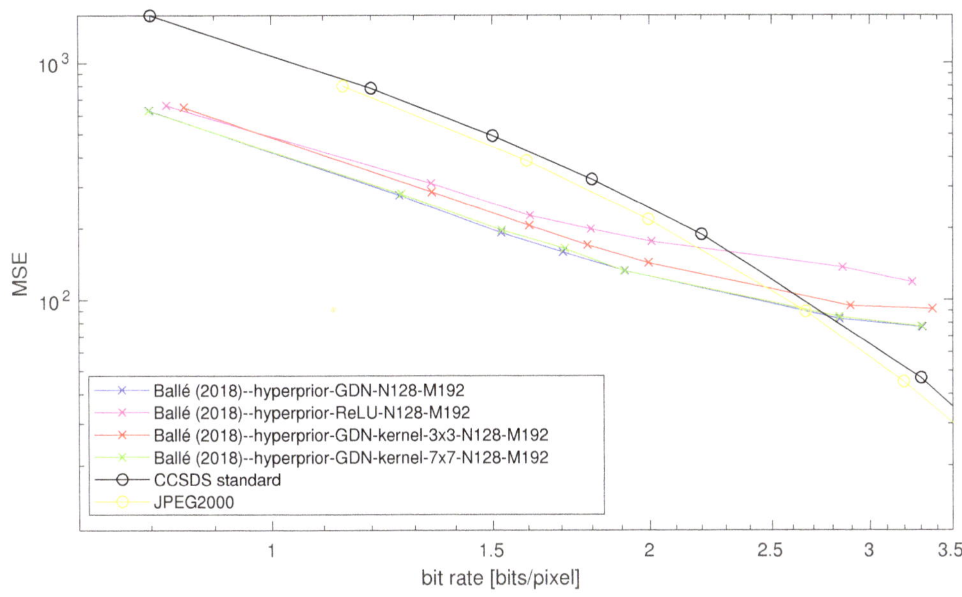

(**a**) Log–log scale. Distortion measure: MSE.

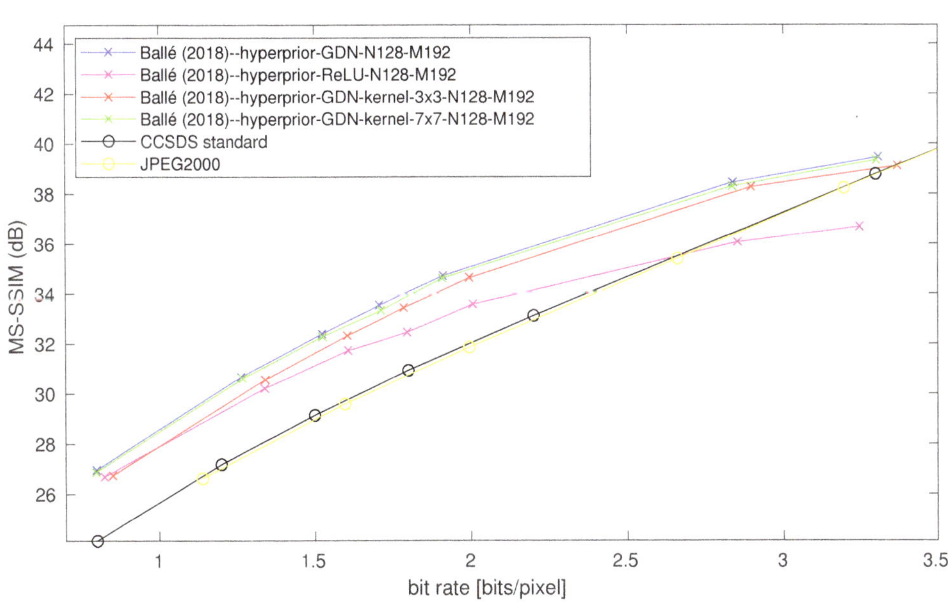

(**b**) Distortion measure: MS-SSIM (dB).

Figure 13. Impact of the GDN/IGDN replacement and of the filter kernel support on performance in terms of MSE and MS-SSIM (dB) (derived as $-10\log_{10}(1-\mathrm{MS\text{-}SSIM})$).

4.6. Impact of the Filter Kernel Support in the Main Autoencoder

The original architecture `Ballé(2018)-hyperprior-N128-M192` of [16] is also compared when the 5×5 filters composing the convolutional layers of the main autoencoder are replaced by 3×3 and 7×7 filters. It is worth mentioning that all the variant architectures considered in this part share the same entropy model obtained through the same auxiliary autoencoder in terms of number of filters and kernel supports, since the objective here is not to assess the impact of the entropy model. According to Figure 13, a kernel support reduction from 5×5 to 3×3 leads to a performance decrease. This result is expected in the sense that filters with a smaller kernel support correspond to a reduced approximation capability. On the other hand, a kernel support increase from 5×5 to 7×7 does not lead to a significant performance improvement. This result indicates that the approximation capability obtained with a kernel support 5×5 is sufficient.

4.7. Impact of the Entropy Model Simplification

4.7.1. At Low Rates

For rates up to 2 bits/pixel, the proposed architectures `Ballé(2018)-laplacian-N128-M192` (with the simplified entropy model) and `Ballé(2018)-s-laplacian-N64-M192` (combining the reduction of the number of filters to $N = 64$ and the simplified Laplacian entropy model) are compared with the non-variational reference method `Ballé(2017)-non-parametric-N192` [13], with the variational reference method `Ballé(2018)-hyperprior-N128-M192` [16], with its version after reduction of the number of filters `Ballé(2018)-s-hyperprior-N64-M192`, with the architecture denoted as `Ballé(2018)-nonparametric-N128-M192` (combining the main auto-encoder in [16] and the non-parametric entropy model in [13]) and its version after reduction of the number of filters `Ballé(2018)-s-non-parametric-N64-M192`. Table 4 shows that the coding part complexity of `Ballé(2018)-s-laplacian-N64-M192` is 13% lower than the one of `Ballé(2018)-s-hyperprior-N64-M192`.

Table 4. Reduction of the encoder complexity induced by simplified entropy model on the coding part-Case of rates up to 2 bits/pixel).

Method	Parameters	FLOPp	Relative
`Ballé(2018)-s-hyperprior-N64-M192`	1,157,696	1.25×10^4	1
`Ballé(2018)-s-laplacian-N64-M192`	526,464	1.09×10^4	0.87

Figure 10 shows the rate-distortion averaged over the validation dataset for the trained models for both MSE and MS-SSIM quality measures. Recall that the architectures were trained for MSE only. The proposed simplified entropy model (`Ballé(2018)-s-laplacian-N64-M192`) achieves an intermediate performance between the variational model (`Ballé(2018)-s-hyperprior-N64-M192`) and the non-variational model (`Ballé (2018)-s-non-parametric-N64-M192`). Obviously, due to the entropy model simplification, `Ballé(2018)-s-laplacian-N64-M192` underperforms the more general and thus more complex `Ballé(2018)-s-hyperprior-N64-M192` model. However, the proposed entropy model, even if simpler, preserves the adaptability to the input image, unlike the models `Ballé(2018)-non-parametric-N128-M192` and `Ballé(2017)-non-parametric-N192` [13]. Please note that the simplified Laplacian entropy model perform close to the hyperprior model at relatively high rates. One possible explanation for this behaviour can be the increased amount of side information required by the hyperprior model [16] for these rates [28].

4.7.2. At High Rates

For high rates (above 2 bits/pixel), the proposed architectures `Ballé(2018)-laplacian-N192-M320` (with the simplified entropy model) and `Ballé(2018)-s-laplacian-N64-M320` (combining the reduction of the number of filters to $N = 64$ and the simplified Laplacian entropy model) are compared with the non-variational reference method `Ballé(2017)-non-parametric-N256` [13], with the variational reference method `Ballé(2018)-hyperprior-N192-M320` [16], with its version after reduction of the number of filters `Ballé(2018)-s-hyperprior-N64-M320`, with the architecture denoted as `Ballé(2018)-nonparametric-N192-M320` (combining the main auto-encoder in [16] and the non-parametric entropy model in [13]) and its version after reduction of the number of filters `Ballé(2018)-s-non-parametric-N64-M320`. Figure 11 displays the rate-distortion averaged over the validation dataset for the trained models in terms of MSE. The proposed simplified entropy method `Ballé(2018)-s-laplacian-N64-M320` achieves an intermediate performance between the variational model (`Ballé(2018)-s-hyperprior-N64-M320`) and the non-variational model `Ballé(2018)-s-non-parametric-N64-M320`, similarly to the models targeting lower rates in Figure 10. Table 5 shows that the coding part complexity of `Ballé(2018)-s-laplacian-N64-M320` is around 16% lower than the one of `Ballé(2018)-s-hyperprior-N64-M320`.

Table 5. Reduction of the encoder complexity induced by simplified entropy model on the coding part-Cas of rates above 2 bits/pixel.

Method	Parameters	FLOPp	Relative
`Ballé(2018)-s-hyperprior-N64-M320`	1,715,008	1.3979×10^4	1
`Ballé(2018)-s-laplacian-N64-M320`	731,392	1.1787×10^4	0.8432

4.7.3. Summary

For either low or high bit rates, the proposed entropy model simplication leads to intermediary performance when compared to the reference architectures [13,16], both in MSE and in MS-SSIM, while it leads to a coding part complexity decrease of more than 10% with respect to [16].

4.8. Discussion About Complexity

According to the previous performance analysis, the computational time complexity of the proposed method is significantly lower than the one of the reference learned compression architecture [16]. However, around 10 kFLOPs/pixel, the attained complexity is at least 2 orders of magnitude higher than the ones of the CCSDS and JPEG2000 [5] standards. Indeed, the complexity of CCSDS 122.0 is around 140 operations per pixel (without optimizations), or 70 MAC (Multiplication Accumulation). The JPEG2000 is 2 to 3 times more complex depending on the optimizations. Note however that the CCSDS 122.0 dates back to 2008 when onboard technologies were limited to radiation-hardened (Rad-Hard) components dedicated to space, with the objectives of 1 Msample/s/W (as specified in the CCSDS 122.0 green book [29]), to process around 50 Mpixels/s. Space technologies, currently developed for the next generation of CNES Earth observation satellites, rather target 5–10 Msample/s/W. Nowadays, the use of commercial off-the-shelf (COTS) components or of dedicated hardware accelerators is envisioned: based on a thinner silicon technology node, they allow higher processing frequencies with consistently lower consumption. For instance, the Movidius Myriade 2 announces 1 TFLOP/s/W. The 10 kFLOP/pixels of the current network would lead to 100 Mpixels/s/W on this component. Therefore, the order of magnitude of the proposed method complexity is not incompatible with an embedded implementation, taking into account the technological leap from the component point of view. Consequently, the complexity increase with respect to the CCSDS one, which we limited as far as possible, is expected to be affordable after computation device upgrading. Please note that, in addition, manufacturers of components dedicated to neural networks provide software suites (for example Xilinx) to optimize the portings. Finally, before on board implementation, a network compression (including pruning, quantization, or tensor decomposition for instance) can be envisioned. However, this is out of the scope of this paper.

5. Conclusions

This paper proposed different solutions to adapt the reference learned image compression models [13,16] to on board satellite image compression, taking into account their computational complexity. We first performed a reduction of the number of filters composing the convolutional layers of the analysis and synthesis transforms, applying a special treatment to the bottleneck. The impact of the bottleneck size, under a drastic reduction of the overall number of filters, was investigated. This study allowed identifying the lowest global number of filters for each rate. For the sake of completeness, we also called into question the other design options of the reference architectures, and especially the parametric activation functions. Second, in order to simplify the entropy model, we also performed a statistical analysis of the learned representation. This analysis showed that most features follow a Laplacian distribution. We thus proposed a simplified parametric entropy model, involving a single parameter. To preserve the adaptivity and thus the performance, this parameter is estimated in the operational phase for each feature of the input image. This entropy model, although far simpler than non-parametric or hyperprior models, brings comparable performance. In a nutshell, by combining the reduction of the global number of filters, and the simplification of the entropy model, we developed a reduced-complexity compression architecture for satellite images that outperforms the CCSDS 122.0-B [6], in terms of rate-distortion trade-off, while maintaining a competitive performance for medium to high rates in comparison with the reference learned image compression models [13,16]. Thereupon, while more complex than traditional CCSDS 122.0 and JPEG 2000 standards, the proposed solutions offer a good compromise between complexity and performance. Thus, we can recommend their use, subject to the availability of suitable on board devices. Besides, future work will be devoted to hardware considerations regarding the on board implementation.

Author Contributions: Conceptualization, M.C. (Marie Chabert), T.O. and C.P.; Formal Analysis, M.C. (Marie Chabert); Investigation, V.A.d.O.; Methodology, V.A.d.O., M.C. (Marie Chabert), T.O. and C.P.; Resources, M.B., C.L., M.C. (Mikael Carlavan), S.H., F.F. and R.C.; Software, V.A.d.O.; Supervision, M.C. (Marie Chabert), T.O. and C.P.; Validation, M.B., C.L., M.C. (Mikael Carlavan), S.H., F.F. and R.C.; Writing—original draft, V.A.d.O. and M.C. (Marie Chabert); Writing—review & editing, T.O. and C.P. All authors have read and agreed to the published version of the manuscript.

Funding: This work has been carried out under the financial support of the French space agency CNES and Thales Alenia Space. Part of this work has been funded by the Institute for Artificial and Natural Intelligence Toulouse (ANITI) under grant agreement ANR-19-PI3A-0004.

Data Availability Statement: Restrictions apply to the availability of these data. Data was obtained from CNES as partner of this study.

Acknowledgments: Experiments presented in this paper were carried out using the OSIRIM platform that is administered by IRIT and supported by CNRS, the Region Midi-Pyrenées, the French Government, ERDF (see http://osirim.irit.fr/site/en).

Conflicts of Interest: The authors declare no conflict of interest.

Abbreviations

The following abbreviations are used in this manuscript:

CCSDS	Consultative committee for space data systems
CNN	Convolutional neural networks
DCT	Discrete cosine transform
GDN	Generalized divisive normalization
IGDN	Inverse generalized divisive normalization
JPEG	Joint photographic experts group
MSE	Mean square error
MS-SSIM	Multi-scale structural similarity index

PCA	Principal component analysis
ReLU	Rectified Linear Unit

Appendix A. Table of Symbols Used

This annex tabulates symbols used in this article.

Table A1. Quantities.

Symbol	Meaning	Reference
$\mathbf{x}$	original image	Section 2.1
$G_a(\mathbf{x})$	analysis transform	Section 2.1
$\mathbf{y}$	learned representation	Section 2.1
$\hat{\mathbf{y}}$	quantized learned representation	Section 2.1
$G_s(\hat{\mathbf{y}})$	synthesis transform	Section 2.1
$\hat{\mathbf{x}}$	reconstructed image	Section 2.1
GDN	generalized divisive normalizations	Section 2.1
IGDN	inverse generalized divisive normalizations	Section 2.1
N	filters composing the convolutional layers	Section 2.1
$n \times n$	kernel support	Section 2.1
M	filters composing the last layer of G_a	Section 2.1
k, l	coordinate indexes of the output of the ith filter	Section 2.1
$v_i(k, l)$	value indexed by (k, l) of the output of the ith filter	Section 2.1
$Q(\mathbf{y})$	quantizer	Section 2.2
J	rate-distortion loss function	Section 2.3.1
$R(\hat{\mathbf{y}})$	rate	Section 2.3.1
$D(\mathbf{x}, \hat{\mathbf{x}})$	distortion between the original image $\mathbf{x}$ and the reconstructed image $\hat{\mathbf{x}}$	Section 2.3.1
λ	parameter that tunes the rate-distortion trade-off	Section 2.3.1
$m(\hat{\mathbf{y}})$	actual discrete probability distribution	Section 2.3.1
$p_{\hat{\mathbf{y}}}(\hat{\mathbf{y}})$	probability model assigned to the quantized representation	Section 2.3.1
$H(\hat{\mathbf{y}})$	bit-rate given by the Shannon cross entropy	Section 2.3.1
$\psi^{(i)}$	distribution model parameter vector associated with each element	Section 2.3.2
$\Delta \mathbf{y}$	i.i.d uniform noise	Section 2.3.2
$\tilde{\mathbf{y}}$	continuous approximated quantized learned representation	Section 2.3.2
$\mathbf{z}$	set of auxiliary random variables	Section 2.3.2
$H_a(\mathbf{y})$	hyperprior analysis transform	Section 2.3.2
$H_s(\hat{\mathbf{z}})$	hyperprior synthesis transform	Section 2.3.2
σ_i	standard deviation of a zero-mean Gaussian distribution	Section 2.3.2
N_{in}	number of features at the considered layer input	Section 3.1.1
N_{out}	number of features at the considered layer output	Section 3.1.1
Param^f	number of parameters associated with the filtering part of the considered layer	Section 3.1.1
δ	term accounting for the bias	Section 3.1.1

Table A1. *Cont.*

Symbol	Meaning	Reference
D	downsampling factor	Section 3.1.1
s_{in}	channel input size	Section 3.1.1
s_{out}	downsampled input channel size	Section 3.1.1
$Operation^f$	number of floating points operations	Section 3.1.1
$Param^g$	number of parameters associated with each IGDN/GDN	Section 3.1.1
$Operation^g$	number of floating points operations of each GDN/IGDN	Section 3.1.1
ζ	random variable that follows a Laplacian distribution	Section 3.2.1
μ	mean value of a Laplacian distribution	Section 3.2.1
b	scale parameter of a Laplacian distribution	Section 3.2.1
$Var(\zeta)$	variance of a laplacian distributed random variable	Section 3.2.1
y_i	feature map elements	Section 3.2.2
I_j	set of indexes covering the jth feature	Section 3.2.2

References

1. Yu, G.; Vladimirova, T.; Sweeting, M.N. Image compression systems on board satellites. *Acta Astronaut.* **2009**, *64*, 988–1005. [CrossRef]
2. Huang, B. *Satellite Data Compression*; Springer Science & Business Media: New York, NY, USA, 2011.
3. Qian, S.E. *Optical Satellite Data Compression and Implementation*; SPIE Press: Bellingham, WA, USA, 2013.
4. Goyal, V.K. Theoretical foundations of transform coding. *IEEE Signal Process. Mag.* **2001**, *18*, 9–21. [CrossRef]
5. Taubman, D.; Marcellin, M. *JPEG2000 Image Compression Fundamentals, Standards and Practice*; Springer Publishing Company: New York, NY, USA, 2013.
6. Book, B. *Consultative Committee for Space Data Systems (CCSDS), Image Data Compression CCSDS 122.0-B-1, Ser. Blue Book*; CCSDS: Washington, DC, USA, 2005.
7. Bengio, Y. Learning deep architectures for AI. *Found. Trends Mach. Learn.* **2009**, *2*, 1–127. [CrossRef]
8. Hu, J.; Shen, L.; Sun, G. Squeeze-and-excitation networks. In Proceedings of the 2018 IEEE Conference on Computer Vision and Pattern Recognition (CVPR), Salt Lake City, UT, USA, 18–22 June 2018; pp. 7132–7141.
9. Redmon, J.; Divvala, S.; Girshick, R.; Farhadi, A. You only look once: Unified, real-time object detection. In Proceedings of the 2016 IEEE Conference on Computer Vision and Pattern Recognition (CVPR), Las Vegas, NV, USA, 27–30 June 2016; pp. 779–788.
10. Kaiser, P.; Wegner, J.D.; Lucchi, A.; Jaggi, M.; Hofmann, T.; Schindler, K. Learning aerial image segmentation from online maps. *IEEE Trans. Geosci. Remote. Sens.* **2017**, *55*, 6054–6068. [CrossRef]
11. Zhang, K.; Zuo, W.; Chen, Y.; Meng, D.; Zhang, L. Beyond a gaussian denoiser: Residual learning of deep CNN for image denoising. *IEEE Trans. Image Process.* **2017**, *26*, 3142–3155. [CrossRef] [PubMed]
12. Wiatowski, T.; Bölcskei, H. A mathematical theory of deep convolutional neural networks for feature extraction. *IEEE Trans. Inf. Theory* **2017**, *64*, 1845–1866. [CrossRef]
13. Ballé, J.; Laparra, V.; Simoncelli, E. End-to-end optimized image compression. In Proceedings of the International Conference on Learning Representations (ICLR), Toulon, France, 24–26 April 2017.
14. Theis, L.; Shi, W.; Cunningham, A.; Huszár, F. Lossy image compression with compressive autoencoders. In Proceedings of the International Conference on Learning Representations (ICLR), Toulon, France, 24–26 April 2017.
15. Rippel, O.; Bourdev, L. Real-Time Adaptive Image Compression. In Proceedings of the International Conference on Machine Learning (ICML), Sydney, Australia, 6–11 August 2017; pp. 2922–2930.
16. Ballé, J.; Minnen, D.; Singh, S.; Hwang, S.J.; Johnston, N. Variational image compression with a scale hyperprior. In Proceedings of the International Conference on Learning Representations (ICLR), Vancouver, BC, Canada, 30 April–3 May 2018.
17. Bellard, F. BPG Image Format. 2015. Available online: https://bellard.org/bpg (accessed on 15 December 2020).
18. Cheng, Z.; Sun, H.; Takeuchi, M.; Katto, J. Deep Residual Learning for Image Compression. In Proceedings of the 2019 IEEE Conference on Computer Vision and Pattern Recognition Workshops, Long Beach, CA, USA, 16–20 June 2019.
19. Ballé, J. Efficient Nonlinear Transforms for Lossy Image Compression. In Proceedings of the 2018 IEEE Picture Coding Symposium (PCS), San Francisco, CA, USA, 24–27 June 2018; pp. 248–252.
20. Lyu, S. Divisive normalization: Justification and effectiveness as efficient coding transform. *Adv. Neural Inf. Process. Syst.* **2010**, *23*, 1522–1530.
21. Rissanen, J.; Langdon, G. Universal modeling and coding. *IEEE Trans. Inf. Theory* **1981**, *27*, 12–23. [CrossRef]

22. Martin, G. Range encoding: An algorithm for removing redundancy from a digitised message. In Proceedings of the Video and Data Recording Conference, Southampton, UK, 24–27 July 1979; pp. 24–27.
23. Van Leeuwen, J. On the Construction of Huffman Trees. In Proceedings of the International Colloquium on Automata, Languages, and Programming (ICALP), Edinburgh, UK, 20–23 July 1976; pp. 382–410.
24. Wang, Z.; Simoncelli, E.P.; Bovik, A.C. Multiscale structural similarity for image quality assessment. In Proceedings of the Thrity-Seventh Asilomar Conference on Signals, Systems & Computers, Pacific Grove, CA, USA, 9–12 November 2003; pp. 1398–1402.
25. Dumas, T.; Roumy, A.; Guillemot, C. Autoencoder based image compression: Can the learning be quantization independent? In Proceedings of the 2018 IEEE International Conference on Acoustics, Speech and Signal Processing (ICASSP), Calgary, AB, Canada, 15–20 April 2018; pp. 1188–1192.
26. Lam, E.Y.; Goodman, J.W. A mathematical analysis of the DCT coefficient distributions for images. *IEEE Trans. Image Process.* **2000**, *9*, 1661–1666. [CrossRef] [PubMed]
27. Pratt, J.W.; Gibbons, J.D. Kolmogorov-Smirnov two-sample tests. In *Concepts of Nonparametric Theory*; Springer: New York, NY, USA, 1981; pp. 318–344.
28. Hu, Y.; Yang, W.; Ma, Z.; Liu, J. Learning End-to-End Lossy Image Compression: A Benchmark. *arXiv* **2020**, arXiv:2002.03711.
29. Book, G. *Consultative Committee for Space Data Systems (CCSDS), Image Data Compression CCSDS 120.1-G-2, Ser. Green Book*; CCSDS: Washington, DC, USA, 2015.

remote sensing

Article

Spectral–Spatial Feature Partitioned Extraction Based on CNN for Multispectral Image Compression

Fanqiang Kong [1], Kedi Hu [1,*], Yunsong Li [2], Dan Li [1] and Shunmin Zhao [1]

[1] College of Astronautics, Nanjing University of Aeronautics and Astronautics, Nanjing 210016, China; kongfq@nuaa.edu.cn (F.K.); danli@nuaa.edu.cn (D.L.); zhaosm@nuaa.edu.cn (S.Z.)

[2] State Key Laboratory of Integrated Service Networks, Xidian University, Xi'an 710071, China; ysli@mail.xidian.edu.cn

* Correspondence: kedi_hu@nuaa.edu.cn; Tel.: +86-137-3516-6766

Abstract: Recently, the rapid development of multispectral imaging technology has received great attention from many fields, which inevitably involves the image transmission and storage problem. To solve this issue, a novel end-to-end multispectral image compression method based on spectral–spatial feature partitioned extraction is proposed. The whole multispectral image compression framework is based on a convolutional neural network (CNN), whose innovation lies in the feature extraction module that is divided into two parallel parts, one is for spectral and the other is for spatial. Firstly, the spectral feature extraction module is used to extract spectral features independently, and the spatial feature extraction module is operated to obtain the separated spatial features. After feature extraction, the spectral and spatial features are fused element-by-element, followed by downsampling, which can reduce the size of the feature maps. Then, the data are converted to bit-stream through quantization and lossless entropy encoding. To make the data more compact, a rate-distortion optimizer is added to the network. The decoder is a relatively inverse process of the encoder. For comparison, the proposed method is tested along with JPEG2000, 3D-SPIHT and ResConv, another CNN-based algorithm on datasets from Landsat-8 and WorldView-3 satellites. The result shows the proposed algorithm outperforms other methods at the same bit rate.

Keywords: spectral–spatial feature; multispectral image compression; partitioned extraction; group convolution; rate-distortion

Citation: Kong, F.; Hu, K.; Li, Y.; Li, D.; Zhao, S. Spectral–Spatial Feature Partitioned Extraction Based on CNN for Multispectral Image Compression. *Remote Sens.* **2021**, *13*, 9. https://dx.doi.org/10.3390/rs13010009

Received: 26 November 2020
Accepted: 17 December 2020
Published: 22 December 2020

Publisher's Note: MDPI stays neutral with regard to jurisdictional claims in published maps and institutional affiliations.

1. Introduction

By capturing digital images of several continuous narrow spectral bands, remote sensors can generate three-dimensional multispectral images that contain rich spectral and spatial information [1]. The abundant information is very useful and has been employed in various applications, such as military reconnaissance, target surveillance, crop condition assessment, surface resource survey, environmental research, and marine applications and so on. However, with the rapid development of multispectral imaging technology, the spectral–spatial resolution of multispectral data becomes higher and higher, resulting in the rapid growth of its data volume. The huge amount of data is not conducive to image transmission, storage, and application, which hinders the development of related technologies. Therefore, it is necessary to find an effective multispectral image compression method to process images before use.

The research of multispectral image compression methods has always received widespread attention. After decades of unremitting efforts, various multispectral image compression algorithms for different application needs have been developed, which can be summarized as follows: predictive coding-based framework [2], vector quantization coding-based framework [3], transform coding-based framework [4,5]. The predictive coding is mainly applied to lossless compression. Its rationale is to use the correlation between pixels to predict the unknown data based on its neighbors, and then to encode the residual between the real value and the predicted value. In [6], Slyz et al. proposed a block-based inter-band lossless multispectral image compression method, in which every image was divided into blocks and the current block was predicted by the corresponding block in the adjacent band. For vector quantization coding, several scalar data sets are formed into a vector, and then the data are quantized as a whole in vector space, so as to be compressed without losing much information. As the performance of the vector quantization coding is closely connected with the codebook, to improve the time efficiency, Qian proposed a fast codebook search method in [7]. In the full search process of the generalized Lloyd algorithm (GLA), if the distance to the partition is better than that of the previous iteration, there is no need to require a search to find the minimum distance partition. Transform coding is an important method in multispectral image compression, which is widely used in lossy compression. This algorithm reduces the correlation between pixels by converting the data to transform domain representation, so that information can be concentrated so as to be quantified and encoded. Karhunen–Loève transform (KLT) [8], discrete cosine transform (DCT) [9] and discrete wavelet transform [10] are all commonly used transform coding algorithms. As we obtain deeper insight into multispectral images, more and more improved algorithms have been developed, such as 3D-SPECK [11], 3D-SPIHT [12], and so on.

The traditional compression methods mentioned above are all effective and obtain great results, but they also have shortcomings. For instance, it is simple to implement the predictive coding algorithm, but the compression ratio is relatively low. Although the vector quantization coding algorithm can achieve a more ideal effect, it is not conducive to implementation due to its computation complexity. To overcome the shortcomings of traditional compression methods and also ensure the compression performance, many multispectral image compression algorithms based on deep learning have been rapidly developed in recent years. Among them, the convolutional neural network (CNN) has emerged as one of the main algorithms in image compression in recent years. The history of CNN started from LeNet-style models, which consist of simple stacks of convolution layers for feature extraction and max-pooling layers for downsampling [13]. In order to extract more features of different scales, AlexNet [14], proposed in 2012, followed this idea and made an improvement by adding several convolutional layers between every two max-pooling layers. To obtain better performance, it is necessary to increase the depth of the network. As a result, VGG [15], GoogLENet [16], ResNet [17] and other excellent network architectures began to emerge one after another. These network frameworks are all milestones in the process of image compression technology and have obtained great grades in past ILSVRC and other competitions.

Inspired by these remarkable network frameworks, many compression methods based on CNN have appeared and showed applicability for visible images. In [18], Ballé proposed an end-to-end optimized image compression method based on CNN with generalized divisive normalization (GDN) joint nonlinearity, by means of the flexible use of linear convolution and nonlinear transformation, the proposed network achieved comparable performance with JPEG2000. To further improve the quality of the reconstructed images, Jiang et al. [19] added CNNs to both encoder and decoder for joint training. The CNN in the encoder produces compact presentation for encoding, and the other CNN in the decoder is to restore the decoded image with high quality, with which block effects can be significantly reduced. It is known that multispectral images are three-dimensional data, in which two dimensions are spatial and one is spectral. As RGB images have three bands as well, it can be seen as special multispectral data. Consequently, many compression methods for visible images can also be applied to multispectral images. In [20], an end-to-end compression framework for multispectral images with optimized residual unit is presented. It is also based on a CNN, and the default architecture of ResNet, which is adopted in the network, was adjusted to better fit for multispectral images. This algorithm has been proven effective and obtains higher PSNR than that of JPEG2000 by about 2 dB. Even so, the methods mentioned above still fail to focus on the strong correlation between spectra of multispectral images, for that is less important for RGB images. However, for multispectral image compression, ignoring spectral correlation may lead to some information loss after compressing. Hence, in this paper, we proposed a novel multispectral image compression method based on partitioned extraction of spectral–spatial feature.

The network is an end-to-end framework based on a CNN and is composed of encoder and decoder. In the encoder, there are two parts for spectral feature extraction and spatial feature extraction, respectively. In the first part, continuous spectral feature extraction modules are adopted to extract spectral features independently. This part does not involve the fusion of spatial information. The second part is for spatial feature extraction, which contains several residual blocks. We use group convolution to separate each channel, so that only spatial features can be extracted without mixing the spectral information within them. Afterwards, all features are fused together and then downsampling is employed to reduce the size of the feature map. Additionally, to make the data more compact, a rate-distortion optimizer is used in the network. After obtaining the intermediate feature data, quantization and lossless entropy encoding are carried out to obtain the compressed binary bit stream. In the decoder, the bit stream first goes through entropy decoding and inverse quantization, and then upsampling helps to restore the image size. Finally, spectral and spatial features are acquired by corresponding deconvolution operations, and the joint feature is used to reconstruct the image. Experimental results demonstrate that our network surpasses JPEG2000 and 3D-SPIHT.

The remainder of this paper is organized as follows. Section 2 introduces our proposed network framework and principal analysis, Section 3 includes experimental parameter settings and the training process, and Section 4 presents the results and comparison with JPEG2000, 3D-SPIHT and the method mentioned in [20] at the same bit rate, which proves the wonderful performance of our network.

2. Proposed Method

In this section, we introduce the proposed multispectral image compression network framework in detail and describe the training flow diagram. We elaborate on several key operations, such as spectral feature extraction module, spatial feature extraction module, rate-distortion optimizer, etc.

2.1. Spectral Feature Extraction Module

2D convolution has been proven with great promise and successfully applied to lots of aspects of image vision and processing, such as target detection, image classification and image compression. However, as multispectral images are three-dimensional, which is more complex, and rich spectral information is even more important, the information loss problem will inevitably be encountered when 2D convolution is used to process multispectral images. Although there have been many precedents of applying deep learning to multispectral image compression, and it has achieved great performance and exceeded some traditional compression methods such as JPEG and JPEG2000, in the process of feature extraction, however, as the convolution kernel is two-dimensional, the spectral redundancy on the third dimension cannot be efficaciously removed, which inhibits the performance of the network.

To deal with this problem, we have come up with the idea of extracting spectral or spatial features separately. Among this, the inspiration of extracting spectral features derives from [21]. Ref. [21] uses three-dimensional kernels for convolution operation, which can maintain the integrity of spectral features in multispectral image data. To avoid the data volume becoming too large, we use a $1 \times 1 \times n$ convolution kernel on the spectral dimension named as 1D spectral convolution to extract spectral features independently. Figure 1 shows the differences between 2D convolution and 1D spectral convolution.

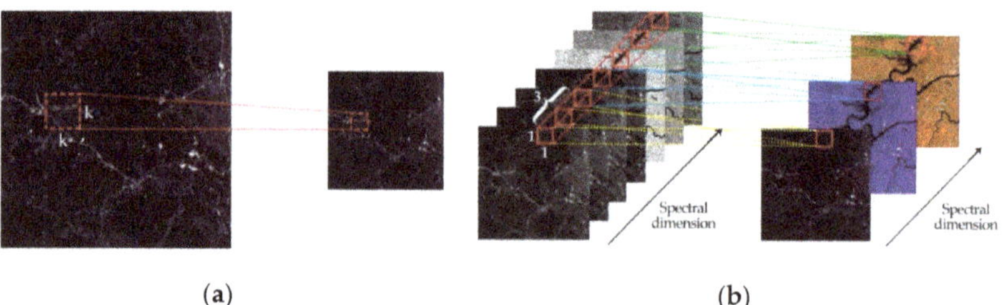

(a) (b)

Figure 1. (a) 2D convolution; (b) 1D spectral convolution.

As shown in Figure 1a, the image is convolved by 2D convolution, whose kernel is two-dimensional, generally followed by activation functions, such as rectified linear units (ReLU) [14], parametric rectified linear units (PReLU) [22], etc. This operation can be expressed as follows:

$$v_{ij}^{xy} = f\left(\sum_{m=1}^{M_{i-1}} \sum_{p=0}^{P_i-1} \sum_{q=0}^{Q_i-1} w_{ijm}^{pq} v_{(i-1)m}^{(x+p)(y+q)} + b_{ij} \right),\tag{1}$$

where i indicates the current layer, j indicates the current feature map of this layer, v_{ij}^{xy} is the output value at (x, y) of the j-th feature map in the i-th layer, $f(\cdot)$ represents the activation function, w_{ijm}^{pq} denotes the weight of the convolution kernel at position (p, q) connected to the m-th feature map (m indexes over the set of feature maps in the $(i-1)$-th layer connected to the current feature map), b_{ij} is the bias of the j-th feature map in the i-th layer, M_{i-1} is the number of feature maps in the $(i-1)$-th layer, P_i and Q_i are the height and width of the convolution kernel, respectively.

Similarly, considering the dimension of the spectrum, 1D spectral convolution operated on 3D images can be formulated as follows:

$$v_{ij}^{xyz} = f\left(\sum_{m=1}^{M_{i-1}} \sum_{p=0}^{P_i-1} \sum_{q=0}^{Q_i-1} \sum_{r=0}^{R_i-1} w_{ijm}^{pqr} v_{(i-1)m}^{(x+p)(y+q)(z+r)} + b_{ij} \right),\tag{2}$$

where R_i is the size of the convolution kernel in the spectral dimension, v_{ij}^{xyz} is the output value at (x, y, z) of the j-th feature map in the i-th layer, and w_{ijm}^{pqr} is weight of the kernel at position (p, q, r) connected to the m-th feature map. As the size of kernel is $1 \times 1 \times n$, by extension, P_i and Q_i are set to 1, Equation (2) can be written as:

$$v_{ij}^{xyz} = f\left(\sum_{m=1}^{M_{i-1}} \sum_{r=0}^{R_i-1} w_{ijm}^{pqr} v_{(i-1)m}^{(x+p)(y+q)(z+r)} + b_{ij} \right),\tag{3}$$

In regard to the activation function, we adopt ReLU as our first choice, as the gradient is usually constant in back propagation when using ReLU, which alleviates the problem of gradient disappearance in deep network training and contributes to network convergence. Additionally, the computation cost is much less when using ReLU than other functions (e.g., sigmoid). In addition, ReLU can make the output of some neurons zero, which ensures the sparsity of the network so as to alleviate the overfitting problem. The ReLU function can be formulated as below:

$$f(x) = max(0, x).\tag{4}$$

In summary, when 2D convolution is operated on three-dimensional images, the output is always two-dimensional, which may cause a large amount of spectral information loss. Therefore, we adopt 1D spectral convolution to retain more feature data of the multi-spectral image.

2.2. Spatial Feature Extraction Module

In order to ensure the spatial information does not mingle with the spectral features, we use group convolution instead of normal 2D convolution in spatial dimensions. Group convolution first appeared in AlexNet, in order to solve the problem of limited hardware resources at that time. Feature maps were distributed to several GPUs for simultaneous processing, and finally concatenated together. Figure 2 shows the differences between normal convolution and group convolution.

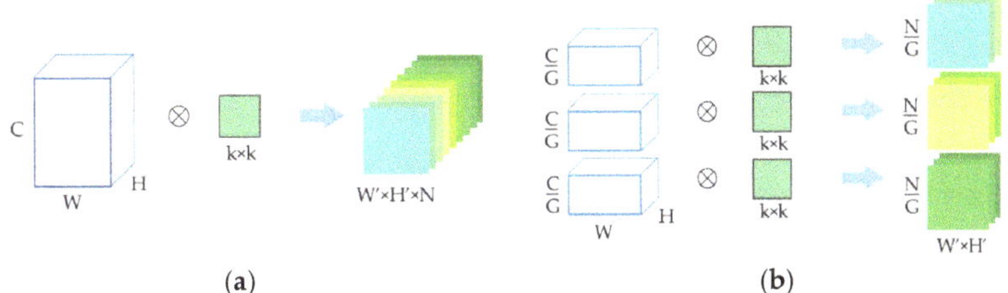

(a)   (b)

Figure 2. (**a**) Normal convolution; (**b**) group convolution.

As shown in Figure 2a, the size of input data is $C \times H \times W$, representing the number of channels, width, and height of the feature map, respectively. The size of the convolution kernel is $k \times k$, and the number of the kernels is N. At this point, the size of the output feature map is $N \times H' \times W'$. The parameter number of N convolution kernels is:

$$params = N \times k \times k \times C. \tag{5}$$

In group convolution, just as its name implies, the input feature maps are divided into several groups, and then convolved separately. Assuming that the size of the input is still $C \times H \times W$ and the number of output feature maps is N. If the input is divided into G groups, the number of input feature maps in each group is C/G, the number of output feature maps in each group is N/G, and the size of convolution kernel is $k \times k$, that is, the amount of convolution kernels remains unchanged and the number of kernels in each group is N/G. Since the feature maps are only convolved by the convolution kernels of the same group, the total number of parameters can be calculated as:

$$params = N \times k \times k \times \frac{C}{G}. \tag{6}$$

By comparing the two Equations (5) and (6), it can be easily known that group convolution can greatly reduce the number of parameters, precisely speaking, it can reduce them to $1/G$. Moreover, as group convolution can increase the diagonal correlation between filters according to [14], filter relationships become sparse after grouping.

Figure 3 shows the correlation matrix between filters of adjacent layers [23], highly correlated filters are brighter, while lower correlated filters are darker. The role of filter groups, namely group convolution, is to take advantage of the block-diagonal sparsity to learn information about the channel dimension. Low correlated filters do not need to be learned, that is to say, they do not need to be given parameters. What is more, as seen in Figure 3, the highly correlated filters can be trained in a more structured way when using group convolution. Therefore, with structured sparsity, group convolution can not only reduce the number of parameters, but also learn more accurately to make a more efficient network.

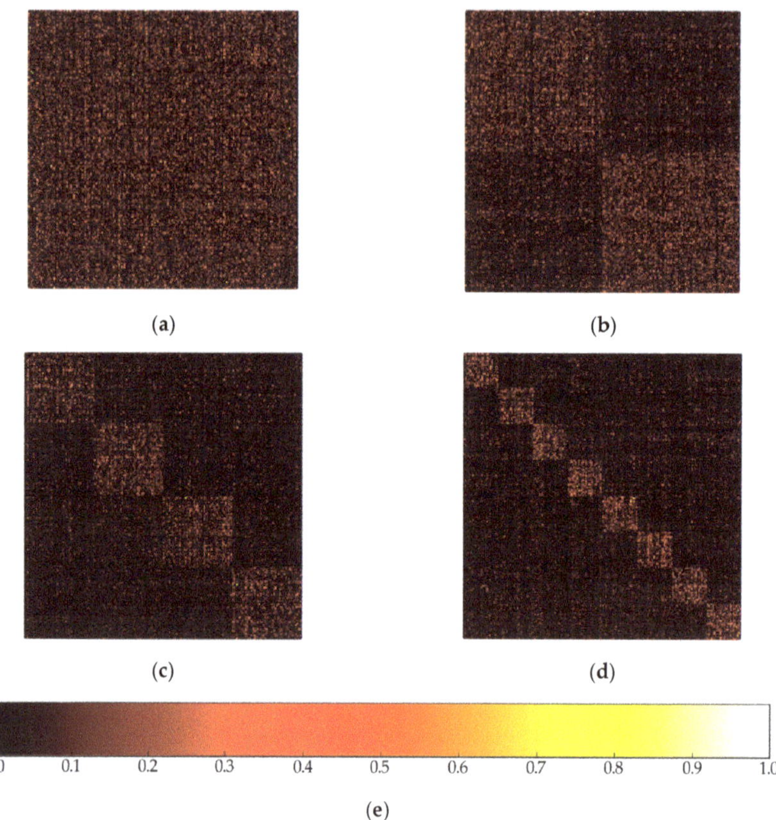

(a) **(b)**

(c) **(d)**

(e)

Figure 3. The correlation matrix between filters of adjacent layers: (**a**) 1 group; (**b**) 2 groups; (**c**) 4 groups; (**d**) 8 groups; (**e**) the correlation illustration.

2.3. Framework of the Proposed Network

The whole framework of the proposed compression network is illustrated in Figure 4. The multispectral images are fed into the forward network first, after feature extraction, the data are then compressed and converted to bit stream successively through quantization and entropy encoder. The structure of the decoder is symmetrical with that of the encoder. As a result, for decoding, the bit stream goes through entropy decoding, inverse quantization, and the backward network, in turn, to restore the images. The detailed architecture of the forward and backward network will be demonstrated in Section 2.3.1.

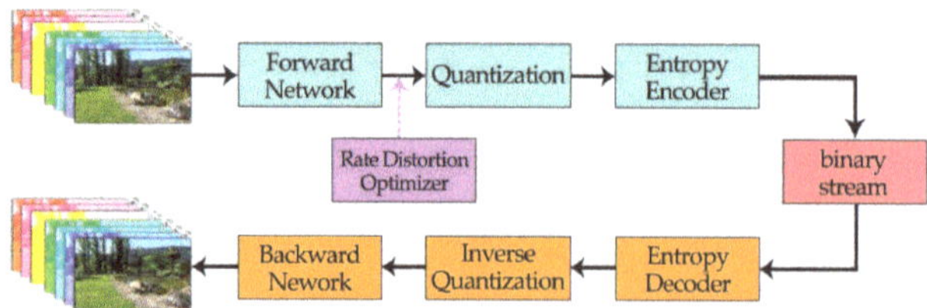

Figure 4. The flow diagram of the proposed network.

2.3.1. The Forward Network and the Backward Network

The architecture of the forward and backward network is shown in Figure 5, the spectral block and the spatial block are shown in Figure 6.

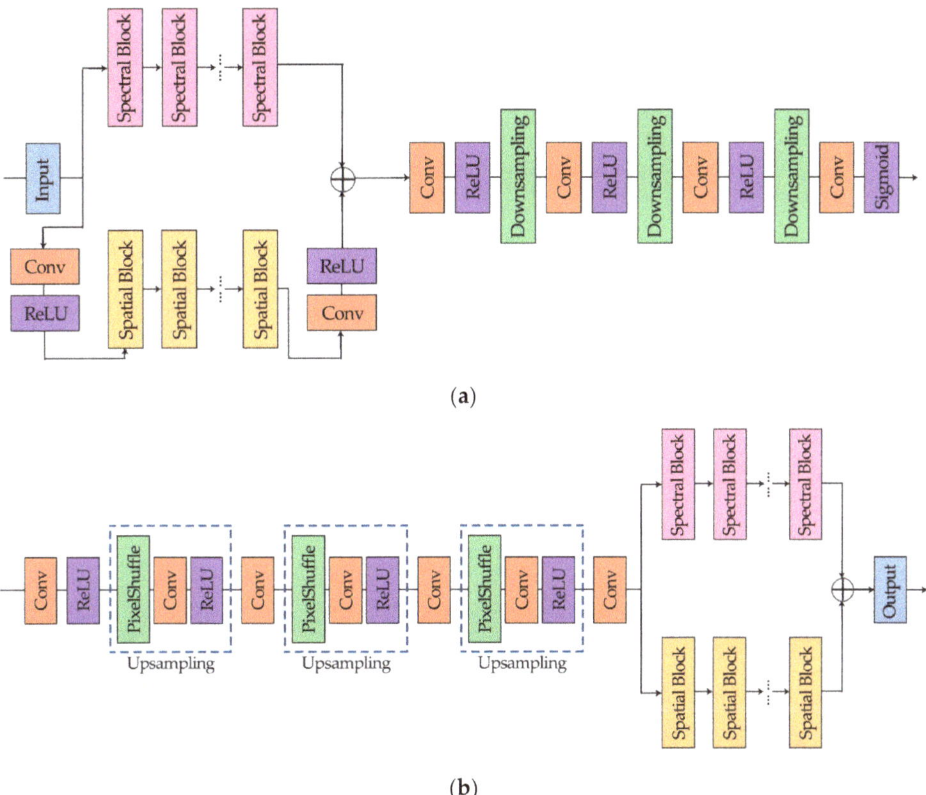

(a)

(b)

Figure 5. (**a**) The forward network; (**b**) the backward network.

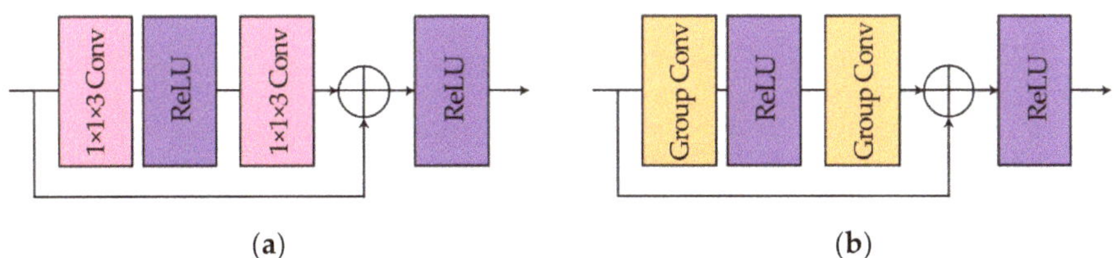

(a)   (b)

Figure 6. (**a**) Spectral block; (**b**) spatial block.

Figure 5 illustrates the detailed process of our network. First of all, the input multispectral images are simultaneously fed into the spectral feature extraction network and the spatial feature extraction network separately, which consist of corresponding function modules. In the spectral part, there are several spectral blocks (Figure 6a), which are based on residual block structure. We replace the convolution layers with 1D spectral convolution as adjusted to meet our expectations, and the size of the kernel is $1 \times 1 \times 3$. Likewise, the spatial part is composed by several spatial blocks with a similar structure, as shown in Figure 6b, and group convolution is used so that each channel will not interact with each other. To be specific, the *GROUP* is set to 7 or 8 as the input multispectral images are of seven or eight bands. Additionally, some convolution layers are added to enhance the ability of the learning features, whose kernel size is 3×3. After extraction, two parts of the features are fused together, and then downsampling is carried out to reduce the size of the feature maps. At the end of the forward network, the sigmoid function plays a role of limiting the value of the intermediate output, in addition, similar to ReLU as well, it introduces nonlinear factors to make the network more expressive to the model.

Symmetric with the forward network, the backward network is formed with upsampling layers, some convolution layers, and the partitioned extraction part. In particular, upsampling is implemented with PixelShuffle [24], which can turn low resolution images into high resolution images using sub-pixel operation.

2.3.2. Quantization and Entropy Coding

After the forward network, the intermediate data are first quantized into a succession of discrete integers by the quantizer. Since the descent gradient is used in the backward propagation to update parameters when training the network, the gradient needs to be passed down. However, the rounding function is not differentiable [25], which will hinder the optimization of the network. Therefore, we relax the function, and it is calculated as:

$$X_Q = round\left[\left(2^Q - 1\right) \times X_s\right],\tag{7}$$

where Q is the quantization level, $X_s \in (0,1)$ is the intermediate datum after sigmoid activation, $round[\cdot]$ is the rounding function, and X_Q is the quantized data. The function rounds the data in the forward network and is skipped during backward propagation, to pass the gradient directly to the previous layer.

Then, we adopt ZPAQ as the lossless entropy coding standard and select "Method-6" as the compression pattern, in order to further process the quantized X_Q and generate the binary bit stream. In the decoder, the bit stream goes through the entropy decoder and de-quantization, and the data $X_Q/(2^Q - 1)$ are finally fed into the backward network to recover the image.

2.4. Rate-Distortion Optimizer

There are two criterions to evaluate a compression method, one is the bit rate, and the other is the quality of the recovered image. To enhance the performance of the network, it is vital to strike a balance between these two criterions. In consequence, rate-distortion optimization is introduced:

$$L = L_D + \lambda L_R,\tag{8}$$

where L is the loss function that should be minimized during training, L_D indicates the distortion loss, L_R represents the rate loss, which can be controlled by the penalty λ. As we use MSE to measure the distortion loss of the recovered image, L_D can be expressed as follows:

$$L_D = \frac{1}{N}\frac{1}{H \times W \times C}\sum_{n=1,x,y,z}^{N}\|I(x,y,z) - \widetilde{I}(x,y,z)\|^2,\tag{9}$$

where N denotes the batch size, I represents the original multispectral image and $\widetilde{I}$ is the recovered image, H, W and C are, respectively, height, width, and spectral band number of the image.

In order to estimate the rate loss, we adopt an Importance-Net to replace the entropy computation with a continuous approximation of the code length. The importance network is used to generate an importance map $P(X)$ learning from the input images [26]. The intention is to assign the bit rate according to the importance of the content of the image, more bits are assigned to complex regions, and fewer bits are assigned to smooth regions. The importance-net is simply composed of four layers, two 1×1 convolution layers and a residual block that consists of two 3×3 convolution layers, which is shown as Figure 7.

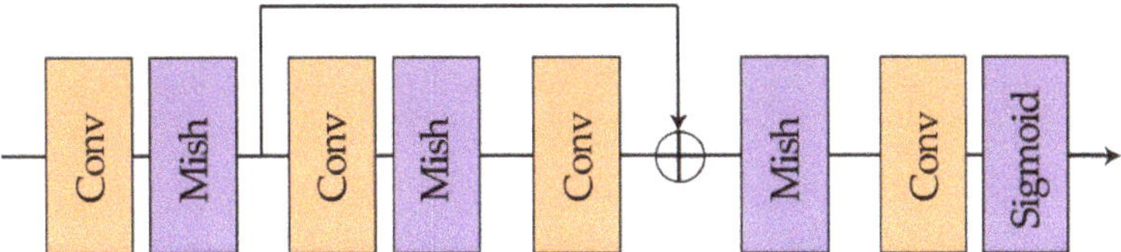

Figure 7. The Importance-Net.

The activation function used in the importance-net is Mish [27], and it has been proven to be smoother than ReLU and achieve better results. Nonetheless, considering the time cost and limited hardware conditions due to the increased complexity of Mish, we only adopt Mish in the importance-net rather than the whole network. Mish can be formulated as:

$$Mish(x) = x \cdot \tanh(\ln(1 + e^x)). \tag{10}$$

After sigmoid activation, the value range of the output is $[0, 1]$. The importance map can be described as below:

$$P(x,y) = \left[\hat{X} \otimes w\right](x,y) + b = \sum_{i=1}^{k}\sum_{j=1}^{k}\left[\hat{X}_n(s_0 x + i, s_0 y + j)w_n(i,j)\right] + b, \tag{11}$$
$$x \in \{0, 1, \ldots, H\}, y \in \{0, 1, \ldots, W\},$$

where $\hat{X}$ represents the un-quantized output after the encoder, w indicates the weight of the importance-net, (x, y) is the spatial location of the pixel, and b, k and s_0 are the bias, size of the kernel and stride, respectively. Unlike [26], we use the mean of $P(X)$ instead of the sum to define the rate loss:

$$P(X) = \sum_N P(x,y), \tag{12}$$

$$L_R = \text{avg}(P(X)), \tag{13}$$

where N is the number of the intermediate channel.

3. Experimental Settings and Training

3.1. Datasets

The 7 band image datasets come from the Landsat-8 satellite. The training dataset contains about 80,000 images. The images we selected include various terrains under different seasons and different weather conditions, which enables the network to learn multiple features, preventing the network training from overfitting. We pick 17 representative images from 80,000 images as a test set and make sure that there are no identical images in the two data sets. The size of training images and test images are 128×128 and 512×512, respectively.

The 8 band image datasets come from the WorldView-3 satellite, which contains about 8700 images of size of 512×512. Likewise, we ensure that the datasets include various terrains under different weather conditions to ensure the diversity of the feature. The test set has 14 images of size of 128×128, and has no identical images with the training set.

3.2. Parameter Settings

We use the Adam optimizer to train the model and update the network. To accelerate the convergence of the network, the initial learning rate is set to 0.0001. Until the loss function drops to a certain degree, then set the learning rate to 0.00001 to seek for the optimal solution. The experimental settings of the training network are listed as Table 1:

Table 1. Parameter settings.

Parameter	Value
Batch Size	16, 32
Learning Rate	$1 \times 10^{-4}, 1 \times 10^{-5}$
Inter-Channels	36, 48
λ	$5 \times 10^{-4}, 1 \times 10^{-3}, 1 \times 10^{-1}, 5 \times 10^{-1}, 1, 5, 8$

3.3. The Training Process

First of all, we initialize the weights of the network randomly, and utilize the Adam optimizer to train the network. In the first stage of training, MSE is introduced into the loss function. Since optimization is a process of restoring the image as close as possible to the original one, we can express it by the following formula:

$$\left(\widetilde{\theta}_1, \widetilde{\theta}_2\right) = \arg \min_{\theta_1, \theta_2} \|Re(En(Se(\theta_1, x) + Sa(\theta_2, x))) - x\|^2, \tag{14}$$

where x is the original image, θ_1 and θ_2 are the parameters of the spectral feature extraction network and spatial feature extraction network, respectively. $Se(\cdot)$ represents the 1D spectral convolution network, $Sa(\cdot)$ is the spatial group convolution network, $En(\cdot)$ denotes quantization coding, and $Re(\cdot)$ denotes the whole decoding and recovering process. To make the loss function decline as soon as possible, θ_1 and θ_2 are disposed to update along with the gradient descent. By fixing θ_1, we can obtain:

$$\widetilde{\theta}_2 = \arg \min_{\theta_2} \|Re(En(Se(\hat{\theta}_1, x) + Sa(\theta_2, x))) - x\|^2, \tag{15}$$

and we can obtain $\widetilde{\theta}_1$ by fixing θ_2:

$$\widetilde{\theta}_1 = \arg \min_{\theta_1} \|Re(En(Se(\theta_1, x) + Sa(\hat{\theta}_2, x))) - x\|^2. \tag{16}$$

During the backward propagation, the quantization needs to be skipped. Accordingly,

$$\left(\hat{\theta}_1, \hat{\theta}_2\right) = \arg \min_{\theta_1, \theta_2} \|Re(Se(\theta_1, x) + Sa(\theta_2, x)) - x\|^2, \tag{17}$$

to simplify the representation of Equation (14), an auxiliary variable x_m is introduced:

$$x_m(\theta_1, \theta_2) = Se(\theta_1, x) + Sa(\theta_2, x), \tag{18}$$

hence, Equation (14) can be written as:

$$\left(\widetilde{\theta}_1, \widetilde{\theta}_2\right) = \arg \min_{\theta_1, \theta_2} \|Re(En(x_m(\theta_1, \theta_2))) - x\|^2. \tag{19}$$

As the first stage of training optimization is completed, we then bring in the rate loss into the loss function. Combining Equations (13) and (19), the final optimization procedure can be formulated as:

$$\left(\widetilde{\theta}_1, \widetilde{\theta}_2\right) = \arg \min_{\theta_1, \theta_2} \left\{ \|Re(En(x_m(\theta_1, \theta_2))) - x\|^2 + \text{avg}[P(\theta_1, \theta_2, x)] \right\}. \tag{20}$$

When the loss function no longer declines, the training reaches the optimal solution. Moreover, in the second stage of the training, a different compression rate can be easily obtained by changing the penalty λ. The value of λ in our experiment is listed in Table 1.

4. Results and Discussion

In this section, we recorded the experimental results, including the performance comparison of our network with other traditional methods at the same bit rate, and the different bit rates have been obtained through adjusting the penalty λ. Meanwhile, to make the results more convincing, the compression method based on CNN using an optimized residual unit in [20] is also added for comparison. For presentation purposes, it is written as ResConv.

4.1. The Evaluation Criterion

To evaluate the performance of the network comprehensively, apart from PSNR measuring the image recovery, we also utilize another metric known as spectral angle (SA) on the spectral dimension to verify the validity of the partitioned extraction method we proposed. SA indicates the angle between two spectra, which can be viewed as two vectors [28], and it can be used to measure the similarity between two spectral dimensions. The formula is written as follows:

$$\mathrm{SA}_{I,\tilde{I}} = \cos^{-1}\left(\frac{\sum\limits_{\lambda}\left(I(x,y,\lambda)\cdot\tilde{I}(x,y,\lambda)\right)}{\sqrt{\sum\limits_{\lambda}I^2(x,y,\lambda)\sum\limits_{\lambda}\tilde{I}^2(x,y,\lambda)}} \right), \tag{21}$$

whose value ranges from -1 to 1. The closer the SA is to zero, the more similar the two vectors are.

4.2. Experimental Results

4.2.1. Spatial Information Recovery

Figure 8 shows the average PSNR of 7 band test sets. As seen from above, our proposed method is about 1 dB better than 3D-SPIHT and exceeds JPEG2000 by 3 dB. Comparing with ResConv, the partitioned extraction method still gains a little advantage of about 0.6 dB. Figure 9 states the detailed comparison of four selected test images, which can show the recovered result comparison of four methods. It is easy to tell that the partitioned extraction algorithm has an obvious superiority when the bit rate is ranging from 0.3–0.4.

For illustrative purposes, Figure 10 shows the visual effects of four test images when the bit rate is around 0.4. To be specific, we display the grayscale image of the third band of the test image to show the differences more clearly. As can be seen from it, with the JPEG2000 and 3D-SPIHT algorithms, the recovered images have obvious block effects and the textures and margins are seriously blurred, whereas the proposed partitioned extraction algorithm performs well under the same bit rate, the same with ResConv, and these two methods preserve more details than any other methods. Figure 11 shows the partial enlarged view of ah_xia for a clearer demonstration. When the bit rate is around 0.4, ResConv and our proposed method both demonstrate impressive performance. However, according to Figure 8, ResConv starts to lose its edge as the bit rate drops to 0.3 or even lower and our method is then more stable.

To further illustrate the advantage of our partitioned extraction method, we augment 8 band test sets into the experiment for better comparison. The average PSNR is shown in Figure 12. As seen from below, our method obtained a higher PSNR than JPEG2000 and 3D-SPIHT, approximately 8 dB and 4 dB, respectively. With regard to ResConv, the proposed method maintains the competitive edge and obtains about 2.5 dB higher than it on average.

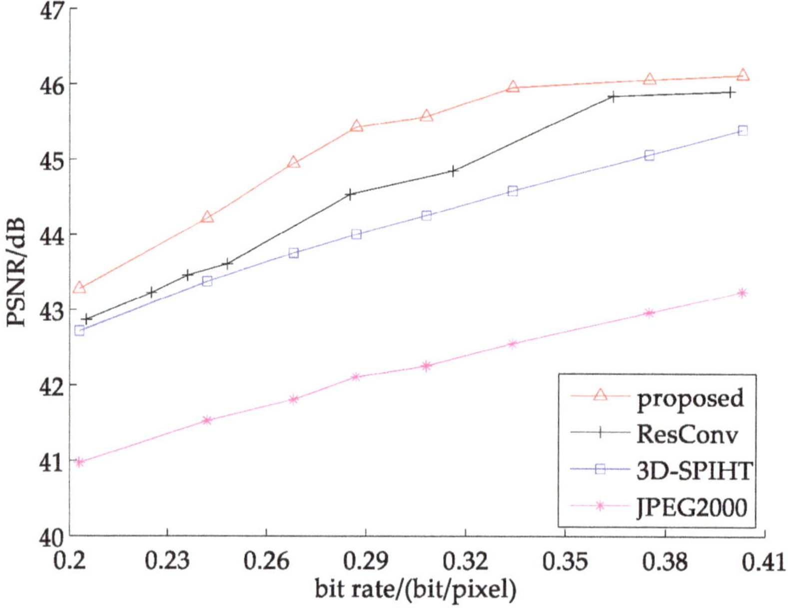

Figure 8. Average PSNR of 7 band test images at different bit rates.

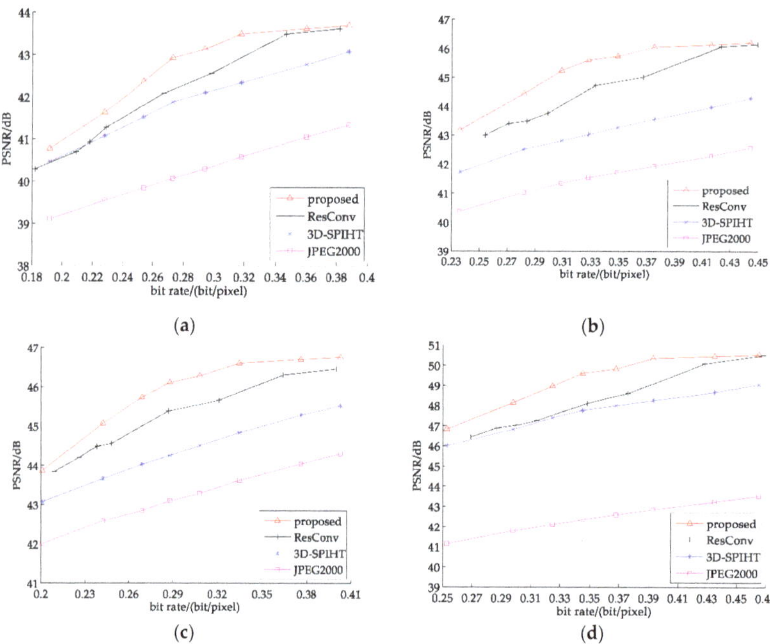

Figure 9. PSNR of recovered images: (**a**) ah_chun; (**b**) ah_xia.; (**c**) hunan_chun; (**d**) tj_dong.

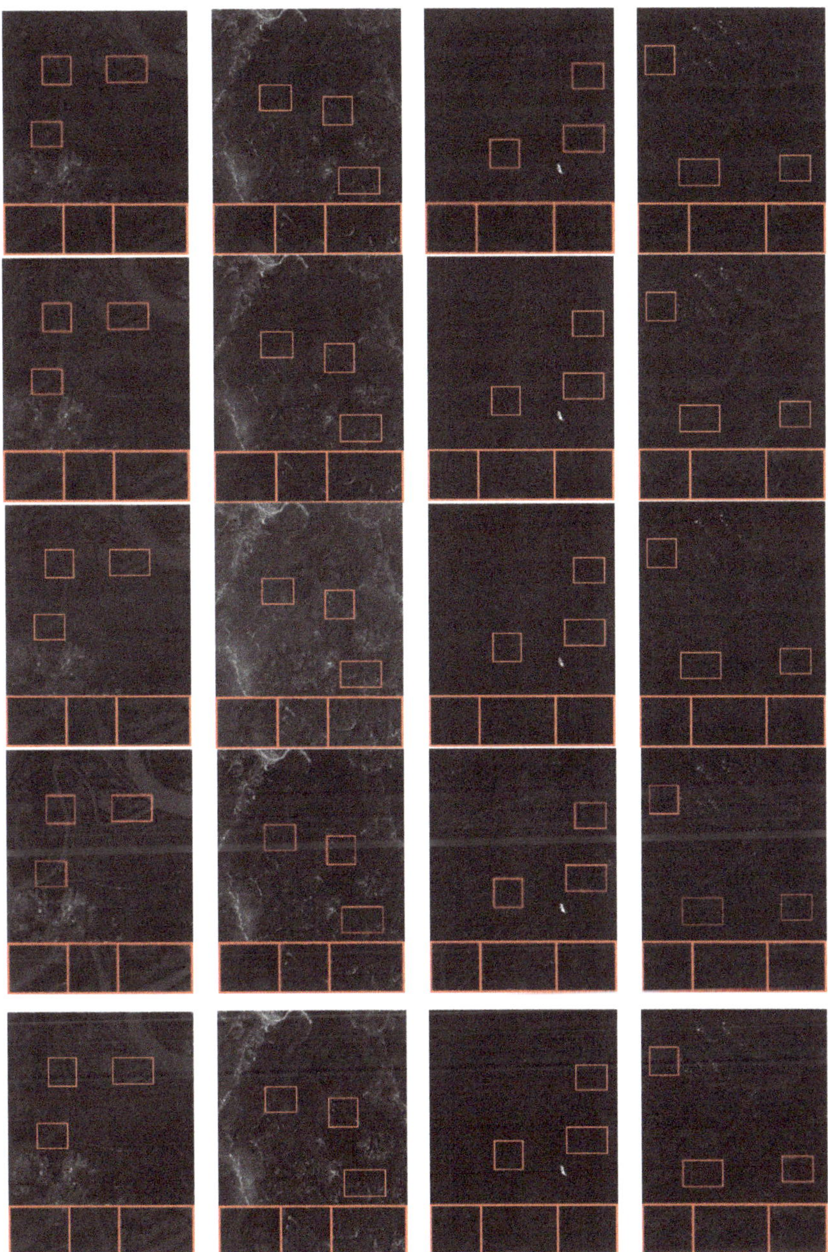

Figure 10. The visual comparison of the recovered images (each column represents the same image).

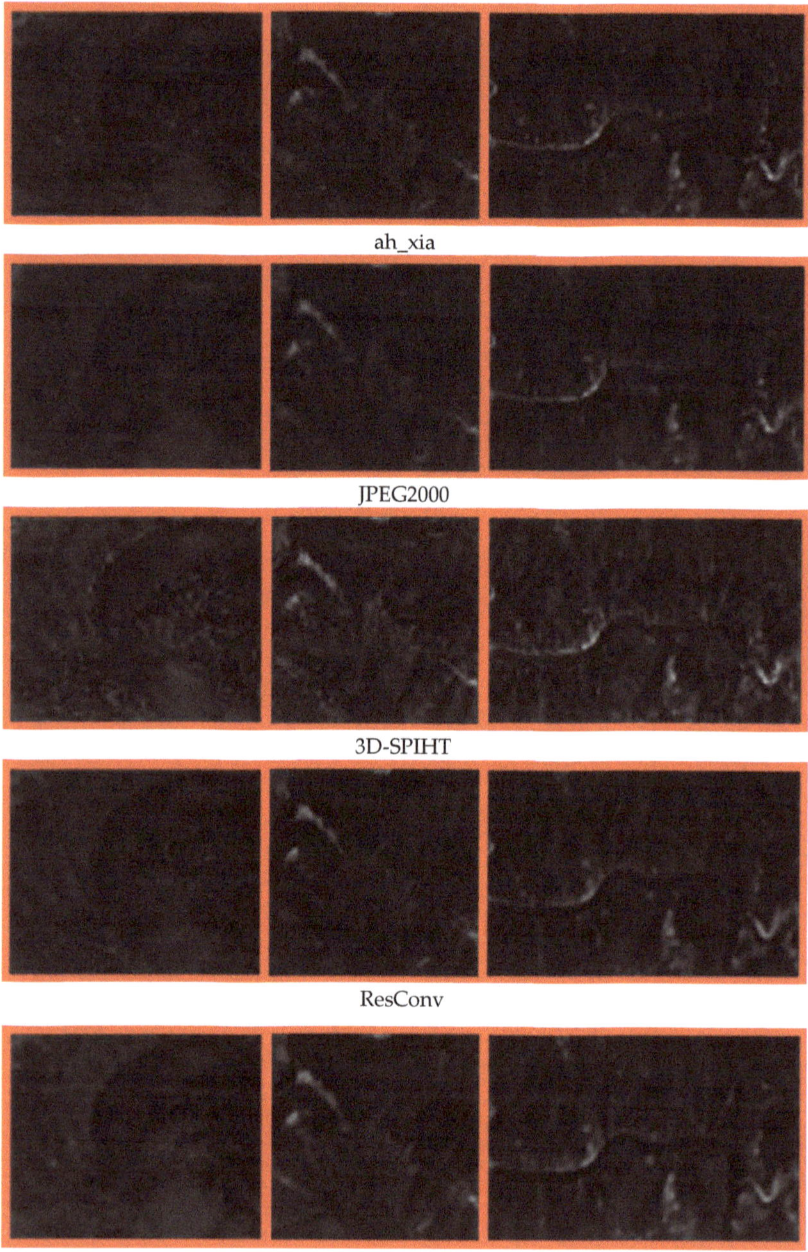

Figure 11. Partial enlarged view of ah_xia.

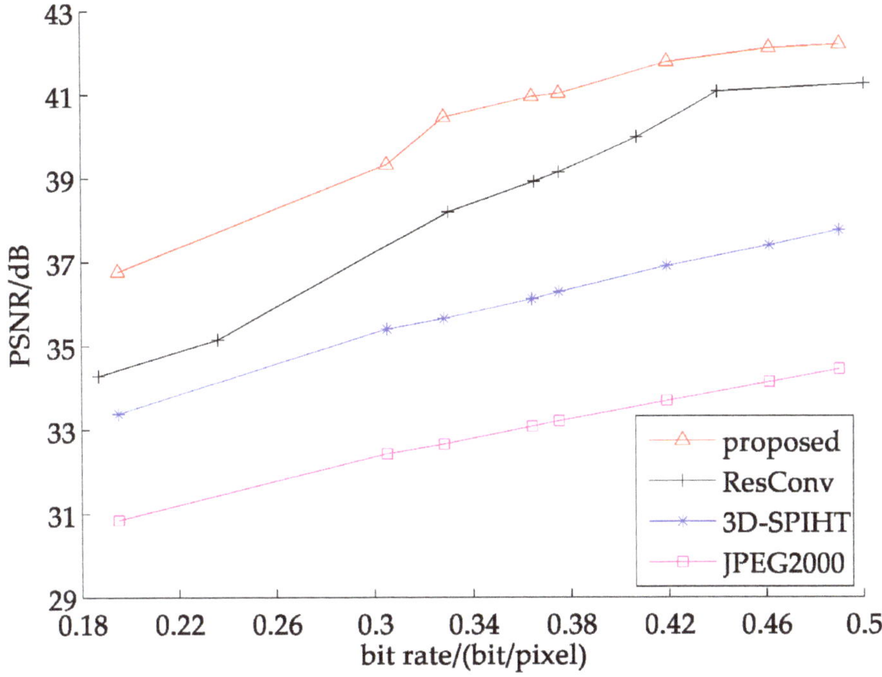

Figure 12. Average PSNR of 8 band test images at different bit rates.

Figure 13 represents the comparison of the PSNR of four test images from 8 band test sets at different bit rates. It can be observed that the advantage of the partitioned extraction method becomes quite prominent, compared with the result of the 7 band test images. Regarding ResConv, in spite of it surpassing JPEG2000 and 3D-SPIHT, its inferiority to our proposed method is more distinct compared with the results of the 7 band test sets. When it comes to processing multispectral images with more bands, or even hyperspectral images, some traditional compression methods will ineluctably be in a more inferior position, as they rarely take the abundant spectral correlation into account.

For visual comparison, as shown in Figures 14 and 15, the JPEG2000 method inevitably generates serious block and ring effects, and detailed texture is ignored too. Furthermore, 3D-SPIHT is relatively better than JPEG2000; however, there are still a lot of blurred texture details in the recovered images. ResConv also obtains recovered images with blurred texture. On the contrary, the algorithm we proposed can retain the detailed texture and edge information of the images to a great extent.

All of the comparison results indicate that these traditional compression methods are not suitable for multispectral image compression, which may cause a lot spatial–spectral information loss. Some CNN-based algorithms may obtain a better result; however, as the bands increase, the inadequacy manifests as well. The partitioned extraction method of spatial–spectral features that we proposed has been proven effective on multispectral image compression, with its higher PSNR and much smoother visual effects.

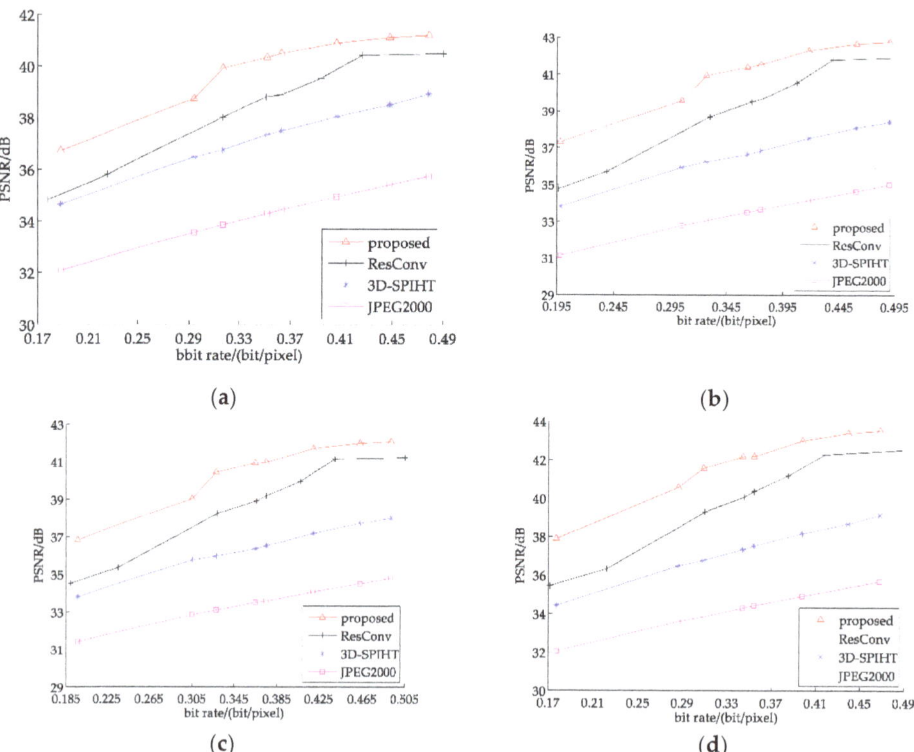

Figure 13. PSNR of recovered images: (**a**) test2; (**b**) test8; (**c**) test14; (**d**) test16.

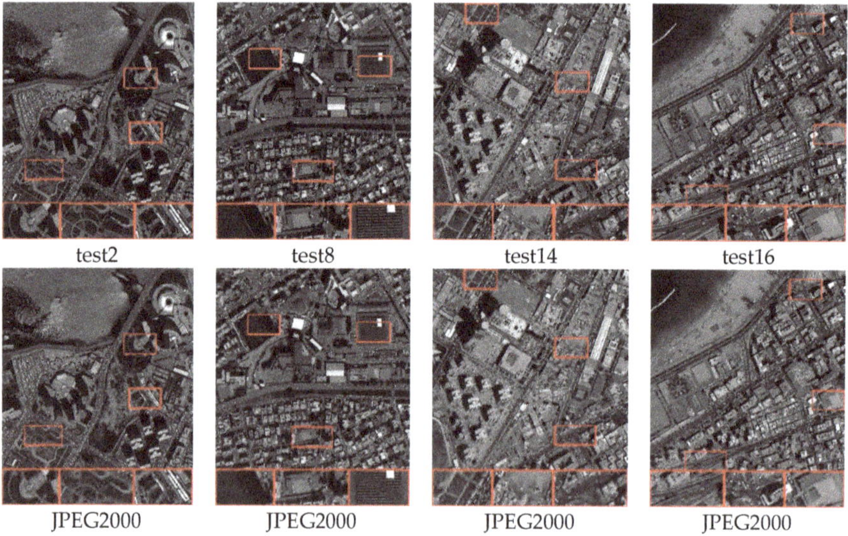

Figure 14. *Cont.*

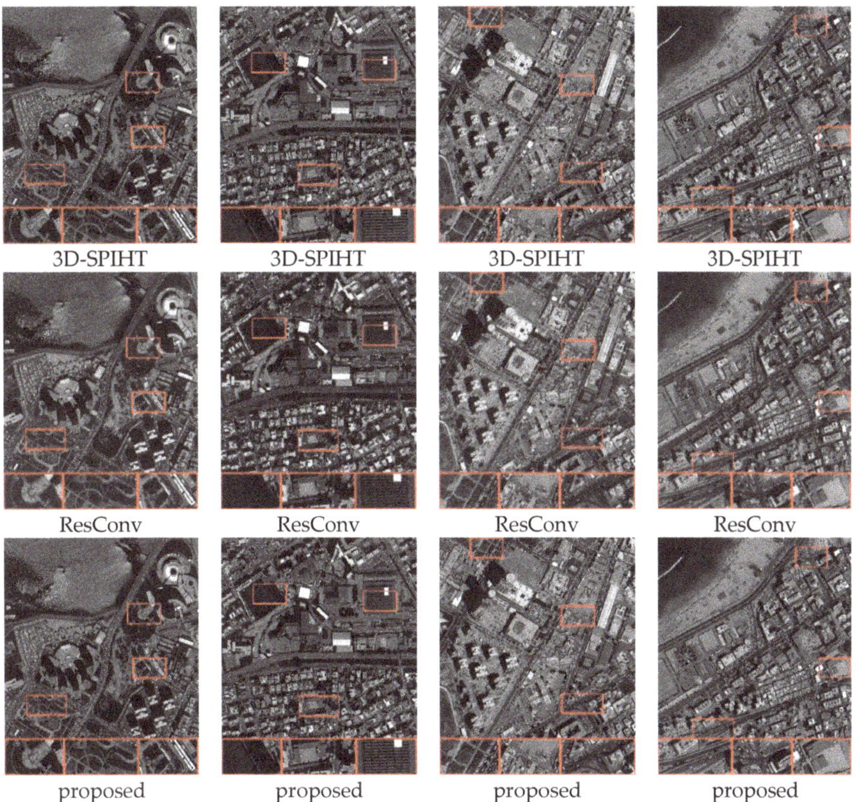

Figure 14. The visual comparison of the recovered images (each column represents the same image).

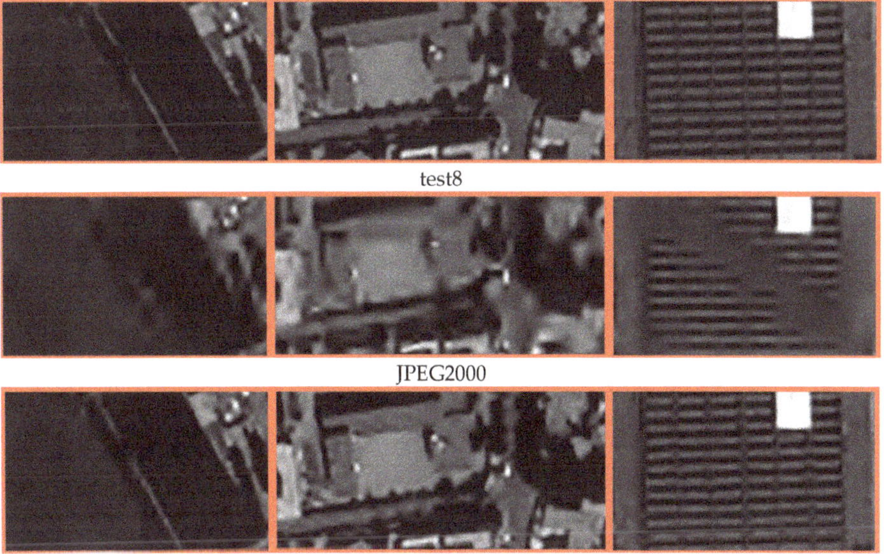

Figure 15. *Cont.*

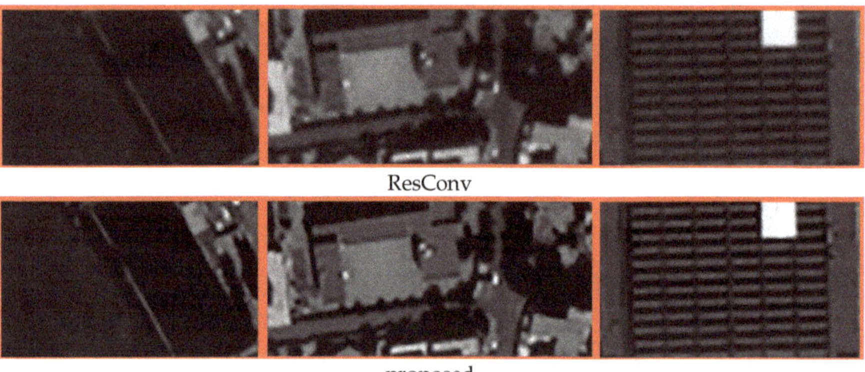

ResConv

proposed

Figure 15. Partial enlarged view of test8.

4.2.2. Spectral Information Recovery

To adapt to our partitioned extraction network structure and verify the effectiveness of spectral information recovery as well, we adopt the SA as the second evaluation criterion. The average SA curves of 7 band and 8 band test sets are shown in Figures 16 and 17, respectively. As seen below, we can find that the SA of the images reconstructed by the partitioned extraction algorithm is always smaller than that of JPEG2000, 3D-SPIHT and ResConv. Tables 2 and 3 list the detailed SA values of four representative test images of 7 band and 8 band, respectively. Supported by the chart and data below, it can be proven that the partitioned extraction algorithm obtains the smallest SA at all bit rates compared with the other three methods, and the smaller SA indicates that the images reconstructed by the proposed partitioned extraction method can obtain better spectral information recovery.

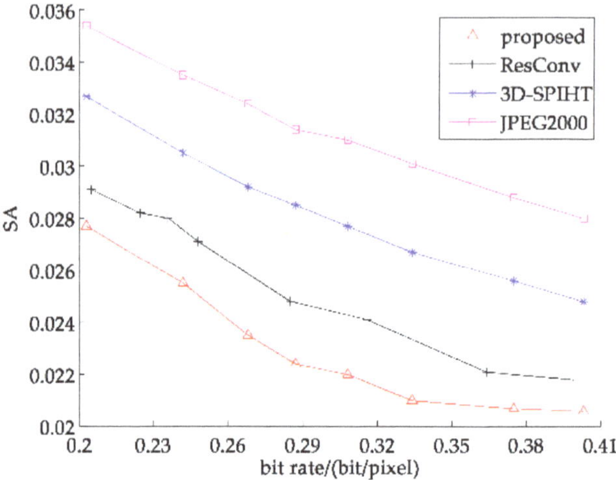

Figure 16. Average spectral angle (SA) curve of 7 band test images.

Table 2. SA of four 7 band test images (around a bit rate of 0.35).

Methods	ah_chun	ah_xia	hunan_chun	tj_dong
Proposed	0.0251	0.0192	0.0182	0.0180
ResConv	0.0265	0.0211	0.0206	0.0201
3D-SPIHT	0.0286	0.0301	0.0249	0.0253
JPEG2000	0.0324	0.0298	0.0255	0.0394

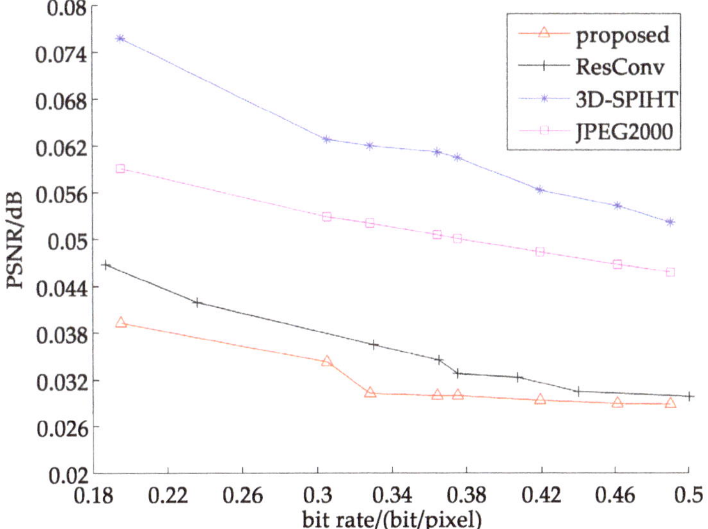

Figure 17. Average SA curve of 8 band test images.

Table 3. SA of four 8 band test images (around a bit rate of 0.35).

Methods	Test2	Test8	Test14	Test16
Proposed	0.0348	0.0300	0.0289	0.0251
ResConv	0.0411	0.0377	0.0327	0.0312
3D-SPIHT	0.0576	0.0645	0.0571	0.0505
JPEG2000	0.0514	0.0517	0.0443	0.0397

5. Conclusions

In this paper, a novel end-to-end framework with partitioned extraction of spatial–spectral features for multispectral image compression is proposed. The algorithm pays close attention to the abundant spectral features of the multispectral images and is committed to preserving the integrity of the spectral–spatial features. The spectral and spatial feature modules extract corresponding features separately, after which the features are fused together for further processing. Likewise, the spectral and spatial features are severally recovered when reconstructing the images, which can help obtain images with high quality. To testify the validity of the framework, experiments are implemented on both 7 band and 8 band test sets. The results show that the proposed algorithm surpasses JPEG2000, 3D-SPIHT and ResConv on PSNR, visual effects and SA as well. The results on the 8 band show that the proposed method has achieved a more obvious superiority, which may prove that spectral information plays an indispensable role in multispectral image processing.

Author Contributions: All the authors made significant contributions to the work. Conceptualization, K.H. and S.Z.; methodology, F.K. and Y.L.; software, K.H.; validation, F.K. and K.H.; formal analysis,

D.L.; investigation, F.K.; resources, F.K.; data curation, Y.L.; writing—original draft preparation, K.H.; writing—review and editing, F.K.; visualization, K.H.; supervision, F.K.; project administration, F.K.; funding acquisition, F.K. All authors have read and agreed to the published version of the manuscript.

Funding: This research was funded by the National Natural Science Foundation of China (grant no. 61801214), and National Key Laboratory Foundation (contract no. 6142411192112).

Acknowledgments: This research was supported by the National Natural Science Foundation of China and National Key Laboratory Foundation, and the authors are also grateful to the editor and reviewers for their constructive comments that helped to improve this work significantly.

Conflicts of Interest: The authors declare no conflict of interest and have no known competing financial interests or personal relationships that could have appeared to influence the work reported in this paper.

References

1. Li, Y.; Zhang, H.K.; Shen, Q. Spectral–Spatial Classification of Hyperspectral Imagery with 3D Convolutional Neural Network. *Remote Sens.* **2017**, *9*, 67. [CrossRef]
2. Gelli, G.; Poggi, G. Compression of multispectral images by spectral classification and transform coding. *IEEE Trans. Image Proc.* **1999**, *8*, 476–489. [CrossRef] [PubMed]
3. Zhou, Z.L. Research on Hyperspectral Image Compression Method. Master's Thesis, Nanjing University of Science and Technology, Nanjing, China, 2008.
4. Li, Y.S.; Wu, C.K.; Chen, J.; Xiang, L.B. Spectral satellite image compression based on wavelet transform. *Acta Opt. Sin.* **2001**, *21*, 691–695.
5. Nian, Y.J.; Liu, Y.; Ye, Z. Pairwise KLT-based compression for multispectral images. *Sens. Imaging* **2016**, *17*, 1–15. [CrossRef]
6. Slyz, M.; Zhang, L. A block-based inter-band lossless hyperspectral image compressor. In Proceedings of the IEEE Data Compression Conference, Snowbird, UT, USA, 29–31 March 2005; pp. 427–436. [CrossRef]
7. Qian, S.E. Hyperspectral data compression using a fast vector quantization algorithm. *IEEE Trans. Geosci. Remote Sens.* **2004**, *42*, 1791–1798. [CrossRef]
8. Hao, P.; Shi, Q. Reversible integer KLT for progressive-to-lossless compression of multiple component images. In Proceedings of the IEEE International Conference on Image Processing, Barcelona, Spain, 14–17 September 2003; p. I-633. [CrossRef]
9. Abousleman, G.P.; Marcellin, M.W.; Hunt, B.R. Compression of hyperspectral imagery using the 3-D DCT and hybrid DPCM/DCT and entropy-constrained trellis coded quantization. In Proceedings of the Conference on Data Compression, Snowbird, UT, USA, 28–30 March 1995. [CrossRef]
10. Sweldens, W. The Lifting Scheme: A Custom-Design Construction of Biorthogonal Wavelets. *Appl. Comput. Harmon. Anal.* **1996**, *3*, 186–200. [CrossRef]
11. Tang, X.L.; Pearlman, W.A. Three-dimensional wavelet-based compression of hyperspectral images. In *Hyperspectral Data Compression*; Springer: Troy, NY, USA, 2006; pp. 273–308. ISBN 978-038-728-579-5.
12. Dragotti, P.L.; Poggi, G.; Ragozini, A.R.P. Compression of multispectral images by three-dimensional SPIHT algorithm. *IEEE Trans. Geosci. Remote Sens.* **2000**, *38*, 416–428. [CrossRef]
13. LeCun, Y.; Jackel, L.; Bottou, L.; Cortes, C.; Denker, J.S.; Drucker, H.; Guyon, I.; Müller, U.A.; Säckinger, E.; Simard, P.; et al. Learning algorithms for classification: A comparison on handwritten digit recognition. In *Neural Networks: The Statistical Mechanics Perspective*; World Scientific: Singapore, 1995; pp. 261–276. [CrossRef]
14. Krizhevsky, A.; Sutskever, I.; Hinton, G.E. ImageNet classification with deep convolutional neural networks. In Proceedings of the 25th International Conference on Neural Information Processing Systems, Red Hook, NY, USA, 3–6 December 2012; pp. 1097–1105.
15. Simonyan, K.; Zisserman, A. Very Deep Convolutional Networks for Large-Scale Image Recognition. *Comput. Sci.* **2014**, arXiv:1409.1556v6.
16. Szegedy, C.; Liu, W.; Jia, Y.Q.; Sermanet, P.; Reed, S.; Anguelov, D.; Erhan, D.; Vanhoucke, V.; Rabinovich, A. Going deeper with convolutions. In Proceedings of the 2015 IEEE Conference on Computer Vision and Pattern Recognition (CVPR), Boston, MA, USA, 7–12 June 2015; pp. 1–9. [CrossRef]
17. He, K.; Zhang, X.; Ren, S.; Sun, J. Deep Residual Learning for Image Recognition. In Proceedings of the 2016 IEEE Conference on Computer Vision and Pattern Recognition (CVPR), Las Vegas, NV, USA, 27–30 June 2016; pp. 770–778. [CrossRef]
18. Ballé, J.; Laparra, V.; Simoncelli, E. End-to-end Optimized Image Compression. In Proceedings of the 2017 International Conference on Learning Representations (ICLR), Toulon, France, 24–26 April 2017.
19. Jiang, F.; Tao, W.; Liu, S.; Ren, J.; Guo, X.; Zhao, D. An End-to-End Compression Framework Based on Convolutional Neural Networks. *IEEE Trans. Circuits Syst. Video Technol.* **2018**, *28*, 3007–3018. [CrossRef]
20. Kong, F.; Zhou, Y.; Shen, Q.; Wen, K. End-to-end Multispectral Image Compression Using Convolutional Neural Network. *Chin. J. Lasers* **2019**, *46*, 1009001-1. [CrossRef]
21. Zhong, Z.; Li, J.; Luo, Z.; Chapman, M. Spectral–Spatial Residual Network for Hyperspectral Image Classification: A 3-D Deep Learning Framework. *IEEE Trans. Geosci. Remote Sens.* **2018**, *56*, 847–858. [CrossRef]

22. He, K.; Zhang, X.; Ren, S.; Sun, J. Delving Deep into Rectifiers: Surpassing Human-Level Performance on ImageNet Classification. In Proceedings of the 2015 IEEE International Conference on Computer Vision (ICCV), Santiago, Chile, 7–13 December 2015; pp. 1026–1034. [CrossRef]
23. A Tutorial on Filter Groups. Available online: https://blog.yani.io/filter-group-tutorial/ (accessed on 9 August 2020).
24. Shi, W.Z.; Caballero, J.; Huszar, F.; Totz, J.; Aitken, A.P.; Bishop, R.; Rueckert, D.; Wang, Z. Real-time single image and video super-resolution using an efficient sub-pixel convolutional neural network. In Proceedings of the 2016 IEEE Conference on Computer Vision and Pattern Recognition (CVPR), Las Vegas, NV, USA, 27–30 June 2016; pp. 1874–1883.
25. Toderici, G.; O'Malley, S.M.; Hwang, S.J.; Vincent, D.; Minnen, D.; Baluja, S.; Covell, M.; Sukthankar, R. Variable rate image compression with recurrent neural networks. *arXiv* **2015**, arXiv:1511.06085v2.
26. Li, M.; Zuo, W.; Gu, S.; Zhao, D.; Zhang, D. Learning Convolutional Networks for Content-Weighted Image Compression. In Proceedings of the 2018 IEEE/CVF Conference on Computer Vision and Pattern Recognition, Salt Lake City, UT, USA, 18–23 June 2018; pp. 3214–3223. [CrossRef]
27. Misra, D. Mish: A Self Regularized Non-Monotonic Activation Function. *arXiv* **2019**, arXiv:1908.08681.
28. Christophe, E.; Leger, D.; Mailhes, C. Quality criteria benchmark for hyperspectral imagery. *IEEE Trans. Geosci. Remote Sens.* **2005**, *43*, 2103–2114. [CrossRef]

 *remote sensing*

Article

Lossy Compression of Multichannel Remote Sensing Images with Quality Control

Vladimir Lukin [1], Irina Vasilyeva [1], Sergey Krivenko [1], Fangfang Li [1], Sergey Abramov [1], Oleksii Rubel [1], Benoit Vozel [2,*], Kacem Chehdi [2] and Karen Egiazarian [3]

[1]   Department of Information and Communication Technologies, National Aerospace University, 61070 Kharkov, Ukraine; lukin@ai.kharkov.com (V.L.); i.vasilieva@khai.edu (I.V.); krivenkos@ieee.org (S.K.); liff@nchu.edu.cn (F.L.); s.abramov@khai.edu (S.A.); o.rubel@khai.edu (O.R.)
[2]   Institut d'Électronique et des Technologies du numéRique, University of Rennes 1, UMR CNRS 6164, 22300 Lannion, France; kacem.chehdi@univ-rennes1.fr
[3]   Computational Imaging Group, Tampere University, 33720 Tampere, Finland; karen.eguiazarian@tuni.fi
*   Correspondence: benoit.vozel@univ-rennes1.fr

Received: 25 October 2020; Accepted: 19 November 2020; Published: 23 November 2020

Abstract: Lossy compression is widely used to decrease the size of multichannel remote sensing data. Alongside this positive effect, lossy compression may lead to a negative outcome as making worse image classification. Thus, if possible, lossy compression should be carried out carefully, controlling the quality of compressed images. In this paper, a dependence between classification accuracy of maximum likelihood and neural network classifiers applied to three-channel test and real-life images and quality of compressed images characterized by standard and visual quality metrics is studied. The following is demonstrated. First, a classification accuracy starts to decrease faster when image quality due to compression ratio increasing reaches a distortion visibility threshold. Second, the classes with a wider distribution of features start to "take pixels" from classes with narrower distributions of features. Third, a classification accuracy might depend essentially on the training methodology, i.e., whether features are determined from original data or compressed images. Finally, the drawbacks of pixel-wise classification are shown and some recommendations on how to improve classification accuracy are given.

Keywords: remote sensing; lossy compression; image quality; image classification; visual quality metrics

---

1. Introduction

Nowadays, remote sensing (RS) is used in numerous applications [1–3] due to the following main reasons. Different types of useful information can be potentially retrieved from RS images, especially high resolution and multichannel data (i.e., a set of co-registered component images of the same territory acquired for different wavelengths, polarizations, even by different sensors [3–6]). The modern RS sensors often offer a possibility of fast and frequent data collection—good examples are multichannel sensors Sentinel-1 and Sentinel-2 which have been launched recently and started to produce a great amount of valuable RS data [7,8].

Therefore, the volume of RS data greatly increases due to the aforementioned factors: better spatial resolution, a larger number of channels, more frequent observations. This causes challenges in RS data processing that relate to all basic stages of their processing: co-registration, calibration, pre- or post-filtering, compression, segmentation, and classification [9,10]. One of the most serious challenges is the compression of multichannel RS images [11–13]. Compression is applied to diminish data size before their downlink transferring from spaceborne (more rarely, airborne, UAV) sensors, to store acquired images in on-land centers of RS data collecting or special depositories, to pass images to potential customers. Lossless compression is often unable to meet requirements to compression ratio (CR) that should be provided, since, even in the most favorable situations of high inter-band correlation of component images [14], CR attained by the best existing lossless compression techniques reaches 4 ... 5 [12].

Lossy compression can provide considerably larger CR values [12,13,15–17] but at the expense of introduced distortions. It is always a problem to reach an appropriate compromise between compressed image quality (characterized in many different ways) and CR [11–13,18–20]. There are several reasons behind this. First, compression providing a fixed CR often used in practice [21,22] leads to compressed images whose quality can vary in wide limits [23]. For a given CR and simpler structure images, introduced distortions are smaller and compressed image quality is higher, whilst for images, having a more complex structure (containing many small-sized details and textures), losses are larger and, thus, image quality can be inappropriate since some important information can be inevitably lost, which is undesired. Certainly, some improvements can be reached due to the employment of a better coder, adaptation to image content [24], use of inter-channel correlation by three-dimensional (3D) compression [12,13,21,22], or some other means. However, the positive effect can be limited, and/or there can be some restrictions, e.g., the necessity to apply image compression standard [12,19,21].

Thus, we are more interested in a different approach that presumes lossy compression of multichannel RS images with a certain control of introduced distortions combined with simultaneous desire to provide a larger CR. Let us explain the advantages of such an approach, when it can be reasonable, and what the conditions are for its realization. Recall here that RS data compressed in a lossy manner can be useful if: (a) introduced losses do not lead to sufficient reduction of solving the final tasks of RS image processing such as segmentation, object detection, spectral unmixing, classification, parameter estimation [17,25–34]; (b) distortions due to lossy compression do not appear themselves as artifacts leading to undesired negative effects (e.g., appearing of ghost artifacts) in solving the aforementioned final task.

Below, we consider the impact of lossy compression on multichannel RS image classification [27,28,33–35]. Specifically, we focus on the case of a limited number of channels, e.g., color, multi-polarization, or multispectral images, due to the following reasons: (a) it is simpler to demonstrate the effects that take place in images due to lossy compression and how these effects influence classification just for the case of a small number of image components; (b) an accurate classification of multichannel images with a small number of components is usually a more difficult task than the classification of hyperspectral data (images with a relatively large number of components) because of a limited number of available features and the necessity to reliably discriminate classes in feature space.

Returning to the advantages of lossy compression of multichannel RS data, we can state the following. First, images compressed in a lossy manner might be classified better than original images [27,34]. Such an effect usually takes place if original images are noisy and coder parameters are adjusted so that the noise removal effect due to lossy compression [13,36–38] is maximal. However, even if the noise in images is absent (or, more exactly, peak signal-to-noise ratio (PSNR) values in original images are high and noise is not seen in visualized component images), the accuracy of image classification may remain practically the same for a relatively wide range of CR variation [27,34,38].

This range width can be different [39]. Its width depends on the following main factors:

- Used classifier, how efficient it is, how it is trained;
- Applied compression technique (coder);
- How well classes are discriminated, and which features are used;
- How many classes are present in each multichannel image;
- How "complex" is a considered image (is it highly textural, does it contain many small-sized or prolonged objects, what is the mean size of fragments belonging to each class).

Therefore, researchers have a wide field of studies intended on understanding how to set a CR or a coder's parameter that controls compression (PCC) for a given multichannel image, used lossy compression technique, and considered classifier. Here we would like to recall some already known aspects and dependencies that will be exploited below. First, for the same CR, considerably larger distortions can be introduced into images with higher complexity [40]. Moreover, larger distortions can be introduced into more complex and/or noisier images for the same PCC, for example, quantization step (QS) of coders based on orthogonal transforms (discrete cosine transform (DCT) or wavelets [40]). Second, the desired quality of a compressed image (a value of a metric that characterizes image quality) can be nowadays predicted and provided with an appropriate accuracy [40,41], at least, for some coders based on DCT as, e.g., the coder AGU [42]. Third, there exist certain dependencies between values of conventional (e.g., PSNR) and/or visual quality metrics and accuracy of correct classification of an entire image or classes [43]. Classification accuracy for classes mainly represented by large size homogeneous objects (water surfaces, meadows) better correlates with the conventional metrics whilst probability of correct classification for classes mainly represented by textures, small-sized and prolonged objects (forests, urban areas, roads, narrow rivers) correlates with visual quality metrics. Besides, classification accuracies for classes can depend on PCC or CR differently [38]. For the latter type of classes, the reduction of probability of correct classification with an increase of CR is usually faster. Fourth, different classifiers produce classification results that can significantly differ both in the sense of obtained classification maps and quantitative characteristics as the total probability of correct classification, probabilities for classes, and confusion matrices [43,44]. Fifth, there are quite successful attempts to predict the total probability of correct classification (TPCC) of compressed RS images based on the fractal models [28].

Aggregating all these, we assume the following:

- There are strict dependencies between distortions due to lossy compression (that can be characterized by certain metrics) and TPCC for multichannel RS images;
- Having such dependences, it is possible to recommend what metrics' values must be provided to ensure practically the same TPCC for compressed data as for original data;
- It is possible to provide the recommended metrics' values, at least, for some modern compression techniques based on DCT;
- Metric's thresholds might depend upon a classifier used and properties of a multichannel image subject to lossy compression;
- Classifier performance might also depend on how it is trained (for example, using an original or compressed multichannel image).

The goal of this paper is to carry out a preliminary analysis are these assumptions valid. The main contributions are the following. First, we show that it is worth applying lossy compression with distortions around their invisibility threshold to provide either some improvement of TPCC (achieved for simple structure images) or acceptable reduction of TPCC (for complex structure images) compared to TPCC for the corresponding original image. Moreover, we show how such compression can be carried out for a particular coder. Second, we demonstrate that classifier training for compressed data usually provides slightly better results than training for original (uncompressed) images.

The paper structure is the following. Section 2 deals with a preliminary analysis of the performance criteria used in lossy compression of RS images. In Section 3, two popular methods of RS data classification based on the maximum likelihood method (MLM) and neural network (NN) are described. Section 4 contains the results of classifier testing for the synthetic three-channel test image. Section 5 provides the results of experiments carried out for real-life three-channel image. Discussion and analysis of some practical peculiarities are presented in Section 6; the conclusions are shown in Section 7.

2. Performance Criteria of Lossy Compression and Their Preliminary Analysis

Consider an image I_{ij}^t where i, j denote pixel indices and I_{Im}, J_{Im} define the processed image size. If one deals with a multichannel image, index q can be used. Let's start with the consideration of the case of a single-component image. To characterize compression efficiency, we have used the following three metrics. The first one, traditional PSNR, is defined by

$$\text{PSNR}_{\text{inp}} = 10\log_{10}\left(\text{DR}^2 / \sum_{i=1}^{I_{Im}}\sum_{j=1}^{J_{Im}}\left(I_{ij}^c - I_{ij}^t\right)^2 / (I_{Im} \times J_{Im})\right) = 10\log_{10}\left(\text{DR}^2/\text{MSE}\right) \tag{1}$$

where DR denotes the range of image representation, I_{ij}^c is the compressed image, MSE is the mean square error of distortions introduced by lossy compression. In general, DR in remote sensing applications might sufficiently differ from 255, which is common for processing RGB color images. Below, we will mainly consider single and multichannel RS images having DR = 255 but will also analyze what should be done if DR is not equal to 255.

PSNR is the conventional metric widely used in many image processing applications. Meanwhile, the drawbacks of PSNR are well known. The main drawback of PSNR is that it is not adequate in characterizing image visual quality [45,46]. Currently, there are numerous other, so-called HVS (human vision system) metrics, that are considered to be more adequate [45–47]. Some of the existing HVS-metrics can be applied only to color images. As we need some metrics that should be good enough and applicable to single component images, we apply the metrics PSNR-HVS and PSNR-HVS-M. The former HVS-metric takes into account less sensitivity of human vision to distortions in high-frequency spectral components, the latter one also takes into consideration a masking effect of image texture and other heterogeneities [47]. These metrics are defined by

$$\text{PHVSM} = 10\log_{10}\left(\text{DR}^2/\text{MSE}_{\text{HVS}}\right) \tag{2}$$

$$\text{PHVSM}_{\text{out}} = 10\log_{10}\left(\text{DR}^2/\text{MSE}_{\text{HVS–M}}\right) \tag{3}$$

where MSE_{HVS} and $\text{MSE}_{\text{HVS–M}}$ are MSEs determined considering the aforementioned effects. If PSNR-HVS is approximately equal to PSNR, then the distortions due to lossy compression have properties similar to an additive white Gaussian noise. If PSNR-HVS is smaller than PSNR, this evidences that distortions are more similar to either non-Gaussian (heavy-tailed), or spatially correlated noise, or both [47] (see also [48] for details). If PSNR-HVS-M is considerably larger than PSNR, then a masking effect of image content is sufficient (most probably, an image is highly textural). In turn, if PSNR-HVS-M is approximately equal or smaller than PSNR, then a masking effect is absent, and/or distortions are like spatially correlated noise.

Figure 1a,c,e present three grayscale test RS images of different complexity—the image Frisco has the simplest structure whilst the image Diego is the most textural. Figure 1b,d,f represent dependences of the considered metrics in Equations (1)–(3) on quantization step which serves as PCC in the coder AGU [42] (this compression method employs 2D DCT in 32 × 32 pixel blocks, the advanced algorithm of coding uniformly quantized DCT coefficients, and embedded deblocking after decompression). As one may expect, all metrics become smaller (worse) if QS increases. Meanwhile, for the same QS, compressed image quality is not the same. As an example, for the case of QS = 20, PSNR and PSNR-HVS-M for the test image Frisco are practically the same (about 40 dB). For the test images Airfield and Diego that have more complex structures, PSNR-HVS-M values are also about 40 dB whilst PSNR values are about 34 dB. This means the following: (1) there are masking effects, i.e., textures and heterogeneities still sufficiently mask introduced distortions; (2) the introduced distortions can be noticed by visual inspection (comparison of original and compressed images). Recall that distortion visibility thresholds are about 35 … 38 dB according to the metric PSNR and about 40 … 42 dB according to the metric PSNR-HVS-M [45].

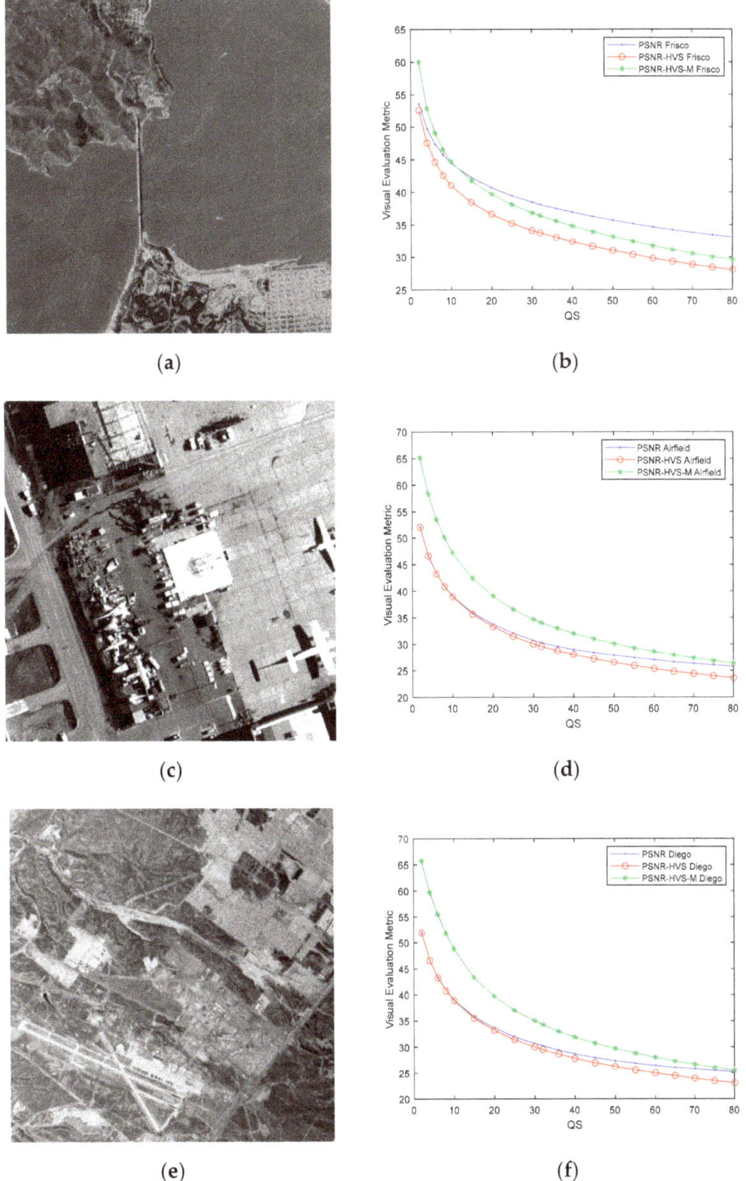

Figure 1. Test grayscale images Frisco (**a**), Airfield (**c**), and Diego (**e**), all of size 512 × 512 pixels, and dependences of the considered metrics (PSNR, PSNR-HVS, PSNR-HVS-M, all in dB) on the quantization step (QS) for the coder AGU for these test images ((**b**), (**d**), and (**f**), respectively).

Note that distortions' invisibility happens if QS is smaller than 18 . . . 20 for grayscale images represented as 8-bit data arrays [23] (more generally, if QS ≤ DR / (12 . . . 13)). One should also keep in mind that the use of the same QS leads to sufficiently different CR. For example, for QS = 20, CR values are equal to 26.3, 4.6, and 4.5 for the considered test images Frisco, Airfield, and Diego, respectively. These facts and dependences are given in Figure 1 confirm the basic statements presented in the Introduction.

Alongside differences in metrics' values for the same QS considered above for QS ≈ 20 (similar differences take place for QS > 20, compare the corresponding plots in Figure 1b,d,f), there are interesting observations for QS ≤ 10 (more generally, QS ≤ DR / 25). In this case, MSE ≈ QS^2 / 12 [20], PSNR exceeds 39 dB and PSNR-HVS-M is not smaller than PSNR and exceeds 45 dB. Thus, the introduced distortions are not visible and the desired $PSNR_{des}$ (or, respectively, MSE_{des}) can be easily provided by a proper setting of QS as QS ≈ $(12MSE_{des})^{1/2}$.

Certainly, there are numerous quality metrics proposed so far. Below we carry our study based on PSNR-HVS-M because of the following reasons. First, it is one of the best component-wise quality metrics [49] that can be calculated efficiently. Second, the cross-correlation between the best existing quality metrics is high (see data in Supplementary Materials Section). Therefore, other quality metrics can be used in our approach under the condition of carrying out a corresponding preliminary analysis of their properties. Note that we have earlier widely used PSNR-HVS-M in image compression and denoising applications [23] utilizing well its main properties.

3. Considered Approaches to Multichannel Image Classification

As has been mentioned above, there are numerous approaches to the classification of multichannel images. Below we consider two of them. The first one is based on the maximum likelihood method (MLM) [50,51] and the second one relies on a neural network (NN) training [50]. In both cases, the pixel-wise classification is studied. There are many reasons behind using the pixel-wise approach and just these classifiers: (a) to simplify the classification task and to use only Q features (q = 1, … , Q), i.e., the values of a given multichannel image in each pixel; (b) to show the problems of pixel-wise classification; (c) MLM and NN based classifiers are considered to be among the best ones [50].

The note that classifiers of RS data classification can be trained in different ways. One option is to train a classifier in advance using earlier acquired images, e.g., uncompressed ones stored for training or other purposes. Another option is to use a part of the obtained compressed image for classifier training and its use for entire image classification [39]. One can expect that classification results would be different where both options have both advantages and drawbacks.

3.1. Maximum Likelihood Classifier

The classification (recognition) function for MLM is based on the calculation of some metric in the feature space

$$\Phi = \begin{cases} 1, & \rho\left(\vec{x}^{\,*}, \vec{x}_s\right) \geq \text{tr}; \\ 0, & \rho\left(\vec{x}^{\,*}, \vec{x}_s\right) < \text{tr}, \end{cases} \tag{4}$$

where $\vec{x}^{\,*}$, $\vec{x}_s$ are feature (attribute) vectors for current and sample objects (image pixel in our case); $\rho(\bullet, \bullet)$ is a used metric of vector similarity; tr denotes the decision undertaking threshold.

As components of the feature vector, in general, one can consider both original ones (pixel (voxel) values of a multichannel image) and "derivative" features calculated on their basis (e.g., ratios of these values).

If $\Phi = 1$, then a current object (pixel) determined by the feature vector $\vec{x}^{\,*}$, is related to a class a_s.

Taking into account the stochastic nature of the features $\vec{x}$, $\vec{x} \in R^c$, information about each object class is contained in parameters $\vec{\theta}$ of joint probability density functions (PDFs) $f_c\left(\vec{x}; \vec{\theta}\right)$.

This information is used for creating statistical decision rules

$$L = \frac{f_c\left(\vec{x}^{\,*}; \vec{\theta} \,|\, a_2\right)}{f_c\left(\vec{x}^{\,*}; \vec{\theta} \,|\, a_1\right)} \geq \text{tr} \tag{5}$$

where L is the likelihood ratio. The threshold tr is determined by a statistical criterion used; for example, the Bayesian criterion provides getting the optimal classifier when the following information is available: PDFs for all sets of patterns, probabilities of each class occurrence, and losses connected with probabilities of misclassifications.

In most practical cases, for object description, statistical sample information is used, i.e., statistical estimates of PDFs obtained at the training stage are employed in likelihood ratio L. Then, sample size, reliability of information about observed objects, and efficiency of using this information in decision rule mainly determine the quality of undertaken decisions.

Training samples are formed using pixels representing a given class; usually, it is some area (or areas, a set of pixels) in an image identified based on some true data for a sensed terrain. The main requirement for the data of training samples is their representativeness—the pixels of the sample must correspond to one class on the ground; such a class should occupy a territory that is fairly well represented by pixels in the image with a given resolution. In other words, the number of pixels in the selected area of the image should ensure the adoption of statistically significant decisions.

When constructing a multi-alternative decision rule (K > 2), the maximum likelihood criterion is used; the threshold tr in the Equation (5) is assumed to be equal to 1. The maximum likelihood criterion makes it possible to eliminate the uncertainty of the solution (when none of the classes can be considered preferable to the others), does not require knowledge of the a priori probabilities of the classes and the loss function, allows evaluating the reliability of the solutions, and can be easily generalized to the case of many classes. In accordance with this criterion, it is believed that the control sample (measured values of features in the current image pixel) belongs to the class, $1 \leq u \leq K$, for which the likelihood function (or rather, its estimate obtained at the training stage) is maximum:

$$f_c\left(\vec{x}^* ; \vec{\theta} \,|\, a_u\right) = \max_{1 \leq k \leq K}\left\{f_c\left(\vec{x}^* ; \vec{\theta} \,|\, a_k\right)\right\} \Rightarrow \vec{x}^* \in a_u. \tag{6}$$

If the assumptions about the normal law of distribution of the feature vector are true, the maximum likelihood method Equation (6) provides optimal recognition. The assumption of normality of features, on the one hand, simplifies the estimation of the parameters of distribution laws (DL) and makes it possible to take into account the presence of mutual correlation relationships between data in different channels in models of classes. On the other hand, it is known that the distributions of real multichannel data are often non-Gaussian. Even with the assumption of the normalization of DL observations due to the central limit theorem, the nonlinearity introduced during the reception and primary processing of information, which manifests itself in a change in the shape and parameters of the observed distribution, cannot be ignored.

To increase the accuracy of approximating empirical non-Gaussian distributions of features and to take into consideration the possible presence of strong correlation relationships, it is advisable to apply multiparameter distribution laws based on non-linear transformations of normal random variables. These transformations include Johnson's transformations.

For describing a class of spectral features, we propose to apply Johnson S_B-distribution [52,53]:

$$f(x) = \frac{\eta \lambda}{\sqrt{2\pi}(x - \varepsilon)(\varepsilon + \lambda - x)} e^{-\frac{1}{2}\left(\gamma + \eta \ln \frac{x-\varepsilon}{\varepsilon+\lambda-x}\right)^2} \tag{7}$$

where η and γ are two shape parameters connected with skewness and kurtosis ($\eta > 0$, $\gamma \in (-\infty, +\infty)$); ε and λ are two scale parameters of a random variable x ($\varepsilon(-\infty, +\infty), \varepsilon \leq x \leq \varepsilon + \lambda$).

Due to a large number of parameters, the model in Equation (4) is quite universal and able to approximate almost any unimodal distribution as well as a wide spectrum of bimodal distributions. Besides, since Johnson distributions are based on nonlinear transformations of normal random variables, their use in describing empirical distributions allows staying within correlation theory framework, which is important for the considered features characterized by non-Gaussian distributions (see examples below) and the presence of strong inter-channel correlation.

The estimates of the parameters for PDFs $\vec{\theta}|a_k = \{\varepsilon_k, \lambda_k, \eta_k, \gamma_k\}$ can be found by a numerical optimization of the loss function minimizing integral MSE of empirical distribution (histogram) representation by the theoretical model in Equation (7).

The multidimensional variant of Equation (7) is the following [52]:

$$f_c(\vec{x}|a) = (2\pi)^{-c/2}\|\Xi\|^{-1/2} \prod_{v=1}^{c} \frac{\eta_v \lambda_v}{(x_v - \varepsilon_v)(\varepsilon_v + \lambda_v - x_v)} \times$$
$$\times \exp\left[-\frac{1}{2} \sum_{v,\kappa=1}^{c} \Xi_{v\kappa}^{-1}\left(\gamma_v + \eta_v \ln \frac{x_v - \varepsilon_v}{\varepsilon_v + \lambda_v - x_v}\right)\left(\gamma_\kappa + \eta_\kappa \ln \frac{x_\kappa - \varepsilon_\kappa}{\varepsilon_\kappa + \lambda_\kappa - x_\kappa}\right)\right]$$

$$(8)$$

where c is the size of the feature vector $\vec{x}$; Ξ is the sample correlation matrix.

Thus, to construct a multidimensional statistical model of correlated data, processing consists of several stages. At the first stage, a traditional statistical analysis is performed by estimating the moments of distributions and histograms of the components of the random vector. Sample estimates of the cross-correlation coefficients of the components $\vec{x}$ are also found. At the second stage, one-dimensional statistical models of the form Equation (7) are constructed for each of the components of multidimensional data. The initial estimation of the distribution density (the histogram of the feature x_v) is used as a target for the desired model $f(x_v)$ described by the Johnson DL parameter vector $\vec{\theta}$. The multidimensional statistical model is built based on multidimensional histograms $\hat{f}_c(\vec{x})$. To construct a multidimensional model as in Equation (8) for a class a_k, one needs to evaluate the matrix of distribution parameters $\Theta_k = [\varepsilon_{ck}, \lambda_{ck}, \eta_{ck}, \gamma_{ck}] : 4 \times c$ and the sample correlation matrix Ξ.

After the process of creation and estimation of the training samples, image pixels are sorted (related) to the classes based on the decision rule (#3). If the pixel-wise classification is applied, data for each pixel with coordinates (i, j) are processed separately (independently). Value vector for a current pixel $\vec{x}_{ij}^{*}$ is put into mathematical models of the class target in Equation (5). The obtained results $\hat{f}_c\left(\vec{x}_{ij}^{*}\right)|a_k, k = 1 \ldots K$ are compared between each other and a maximal estimate of the likelihood function is chosen; its number is the number of the class the current pixel (i, j) is referred to.

3.2. NN Classifier Description

As the classifier is based on a neural network for image processing, a feedforward neural network is used. Nowadays there exist many NNs and approaches to their training and optimization [54]. We have chosen a simple but efficient NN for our application that can be easily implemented or placed on different platforms and devices. The employed classifier is an all-to-all connected perceptron combined with a self-organizing map to treat obtained weights or probabilities of the pixel belonging to one of the classes. The architecture details are the following: input layer for incoming data of three-color components, one hidden neuron layer with 30 fully connected neurons (this number has shown to perform equally well with larger possible numbers of hidden layer neurons). For the hidden layer, the tanh activation function is exploited. Scaled conjugate gradient backpropagation is used as a global training function. As has been mentioned above, the output layer of the used NN is a self-organized map fully connected with neurons of the hidden layer. It provides the mapping of output probabilities of the decision to the list of classes as vectors.

Training and validation processes have been performed in the following way. For this purpose, a certain image is divided into a set of pixel color value vectors. Hereby, the obtained set of vectors is permuted to introduce non-ordinary allocation for each training. For NN training, we have taken 70% of the produced set, the other 30% of data is used for validation. The self-dataset validation at this first stage has to be performed to prove the chosen NN architecture and NN parameters, and other training process peculiarities. We have considered different configurations of NN-based classifiers, especially varying the number of hidden layers, neurons, and connections among them. It has been established that complex architectures, like multi-layer perceptron, do not provide better efficiency of the classification and even cause over-fitting using the same training dataset. As the result, the chosen architecture was NN with one hidden layer with a fully optimal number of training epochs equal to 50 for the given NN. The proposed classifier is easy to use, and it is fast (which is the crucial point for the application of classification to remotely sensed data). The overall training process was repeated 100 times with full permutation of the dataset and the obtained classifier has been applied to test images.

4. Analysis of Synthetic Three-Channel Image

4.1. MLM Classifier Results

Creating a test multichannel image, we have taken into account the following typical properties of classes that might take place in real-life RS images:

Property 1. *Features rarely have distributions close to Gaussian; since in our case, color component values (see the test color image and its components in Figure 2a–d, respectively) are the used features, then this property relates to them;*

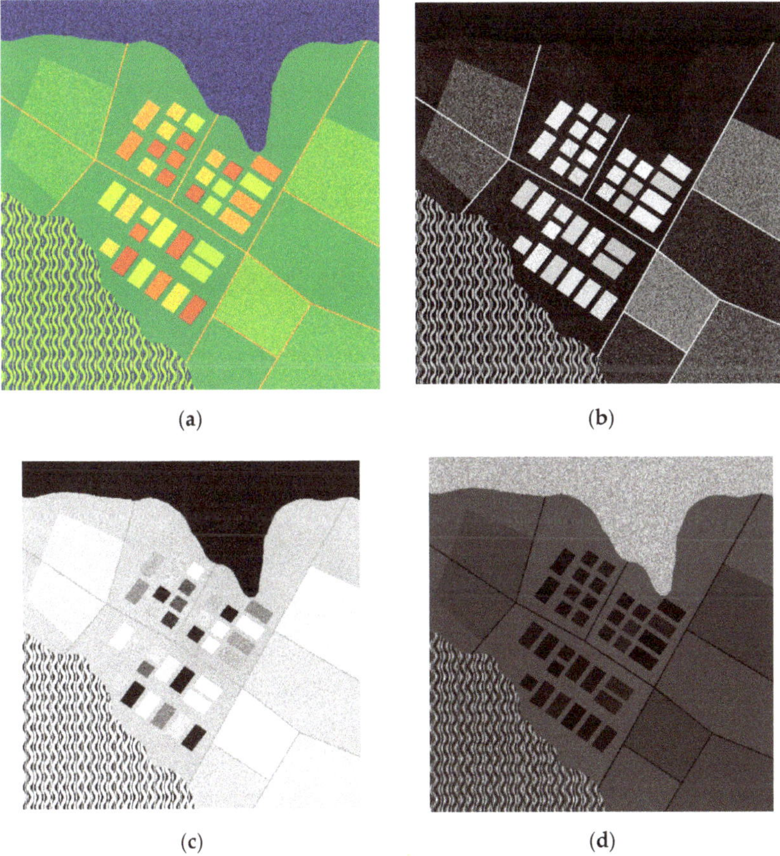

(a)   (b)

(c)   (d)

Figure 2. *Cont.*

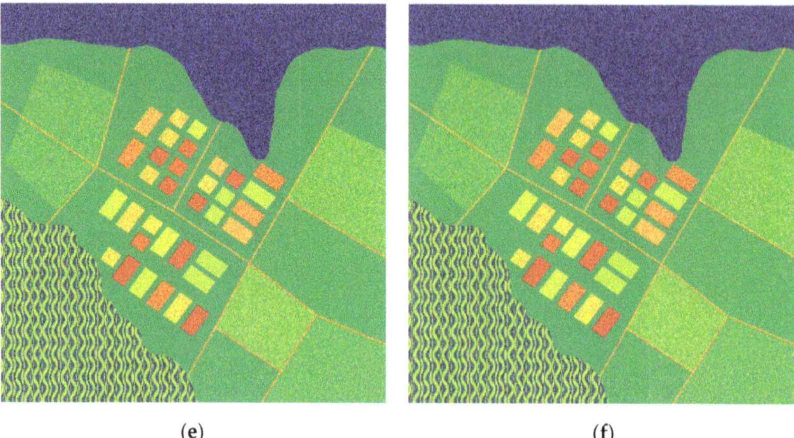

(e)   (f)

Figure 2. Three-channel test image: original (**a**), its three components (**b–d**), compressed with providing PSNR-HVS-M = 36 dB (**e**), compressed with providing PSNR-HVS-M = 30 dB (**f**).

Property 2. *Objects that relate to a certain class are rarely absolutely homogeneous (even water surface usually cannot be considered as absolutely homogeneous); many factors lead to diversity (variations) of pixel values as variations of physical and chemical characteristics of reflecting or irradiating surface, noise, etc.*

Property 3. *Features overlap (intersect) in the feature space; just this property usually leads to misclassifications especially if such overlapping takes place for many features;*

Property 4. *The numbers of pixels belonging to different classes in each RS image can differ a lot, it might be so that a percentage of pixels belonging to one class is less than 1% whilst for another class, there can be tens of percent of the corresponding pixels.*

Property 5. *There can be objects of two or even more classes in one pixel [54,55] that requires the application of unmixing methods but we do not consider such, more complex, situations in our further studies (usually more than three components of multichannel RS data are needed to carry out efficient unmixing).*

Let us show that we have, in a more or less extent, incorporated these properties in our test image. Table 1 contains data about seven classes, histograms for them, and approximations using Johnson S_B-distribution. Class 1 can be conditionally treated as "Roads", Class 2—as agricultural fields (Field-Y), Class 3—as agricultural fields of another color (Field-R), Class 4—as "Water", Class 5—as "Grass", Class 6—as "Trees", Class 7—as some very textural or heterogeneous class like urban area or vineyard ("Texture").

Table 1. Information about classes.

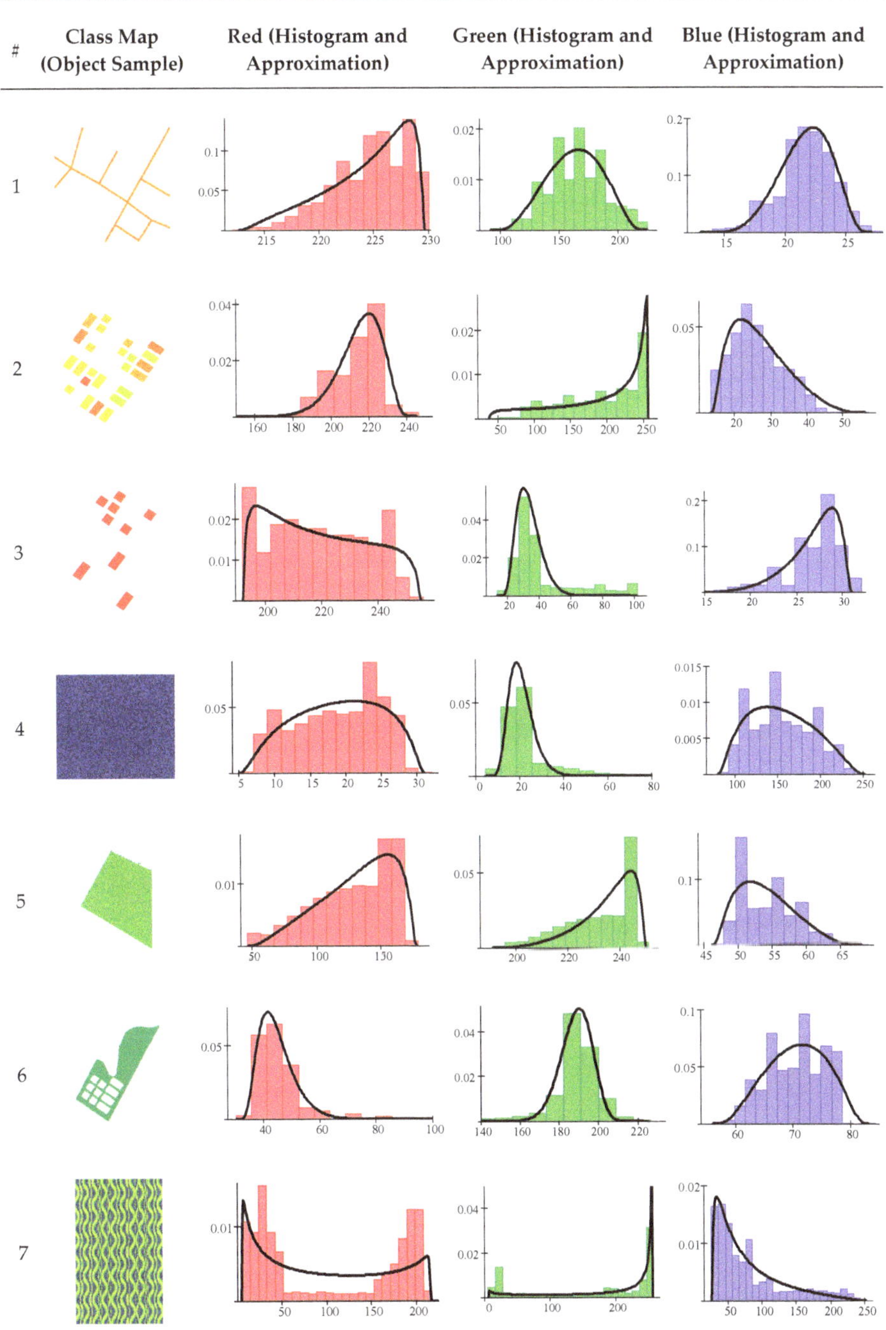

As one can see, Property 1 (non-Gaussian distribution) holds for practically all particular distributions. Property 2 (internal heterogeneity) takes place for all classes, especially for Classes 2 and 7. Property 3 (intersection of features) takes place for many classes—all three features for Classes 1 and 2, all three features for Classes 5 and 6; features of Class 7 overlap practically with all other classes (intersection of features can be also seen from a comparison of component images where some classes can be hardly distinguished, e.g., Classes 2 and 4 in Green component, Figure 2c). Property 4 is also observed—there is a small percentage of pixels belonging to Class 1 whilst quite many pixels belong to Classes 4, 5, and 7.

Figure 3a presents the true map of classes where Class 1 is shown by light brown, Class 2—by yellow; Class 3—by red, Class 4—by blue; Class 5—by light green, Class 6—by dark green, Class 7—by white.

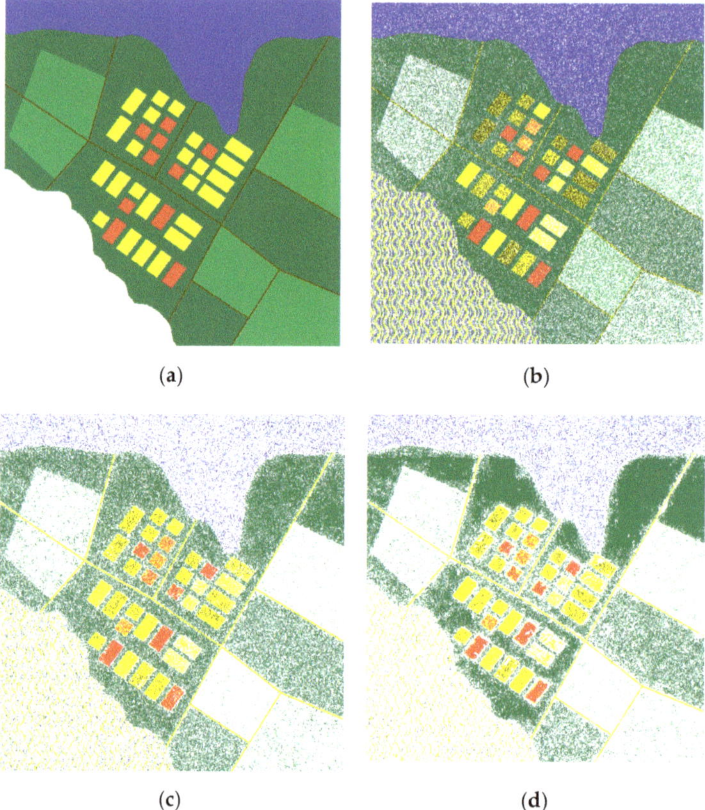

Figure 3. Classification results: true map (**a**); classification map for original image (**b**); classification map for image compressed with PSNR-HVS-M = 36 dB (**c**); classification map for image compressed with PSNR-HVS-M = 30 dB (**d**).

MLM classifier applied to original image pixel-wise (Figure 3a) produces a classification map given in Figure 3b. As one can see, there are quite many misclassifications. Classes 4 and 6 are classified in the best way (probabilities of correct classification P_{44} and P_{66} are equal to 0.85 and 0.82, respectively) whilst Class 5 is recognized in the worst manner (P_{55} = 0.19). Classes 1 and 7 are not recognized well (P_{11} and P_{77} are equal to 0.48 and 0.42, respectively). Classes 2 and 3 are characterized with P_{22} = 0.67 and P_{33} = 0.77, respectively.

Quite low probabilities of classification can be explained by several factors. Some of them have been already mentioned above—intersections in feature space are sufficient. One more possible reason is that the MLM classifier is based on distribution approximations and these approximations can be not perfect (see histograms and their approximations in Table 1).

Quite many pixels are erroneously related to Class 7 ("Texture") for which distributions are very wide and they intersect with distributions for other classes.

Figure 2e,f presents images compressed providing PSNR-HVS-M $\approx$ 36 dB and PSNR-HVS-M $\approx$ 30 dB. In these cases, CR values are about 4.5 and 8.9, respectively (for component-wise lossy compression). For the case of PSNR-HVS-M$\approx$36 dB, introduced distortions are visible [45]. They mostly appear themselves as smoothing of variations in quasi-homogeneous regions (consider the fragments for Class 6 (dark blue) for images in Figure 2a,e). The effects of such suppression of noise or high-frequency variations due to lossy compression are known for lossy compression [36,37] and they might have even a positive effect for classification (under certain conditions). For the case of PSNR-HVS-M $\approx$ 30 dB, the distortions due to lossy compression (see the compressed image in Figure 2f) appear themselves in variations' smoothing and edge/object smearing (this can be seen well for Class 1). Clearly, such effects might have an impact on classification.

Probabilities of correct classification for classes are given in Table 2. Note that the MLM classifier has been "trained" using distribution approximations obtained for the original (uncompressed) three-channel image. As one can see, probabilities for particular classes depend on compressed image quality and compression ratio in a different manner. P_{11} steadily decreases and becomes close to zero. This happens because of two reasons. First, features for Class 1 sufficiently intersect with features for Class 2. Second, due to lossy compression, feature distribution after compression differs from feature distribution for the original image (for which the MLM classifier has been trained) where this difference increases with the reduction of PSNR-HVS-M and CR increase. This is illustrated by distributions presented in Table 3 for Class 1 and, partly, in Table 4 for Class 2.

Table 2. Probabilities of correct classifications for particular classes depending on image quality.

PSNR-HVS-M, dB	P_{11}	P_{22}	P_{33}	P_{44}	P_{55}	P_{66}	P_{77}
∞ (original image)	0.475	0.667	0.773	0.854	0.193	0.822	0.415
48	0.190	0.781	0.686	0.384	0.093	0.809	0.579
45	0.147	0.795	0.652	0.308	0.077	0.777	0.623
42	0.101	0.804	0.615	0.253	0.058	0.712	0.665
36	0.031	0.802	0.568	0.163	0.040	0.545	0.777
30	0.001	0.753	0.478	0.177	0.042	0.556	0.843

Table 3. Illustration of distribution changes due to lossy compression for Class 1.

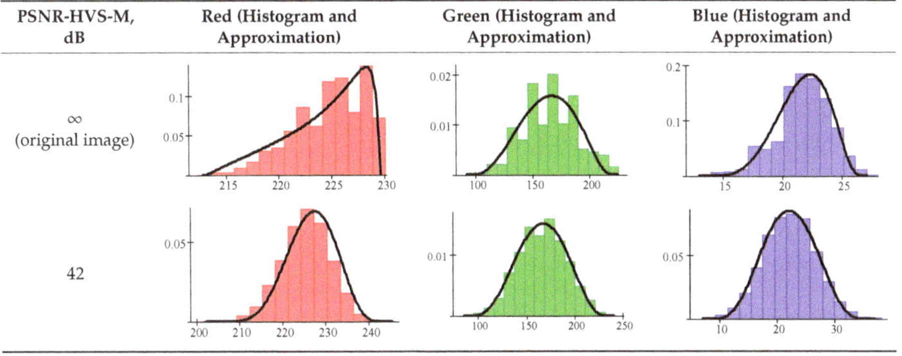

PSNR-HVS-M, dB	Red (Histogram and Approximation)	Green (Histogram and Approximation)	Blue (Histogram and Approximation)
∞ (original image)			
42			

Table 3. *Cont.*

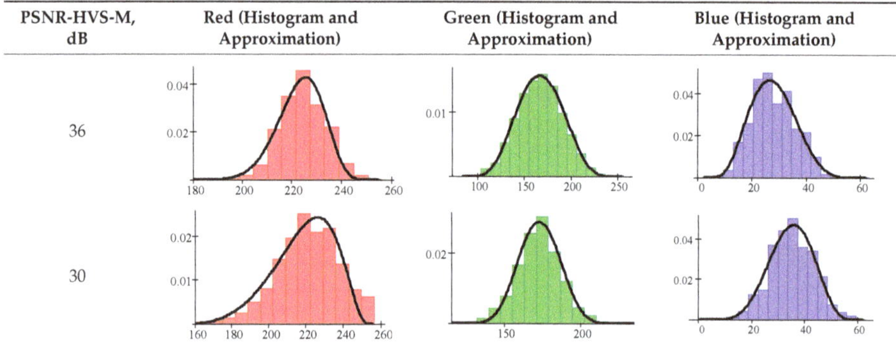

Table 4. Illustration of distribution changes due to lossy compression for Class 2.

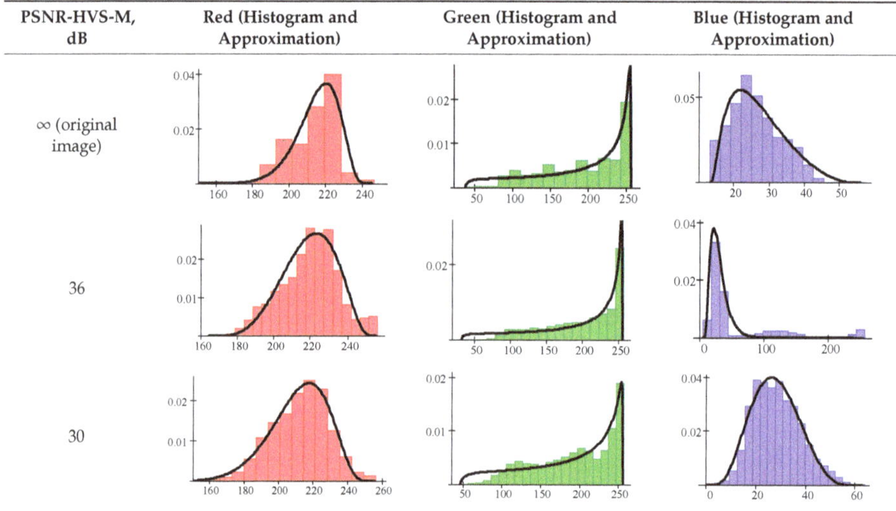

Analysis of data in Table 2 also shows the following. There are some classes (Classes 3–6) for which a larger CR and smaller PSNR-HVS-M (that correspond to more sufficient lossy compression) results in a reduction of probabilities of correct classification (analyze data in columns for P_{33}, P_{44}, P_{55}, and P_{66}). Meanwhile, there are classes, for which probabilities of correct classification increase—see data for P_{77}. This is because of the decrease in misclassifications in the texture area (see the maps in Figure 3c,d). There is also one class (Class 2) for which P_{22} is the largest for PSNR-HVS-M about 40 dB (due to noise filtering effect) but it reduces for smaller PSNR-HVS-M (larger CR).

One probable reason why the MLM classifier trained for original images and applied to compressed ones loses efficiency is that feature distributions observed for compressed data differ from those ones observed for original RS data and used for classifier training.

Then, one can expect that the MLM classifier has to be trained for compressed images. We have carried out the corresponding study and the results are presented in Table 5. The conclusion that follows from the analysis of data in Table 5 might seem trivial—it is needed to train the MLM classifier for compressed images with the same characteristics of compression. Following this rule leads to a considerable improvement in classification accuracy. For example, if original data are used in training and then the MLM classifier is applied to data compressed providing PSNR-HVS-M = 42 dB, then P_{cc} = 0.501. At the same time, if the image compressed providing PSNR-HVS-M = 42 dB was used in training, P_{cc} radically increases and becomes equal to 0.579. Even more surprising results are observed for cases of images compressed producing PSNR-HVS-M equal to 36 and 30 dB. P_{cc} for the corresponding training reaches 0.597 and 0.527, respectively. Note that P_{cc} = 0.627 for original image classified using training for the original image. Then, it occurs that for images compressed with PSNR-HVS-M about 40 dB almost the same P_{cc} is observed if they are trained for the corresponding compressed images.

Table 5. Total probabilities of correct classification Pcc for different images depending upon "training" data.

	Image Used for Training	Total Probabilities of Correct Classification for Compressed Images		
		PSNR-HVS-M = 42 dB	PSNR-HVS-M = 36 dB	PSNR-HVS-M = 30 dB
0	Original (uncompressed)	0.501	0.434	0.442
1	Compressed with providing PSNR-HVS-M = 42 dB	0.579	0.566	0.563
2	Compressed with providing PSNR-HVS-M = 36 dB	0.519	0.597	0.599
3	Compressed with providing PSNR-HVS-M = 30 dB	0.408	0.446	0.527

Classification maps obtained for this methodology of training are presented in Figure 4a,b. They can be compared to the maps in Figure 3c,d, respectively. This comparison clearly demonstrates that training for the corresponding compressed data is preferable. Many classes are recognized sufficiently better. One interesting point is that many classifications might appear in the neighborhoods of sharp edges in multichannel images (near edges of Class 4). This is because such edges are smeared by lossy compression and pixels with "intermediate values" that can be referred to as "wrong classes" appear.

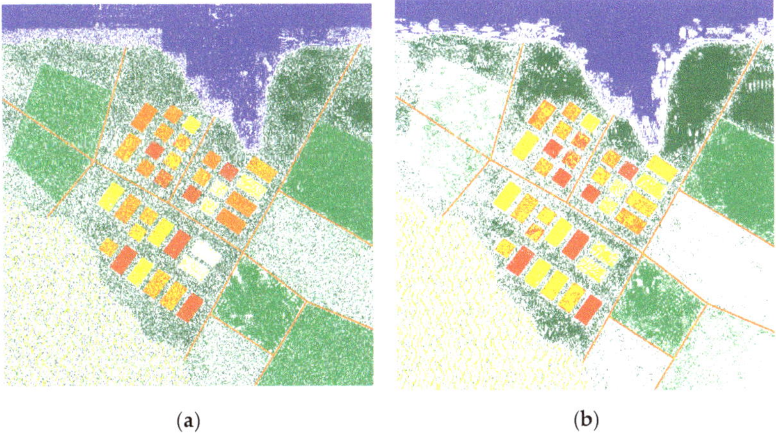

(a)   (b)

Figure 4. Classification maps for the compressed image with providing PSNR-HVS-M = 36 dB using maximum likelihood method (MLM) classifier trained for the same compressed image (**a**); compressed image with providing PSNR-HVS-M = 30 dB using MLM classifier trained for the same compressed image (**b**).

4.2. NN-Based Classification

Let us start by analyzing data for classifying the original three-channel image using NN training for this image. The obtained confusion matrix is presented in Table 6. The corresponding map is given in Figure 5a. As one can see, several classes are recognized well: Field-R (Class 3), Water (class 4), Trees (Class 6). There are many misclassifications for Classes "Roads" (with Class Field-Y that has practically the same colors) and "Texture" (its pixels can be erroneously related to classes "Water", "Trees", "Grass", and, more rarely, Field-Y). The results are relatively bad for Class 2 and Class 5. Thus, even for this almost ideal case (classified image is the same as that one used for training, the NN-based classifier able to work with non-Gaussian distributions of features is applied), the classification results are far from being perfect.

Table 6. Confusion matrix for the original image classified by neural networks (NN) trained for this image.

Class	Probability of Decision						
	Road	Field-Y	Field-R	Water	Grass	Trees	Texture
Road	**0.356**	0.645	0	0	0	0	0
Field-Y	0.109	**0.608**	0.002	0	0	0	0.281
Field-R	0	0.004	**0.996**	0	0	0	0
Water	0	0	0	**0.986**	0	0	0.014
Grass	0	0	0	0	**0.744**	0.245	0.012
Tree	0	0	0	0	0.072	**0.923**	0.005
Texture	0	0.03	0	0.26	0.116	0.153	**0.441**

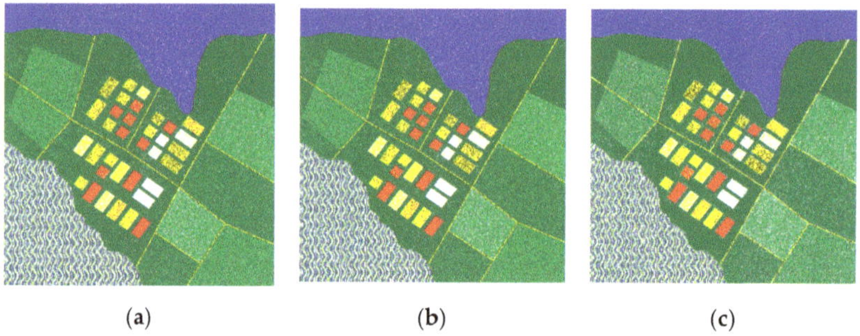

(a)   (b)   (c)

Figure 5. Classification results: for compressed image (PSNR-HVS-M = 42 dB) (**a**), for compressed image (PSNR-HVS-M = 36 dB) (**b**); for compressed (PSNR-HVS-M = 30 dB) (**c**).

Taking into account the results for the MLM classifier, we also considered the case when the NN classifier has been trained for the original image and then applied to compressed images. The data obtained for the cases of PSNR-HVS-M equal to 42 dB, 36 dB, and 30 dB are given in Tables 7–9, respectively. Analysis of data in Table 7 shows the following. As was expected, all probabilities have changed. Let us consider the diagonal elements marked by Bold that correspond to probabilities of correct classification of particular classes. In this sense, there are interesting observations (compare the corresponding data in Tables 6 and 7): some probabilities have decreased sufficiently (see P_{11}), some decreased slightly (P_{33}, P_{44}, P_{66}, P_{77}). Meanwhile, some probabilities have slightly increased (P_{22} and P_{55}).

Table 7. Confusion matrix for compressed image classified by NN trained for the original image and applied for the compressed image (PSNR-HVS-M = 42 dB).

Class	Probability of Decision						
	Road	Field-Y	Field-R	Water	Grass	Trees	Texture
Road	**0.299**	0.699	0	0	0	0	0
Field-Y	0.089	**0.627**	0	0	0	0	0.276
Field-R	0	0.018	**0.982**	0	0	0	0
Water	0	0	0	**0.980**	0	0	0.020
Grass	0	0	0	0	**0.753**	0.219	0.028
Tree	0	0	0	0	0.084	**0.911**	0.006
Texture	0	0.03	0	0.26	0.116	0.153	**0.441**

Table 8. Confusion matrix for compressed image classified by NN trained for original image and applied for the compressed image (PSNR-HVS-M = 36 dB).

Class	Probability of Decision						
	Road	Field-Y	Field-R	Water	Grass	Trees	Texture
Road	**0.271**	0.728	0.001	0	0	0	0
Field-Y	0.078	**0.631**	0.012	0	0.001	0	0.279
Field-R	0	0.027	**0.973**	0	0	0	0
Water	0	0	0	**0.979**	0	0	0.021
Grass	0	0	0	0	**0.721**	0.22	0.059
Tree	0	0	0	0	0.078	**0.915**	0.007
Texture	0	0.049	0	0.253	0.123	0.148	**0.426**

Table 9. Confusion matrix for compressed image classified by NN trained for the original image and applied for the compressed image (PSNR-HVS-M = 30 dB).

Class	Probability of Decision						
	Road	Field-Y	Field-R	Water	Grass	Trees	Texture
Road	**0.217**	0.766	0	0	0	0	0.017
Field-Y	0.067	**0.646**	0.007	0	0.005	0	0.275
Field-R	0	0.05	**0.946**	0.004	0	0	0
Water	0	0	0	**0.992**	0	0	0.008
Grass	0	0	0	0	**0.658**	0.21	0.131
Tree	0	0	0	0	0.051	**0.925**	0.025
Texture	0	0.067	0	0.243	0.141	0.131	**0.418**

Analysis of data in Tables 8 and 9 shows that lossy compression negatively influences P_{11}, i.e., Probability of correct classification for the Class "Road". Recall that the same effect has been observed for the MLM. With a larger CR (smaller provided PSNR-HVS-M), probabilities P_{33} and P_{77} continue to decrease. P_{22} slowly increases. P_{44} remains practically the same and very high. P_{66} also remains practically the same. Finally, P_{55} decreases. Thus, we can state that being trained for the original image, the NN classifier continues working well for PSNR-HVS-M about 40 dB and then its performance starts making worse faster (with further increase of CR).

Figure 5 shows classification results for the test image compressed providing three different values of PSNR-HVS-M. The results are quite similar. However, some differences can be noticed:

1. Less misclassifications are observed for PSNR-HVS-M = 30 dB for the class "Water".

2. There are quite many misclassifications around the edge of Class "Water".

3. More misclassifications appear for the class "Grass" when CR increases.

4. Many pixels and even some objects that belong to Class 2 (Field-Y) are classified as "Texture" (white color). This example one more time shows the problems in image classification if an image

to be classified contains a class like "Texture". The problems occur for the class "Texture" itself and the classes for which there is an intersection of features with features of the class "Texture".

Let us now check what happens if compressed images are used for training. Table 10 presents the confusion matrix obtained for the case of NN training using image compressed providing PSNR-HVS-M = 42 dB. These data can be compared to the corresponding data in Table 7. If compressed image is employed for training, probabilities P_{11}, P_{55}, and P_{77} are sufficiently better, probability P_{22} is sufficiently worse, other probabilities are slightly worse. Thus, the quality of classification, in general, slightly improves. The classification map is presented in Figure 6b.

Table 10. Confusion matrix for compressed image classified by NN trained for this image (PSNR-HVS-M = 42 dB).

Class	Probability of Decision						
	Road	Field-Y	Field-R	Water	Grass	Trees	Texture
Road	**0.455**	0.545	0.001	0	0	0	0
Field-Y	0.137	**0.561**	0.014	0	0	0	0.289
Field-R	0	0.026	**0.974**	0	0	0	0
Water	0	0	0	**0.961**	0	0	0.039
Grass	0	0	0	0	**0.820**	0.165	0.015
Tree	0	0	0	0	0.068	**0.929**	0.004
Texture	0	0.029	0	0.234	0.124	0.153	**0.461**

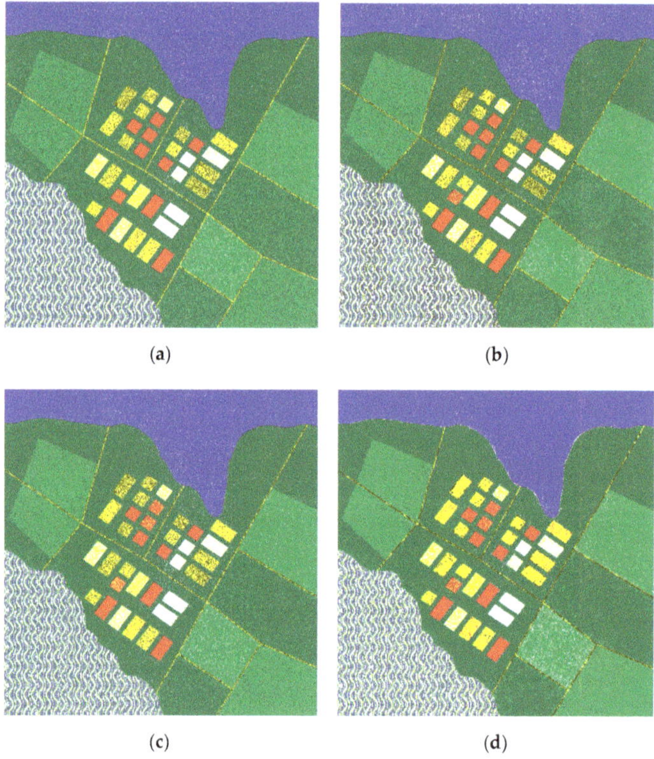

Figure 6. Results of image classification by NN trained for the same image: original (**a**), compressed with PSNR-HVS-M = 42 dB (**b**), compressed with PSNR-HVS-M = 36 dB (**c**), compressed with PSNR-HVS-M = 30 dB (**d**).

The next considered case is the classification of the image compressed producing PSNR-HVS-M = 36 dB. The obtained probabilities are given in Table 11, the classification map is presented in Figure 6c. The probabilities of correct classification in Table 11 are almost the same as the corresponding values in Table 10. Comparing the data in Tables 8 and 11, it is possible to state that it is worth using a compressed image for training (most probabilities of correct classification for classes are better in this case). The classification map is given in Figure 6c and it is quite similar to those in Figure 6a,b. This means that there is no considerable degradation of classification accuracy compared to the previous two cases.

Table 11. Confusion matrix for compressed image classified by NN trained for this image (PSNR-HVS-M = 36 dB).

Class	Probability of Decision						
	Road	Field-Y	Field-R	Water	Grass	Trees	Texture
Road	**0.507**	0.492	0.002	0	0	0	0
Field-Y	0.127	**0.562**	0.005	0	0.001	0	0.305
Field-R	0	0.069	**0.931**	0	0	0	0
Water	0	0	0	**0.989**	0	0	0.01
Grass	0	0	0	0	**0.901**	0.087	0.013
Tree	0	0	0	0	0.039	**0.956**	0.005
Texture	0	0.029	0	0.236	0.142	0.141	**0.452**

Finally, the last case is the classification of the image compressed providing PSNR-HVS-M = 30 dB, i.e., with considerable distortions. The confusion matrix is given in Table 12, the classification map is represented in Figure 6d. Compared to the corresponding data in Table 9, a sufficient increase of P_{11}, P_{55}, P_{66}, and P_{77} is observed, P_{22}, P_{33}, and P_{44} have increased a little too. This means that it is worth using the compressed image for training, especially if compression is carried out with a large CR.

Table 12. Confusion matrix for compressed image classified by NN trained for this image (PSNR-HVS-M = 30 dB).

Class	Probability of Decision						
	Road	Field-Y	Field-R	Water	Grass	Trees	Texture
Road	**0.692**	0.304	0	0	0.002	0.001	0.01
Field-Y	0.055	**0.654**	0.009	0	0.019	0	0.263
Field-R	0	0.033	**0.967**	0.004	0	0	0
Water	0	0	0	**0.994**	0	0	0.006
Grass	0	0	0	0	**0.939**	0.035	0.026
Tree	0	0	0	0	0.023	**0.972**	0.005
Texture	0	0.049	0	0.221	0.113	0.131	**0.486**

Attentive analysis of the classification map in Figure 6d allows noticing quite many misclassifications near the edge between classes "Water" and "Trees" where these misclassified pixels are related to "Texture". We associate this effect with edge smearing and other effects that happen near high contrast edges if the compression ratio is quite high.

4.3. Preliminary Conclusions

Summarizing the obtained results and conclusions, we can state the following:

1. Classification by any classifier is not perfect if features intersect in feature space and/or if not all possible features are used; here we mean that pixel-wise classification is unable to exploit spatial information that can be useful for improving classification accuracy;
2. If a classifier is trained for original (uncompressed) data and then applied to compressed images, then more compression (a larger CR, a smaller PSNR or PSNR-HVS-M) leads to classification worsening (in general); meanwhile, the difference in Pcc and probabilities of correct classification for particular classes can be negligible; classification accuracy for some particular classes can even improve due to noise/variation suppression (if noise or high frequency variations are present);
3. It is worth training a classifier (at least, MLM and NN-based classifiers) for "conditions" they will be applied; these conditions can be described by quality metrics' values that are planned to be provided at compression stage or QS value that will be used;
4. Different classifiers can produce sufficiently different classification maps for the same RS data; quantitative parameters that characterize classification can be considerably different as well; a good example are data for Class 6—the MLM classifier does not recognize it well enough whilst the NN classifier performs sufficiently better, the situation is the opposite for Class 7;
5. Lossy compression negatively influences classes that are mainly associated with small-sized and prolonged objects such as roads, narrow rivers, and cricks, small houses; the reason is that lossy compression smears such kinds of objects and changes feature distributions; in this sense, the use of visual quality metrics and setting desired values for them at compression stage can be useful to avoid sufficient degradation of classification accuracy for such classes;
6. For pixel-wise classification, the misclassifications usually appear as isolated pixels; if distortions due to compression are larger, the misclassifications might appear as groups of pixels, especially in the neighborhoods of high contrast edges in RS images; this is undesired and it is worth avoiding such situations that mostly happen if distortions due to lossy compression are visible (in our experiments, if PSNR-HVS-M = 30 dB and, more rarely, if PSNR-HVS-M = 36 dB);
7. This runs us into the idea that, possibly, it is worth carrying out lossy compression either without visually noticeable distortions or with distortions relating to partial suppression of noise in quasi-homogeneous image regions; this can be controlled by either setting the desired PSNR-HVS-M ($\approx$40 dB) or the corresponding PCC for the used coder (e.g., QS for AGU).

Certainly, these conclusions are based on analysis of only one test image, and more studies are needed. One opportunity is to check them for real-life RS data.

5. Analysis of Real-Life Three-Channel Image

Our experiments with real-life images have been done using a three-channel Landsat TM image earlier used in our studies [38,39]. This test image (shown in pseudo-color representation in Figure 7a) contains three component images acquired in optical bands with central wavelengths equal to 0.66 µm, 0.56 µm, and 0.49 µm, respectively. They have been associated with R, G, and B components of the color image that has a size of 512 × 512 pixels. The image has five recognizable classes, namely, "Soil" (Class 1), "Grass" (Class 2), "Water" (Class 3), "Urban (Roads and Buildings)" (Class 4), and "Bushes" (Class 5). The image fragments used in classifier training are shown in Figure 7b whilst the pixel used for verification of classifiers are marked by the same colors in Figure 7c. Details concerning the numbers of pixels are given in [39]. Class 1 is marked by red color, Class 2—by green, Class 3—by dark blue; Class 4—by yellow, Class 5—by azure.

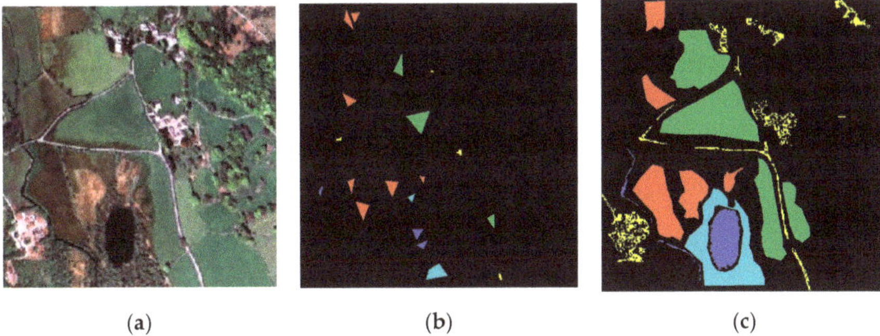

(a)   (b)   (c)

Figure 7. The three-channel image in pseudo-color representation (**a**); pixel groups used for training (**b**), pixel groups used for verification (**c**).

As one can see, the total number of classes is less than for the test image in the previous section, but some classes are the same or similar. Class 1 is similar to Field-R, Class 5—to Class "Texture". As in test RS data, color features for many classes intersect (consider Classes "Grass" and "Soil", Classes "Soil" and Bushes"). This means that classification is not a simple task.

A part of data concerning lossy compression of this image component-wise are presented in Table 13 (more details can be found in [39]). As one can see, to provide PSNR-HVS-M$_{des}$ = 42 dB, one has to use QS ≈ 17.3. To provide PSNR-HVS-M$_{des}$ ≈ 42 dB, QS ≈ 29.3. This is in good agreement with the data in Figure 1 for other test images. Thus, it is possible to give recommendations on what QS to use for providing PSNR-HVS-M$_{des}$ (see details in [23]).

Table 13. Compression parameters.

Component	PSNR-HVS-M Desired	QS	PSNR-HVS-M Provided	CR
	42 dB	17.372	41.941	6.008
Pseudo-red	36 dB	29.825	36.059	8.952
	30 dB	55.558	30.007	16.503
	42 dB	17.285	41.936	5.994
Pseudo-green	36 dB	29.856	35.992	8.968
	30 dB	55.134	29.996	16.445
	42 dB	17.227	41.847	6.016
Pseudo-blue	36 dB	29.729	35.998	8.993
	30 dB	55.014	30.0008	16.577

Note that the provided CR can be quite large. It is about 6 for PSNR-HVS-M$_{des}$ = 42 dB and reaches 16.5 for PSNR-HVS-M$_{des}$ = 30 dB.

We have carried out the analysis for six values of PSNR-HVS-M (45, 42, 39, 36, 33, and 30 dB). Figure 8 presents the considered three-channel image (in pseudo-color representation) compressed providing PSNR-HVS-M equal to 42 dB, 36 dB, and 30 dB. They are all very similar between each other and similar to the original image. A more attentive analysis allows finding differences for images in Figure 8b,c that mainly appear themselves in the neighborhoods of small-sized objects and high contrast edges.

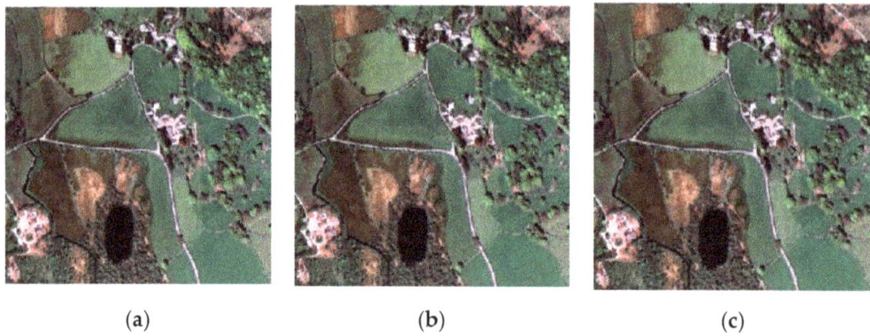

(a)   (b)   (c)

Figure 8. Three-channel image in pseudo-color representation: compressed with providing PSNR-HVS-M = 42 dB (**a**), compressed with providing PSNR-HVS-M = 36 dB (**b**); compressed with providing PSNR-HVS-M = 30 dB (**c**).

5.1. MLM Classifier Results

Let us start by considering the results of applying the MLM classifier trained for the original image. Tables 14–17 present the obtained data in the form of confusion matrices. As it is seen, probabilities of correct classification for particular classes vary from 0.75 to 0.99 and the smallest probabilities take place for rather heterogeneous classes "Soil" and "Bushes".

Table 14. Classification probabilities for the MLM method trained for the original image and applied to it.

Class	Probability of Decision				
	Soil	Grass	Water	Urban	Bushes
Soil	**0.747**	0.027	0	0.037	0.189
Grass	0.173	**0.812**	0	0.008	0.007
Water	0	0	**0.967**	0	0.033
Urban	0.004	0	0	**0.989**	0.007
Bushes	0.091	0.068	0.013	0.016	**0.812**

Table 15. Classification probabilities (confusion matrix) for the MLM method trained for the original image and applied to the compressed image (PSNR-HVS-M = 42 dB).

Class	Probability of Decision				
	Soil	Grass	Water	Urban	Bushes
Soil	**0.732**	0.025	0	0.043	0.199
Grass	0.177	**0.808**	0	0.008	0.006
Water	0	0	**0.963**	0	0.037
Urban	0.005	0	0	**0.987**	0.008
Bushes	0.118	0.07	0.014	0.018	**0.779**

If the same classifier is applied to a compressed image (PSNR-HVS-M=42 dB), the results are slightly worse (see data in Table 15). Reduction of probabilities of correct classification by 0.002 … 0.033 takes place. The largest reduction is observed for the most heterogeneous class "Bushes".

Table 16. Classification probabilities (confusion matrix) for the MLM trained for the original image and applied to the compressed image (PSNR-HVS-M = 36 dB).

Class	Probability of Decision				
	Soil	Grass	Water	Urban	Bushes
Soil	**0.744**	0.028	0	0.043	0.185
Grass	0.185	**0.799**	0	0.009	0.007
Wate»	0	0	**0.959**	0	0.041
Urban	0.005	0	0	**0.985**	0.010
Bushes	0.133	0.074	0.016	0.018	**0.758**

Table 17. Classification probabilities (confusion matrix) for the MLM trained for the original image and applied to the compressed image (PSNR-HVS-M = 30 dB).

Class	Probability of Decision				
	Soil	Grass	Water	Urban	Bushes
Soil	**0.724**	0.026	-	0.044	0.206
Grass	0.178	**0.809**	0	0.006	0.007
Water	-	0	**0.942**	0	0.058
Urban	0.007	-	0	**0.982**	0.011
Bushes	0.145	0.079	0.016	0.019	**0.741**

Consider now the data obtained if the MLM classifier is applied to the image compressed with visible distortions (PSNR-HVS-M = 36 dB). The confusion matrix is given in Table 16. Most probabilities of correct classification for particular classes continue to decrease although this reduction is not essential—up to 0.021 compared to the previous case (Table 15).

Finally, let us analyze data for a compressed image (PSNR-HVS-M = 30 dB) to which the MLM classifier has been applied. The data are presented in Table 17. The tendency is the same—most probabilities of correct classification for particular classes continue to decrease but, again, the reduction is not large.

Classification maps for three images are presented in Figure 9. They are quite similar although some differences can be found. Compression does not lead to radical degradation of classification accuracy.

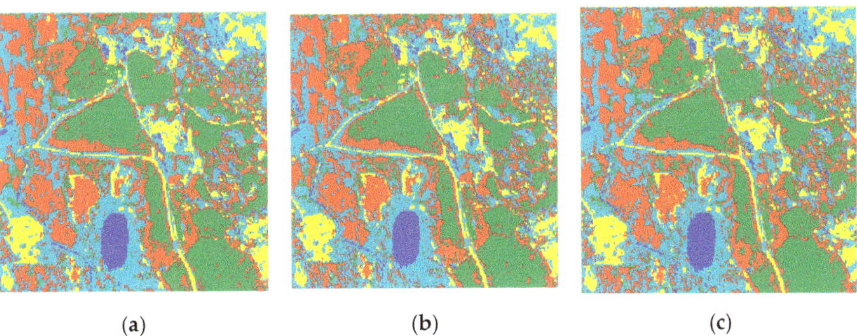

(a)   (b)   (c)

Figure 9. Classification results: for original image (**a**), for compressed image (PSNR-HVS-M = 36 dB) (**b**), for compressed (PSNR-HVS-M = 30 dB) (**c**).

One can be interested in the behavior of P_{cc} depending on compression. For original image P_{cc} = 0.865; for compressed images it equals to 0.853, 0.854, 0.853, 0.849, 0.842, and 0.839 for PSNR-HVS-M equal to 45, 42, 39, 36, 33, and 30 dB, respectively. So, it is possible to state that lossy compression providing PSNR-HVS-M about 40 dB does not lead to sufficient reduction of P_{cc} for the considered case. Moreover, if training is done for the image compressed with the same conditions as an image subject to classification, classification results can improve. For example, if training has been done for the compressed image (PSNR-HVS-M = 36 dB) and then applied to this image (to verification set of pixels), P_{cc} increases to 0.855.

We have analyzed distributions of features before and after compression. One reason why P_{cc} does not radically reduce with CR increase is that the corresponding distributions do not differ a lot. The largest differences are observed for the classes "Soil" and "Bushes". It might be slightly surprising that the class "Urban" is recognized so well. The reason is that, for the considered image, features for this class do not sufficiently overlap with features for other classes.

5.2. NN Classifier Results

Consider now the results of the NN-based classification of the real-life image. The ideal case data (NN trained for original image is applied to the original image) are given in Table 18. The results can be compared to the data in Table 14. The NN classifier better recognizes the classes "Soil" and "Grass", the results for the class "Water" are approximately the same.

Table 18. Classification probabilities for the NN-based method trained for the original image and applied to it.

Class	Probability of Decision				
	Soil	Grass	Water	Urban	Bushes
Soil	**0.950**	0	0	0	0.05
Grass	0	**1**	0	0	0
Water	0	0	**0.952**	0	0.048
Urban	0.098	0	0	**0.767**	0.135
Bushes	0.187	0.009	0.02	0.009	**0.774**

Suppose now that this classifier (NN trained for the original image) is applied to compressed images. The results for different qualities of compressed data are presented in Tables 19–21. If PSNR-HVS-M is equal to 42 dB or 36 dB, the results keep practically the same. Only the probability of correct classification for "Bushes" steadily decreases. Reduction of classification accuracy occurs to be larger for the image compressed with PSNR-HVS-M equal to 30 dB. Mainly, reduction takes place for the classes "Water" and "Urban".

Table 19. Classification probabilities (confusion matrix) for the NN-based method trained for the original image and applied to the compressed image (PSNR-HVS-M = 42 dB).

Class	Probability of Decision				
	Soil	Grass	Water	Urban	Bushes
Soil	**0.942**	0	0	0	0.058
Grass	0	**0.999**	0	0	0.001
Water	0	0	**0.942**	0	0.058
Urban	0.01	0	0	**0.765**	0.136
Bushes	0.207	0.016	0.018	0.009	**0.751**

Table 20. Classification probabilities (confusion matrix) for the NN-based method trained for original image and applied to the compressed image (PSNR-HVS-M = 36 dB).

Class	Probability of Decision				
	Soil	**Grass**	**Water**	**Urban**	**Bushes**
Soil	**0.944**	0	0	0	0.057
Grass	0	**1**	0	0	0
Water	0	0	**0.941**	0	0.059
Urban	0.108	0	0	**0.761**	0.131
Bushes	0.236	0.022	0.017	0.007	**0.718**

Table 21. Classification probabilities (confusion matrix) for the NN-based method trained for the original image and applied to the compressed image (PSNR-HVS-M = 30 dB).

Class	Probability of Decision				
	Soil	**Grass**	**Water**	**Urban**	**Bushes**
Soil	**0.941**	0	0	0	0.059
Grass	0	**0.999**	0	0	0.001
Water	0.001	0	**0.919**	0	0.08
Urban	0.099	0	0	**0.747**	0.153
Bushes	0.247	0.028	0.014	0.004	**0.707**

Some classification maps are presented in Figure 10. They do not differ a lot from each other. Some pixels that belong to the class "Water" for the narrow river (see the left low corner in Figure 10c) "disappear" (become misclassified). This is because of the effects of smearing prolonged objects due to lossy compression. The MLM classifier training for compressed images was performed using training samples for fragments shown in Figure 7b. Due to the usage of the same data for both training and validation, noticeable classification improvement can be observed (at least, for several classes).

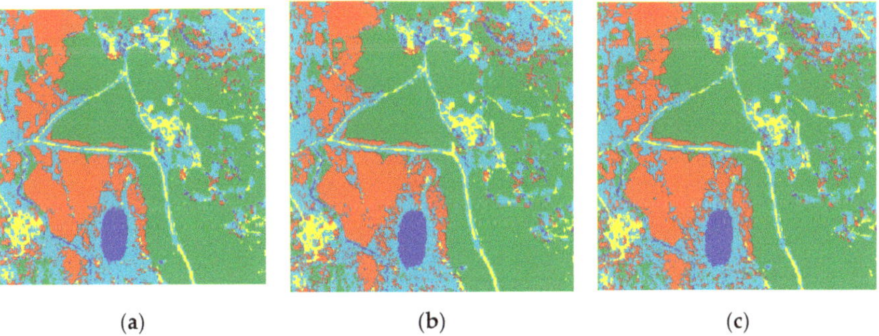

(a)   (b)   (c)

Figure 10. Classification results provided by NN-based method: for original image (**a**), for compressed image (PSNR-HVS-M = 36 dB) (**b**); for compressed (PSNR-HVS-M = 30 dB) (**c**).

5.3. Brief Analysis for Sentinel-2 Three-Channel Images

One can be interested whether the observed dependences hold for other images or their fragments, other imagers, and other compression techniques. Sentinel-2 offers wide possibilities to check this since it provides a huge amount of data that can be exploited for different purposes. For our study, we have taken three-channel images of the Kharkiv region (Ukraine) in visible range acquired on 30 August 2019 when there were practically no clouds (images are available at [56]). The analyzed 512 × 512 pixel fragments are for the neighborhood of Staryi Saltiv (45 km north-east from Kharkiv, Ukraine, set 1) and north part of Kharkiv (set 2)—see Figure 11. The main reason for choosing these fragments is the availability of ground truth data that allow easy marking of four typical classes: Urban, Water,

Vegetation, and Bare Soil. One more reason is that the image in Figure 11a is considerably less complex (textural) than the image in Figure 11b.

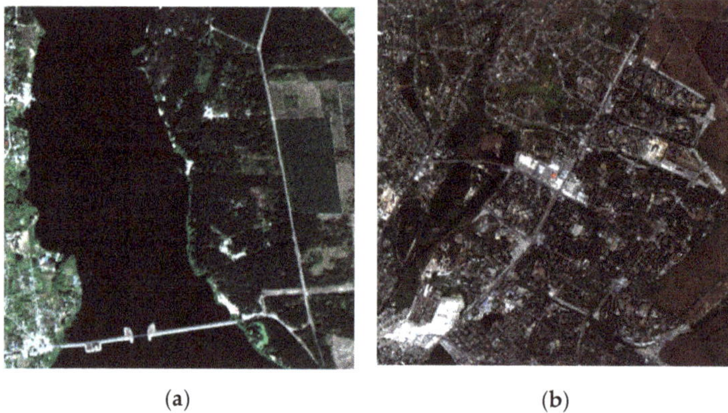

(a)   (b)

Figure 11. Fragments of Sentinel-2 images of Staryi Saltiv (**a**) and north part of Kharkiv (**b**).

First, these fragments have been compressed component-wise with providing a set of PSNR-HVS-M values using three coders: AGU used in the previous part of this paper, SPIHT [57] which can be considered as an analog of JPEG2000 standard, and Advanced DCT coder (ADCTC) [58] that uses partition scheme optimization for better adaptation to image content. Basic data on compression performance are given in Table 22. The observed dependences are predictable. CR for all coders increases if desired PSNR-HVS-M reduces. CR values for different components for the same desired PSNR-HVS-M and set differ but not considerably. CR for AGU is usually slightly larger than for SPIHT (for the same conditions), ADCTC outperforms both coders. CR values for set 2 images are several times smaller than for the corresponding set 1 images due to the higher complexity of set 2 RS data.

Table 22. CR comparison for real-life data for three coders, both sets.

Color	Desired PSNR-HVS-M	CR for AGU	CR for SPIHT	CR for ADCTC
	45.000	10.338	9.467	11.306
Pseudo-Red, set 1	39.000	18.383	16.631	20.362
	33.000	32.929	29.303	36.465
	45.000	3.317	3.283	3.498
Pseudo-Red, set 2	39.000	4.520	4.762	4.886
	33.000	7.121	7.360	7.986
	45.000	7.179	5.957	7.513
Pseudo-Green, set 1	39.000	12.666	10.914	13.513
	33.000	23.907	20.833	26.030
	45.000	3.214	3.150	3.363
Pseudo-Green, set 2	39.000	4.356	4.564	4.684
	33.000	6.784	7.105	7.573
	45.000	6.665	5.727	7.216
Pseudo-Blue, set 1	39.000	11.942	10.336	12.740
	33.000	24.914	22.346	27.191
	45.000	3.212	3.252	3.418
Pseudo-Blue, set 2	39.000	4.314	4.698	4.785
	33.000	6.882	7.380	7.817

Second, classifier training has been performed and probabilities of correct classification for all four classes as well as the total probability of correct classification have been obtained. For the NN classifier and AGU coder, the data are presented in Tables 23 and 24. Here and below abbreviation SS1 means that classification has been done for an original (uncompressed) image 1 whilst, e.g., SS1_45 means that classification has been done for the image compressed with providing PSNR-HVS-M = 45 dB. Training has been done for the original image. As one can see, compression results for the Set 1 image practically do not depend on compressed image quality. Even for PSNR-HVS-M = 30 dB probabilities of correct classification are practically the same as for original data. Meanwhile, for the Set 2 image, the situation is another. There is a tendency for classification accuracy degradation if compressed image quality becomes worse. This is mainly due to the reduction of correct classification probabilities for more heterogeneous classes, e.g., vegetation. For this image, it is possible to recommend compression with providing PSNR-HVS-M about 42 dB to avoid considerable reduction of classification accuracy.

Table 23. Probabilities of correct classification depending on compressed image quality for Set 1, AGU coder, NN classifier.

Classes	SS1	SS1_45	SS1_42	SS1_39	SS1_36	SS1_33	SS1_30
Urban	0.774	0.777	0.775	0.777	0.778	0.773	0.786
Water	0.998	0.998	0.998	0.999	0.999	0.999	0.998
Vegetation	0.915	0.913	0.914	0.915	0.918	0.921	0.927
Bare soil	0.809	0.811	0.809	0.809	0.812	0.798	0.801
P_{total}	0.874	0.875	0.874	0.875	0.877	0.873	0.878

Table 24. Probabilities of correct classification depending on compressed image quality for Set 2, AGU coder, NN classifier.

Classes	SS2	SS2_45	SS2_42	SS2_39	SS2_36	SS2_33	SS2_30
Urban	0.877	0.872	0.871	0.865	0.861	0.861	0.864
Water	0.654	0.659	0.662	0.661	0.654	0.644	0.647
Vegetation	0.857	0.825	0.815	0.801	0.788	0.797	0.788
Bare soil	0.890	0.867	0.865	0.869	0.866	0.860	0.870
P_{total}	0.820	0.810	0.803	0.799	0.792	0.791	0.792

We have also analyzed the possibility of using compressed images for training. The probability of correct classification has improved by about 0.01 for Set 1 image and remained practically the same for Set 2 image.

Another part of our study relates to classification by MLM (the same maps have been used for training the NN and MLM). MLM has been applied to data compressed by three aforementioned coders. For the AGU coder, the obtained data are presented in Tables 25 and 26.

Table 25. Probabilities of correct classification depending on compressed image quality for Set 1, AGU coder, ML classifier.

Classes	SS1	SS1_45	SS1_42	SS1_39	SS1_36	SS1_33	SS1_30
Urban	0.695	0.727	0.727	0.735	0.745	0.743	0.771
Water	0.979	0.99	0.992	0.992	0.993	0.992	0.988
Vegetation	0.9	0.896	0.896	0.895	0.899	0.901	0.902
Bare soil	0.85	0.852	0.851	0.849	0.852	0.831	0.826
P_{total}	0.856	0.866	0.866	0.868	0.872	0.867	0.872

Table 26. Probabilities of correct classification depending on compressed image quality for Set 2, AGU coder, ML classifier.

Classes	SS2	SS2_45	SS2_42	SS2_39	SS2_36	SS2_33	SS2_30
Urban	0.917	0.914	0.913	0.913	0.91	0.913	0.918
Water	0.923	0.902	0.893	0.885	0.873	0.873	0.857
Vegetation	0.691	0.667	0.647	0.645	0.634	0.631	0.619
Bare soil	0.856	0.819	0.815	0.817	0.812	0.802	0.812
P_{total}	0.847	0.826	0.817	0.815	0.807	0.805	0.801

It is seen that lossy compression leads to a positive effect for the Set 1 image. For all classes except Bare Soil, the probabilities of correct classification improve or remain the same. Total probability also improves and remains at approximately the same level for all considered qualities of the compressed image (Table 25). Meanwhile, for the Set 2 image, larger CR leads to a steady reduction of total probability and decreasing of probabilities for most classes Table 26).

We have also checked whether or not it is worth carrying out training for compressed images instead of uncompressed ones. The answer, as earlier, is yes. In particular, for images compressed with providing PSNR-HVS-M = 42 dB. The total probabilities equal to 0.893 for the Set 1 data and 0.858 for the Set 2 image. Thus, it is worth using those data in training that have been obtained with the same compression conditions as images subject to classification.

Data obtained for the SPIHT coder are presented in Tables 27 and 28. Their analysis shows the following. Again, lossy compression has a small impact on the classification of the Set 1 data. An optimum is observed for PSBR-HVS-M about 40 dB. The impact of compression for the Set 2 data is also small. For some classes, probabilities of correct classification improve, for others become slightly worse. In aggregate, the total probability remains almost the same.

Table 27. Probabilities of correct classification depending on compressed image quality for Set 1, SPIHT coder, ML classifier.

Classes	SS1	SS1_45	SS1_42	SS1_39	SS1_36	SS1_33	SS1_30
Urban	0.695	0.721	0.724	0.728	0.735	0.742	0.759
Water	0.979	0.991	0.991	0.993	0.993	0.992	0.985
Vegetation	0.9	0.899	0.901	0.9	0.904	0.908	0.906
Bare soil	0.85	0.856	0.864	0.858	0.858	0.848	0.816
P_{total}	0.856	0.866	0.87	0.87	0.872	0.873	0.867

Table 28. Probabilities of correct classification depending on compressed image quality for Set 2, SPIHT coder, ML classifier.

Classes	SS2	SS2_45	SS2_42	SS2_39	SS2_36	SS2_33	SS2_30
Urban	0.917	0.915	0.915	0.917	0.916	0.92	0.924
Water	0.923	0.918	0.913	0.903	0.892	0.884	0.875
Vegetation	0.691	0.686	0.688	0.692	0.697	0.699	0.707
Bare soil	0.856	0.847	0.858	0.862	0.853	0.861	0.859
P_{total}	0.847	0.841	0.843	0.843	0.839	0.841	0.841

Consider now the data obtained for ADCTC. They are presented in Tables 29 and 30. Their analysis shows the same tendencies as those observed for the SPIHT coder.

Table 29. Probabilities of correct classification depending on compressed image quality for Set 1, ADCTC, ML classifier.

Classes	SS1	SS1_45	SS1_42	SS1_39	SS1_36	SS1_33	SS1_30
Urban	0.695	0.723	0.725	0.734	0.736	0.751	0.76
Water	0.979	0.992	0.994	0.995	0.995	0.995	0.995
Vegetation	0.9	0.898	0.9	0.901	0.901	0.91	0.913
Bare soil	0.85	0.863	0.855	0.85	0.864	0.84	0.827
P_{total}	0.856	0.869	0.869	0.87	0.874	0.874	0.874

Table 30. Probabilities of correct classification depending on compressed image quality for Set 2, ADCTC coder, ML classifier.

Classes	SS2	SS2_45	SS2_42	SS2_39	SS2_36	SS2_33	SS2_30
Urban	0.917	0.914	0.911	0.91	0.911	0.912	0.915
Water	0.923	0.918	0.912	0.906	0.909	0.901	0.898
Vegetation	0.691	0.674	0.688	0.682	0.664	0.67	0.693
Bare soil	0.856	0.848	0.848	0.847	0.843	0.859	0.868
P_{total}	0.847	0.838	0.84	0.836	0.832	0.836	0.844

6. Discussion

Above, we have considered tendencies for one test and three real-life three-channel images presented as 8-bit 2D data for each component. In practice, images can be presented differently, for example, by 16-bit data [13] or by 10-bit data after certain normalization [59]. Then, a question arises how to provide the desired PSNR-HVS-M (e.g., 40 dB) for a given multichannel image to be compressed. To answer this question, let us consider some data. First, Figure 11 taken from [60] presents the dependences of PSNR-HVS-M on QS for nine 8-bit grayscale test images of different complexity for AGU. The average curve obtained for these nine test images is also given. It is seen that this average curve allows the approximate setting of QS to provide the desired PSNR-HVS-M. For example, to provide PSNR-HVS-M $\approx$ 43 dB, one has to set $QS_{rec} \approx$ 15. To provide PSNR-HVS-M $\approx$ 40 dB, it is possible to set $QS_{rec} \approx$ 20. If D is not equal to 255, then the recommended QS is

$$QS_{recD} = QS_{rec}D/255, \qquad (9)$$

where QS_{rec} is determined from the average curve in Figure 11. As it follows from the analysis of data in Figure 11, the use of QS_{rec} or QS_{recD} provides PSNR-HVS-M approximately, errors can be up to 1 ... 2 dB depending upon the complexity of an image to be compressed. Such accuracy can be treated as acceptable, since, as it is shown in the previous Section, change of PSNR-HVS-M by even 2 dB does not lead to radical changes of P_{cc} and probabilities of correct classification for particular classes. Second, if errors in providing the desired PSNR-HVS-M are inappropriate, accuracy can be improved by applying a two-step procedure proposed in [60]. As it has been shown in Table 13, QS that should be used in component-wise compression is practically the same for all components. This means that it is enough to determine QS_{rec} or QS_{recD} for one component image and then apply it for compressing other components (this can save time and resources at the data compression stage). Moreover, QS_{rec} or QS_{recD} determined according to recommendations given above can be used in joint compression of all components of multichannel images or groups of components [59] by the 3D version of AGU. In this case, the positive effect is twofold. First, a larger CR is provided compared to the component-wise compression. Second, a slightly larger quality of the compressed image can be ensured.

Figure 12 demonstrates the dependences PSNR-HVS-M on QS for different test images. Figure 13 shows the RS data processing flowchart. Acquired images (e.g., on-board) are subject to "careful" lossy compression in Quality Control Compression Unit where PCC (e.g., QS) is determined using the desired threshold for a chosen quality metric (e.g., 40 dB for PSNR-HVS-M) and rate/distortion curves obtained in advance (like those in Figure 12). PCC corrections can be done if, e.g., images are normalized before compression. In the Image Classification Unit, RS data are subject to classification where the classifier can be trained using either earlier processed images or data that have been just received. If time limitations are not strict, the second option seems preferable (according to results obtained in our analysis).

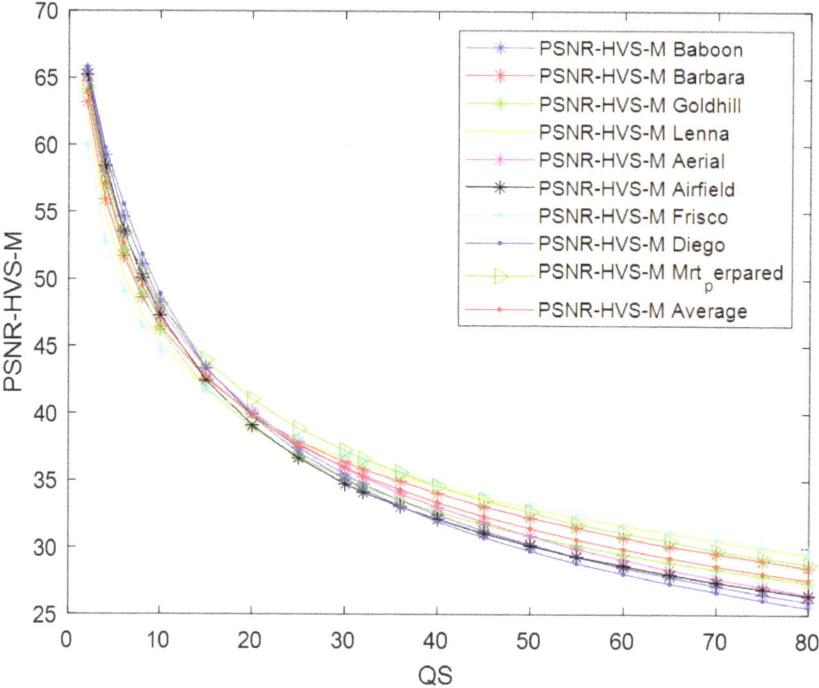

Figure 12. Dependences PSNR-HVS-M on QS for nine test images (see the list in the upper part of the plot) and the average curve for AGU.

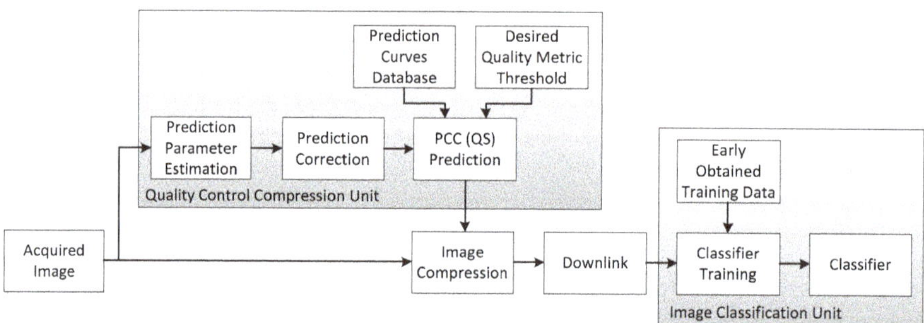

Figure 13. Flowchart of the proposed processing approach.

Another question that has arisen several times is the following—is it possible to improve classification accuracy? One answer that easily follows from our analysis is that the pixel-wise classification (at least, its simple version used by us) does not allow exploiting information from neighboring pixels. Such information can be of different types and it can be used differently. For example, a pixel can be preliminarily classified as belonging to a homogeneous, textural, or locally active area (by locally active, we mean pixel belonging to a small-sized object or edge or their neighborhood). Then, different features can be determined and used for such pre-classified pixels. However, this approach is out of the scope of this paper.

There are also two other approaches possible. They have been proposed in [39]. First, several elementary classifiers (for example, MLM, NN, and SVM ones) can be applied in parallel, and then their outputs can be combined. Second, post-processing of classification results with preliminary detection of edges can be performed. Note that these approaches can lead to a sufficient (up to 0.1 ... 0.2) increase of P_{cc}, especially if it is low for originally classified data. Meanwhile, more experiments for verifying the methods developed in [39] are needed.

Finally, the last question is how to improve CR without losing classification performance for multichannel RS data? In our opinion, 3D compression should be applied. However, in this case, additional studies are also desired.

7. Conclusions

We have considered the task of lossy compression of multichannel images by introducing different levels of distortions to analyze their influence on classification accuracy. Two classifiers, namely MLM and NN, have been studied. The DCT-based coder AGU has been used. In addition, two other transform-based coders have been employed with a brief analysis of their performance. One artificial and three real-life three-channel images have been thoroughly analyzed. Component-wise compression has been addressed with quality controlled (characterized) by visual quality metric PSNR-HVS-M.

The following has been shown for the case of classifier training for original (uncompressed) data;

- there is no sufficient decrease in classification accuracy if images are compressed without visible distortions (PSNR-HVS-M exceeds 40 ... 42 dB); in some cases, TPCC can even increase compared to TPCC for original (uncompressed) data;
- lossy compression might have no considerable negative impact on classification accuracy even if distortions are visible (PSNR-HVS-M < 40 dB) but this happens if a compressed image has a simple structure;
- distortions of the aforementioned level result in different CR depending upon image complexity; this means that images with complex structure have to be compressed with smaller CR to avoid considerable reduction of P_{cc};
- dependences of probabilities of correct classification of particular classes on compression parameters are different; for classes mostly represented by large-size objects, lossy compression is not so crucial; it might even happen that lossy compression can improve classification accuracy if noise or high-frequency variations are partly suppressed due to it;
- for classes represented by small-sizes and prolonged objects as well as some textures, lossy compression might lead to sufficient reduction of classification accuracy. This is due to the image smearing effect that transforms features.

The situation is slightly different if the training is done for compressed data. Then, it is worth using for training such compressed data that are characterized by the same quality (for example, approximately the same PSNR-HVS-M value) as data to which classification will be applied. Then, even for PSNR-HVS-M about 36 dB, the classification can be almost as good enough as for the original image classified by the classifier trained for original data.

It has been also shown that the main dependencies are observed for different coders (in particular, compression techniques based on DCT and wavelets) and for data provided by different multispectral systems. We also believe that other than PSNR-HVS-M visual quality metrics can be applied within the proposed approach to lossy compression. Thus, the approach is quite general.

Finally, practical recommendations on how to set coder parameters to avoid sufficient reduction of classification accuracy are given. Possibilities of classification improvement are briefly mentioned and directions of future research are described.

Supplementary Materials: The following are available online at http://www.mdpi.com/2072-4292/12/22/3840/s1, The Spearman Rank-Order Correlations Coefficient (SROCC) values for 49 quality metrics for TID 2013 database.

Author Contributions: I.V. created and tested MLM classifier, O.R. created and tested NN-based classifier, F.L. performed compression with desired characteristics; S.K. participated in model data design, S.A. was responsible for simulation results analysis, V.L. put forward the idea of coder parameter selection and prepared the paper draft, B.V. was responsible for real-life image result analysis, K.C. proposed methodology of investigation, K.E. carried out editing and supervision. All authors have read and agreed to the published version of the manuscript.

Funding: This research received no external funding.

Conflicts of Interest: The authors declare no conflict of interest.

References

1. Mielke, C.; Boshce, N.K.; Rogass, C.; Segl, K.; Gauert, C.; Kaufmann, H. Potential Applications of the Sentinel-2 Multispectral Sensor and the ENMAP hyperspectral Sensor in Mineral Exploration. In Proceedings of the EARSeL eProceedings, Warsaw, Poland, 13 February 2014; Volume 13, pp. 93–102.

2. Schowengerdt, R.A. *Remote Sensing: Models and Methods for Image Processing*, 3rd ed.; Academic Press: San Diego, CA, USA, 2007; p. 515.

3. Joshi, N.; Baumann, M.; Ehammer, A.; Fensholt, R.; Grogan, K.; Hostert, P.; Jepsen, M.R.; Kuemmerle, T.; Meyfroidt, P.; Mitchard, E.T.A.; et al. A Review of the Application of Optical and Radar Remote Sensing Data Fusion to Land Use Mapping and Monitoring. *Rem. Sens.* **2016**, *8*, 23. [CrossRef]

4. Deledalle, C.; Denis, L.; Tabti, S.; Tupin, F. MuLoG, or How to Apply Gaussian Denoisers to Multi-Channel SAR Speckle Reduction? *IEEE Trans. Image Process.* **2017**, *26*, 4389–4403. [CrossRef] [PubMed]

5. Ponomaryov, V.; Rosales, A.; Gallegos, F.; Loboda, I. Adaptive vector directional filters to process multichannel images. *Ieice Trans. Commun.* **2007**, *90*, 429–430. [CrossRef]

6. Zhong, P.; Wang, R. Multiple-Spectral-Band CRFs for Denoising Junk Bands of Hyperspectral Imagery. *IEEE Trans. Geosci. Remote Sens.* **2013**, *51*, 2269–2275. [CrossRef]

7. First Applications from Sentinel-2A. Available online: http://www.esa.int/Our_Activities/Observing_the_Earth/Copernicus/Sentinel-2/First_applications_from_Sentinel-2A (accessed on 10 January 2019).

8. Kussul, N.; Lemoine, G.; Gallego, F.J.; Skakun, S.; Lavreniuk, M.; Shelestov, A. Parcel-Based Crop Classification in Ukraine Using Landsat-8 Data and Sentinel-1A Data. *IEEE J. Sel. Top. Appl. Earth Obs. Remote Sens.* **2016**, *9*, 2500–2508. [CrossRef]

9. Bioucas-Dias, J.M.; Plaza, A.; Camps-Valls, G.; Scheunders, P.; Nasrabadi, N.; Chanussot, J. Hyperspectral Remote Sensing Data Analysis and Future Challenges. *IEEE Geosci. Remote Sens. Mag.* **2013**, *1*, 6–36. [CrossRef]

10. Chi, M.; Plaza, A.; Benediktsson, J.A.; Sun, Z.; Shen, J.; Zhu, Y. Big Data for Remote Sensing: Challenges and Opportunities. *Proc. IEEE* **2016**, *104*, 2207–2219. [CrossRef]

11. Christophe, E. Hyperspectral Data Compression Tradeoff. *Opt. Remote Sens. Adv. Signal Process. Exploit. Tech.* **2011**, *7*, 9–29.

12. Blanes, I.; Magli, E.; Serra-Sagrista, J. A Tutorial on Image Compression for Optical Space Imaging Systems. *IEEE Geosci. Remote Sens. Mag.* **2014**, *2*, 8–26. [CrossRef]

13. Zemliachenko, A.N.; Kozhemiakin, R.A.; Uss, M.L.; Abramov, S.K.; Ponomarenko, N.N.; Lukin, V.V. Lossy compression of hyperspectral images based on noise parameters estimation and variance stabilizing transform. *J. Appl. Remote Sens.* **2014**, *8*, 25. [CrossRef]

14. Manolakis, D.; Lockwood, R.; Cooley, T. On the Spectral Correlation Structure of Hyperspectral Imaging Data. In Proceedings of the IGARSS 2008—2008 IEEE International Geoscience and Remote Sensing Symposium, Boston, MA, USA, 7–11 July 2008; pp. II-581–II-584.

15. Lam, K.W.; Lau, W.; Li, Z. The effects on image classification using image compression technique. *Amsterdam, Int. Arch. Photogramm. Remote Sens.* **2000**, *33*, 744–751.

16. Penna, B.; Tillo, T.; Magli, E.; Olmo, G. Transform Coding Techniques for Lossy Hyperspectral Data Compression. *IEEE Trans. Geosci. Remote Sens.* **2007**, *45*, 1408–1421. [CrossRef]

17. Kaur, A.; Wasson, V. Compressing Land Images using Fractal Lossy Compression. *Int. J. Eng. Comput. Sci.* **2014**, *3*, 8806–8811.

18. Taubman, D.; Marcellin, M. *JPEG2000 Image Compression Fundamentals, Standards and Practice*; Kluwer Academic Publishers: Boston, MA, USA, 2002; p. 777.

19. Santos, L.; Lopez, S.; Callico, G.; Lopez, J.; Sarmiento, R. Performance evaluation of the H.264/AVC video coding standard for lossy hyperspectral image compression. *IEEE J. Sel. Topics Appl. Earth Observ. Remote Sens.* **2012**, *5*, 451–461. [CrossRef]

20. Krivenko, S.; Krylova, O.; Bataeva, E.; Lukin, V. Smart Lossy Compression of Images Based on Distortion Prediction. *Telecommun. Radio Eng.* **2018**, *77*, 1535–1554. [CrossRef]

21. Khelifi, F.; Bouridane, A.; Kurugollu, F. Joined spectral trees for scalable SPIHT-based multispectral image compression. *IEEE Trans. Multimed.* **2008**, *10*, 316–329. [CrossRef]

22. Zemliachenko, A.; Kozhemiakin, R.; Abramov, S.; Lukin, V.; Vozel, B.; Chehdi, K.; Egiazarian, K. Prediction of compression ratio for DCT-based coders with application to remote sensing images. *J. Sel. Top. Appl. Earth Obs. Remote Sens.* **2018**, *11*, 257–270. [CrossRef]

23. Zemliachenko, A.; Ponomarenko, N.; Lukin, V.; Egiazarian, K.; Astola, J. Still Image/Video Frame Lossy Compression Providing a Desired Visual Quality. *Multidimens. Syst. Signal Process.* **2015**, *27*, 697–718. [CrossRef]

24. Shi, C.; Wang, L.; Zhang, J.; Miao, F.; He, P. Remote Sensing Image Compression Based on Direction Lifting-Based Block Transform with Content-Driven Quadtree Coding Adaptively. *Rem. Sens.* **2018**, *10*, 999. [CrossRef]

25. Balasubramanian, R.; Ramakrishnan, S.S. Wavelet application in compression of a remote sensed image. In Proceedings of the 2013 the International Conference on Remote Sensing, Environment and Transportation Engineering (RSETE 2013), Nanjing, China, 26–28 July 2013; pp. 659–662.

26. Jagadeesh, P.; Kumar, S.; Nedumaan, J. Accepting the Challenges in Devising Video Game Leeway and Contrivance. *Int. J. Inf. Secur.* **2015**, *14*, 582–590.

27. Ozah, N.; Kolokolova, A. Compression improves image classification accuracy. In Proceedings of the 32nd Canadian Conference on Artificial Intelligence, Canadian AI 2019, Kingston, ON, Canada, 28–31 May 2019; pp. 525–530.

28. Chen, Z.; Ye, H.; Yingxue, Z. Effects of Compression on Remote Sensing Image Classification Based on Fractal Analysis. *IEEE Trans. Geosci. Remote Sens.* **2019**, *57*, 4577–4589. [CrossRef]

29. García-Sobrino, J.; Laparra, V.; Serra-Sagristà, J.; Calbet, X.; Camps-Valls, G. Improved Statistically Based Retrievals via Spatial-Spectral Data Compression for IASI Data. *IEEE Trans. Geosci. Remote Sens.* **2019**, *57*, 5651–5668. [CrossRef]

30. Gimona, A.; Poggio, L.; Aalders, I.; Aitkenhead, M. The effect of image compression on synthetic PROBA-V images. *Int. J. Remote Sens.* **2014**, *35*, 2639–2653. [CrossRef]

31. Perra, C.; Atzori, L.; De Natale, F.G.B. Introducing supervised classification into spectral VQ for multi-channel image compression. In Proceedings of the IGARSS 2000, IEEE 2000 International Geoscience and Remote Sensing Symposium. Taking the Pulse of the Planet: The Role of Remote Sensing in Managing the Environment (Cat. No.00CH37120), Honolulu, HI, USA, 24–28 July 2000; pp. 597–599. [CrossRef]

32. Izquierdo-Verdiguier, E.; Laparra, V.; Gómez-Chova, L.; Camps-Valls, G. Encoding Invariances in Remote Sensing Image Classification With SVM. *IEEE Geosci. Remote Sens. Lett.* **2013**, *10*, 981–985. [CrossRef]

33. Zabala, A.; Pons, X. Impact of lossy compression on mapping crop areas from remote sensing. *Int. J. Remote Sens.* **2013**, *34*, 2796–2813. [CrossRef]

34. Zabala, A.; Pons, X.; Diaz-Delgado, R.; Garcia, F.; Auli-Llinas, F.; Serra-Sagrista, J. Effects of JPEG and JPEG2000 Lossy Compression on Remote Sensing Image Classification for Mapping Crops and Forest Areas. In Proceedings of the 2006 IEEE International Symposium on Geoscience and Remote Sensing, Denver, CO, USA, 31 July–4 August 2006; pp. 790–793. [CrossRef]
35. Popov, M.A.; Stankevich, S.A.; Lischenko, L.P.; Lukin, V.V.; Ponomarenko, N.N. Processing of hyperspectral imagery for contamination detection in urban areas. In *Environmental Security and Ecoterrorism*; Springer: Berlin/Heidelberg, Germany, 2010; pp. 25–26.
36. Ponomarenko, N.; Krivenko, S.; Lukin, V.; Egiazarian, K.; Astola, J. Lossy Compression of Noisy Images Based on Visual Quality: A Comprehensive Study. *Eurasip. Jasp.* **2010**, *2010*, 976436. [CrossRef]
37. Al-Shaykh, O.K.; Mersereau, R.M. Lossy compression of noisy images. *IEEE Trans. Image Process* **1998**, *7*, 1641–1652. [CrossRef]
38. Lukin, V. Processing of Multichannel RS data for Environment Monitoring. *Proc. Nato Adv. Res. Workshop Geogr. Inf. Process. Vis. Anal. Env. Secur.* **2009**, *1*, 129–138.
39. Proskura, G.; Vasilyeva, I.; Fangfang, L.; Lukin, V. Classification of Compressed Multichannel Images and Its Improvement. In Proceedings of the 2020 30th International Conference Radioelektronika (RADIOELEKTRONIKA), Bratislava, Slovakia, 15–16 April 2020; pp. 62–67.
40. Lukin, V.; Zemliachenko, A.; Krivenko, S.; Vozel, B.; Chehdi, K. *Lossy Compression of Remote Sensing Images with Controllable Distortions. Satellite Information Classification and Interpretation*; Rustam, B.R., Ed.; IntechOpen: Londone, UK, 2018. [CrossRef]
41. Krivenko, S.S.; Abramov, S.K.; Lukin, V.V.; Vozel, B.; Chehdi, K. Lossy DCT-based compression of remote sensing images with providing a desired visual quality. In *Image and Signal Processing for Remote Sensing XXV*; International Society for Optics and Photonics: Bellingham, WA, USA, 2019; p. 11155. [CrossRef]
42. Ponomarenko, N.N.; Lukin, V.V.; Egiazarian, K.; Astola, J. DCT Based High Quality Image Compression. In *Scandinavian Conference on Image Analysis*; Springer: Berlin/Heidelberg, Germany, 2005; pp. 1177–1185.
43. Lukin, V.; Abramov, S.; Krivenko, S.; Kurekin, A.; Pogrebnyak, O. Analysis of classification accuracy for pre-filtered multichannel remote sensing data. *J. Expert Syst. Appl.* **2013**, *40*, 6400–6411. [CrossRef]
44. Congalton, R.G.; Green, K. *Assessing the Accuracy of Remotely Sensed Data: Principles and Practices*; CRC Press: Florida, FL, USA, 1999.
45. Lukin, V.; Ponomarenko, N.; Egiazarian, K.; Astola, J. Analysis of HVS-Metrics' Properties Using Color Image Database TID2013. *Proc. Acivs* **2015**, *9386*, 613–624.
46. Lin, W.; Jay Kuo, C.C. Perceptual visual quality metrics: A survey. *J. Vis. Commun. Image Represent.* **2011**, *22*, 297–312. [CrossRef]
47. Ponomarenko, N.; Silvestri, F.; Egiazarian, K.; Carli, M.; Astola, J.; Lukin, V. On between-coefficient contrast masking of DCT basis functions. In Proceedings of the Third International Workshop on Video Processing and Quality Metrics, Scottsdale, AZ, USA, 25–26 January 2007; Volume 7.
48. Available online: http://ponomarenko.info/psnrhvsm.htm (accessed on 20 November 2020).
49. Ieremeiev, O.; Lukin, V.; Okarma, K.; Egiazarian, K. Full-Reference Quality Metric Based on Neural Network to Assess the Visual Quality of Remote Sensing Images. *Remote Sens.* **2020**, *12*, 2349. [CrossRef]
50. Sun, J.; Yang, J.; Zhang, C.; Yun, W.; Qu, J. Automatic remotely sensed image classification in a grid environment based on the maximum likelihood method. *Math. Comput. Model.* **2013**, *58*, 573–581. [CrossRef]
51. Sisodia, P.S.; Tiwari, V.; Kumar, A. Analysis of Supervised Maximum Likelihood Classification for remote sensing image. In *International Conference on Recent Advances and Innovations in Engineering (ICRAIE-2014)*; IEEE: Piscataway, NY, USA, 2014; pp. 1–4. [CrossRef]
52. Parresol, B.R. *Recovering Parameters of Johnson's S_B Distribution*; US Department of Agriculture, Forest Service, Southern Research Station: Asheville, NC, USA, 2003; Volume 9.
53. Lemonte, A.J.; Bazán, J.L. New class of Johnson SB distributions and its associated regression model for rates and proportions. *Biom. J.* **2016**, *58*, 727–746. [CrossRef]
54. Bishop, C.M. *Pattern Recognition and Machine Learning*; Springer: New York, NY, USA, 2006.
55. Bioucas-Dias, J.M. Hyperspectral Unmixing Overview: Geometrical, Statistical, and Sparse Regression-Based Approaches. *IEEE J. Sel. Top. Appl. Earth Obs. Remote Sens.* **2012**, *5*, 354–379. [CrossRef]

56. Available online: https://apps.sentinel-hub.com/eo-browser/?lat=46.45&lng=34.12&zoom=6&time=2019-11-03&preset=3_NDVI&datasource=Sentinel-2%20L1C (accessed on 20 November 2020).

57. Taubman, D.; Marcellin, M.W. *JPEG2000 Image Compression: Fundamentals, Standards and Practice*; Kluwer Academic Publishers: Dordrecht, The Netherlands, 2002. [CrossRef]

58. Ponomarenko, N.N.; Egiazarian, K.O.; Lukin, V.V.; Astola, J.T. High-Quality DCT-Based Image Compression Using Partition Schemes. *IEEE Signal Process. Lett.* **2007**, *14*, 105–108. [CrossRef]

59. Zemliachenko, A.; Lukin, V.; Ieremeiev, O.; Vozel, B. Peculiarities of Hyperspectral Image Lossy Compression for Sub-band Groups. In Proceedings of the 2019 IEEE 2nd Ukraine Conference on Electrical and Computer Engineering (UKRCON), Lviv, Ukraine, 2–6 July 2019; pp. 918–923.

60. Li, F.; Krivenko, S.; Lukin, V. A Two-step Approach to Providing a Desired Visual Quality in Image Lossy Compression. In Proceedings of the 2020 IEEE 15th International Conference on Advanced Trends in Radioelectronics, Telecommunications and Computer Engineering (TCSET), Lviv-Slavske, Ukraine, 25–29 February 2020; pp. 502–506. [CrossRef]

Publisher's Note: MDPI stays neutral with regard to jurisdictional claims in published maps and institutional affiliations.

Article

FPGA-Based On-Board Hyperspectral Imaging Compression: Benchmarking Performance and Energy Efficiency against GPU Implementations

Julián Caba [1,*,†], María Díaz [2,†], Jesús Barba [1], Raúl Guerra [2] and Jose A. de la Torre [1] and Sebastián López [2]

[1] School of Computer Science, University of Castilla-La Mancha (UCLM), 13071 Ciudad Real, Spain; jesus.barba@uclm.es (J.B.); joseantonio.torre@uclm.es (J.A.d.l.T.)

[2] Institute for Applied Microelectronics (IUMA), University of Las Palmas de Gran Canaria (ULPGC), 35001 Las Palmas de Gran Canaria, Spain; mdmartin@iuma.ulpgc.es (M.D.); rguerra@iuma.ulpgc.es (R.G.); seblopez@iuma.ulpgc.es (S.L.)

* Correspondence: julian.caba@uclm.es

† These authors contributed equally to this work.

Received: 17 September 2020; Accepted: 9 November 2020; Published: 13 November 2020

Abstract: Remote-sensing platforms, such as Unmanned Aerial Vehicles, are characterized by limited power budget and low-bandwidth downlinks. Therefore, handling hyperspectral data in this context can jeopardize the operational time of the system. FPGAs have been traditionally regarded as the most power-efficient computing platforms. However, there is little experimental evidence to support this claim, which is especially critical since the actual behavior of the solutions based on reconfigurable technology is highly dependent on the type of application. In this work, a highly optimized implementation of an FPGA accelerator of the novel *HyperLCA* algorithm has been developed and thoughtfully analyzed in terms of performance and power efficiency. In this regard, a modification of the aforementioned lossy compression solution has also been proposed to be efficiently executed into FPGA devices using fixed-point arithmetic. Single and multi-core versions of the reconfigurable computing platforms are compared with three GPU-based implementations of the algorithm on as many NVIDIA computing boards: Jetson Nano, Jetson TX2 and Jetson Xavier NX. Results show that the single-core version of our FPGA-based solution fulfils the real-time requirements of a real-life hyperspectral application using a mid-range Xilinx Zynq-7000 SoC chip (XC7Z020-CLG484). Performance levels of the custom hardware accelerator are above the figures obtained by the Jetson Nano and TX2 boards, and power efficiency is higher for smaller sizes of the image block to be processed. To close the performance gap between our proposal and the Jetson Xavier NX, a multi-core version is proposed. The results demonstrate that a solution based on the use of various instances of the FPGA hardware compressor core achieves similar levels of performance than the state-of-the-art GPU, with better efficiency in terms of processed frames by watt.

Keywords: hyperspectral imaging; lossy compression; on-board processing; FPGA; GPU; real-time performance; UAV; parallel computing

---

1. Introduction

Hyperspectral technology has experienced a steady surge in popularity in recent decades. Among the main reasons that have given hyperspectral imaging greater visibility is the richness of spectral information collected by this kind of sensor. This feature has positioned hyperspectral analysis techniques as the mainstream solution for the analysis of land areas and the identification and discrimination of visually similar surface materials. As a consequence, this technology has acquired increasing relevance, being widely used for a variety of applications, such as precision agriculture, environmental monitoring, geology, urban surveillance and homeland security, among others. Nevertheless, hyperspectral image processing is accompanied by the management of large amounts of data, which affects, on one hand, its real-time performance and, on the other hand, the requirements of the on-board storage resources. Additionally, the latest technological advances are promoting to market hyperspectral cameras with higher spectral and spatial resolutions. All of this makes the efficient data handling, from an on-board processing, communication, and storage point of view, even more challenging [1,2].

Traditionally, images sensed by spaceborne Earth-observation missions are not on-board processed. The main rationale behind this is the limited on-board power capacity that forces the use of low-power devices, which are normally not as highly performing as their commercial counterparts [3–8]. In this regard, images are subsequently downloaded to the Earth surface where they are off-line processed on high-performance computing systems based on Central Processing Units (CPUs), Graphics Processing Units (GPUs), Field-Programmable Gate Arrays (FPGAs), or heterogeneous architectures. In the case of airborne capturing platforms, images are normally on-board stored and hence the end user could not access them until the flight mission is over [9]. Additionally, unmanned aerial vehicles (UAVs) have gained momentum in recent years. They have become a very popular solution for inspection, surveillance and monitoring since they represent a lower-cost approach with a more flexible revisit time than the aforementioned Earth-observation platforms. In this context, hyperspectral image management is addressed similarly to how it is done in airborne platforms, although a lot of efforts have been made recently to transmit the images to the ground segment as soon as they are captured [10,11].

Regrettably, the data transmission from the aforementioned remote-sensing observation platforms introduces important delays. They are mainly related to the transference of large volumes of data and the limited communication bandwidths between the source and the final target, which has also kept relatively stable over the years [6,12,13]. Consequently, it reveals a bottleneck in the downlink systems that can seriously affect to the effective performance of real-time or nearly real-time applications. However, the steadily growing data-rate of the latest-generation sensors makes it compulsory to reach higher compression ratios and to carry out a real-time compression performance in order to prevent the unnecessary accumulation of high amounts of uncompressed data on-board and to facilitate efficient data transfers [14].

In this scenario of limited communication bandwidths and increasing data volumes, it is becoming necessary to move from lossless or near-lossless compression approaches to lossy compression techniques. Despite most of the state-of-the-art lossless compressors bringing a quite satisfactory rate-distortion performance, the former approaches provide very moderate compression ratios of about 2~3:1 [15,16], which presently are not sufficient to handle the high input data-rate of the newest-generation sensors. For this reason, a research effort targeting lossy compression is currently being made [17–21]. Due to the limited on-board computational capabilities of remote-sensing hyperspectral acquisition systems, low-complexity compression schemes stand as the most practical solution for such restricted environments [21–24]. Nevertheless, most of the state-of-the-art lossy compressors are generalizations of existing 2D images or video compression algorithms [25]. For this reason, they are normally characterized by high computational burden, intensive memory requirements and a non-scalable nature. These features prevent their use in power-constrained applications with limited hardware resources, such as on-board compression [26,27].

In this context, the Lossy Compression Algorithm for Hyperspectral Image Systems (HyperLCA) [28] was developed as a novel hardware-friendly lossy compressor for hyperspectral images. HyperLCA was introduced as a low-complexity alternative that provides a good compression performance at high compression ratios with a reasonable computational burden. Additionally, this algorithm permits compressing blocks of image pixels independently which promotes, on the one hand, the reduction of the data to be managed at once besides the hardware resources to be allocated. On the other hand, the HyperLCA algorithm becomes a very competitive solution for most applications based on pushbroom/whiskbroom scanners, paving the way for real-time compression performance. The flexibility and the high level of parallelism intrinsic to the HyperLCA algorithm has been previously evaluated in earlier publications. In particular, its suitability for real-time performance in applications characterized by high data-rates with restrictions in computational resources due to power, weight or space was tested in [29].

In this work, we focus on a use case where a visible-near-infrared (VNIR) hyperspectral pushbroom scanner is mounted onto a UAV. In particular, we have analyzed the performance of FPGAs for the lossy compression of hyperspectral images against low-power GPUs (LPGPUs) in order to establish the benefits and barriers of using each one of these hardware devices for the on-board compression task. Specifically, we have implemented the HyperLCA algorithm in a Xilinx Zynq-7000 programmable System on Chip (SoC). We have selected this SoC because it can be found in low-cost, low-weight and compact-size development boards, such as MicroZed^{TM} and PYNQ boards. Although UAVs have been consolidated as trending aerial observation platforms, their acquisition costs are still not accessible for many end customers, not only those who want to purchase them but also those who lease their services. For this reason, we must also aim to solve the economic implications that comes along with these devices. On this basis, in this work we have focused on the on-board computing platform, searching for a less expensive alternative that, in exchange, cannot offer the same level of both performance and functionality than other costly commercial products. Additionally, it is important to note that while experiments carried out in this work are oriented to the current necessities imposed by an application based on drones, all drawn conclusions can be extrapolated to other fields in which remotely sensed hyperspectral images have to be compressed in real time, such as spaceborne missions that employs next-generation space-grade FPGAs.

The rest of this paper is organized as follows. Section 2 gives outlines of the majority issues around the operations involved in the HyperLCA algorithm. In addition, it also includes a detailed explanation about the implementation model developed for the execution of the HyperLCA compressor on the selected FPGA SoC. Section 3 analyzes the obtained experimental results in terms of both quality of the compression process and hardware implementation points of view. Section 4 discusses the strengths and limitations of the proposed solution and includes a comparison evaluation of the obtained results with those achieved by a selection of LPGPUs embedded systems, following the implementation model shown in [29]. Finally, Section 5 draws the main conclusions of this work.

2. Materials and Methods

2.1. HyperLCA Algorithm

The HyperLCA algorithm is a novel lossy compressor for hyperspectral images especially designed for applications based on pushbroom/whiskbroom sensors. Concretely, this transform-based solution can independently compress blocks of image pixels regardless of any spatial alignment between pixels. Consequently, it facilitates the parallelization of the entire compression process and makes it relatively simple to stream the compression of hyperspectral frames collected by a pushbroom or whiskbroom sensor. Additionally, the HyperLCA compressor also allows fixing a desired minimal compression ratio and guaranties that at least this minimum will be achieved for each block of image pixels. This makes it possible to know in advance the maximum data rate that will be attained after the acquisition and compression processes in order to efficiently manage the data transfers and/or storage. Additionally, the HyperLCA compressor provides quite satisfactory rate-distortion results for higher compression ratios than those achievable by lossless compression approaches. Furthermore, the HyperLCA algorithm preserves the most characteristic spectral features of image pixels that are potentially more useful for ulterior hyperspectral analysis techniques, such as target detection, spectral unmixing, change detection, anomaly detection, among others [30–34].

Figure 1 shows a graphic representation of the main computing stages involved by the HyperLCA compressor. First, the *Initialization* stage configures the same parameters (p_{max}) employing the input parameters CR, N_{bits} and BS. Since the HyperLCA compressor follows an unmixing-like strategy, the p_{max} most representative pixels within a block of image pixels to be independently processed are compressed with the highest precision. Then, other image pixels are reconstructed using their scalar projections, V, over the previously selected pixels. Therefore, p_{max} is estimated according to the minimum desired compression ratio to be reached, CR, the number of hyperspectral pixels within each image block, BS and the number of bits used for representing the values of V vectors. Second, these p_{max} characteristic pixels are selected using orthogonal projection techniques in the *HyperLCA Transform* stage. It results in a spectral uncorrelated version of the original hyperspectral data. Thirdly, the outputs of the *HyperLCA Transform* stage are adapted in the following *HyperLCA Preprocessing* stage for being later more efficiently codified in the fourth and last *HyperLCA Entropy Coding* stage. In the following lines, the operations involved in these four algorithm stages are further analyzed in order to give a glance of the decisions taken for the acceleration of the HyperLCA algorithm in hardware devices based on FPGAs.

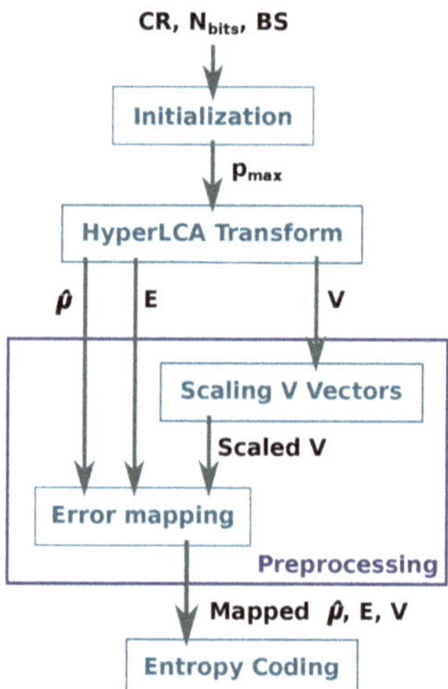

Figure 1. Data flow among the different computing stages of the HyperLCA compressor.

2.1.1. General Notation

Before starting the algorithm description, it is necessary to introduce the general notation followed thorough the rest of this manuscript. In this sense, $\mathbf{HI} = \{\mathbf{F}_i, i = 1, ..., nr\}$ is a sequence of nr scanned cross-track lines of pixels acquired by a pushbroom sensor, $\mathbf{F}_i$, comprised by nc pixels with nb spectral bands. Pixels within $\mathbf{HI}$ are grouped in blocks of BS pixels, $\mathbf{M}_k = \{\mathbf{r}_j, j = 1, ..., BS\}$, being normally BS equal to nc, or multiple of it, and k spans from 1 to $\frac{nr \cdot nc}{BS}$. $\hat{\mu}$ represents the average pixel of each image block, $\mathbf{M}_k$. Therefore, $\mathbf{C}$ is defined as the centralized version of $\mathbf{M}_k$ after being subtracted $\hat{\mu}$ from all image pixels. $\mathbf{E} = \{\mathbf{e}_n, n = 1, ..., p_{max}\}$ saves the p_{max} most different hyperspectral pixels extracted from each $\mathbf{M}_k$ block. $\mathbf{V} = \{\mathbf{v}_n, n = 1, ..., p_{\max}\}$ comprises $p_{\max}$ vectors of BS elements where each $\mathbf{v}_n$ vector corresponds to the projection of the BS pixels within $\mathbf{M}_k$ onto the corresponding n extracted pixel, $\mathbf{e}_n$. $\mathbf{Q} = \{\mathbf{q}_n, n = 1, ..., p_{\max}\}$ and $\mathbf{U} = \{\mathbf{u}_n, n = 1, ..., p_{\max}\}$ save $p_{\max}$ pixels of nb bands that are orthogonal among them.

2.1.2. HyperLCA Initialization

The HyperLCA compressor must first determine the number of pixels, $\mathbf{e}_n$, and projection vectors, $\mathbf{v}_n$, ($p_{\max}$) to be extracted from each image block, $\mathbf{M}_k$, according to the input parameters CR, N_{bits} and BS, as shown in Equation (1). In it, DR refers to the number of bits per pixel per band used for representing the hyperspectral image elements to be compressed. As can be deduced, the number of selected pixels, $p_{\max}$, directly determines the maximum compression ratio to be reached with the selected algorithm configuration. Furthermore, bigger $p_{\max}$ results in better reconstructed images but lower compression ratios.

$$p_{\max} \leq \frac{DR \cdot nb \cdot (BS - CR)}{CR \cdot (DR \cdot nb + N_{\text{bits}} \cdot BS)} \tag{1}$$

2.1.3. HyperLCA Transform

The HyperLCA compressor is a transform-based algorithm that employs a modified version of the well-known Gram-Schmidt orthogonalization method to obtain a compressed and uncorrelated image. In this regard, the *HyperLCA Transform* stage oversees carrying out the spectral transform. For this reason, it is the algorithm stage that involves the most demanding computing operations and thus, which are more prone to be accelerated in parallel dedicated hardware devices.

The *HyperLCA Transform* is described in detail in Algorithm 1. As can be seen, the *HyperLCA Transform* stage receives as inputs the hyperspectral image block to be compressed, $\mathbf{M}_k$, and the number of hyperspectral pixels to be extracted, $p_{\max}$. As outputs, the *HyperLCA Transform* states the set of the $p_{\max}$ most different hyperspectral pixels, $\mathbf{E}$, and their corresponding projection vectors, $\mathbf{V}$. For doing so, the hyperspectral image block, $\mathbf{M}_k$, is first centralized by subtracting the average pixel, $\hat{\mu}$, from all pixels within $\mathbf{M}_k$, which results in the centralized version of the data, $\mathbf{C}$, in line 3. Second, the $p_{\max}$ most characteristic pixels are sequentially extracted from lines 4 to 15. In this iterative process, the brightness of each image pixel is calculated in lines 5 to 7. It is defined as the dot product of each image pixel with itself (see line 6). The extracted pixels, $\mathbf{e}_n$, are those pixels within the original image block, $\mathbf{M}_k$, with the highest brightness in each iteration, n (see line 8 and 9). Afterwards, the orthogonal vectors $\mathbf{q}_n$ and $\mathbf{u}_n$ are accordingly defined as shown in lines 10 and 11, respectively. $\mathbf{u}_n$ is employed to estimate the projection of each image pixel over the direction spanned by the selected pixel, $\mathbf{e}_n$, deriving in the projection vector, $\mathbf{v}_n$ in line 12. Finally, this information is subtracted from $\mathbf{C}$ in line 13.

Accordingly, $\mathbf{C}$ retains in next iterations the spectral information that cannot be represented by the already selected pixels $\mathbf{e}_n$. For this reason, once the $p_{\max}$ pixels have been extracted, $\mathbf{C}$ contains the spectral information that will be not recovered after the decompression process. Hence, it is represented of the losses introduced by the HyperLCA compressor. Consequently, the remaining information in $\mathbf{C}$ could be used to establish extra stopping conditions based on quality metrics, such as the *Signal-to-Noise Ratio* (SNR) or the *Maximum Absolute Difference* (MAD). In this work, these extra stopping conditions have been discarded, and thus, a fixed number of $p_{\max}$ iterations is executed for all image blocks, $\mathbf{M}_k$.

Algorithm 1 HyperLCA Transform.

Inputs:
$\mathbf{M}_k = [\mathbf{r}_1, \mathbf{r}_2, ..., \mathbf{r}_{BS}], p_{max}$
Outputs:
$\hat{\mu}; \hat{\mu}; \mathbf{E} = [\mathbf{e}_1, \mathbf{e}_2, ..., \mathbf{e}_{p_{max}}]; \mathbf{V} = [\mathbf{v}_1, \mathbf{v}_2, ..., \mathbf{v}_{p_{max}}]$
Algorithm:
1: {Additional stopping condition initialization.}
2: Average pixel: $\hat{\mu}$;
3: Centralized image: $\mathbf{C} = \mathbf{M}_k - \hat{\mu} = [\mathbf{c}_1, \mathbf{c}_2, ...\mathbf{c}_{BS}]$;
4: **for** $n = 1$ **to** p_{max} **do**
5:     **for** $j = 1$ **to** BS **do**
6:         Brightness Calculation: $\mathbf{b}_j = \mathbf{c}'_j \cdot \mathbf{c}_j$;
7:     **end for**
8:     Maximum Brightness: $j_{max} = \text{argmax}(\mathbf{b}_j)$;
9:     Extracted pixels: $\mathbf{e}_n = \mathbf{r}_{j_{max}}$;
10:     $\mathbf{q}_n = \mathbf{c}_{j_{max}}$;
11:     $\mathbf{u}_n = \mathbf{q}_n / b_{j_{max}}$;
12:     Projection vector: $\mathbf{v}_n = \mathbf{u}'_n \cdot \mathbf{C}$;
13:     Information Subtraction: $\mathbf{C} = \mathbf{C} - \mathbf{q}_n \cdot \mathbf{v}_n$;
14:     {Additional stopping condition checking.}
15: **end for**

2.1.4. HyperLCA Preprocessing

To ensure an efficient entropy coding of the *HyperLCA Transform* outputs, they must be first adapted in the *HyperLCA Preprocessing* stage. This transformation process is performed in two steps:

Scaling V Vectors

$\mathbf{V}$ vectors represent the projection of each pixel within $\mathbf{M}_k$ over the direction spanned by each orthogonal vector, $\mathbf{u}_n$, in each iteration. Therefore, potential values of each element of $\mathbf{V}$ vectors are in the range of $(-1, 1]$. However, the later *Entropy-coding* stage works exclusively with integers. Consequently, elements within $\mathbf{V}$ must be scaled to fully exploit the dynamic range offered by the input parameter N_{bits}, as shown in Equation (2). After doing so, the scaled $\mathbf{V}$vectors are rounded up to the closest integer values.

$$v_{j_{scaled}} = (v_j + 1) \cdot (2^{N_{bits}-1} - 1) \tag{2}$$

Error Mapping

The codification stage also exploits the redundancies within the data in the spectral domain to assign the shortest word length to the most common values. With it, the compression ratio achieved by the *HyperLCA Transform* could be even increased. To this end, the output vectors of the *HyperLCA Transform* ($\hat{\mu}$, $\mathbf{E}$ and scaled $\mathbf{V}$) are lossless pre-processed in the *Error Mapping* stage. To do so, the prediction error mapper described in the Consultative Committee for Space Data Systems (CCSDS) Blue Books is employed [35]. In this regard, each output vector, ($\hat{\mu}$, $\mathbf{e}_n$ and scaled $\mathbf{v}_n$) is individually processed and transformed to be exclusively composed of positive integer values closer to zero.

2.1.5. HyperLCA Entropy Coding

The *Entropy Coding* is the last stage of the HyperLCA compressor. It follows a lossless entropy-coding strategy based on the Golomb–Rice algorithm [36]. As in the *Error Mapping* stage, each single output vector is independently coded. For doing so, the compression parameter, M, is estimated as the average value of the targeted vector. Afterwards, each of its elements is divided by M in order to obtain the results of the division, the quotient (q) and the remainder (r). On one hand, the quotient, q, is codified using unary code. On the other hand, the remainder, r, could be coded using $b = log_2(M) + 1$ bits if M is power of 2. Nevertheless, M can actually be any positive integer. For this reason, the remainder, r is coded as plain binary using $b - 1$ bits for r values smaller than $2^b - M$, otherwise it is coded as $r + 2^b - M$ using b bits.

2.1.6. HyperLCA Data Types and Precision Evaluation

Most of the compression performance achieved by the HyperLCA algorithm is obtained in the *HyperLCA Transform* stage, originally designed to use floating-point precision. However, FPGA devices are, in general, more efficient dealing with integer operations. Additionally, the execution of floating-point operations in different devices may produce slightly different results. For this reason, the performance of the HyperLCA algorithm using integer arithmetic was largely drawn in [37] in order to adapt it for being more suitable for this kind of device. In particular, it was used the fixed-point concept in a custom way employing integer arithmetic and bits shifting [38]. In this previous work, two different versions of the *HyperLCA Transform* were considered employing 32 and 16 bits, respectively, for representing the image values stored in the centralized version of the hyperspectral frame to be processed, **C**.

It is important to note that the aforementioned proposed versions were developed for working with hyperspectral images whose element values could be represented with up to 16 bits per pixel per band as maximum. In this context, the quality of the compression results is very similar between the 32-bits fixed point and the single precision floating-point versions. Nevertheless, the compression performance obtained by the 16-bit version is not as competitive as the other two solutions. It is shown that this 16-bit approach provides its best performance for very high compression ratios, small *BS* and images packaged using less than 16 bits per pixel per band. This makes the 16-bits version into a very interesting option for applications with limited hardware resources, especially the availability of in-chip memory, one of the weaknesses of FPGAs.

On this basis, we have made some performance-enhancing improvements to the 16-bit version to obtain better compression results. In this context, we have assumed that the available capturing system codes hyperspectral images with a maximum of 12 bits, which is also the most common scenario in remote-sensing applications [39,40] and in the one outlined in this manuscript. Table 1 summarizes the number of bits used by the two algorithm versions introduced in [37], known as I32 and I16, and the one described in this work, referred as to I12, for some algorithm variables. It should be pointed out that when integer arithmetic is used, **V** vectors are directly represented using integer data types and do not need to be scaled and rounded to integer values.

As described in Section 2.1.3, the information presents in matrix **C** decreases with every iteration, n, though some specific values may increase in some quite unlikely situations. For this reason, initial values of **C** were divided by 2 in the previous 16-bit version in order to avoid overflowing, what directly decreases the precision in one bit. In the new version proposed in this work, this fact is discarded since we have 2 bits extra for these improbable situations. As can be noticed from Table 1, image **C** is stored employing 16 bits in the new I12 version as well as in the I16 version. However, it is assumed that images values are coded employing 12 bits as maximum instead of 16 bits. It permits having 2 extra bits for representing the integer part of the fixed-point values of **C** elements. Consequently, image precision is not altered as in previous 16-bit version for dealing with possible overflowing scenarios. Additionally, this new version also allows having 2 bits for representing the decimal part of the fixed-point values of matrix **C**, which is not possible in the I16 version.

Table 1. Number of bits used for the integer and decimal parts used to represent the variables involved in the *HyperLCA Transform*. Three versions of the algorithm were developed to use integer arithmetic (I32, I16, I12).

Variable	Integer Part			Decimal Part			Total		
	I32	I16	I12	I32	I16	I12	I32	I16	I12
C	20	16	14	12	0	2	32	16	16
b	48	48	48	16	16	16	64	64	64
q	20	20	20	12	12	12	32	32	32
u	2	2	2	30	30	30	32	32	32
v	2	2	2	30	30	30	32	32	32
μ	16	16	12	0	0	0	16	16	12

2.2. FPGA Implementation of the HyperLCA Algorithm

An FPGA (Field-Programmable Gate Array) can be seen as a whiteboard for designing specialized hardware accelerators (HWacc), by composition of predefined memory and logic blocks that are available in the platform. Therefore, a HWacc is a set of architectural FPGA resources, connected and configured to carry out a specific task. Each vendor proposes its own reconfigurable platform, instantiating a particular mix of such resources, around a particular interconnection architecture.

FPGAs provide flexibility to designers, since the silicon adapts to the solution, instead of fitting the solution to the computing platform as is the case of GPU-based solutions. On top of that, FPGA-based implementations can take advantage of the fine-grain parallelism of their architecture (operation level) as well as task-level concurrency. In this paper, the HyperLCA lossy compressor has been implemented onto a heterogeneous Zynq-7000 SoC (System-on-a-Chip) from Xilinx that combines a Processor System (PS), based on a dual core ARM processor, and a Programmable Logic (PL), based on a Artix-7/Kintex-7 FPGA architecture.

The development process followed a productive methodology based on High-Level Synthesis (HLS). This methodology focuses the effort on the design and verification of the HWacc, as well as the exploration of the solution space that helps to speed up the search for value-added solutions. The starting point of a methodology based on HLS is a high-abstraction model of the functionality to be deployed on the FPGA, usually described by means of high-level programming languages, such as C or C++. Then, HLS tools can generate the corresponding RTL (Register Transfer Level) implementation, functionally equivalent to the C or C++ model [41,42].

Productivity is the strongest point of HLS technology, and one of the main reasons why hardware architects and engineers have been recently attracted to it. However, HLS technology (more specifically, the tools that implement the synthesis process) has some weaknesses. For example, despite the fact that the designer can describe a modular and hierarchical implementation of a HWacc (i.e., grouping behavior via functions or procedures), all sub-modules are orchestrated by a global clock due to the way the translation to RTL from C is done. Another example is the rigid semantic when specifying dataflow architectures, allowing a reduced number of alternatives. This prevents the designer from obtaining optimal solutions for certain algorithms and problems [43,44], as was the case of the HyperLCA compressor. Therefore, to overcome the limitations of current HLS tools, a hybrid solution that combines modules developed using VHDL and HLS-synthesized C or C++ blocks has been selected. The result allows the achievement of the maximum performance. Otherwise, it could not be possible to realize either the synchronization mechanisms or the parallel processing proposed in this work. On top of that, this approach makes it also possible to optimize the necessary resources because of the use of custom producer-consumer data exchange patterns that are not supported by HLS tools.

The four main stages of the HyperLCA compressor, described in Section 2.1 and shown in Figure 1, have been restructured in order to better adapt to devices with a high degree of fine-grain parallelism such as FPGAs. Thus, the operations involved in these parts of the HyperLCA algorithm have been grouped in two main stages: *HyperLCA Transform* and *HyperLCA Coder*. These new stages can run in parallel, which improves the performance.

The *Initialization* stage (Section 2.1.2), where the calculation of the p_{max} parameter is done, is considered only at design time. This is because several hardware components, such as internal memories or FIFOs, must be configured with the appropriate size. In this sense, it is worth mentioning that the p_{max} value (Equation (1)) depends on other parameters also known at design time (the minimum desired compression ratio (*CR*), block size (*BS*) and the number of bits (N_{bits}) used for scaling the projection vectors, **V**, and, therefore, can be fixed for the HWacc.

2.2.1. HyperLCA Transform

HyperLCA Transform stage performs the operations of the HyperLCA transform itself (described in Section 2.1.3), and also the computation of the average pixel and the scaling of **V** vector. These are the most computational demanding operations of the algorithm and, together with the fact that they are highly parallelizable by nature, are good candidates for acceleration.

Figure 2 shows an overview of the hardware implementation of the *HyperLCA Transform* stage. *Avg_Cent*, *Brightness* and *Proj_Sub* modules have been modeled and implemented using the aforementioned HLS tools, while the memory buffers and custom logic that integrates and orchestrates all the components in the design have been instantiated and implemented using VHDL language, respectively. This HWacc has a single entry corresponding to a hyperspectral block ($\mathbf{M}_k$) that will be compressed, while the output is composed of three elements (which differs to the output of Algorithm 1) in order to achieve a high level of parallelism. It must also be mentioned that the output of the *HyperLCA Transform* stage feeds the input of the *HyperLCA Coder* stage that performs the error mapping and entropy coding in parallel. Thus, the centroid, $\hat{\mu}$, is obtained as depicted in Algorithm 1, while the p_{max} most different hyperspectral pixels, **E**, are not directly obtained. In this regard, the HWacc provides the indexes of such pixels, j_{max}, in each loop iteration of Algorithm 1 (outer loop), while the *HyperLCA Coder* is the responsible to obtaining each hyperspectral pixel, $\mathbf{e}_n$, from the external memory, in which is stored the hyperspectral image, to build the vector of most different hyperspectral pixels, **E**. Finally, the projection of pixels within each image block, **V**, is provided by the HWacc in a batch mode, i.e., each loop iteration (outer loop of Algorithm 1) obtains a projection (v_n) that forms part of the **V** vector.

The architecture of the *HLCA Transform* HWacc can be divided into two main modules: *Avg_Cent*, which corresponds to lines 2 and 3 of Algorithm 1, and *Loop_Iter* (the main loop body, lines from 5 to 14). These modules are connected by a bridge buffer (*BBuffer*), whose depth is only 32 words, and also share a larger buffer (*SBuffer*) with capacity to store a complete hyperspectral block. The size of the *SBuffer* depends on the *BS* algorithm parameter and its depth is determined at design time. The role of this *SBuffer* is to avoid the costly access to external memory, such it is the case of double data rate (DDR) memory in the Zynq-7000 SoCs.

The implementation of the *SBuffer* is based on a first-in, first-out (FIFO) memory that is written and read by different producers (i.e., *Avg* and *Brightness*) and consumers (i.e., *Cent* and *Projection*) of the original hyperspectral block and the intermediate results (transformations). Since there are more than one producer and consumer for the *SBuffer*, a dedicated synchronization and control access logic has been developed in VHDL (not illustrated in Figure 2). The use of a FIFO contributes to reduce the on-chip memory resources in the FPGA fabric, being its use feasible because of the linear pattern access of the producers and consumers. However, this type of solution would not have been possible with HLS because the semantic of stream-based communication between stages in a dataflow limits the number of producers and consumer to one. Also, it is not possible, with the used HLS technology, to exploit inter-loop parallelism as is done in the proposed solution. Notice that there is a data dependency between iterations in Algorithm 1 (centralized block, **C**) and, therefore, the HLS tool infers that the next iteration must wait for the previous one to finish, resulting in fact in sequential computation. However, a deeper analysis of the behavior of the algorithm shows that the computation of the brightest pixel for iteration $n + 1$ can be performed as it has received the output of the subtraction stage, which will be still processing iteration n.

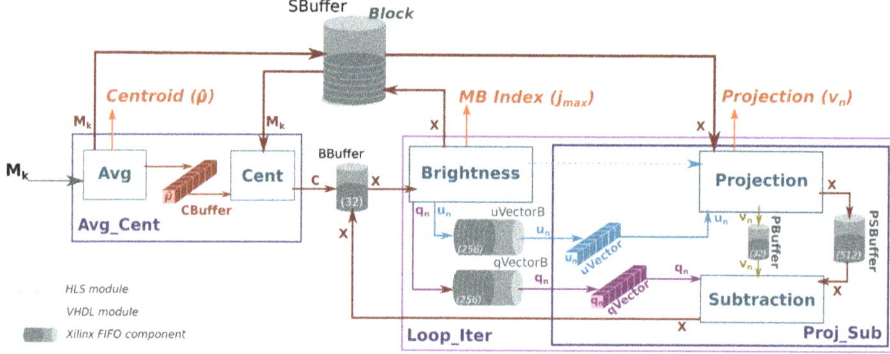

Figure 2. Overview of the *HyperLCA Transform* hardware accelerator. Light blue and white boxes represent modules implemented using HLS. Light red boxes and FIFOs represent glue logic and memory elements designed and instantiated using VHDL language.

The *Avg_Cent* module has been developed using HLS technology and contains two sub-modules, *Avg* and *Cent*, that implement lines 2 and 3 of Algorithm 1, respectively.

Avg This sub-module computes the centroid or average pixel, $\hat{\mu}$, of the original hyperspectral block, $\mathbf{M}_k$, following line 2 of Algorithm 1, and stores it in *CBuffer*, a buffer that shares with *Cent* sub-module. During this operation, *Avg* forwards a copy of the centroid to the *HyperLCA Coder* via a dedicated port (orange arrow). At the same time, the *Avg* sub-module writes all the pixels of the $\mathbf{M}_k$ into the *SBuffer*. A copy of the original hyperspectral block, $\mathbf{M}_k$, will be available once *Avg* finishes, ready to be consumed as a stream by *Cent*, which reduces the latency.

Figure 3 shows in detail the functioning of the *Avg* stage. The main problem in this stage is the way the hyperspectral data is stored. In our case, the hyperspectral block, $\mathbf{M}_k$, is ordered by the bands that make up a hyperspectral pixel. However, to obtain the centroid, $\hat{\mu}$, the hyperspectral block must be read by bands (in-width reading) instead of by pixels (in-depth reading). We introduce an optimization that handles the data as it is received (in-depth), avoiding the reordering of the data to maintain a stream-like processing. This optimization consists of an accumulate vector, whose depth is equal to the number of bands that stores partial results of the summation for each band, i.e., the first position of this vector contains the partial results of the first band, the second position the partial results of the second band and so on. The use of this vector removes the loop-carry dependency in the HLS loop that models the behavior of the *Avg* sub-module, saving processing cycles. The increase in resources is minimal, which is justified by the gain in performance.

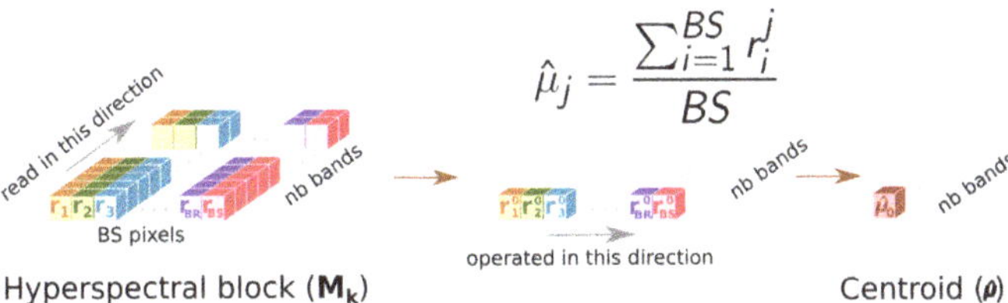

$$\hat{\mu}_j = \frac{\sum_{i=1}^{BS} r_i^j}{BS}$$

Hyperspectral block ($\mathbf{M_k}$)   Centroid ($\boldsymbol{\mu}$)

Figure 3. Overview of *Avg* stage.

Cent This sub-module reads the original hyperspectral block, $\mathbf{M}_k$, from the *SBuffer* to centralize it
(**C**). This operation consists of subtracting the average pixel, calculated in the previous stage,
from each hyperspectral pixel of the block (line 3 of Algorithm 1). Figure 4 shows this process,
highlighting the elements that are involved in the centralization of the first hyperspectral pixel. Thus,
the *Cent* block reads the centroid, $\hat{\mu}$, which is stored in the *CBuffer*, as many times hyperspectral
pixels have the original block (i.e., *BS* times in the example illustrated in Figure 4). Therefore,
CBuffer is an addressable buffer that permanently stores the centroid of the current hyperspectral
block that is being processed. The result of this stage is written into the *BBuffer* FIFO, which
makes unnecessary an additional copy of the centralized image, **C**. As soon as the centralized
components of the hyperspectral pixels are computed, the data is ready at the input of the *Loop_Iter*
module and, therefore, it can start to perform its operations without waiting for the block to be
completely centralized.

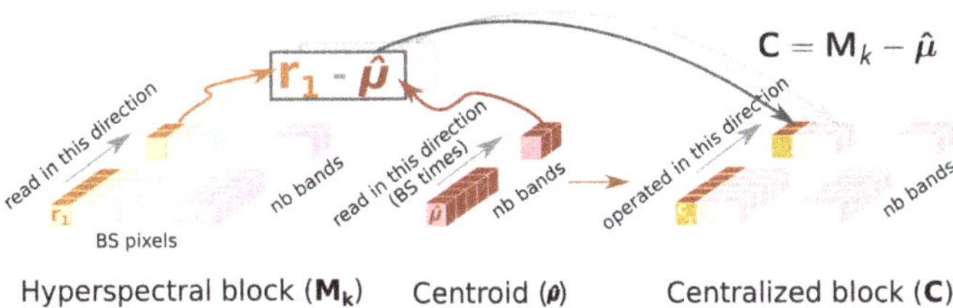

Hyperspectral block ($\mathbf{M_k}$)   Centroid ($\boldsymbol{\mu}$)   Centralized block (**C**)

Figure 4. Overview of *Cent* stage.

The *Loop_Iter* module instantiates the *Brightness* and *Proj_Sub* sub-modules which have been designed
and implemented using HLS technology. Both modules are connected by two FIFO components (*uVectorB*
and *qVectorB*) using customized VHDL code to link them. Unlike *Avg_Cent* module, *Loop_Iter* module is
executed several times (specifically p_{max} times, line 4 of Algorithm 1) for each hyperspectral block that is
being processed.

Brightness This sub-module starts as soon as there is data in the *BBuffer*. In this sense, *Brightness*
sub-module works in parallel with the rest of the system; the input of the *Brightness* module is the
output of *Cent* module in the first iteration, i.e., the centralized image, **C**, while the input for all other
iterations is the output of *Subtraction* sub-module that corresponds to the image for being subtracted,
depicted as **X** for the sake of clarity (see brown arrows in Figure 2).

Brightness sub-module has been optimized to achieve a dataflow behavior that takes the same time
regardless of the location of the brightest hyperspectral pixel. Figure 5 shows how the orthogonal
projection vectors $\mathbf{q}_n$ and $\mathbf{u}_n$ are obtained by the three sub-modules in *Brightness*. First, the *Get
Brightness* sub-module reads in order the hyperspectral pixel of the block from the *BBuffer* (**C** or **X**,
it depends on the loop iteration) and calculates its brightness (b_j) as specified in line 6 of Algorithm 1.

Get Brightness also makes a copy of the hyperspectral pixel in an internal buffer (*actual_pixel*) and in *SBuffer*. Thus, *actual_pixel* contains the current hyperspectral pixel whose brightness is being calculated, while *SBuffer* will contain a copy of the hyperspectral block with transformations (line 6 and assignment in line 13 of Algorithm 1).

Once the brightness of the current hyperspectral pixel is calculated, the *Update Brightness* sub-module will update the internal vector *brightness_pixel* if the current brightness is greater than the previous one. Regardless of such condition, the module will empty the content of *actual_pixel* in order to keep the dataflow with the *Get Brightness* sub-module. The operations of both sub-modules are performed until all hyperspectral pixels of the block are processed (inner loop, lines 5 to 7 of Algorithm 1). The reason to use a vector to store the brightest pixel instead of a FIFO is because the HLS tool would stall the dataflow otherwise.

Finally, the orthogonal projection vectors $\mathbf{q}_n$ and $\mathbf{u}_n$ are accordingly obtained from the brightest pixel (lines 10 and 11 of Algorithm 1) by the module *Build quVectors*. Both are written in separate FIFOs: *qVectorB* and *uVectorB*, respectively. Furthermore, the contents of these FIFOs are copied in *qVector* and *uVector* arrays in order to get a double space memory that does not deadlock the system and allows **Proj_Sub** sub-module to read several times (concretely *BS* times) the orthogonal projection vectors $\mathbf{q}_n$ and $\mathbf{u}_n$ to obtain the projected image vector, $\mathbf{v}_n$, and transform the current hyperspectral block. This module also returns the index of the brightest pixel, j_{max}, so that the *HyperLCA Coder* stage reads the original pixel from the external memory, such as DDR, where the hyperspectral image is stored in order to build the compressed bitstream.

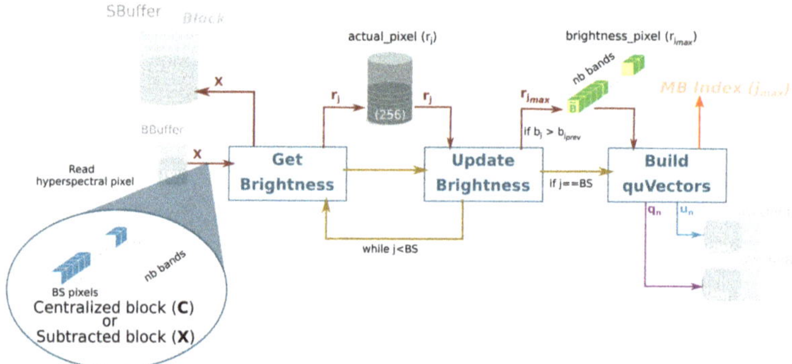

Figure 5. Overview of *Brightness* stage.

Proj_Sub Although this sub-module is represented by separate *Projection* and *Subtraction* boxes in Figure 2, it must be mentioned that both perform their computations in parallel. The *Proj_Sub* sub-module reads just once the hyperspectral block that was written in *SBuffer* by the *Brightness* sub-module (**X**). Figure 6 shows an example of *Projection* and *Subtraction* stages. First, each hyperspectral pixel of the block is read by the *Projection* sub-module to obtain the projected image vector according to line 12 of Algorithm 1. At the same time, the hyperspectral pixel is written in *PSBuffer*, which can store two hyperspectral pixels. Two is the number of pixels because the *Subtraction* stage begins right after the first projection on the first hyperspectral pixel is ready, i.e., the executions of *Projection* and *Subtraction* are shifted by one pixel. Figure 6 shows such behavior. While pixel $\mathbf{r}_1$ is being consumed by *Subtraction*, pixel $\mathbf{r}_2$ is being written in *PSBuffer*. During the projection of the second hyperspectral pixel ($\mathbf{r}_2$), the subtraction of the first one ($\mathbf{r}_1$) can be performed since all the input operands, included the projection $\mathbf{v}_n$, are available, following the expression in line 13 of Algorithm 1.

The output of the *Projection* sub-module is the projected image vector, $\mathbf{v}_n$, which is forwarded to the *HyperLCA Coder* accelerator (through the *Projection* port) and to the *Subtration* sub-module (via the *PBuffer* FIFO). At the same time, the output of the *Subtraction* stage feeds the *Loop_Iter* block (see purple arrow, labeled as **X**, in Figure 2) with the pixels of the transformed block in the ith iteration. It means that *Brightness* stage can start the next iteration without waiting to get the complete image subtracted. Thus, the initialization interval between loop-iterations is reduced as much as possible because the *Brightness* starts when the first subtracted data is ready.

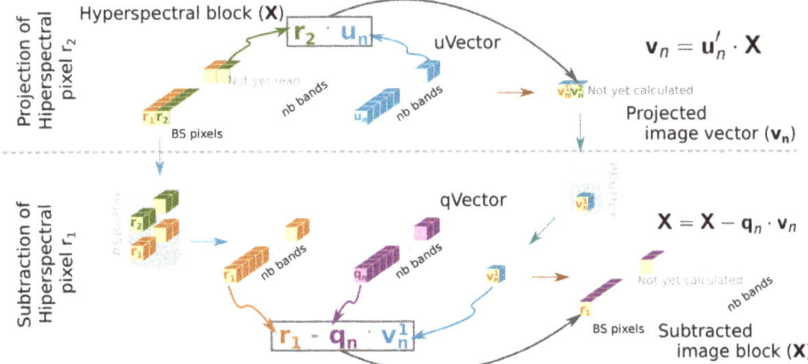

Figure 6. Example of Projection and Subtraction stages.

The FPGA-based solution described above highlights how the operations are performed in parallel to make the most out of such technology. Also, it has been spotted specific synchronization scenarios that are not possible to design using solely HLS. due to the current communication semantics supported by the synthesis tools. For example, is not possible for current synthesis tools to perform an analysis of the data and control dependencies such as the one done in this work. However, a hybrid solution based on HLS and hand-written VHDL code to glue the RTL models synthesized from C/C++ models, brings to life an efficient dataflow (see Figure 7). In this regard, the use of optimal sized FIFOs to interconnect the modules is key. For example, while the *Brightness* sub-module is filling the *SBuffer*, the *Projection* sub-module is draining it, and at the same time this sub-module supplies to the *Subtraction* sub-module with the same data read from *SBuffer*. Finally, *Subtraction* sub-module feeds back the *Brightness* sub-module through the *BBuffer* FIFO. The *Brightness* sub-module fills, in turn, the *SBuffer*, with the same data, closing the circle; this loop is repeated p_{max} times.

Furthermore, the initialization interval between image blocks has been reduced. The task performed by the *Avg* sub-module for block $k + 1$ (see Figure 7) can be scheduled at the same time that the *Projection* and *Subtraction* sub-modules are computing their outputs for block k, and right after the completion of the *Brightness* sub-module for block k. This is possible since the glue logic discards the output of the *Subtraction* sub-module during the last iteration. This logic ensures that the *BBuffer* is filled with the output from the *Avg* sub-module that feeds the first execution of *Brightness* for the first iteration of block $k + 1$, resulting in an overlapped execution of the computation for blocks k and $k + 1$.

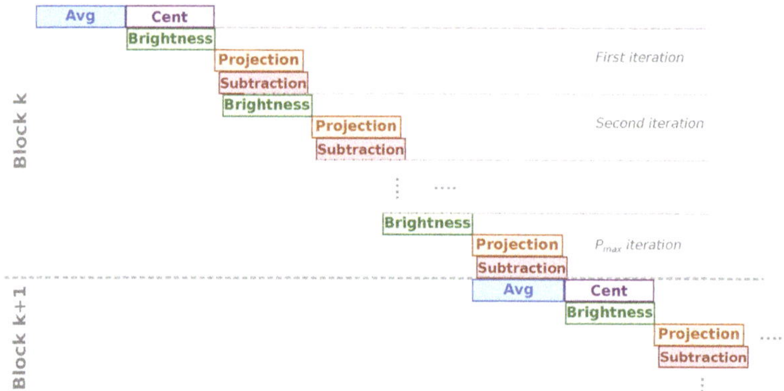

Figure 7. Dataflow of *HyperLCA Transform* hardware accelerator. Overlapping of operations within inter-loops and inter-blocks.

Unfortunately, despite the optimizations introduced in the dataflow architecture, the HWacc is not able to reach the performance target as shown by the experimental results (see Section 3). The value in the column labeled as 1 PE (Processing Element) is clearly below the standard frame rates provided by contemporary commercial hyperspectral cameras. However, the number of FPGA resources required by the one-PE version of the HWacc is very low (see Section 3.2) which makes it possible to implement further parallelism strategies to speed up the compression process. Thus, three options open up for solving the performance problem.

First, task-level parallelism approach is possible by means of the use of several instances of the HWacc working concurrently. Second, increase the intra-HWacc parallelism using multiple operators, computing several pixel bands at the same time. Thirdly, a combination of the two previous approached. Independently of the strategy chosen, the limiting factor is the version of the Xilinx Zynq-7000 programmable SoC, that would have enough resources to support the design. In Section 4, a detailed analysis of several single-HWacc (with variations in the number of PEs) and multi-HWacc versions of the design is drawn.

So far, it has been described the inner architecture of a HWacc that only performs a computational operation over a single band component of a hyperspectral pixel. However, it can be modified to increase the number of bands that are processed in parallel. Thus, the HWacc of HyperLCA compressor turns from a single-PE to multiple PEs. This fact opens two new challenges. First challenge is to increase the width of the input and output ports of the modules, in accordance with the number of bands that would be processed in parallel. It must be mentioned that it is technologically possible because HLS-based solutions allow designers to build their own data types. For example, if a band component of a hyperspectral pixel is represented by an unsigned integer of 16-bits, we could define our own data type consisting of an unsigned integer of 160-bits packing ten bands of a hyperspectral pixel (see Figure 8). The second challenge has to do with the strategy to process the data in parallel. In this regard, a solution based on the map-reduce programming model has been followed.

Figure 8 shows an example of the improvements applied to the *Cent* stage, following the above-mentioned optimizations. The input of this stage is the hyperspectral block, $\mathbf{M}_k$, and the average pixel, $\hat{\mu}$, which are read in blocks of N-bands. The example assumes that the block is composed of ten bands and uses an user data type, specifically an unsigned int of 160-bits (10 bands by 16-bits to represent each band). Then, both blocks are broken down into the individual components that feed the PEs in an orderly fashion. This process is also known as *map* phase in the map-reduce programming model [45]. It must be mentioned that the design needs as many PEs as number of divisions made in the block.

Once the PEs have performed the assigned computational operation, the *reduce* phase of the map-reduce model is executed. For *Cent* sub-module, this stage consists of gathering in a block the partial results produced by each one of the PEs. Thus, a new block of N-bands is built, which in turn is part of the centralized block (**C**), which is the output of *Cent* sub-module.

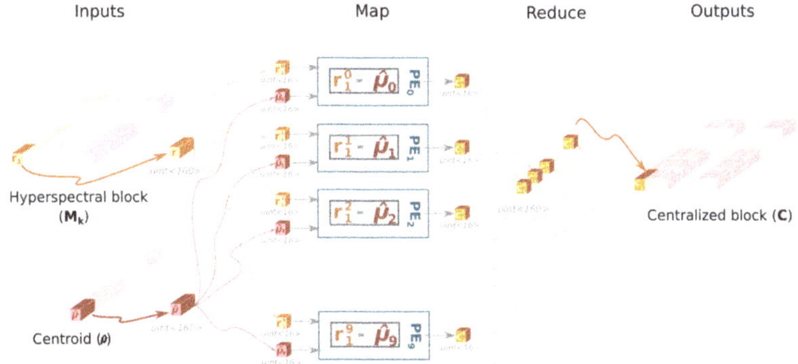

Figure 8. Map-reduce programming model and data packaging on *Cent* stage.

2.2.2. Coder

The *HyperLCA Coder* is the second of the HWacc developed, responsible for the error mapping and entropy-coding stages of the HyperLCA compression algorithm (Figure 1). The coder module teams up with the transform module to perform in parallel the CCSDS prediction error mapping [35] and the Golomb–Rice [36] entropy-coding algorithms as the different vectors are received from the *HyperLCA Transform* block.

The *HyperLCA transform* block generates the centroid, $\hat{\mu}$, extracted indexes of pixels, j_{max}, and projection vectors, $\mathbf{v}_n$, for an input hyperspectral block, $\mathbf{M}_k$. These arrays are consumed as they are received, reducing the need for large intermediate buffers. To minimize the necessary bandwidth to the memory that stores the hyperspectral image, only the indexes of the selected pixels in each iteration of the transform algorithm (line 8 of Algorithm 1) are provided to the coder (*MB index* port).

The operation of the transform and coder blocks overlaps in time. Since the coder takes approximately half of the time the transform needs to generate each vector (see Section 3.2) for the maximum number of PEs (i.e., the maximum performance achieved), a contention situation is not taking place, reducing the pressure over the FIFOs that connect both blocks and, therefore, requiring less space for these communication channels.

Figure 9 sketches the internal structure of the *HyperLCA Coder* that has been modeled entirely using Vivado HLS. It is a dataflow architecture comprising three steps. During the first step, the prediction mapping and entropy coding of all the input vectors are performed by the *coding command generator*. The result of this step is a sequence of commands that are subsequently interpreted by the *compressed bitstream generator*.

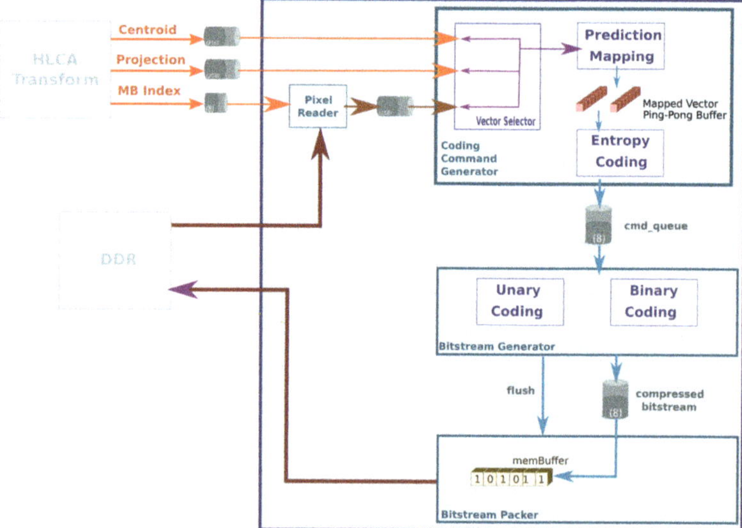

Figure 9. Overview of the *HyperLCA Coder* hardware accelerator.

The generation of the bitstream was extracted from the entropy-coding original functionality, which enabled a more efficient implementation of the latter, which could be re-written as a perfect loop and, therefore, Vivado HLS was able to generate a pipelined datapath with the minimum initiation internal (II = 1).

The generation of the compressed bitstream is simple. This module is continuously reading the *cmd_queue* FIFO for a new command to be processed. A command contains the operation (unary of binary coding) as well as the word (quotient or reminder, see Section 2.1.5) to be coded, and the number of bits to generate. Unary and binary coding functions simply iterate over the word to be coded and produces a sequence if bits which corresponds to the compressed form of the hyperspectral block.

Finally, the third step packs the compressed bitstream in words and written to memory. For this implementation, the width of the memory word is 64 bits, the maximum allowed by the *AXI Master* interface port for the Zynq-7020 SoC. The *bitstream packer* module instantiates a small buffer (64 words) that is flushed to DDR memory once it has been filled. This way, average memory access cycles per word is optimized by means of the use of burst requests.

As mentioned above, the *HyperLCA Transform* block feeds the coder with the indexes of the extracted pixels $\mathbf{e}_n$, which correspond to the highest brightness as the hyperspectral block is processed. Hence, it is necessary to retrieve the *nb* spectral bands from memory before the coder could start generating the compressed stream for the pixel vector. This is the role played by the *pixel reader* module. As in the case of the *bitstream packer* step, the *pixel reader* makes it use of a local buffer and issue burst requests to read the bands in the minimum number of cycles.

While the computing complexity of the coder module is low, the real challenge when it comes to the implementation of its architecture is to write a C++ HLS model that is consistent through the whole design, verification and implementation processes. To achieve this goal, it has been provided the communication channels with extra semantic so as to keep the different stages sync, despite the fact the model of computation of the architecture falls into the category of GALS (Globally Asynchronous, Locally Synchronous) systems. Side-channel information is embedded in the *cmd_queue* and *compressed_stream* FIFOs that connects the different stages of the coder. This information is used by the different modules to reset their internal states or stop and resume their operation (i.e., special commands to be interpreted by the bitstream generator or a flush signal as input to the packing module). This way, it is possible to integrate under a single HLS design all the functionality of the coder, which simplifies and speeds up the design process. On top of that, this strategy allowed avoiding the need to tailor the C++ HLS specification to make it usable in different steps of the design process. For example, it is common to write variations of the model depending on whether functional (C simulation) or RTL verification (co-simulation) is taking place due to the fact that the former is based on an untimed model and the latter introduces timing requirements [46,47].

Designing for parallelism is key to obtain the maximum performance. The dataflow architecture of the coder ensures an inter-module level of concurrency. However, the design must be balanced as to the latency of the different stages in the dataflow. Otherwise, the final result could be jeopardized. As mentioned before, decoupling the generation of the compressed bitstream from the entropy-coding logic, led to a more efficient implementation of the latter by the HLS synthesis tool. Also, this change helped to redistribute the computing effort, achieving a more balanced implementation.

In the first stage of the dataflow (*coding command generator*) a simple logic that controls the encoding of each input vector plus a header is implemented. It is an iterative process that performs the error mapping and error coding over the centroid, and p_max times over the extracted pixels and projections vectors. The bulk of this process is, thus, the encoding algorithm. The encoding is delegated in another module that implements an internal dataflow itself. In this way, it is possible to reduce the interval between two encoding operations. As can be seen in Figure 9, the prediction mapping and entropy-coding sub-modules communicates through a ping-pong buffer for the *mapped* vector.

To conclude this section, it is worth mentioning a couple of optimizations carried out related to the operations *pow* and *log*, which are used by the prediction mapping and entropy-coding algorithms. This type of arithmetic operation is costly to implement in FPGAs since the generated hardware is based on tables that consume on-chip resources (mainly BRAM memories), and an iterative processes that boosts latency. Since the base used in this application is 2, it can be largely simplified. Thus, we can substitute the *pow* and *log* operations by a logical shift instruction and the GCC compiler *__builtin_clz(x)* built-in function, respectively. This change is part of the refactoring process of the reference code implementation (golden model) that is almost mandatory at the beginning of any HLS project. The *__builtin_clz(x)* function is synthesizable and counts the leading zeros of the integer x. Therefore, the computation of the lowest power of 2 higher than M, performed during the entropy coding, is redefined as follows:

Listing 1: FPGA implementation of costly arithmetic operations during entropy coding.

```
1  //Original code
2  b = log2 (M) + 1;
3  difference =  pow(2,b) – M;
4
5  //FPGA optimization
6  b = (32 – __builtin_clz (M) );
7  difference = (1<<b) – M;
```

2.3. Reference Hyperspectral Data

In this section, we introduce the hyperspectral imagery used in this work to evaluate the performance of the proposed computing approach using reconfigurable logic devices. This data set is composed of 4 hyperspectral images that were also employed in [29], where the HyperLCA algorithm was implemented in low-power consumption embedded GPUs. We have kept the same data set in order to compare in Section 3 the performance of the developed FPGA-based solution with the results obtained by the GPU accelerators.

In particular, the test bench was sensed by the acquisition system extensively analyzed in [48]. This aerial platform mounts a *Specim FX10* pushbroom hyperspectral camera on a DJI Matrice 600 drone. The image sensor covers the range of the electromagnetic spectrum between 400 and 1000 nm using 1024 spatial pixels per scanned cross-track line and 224 spectral bands. Nevertheless, the hyperspectral images used in the experiments only retain the spectral information of 180 spectral bands. Concretely, the first 10 spectral bands and the last 34 bands have been discarded due to the low spectral response of the hyperspectral sensor at these wavelengths.

The data sets were collected over some vineyard areas in the center of Gran Canaria island (Canary Islands, Spain) and in particular, in a village called Tejeda, during two different flight campaigns. Figure 10 and Figure 11 display some Google Maps pictures of the scanned terrain in both flight missions, whose exact coordinates are 27°59′35.6″N 15°36′25.6″W (green point in Figure 10) and 27°59′15.2″N 15°35′51.9″W (red point in Figure 11), respectively. Both flight campaigns were performed at a height of 45 m over the ground, which results in a ground sampling distance in line and across line of approximately 3 cm.

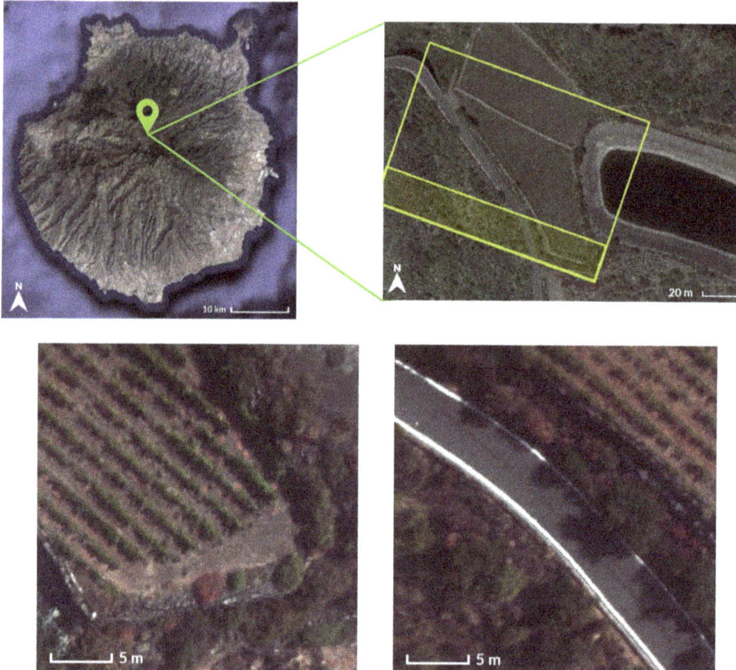

Figure 10. Google Maps pictures of the vineyard areas sensed during the first flight campaign. False RGB representations of the hyperspectral images employed in the experiments.

The former was carried out with a drone speed of 4.5 m/s and the camera frame rate set to 150 frames per second (FPS). In particular, this flight mission consisted of 12 waypoints that led to a total of 6 swathes, but only one was used for the experiments, which has been highlighted in green in Figure 10. Two portions of 1024 hyperspectral frames with all their 1024 hyperspectral pixels were selected from this swath to generate two of the data sets that compose the test bench. A closer view of these selected areas can be also seen in Figure 10. These images are false RGB representations extracted from the acquired hyperspectral data.

The latter was carried out with a drone speed of 6 m/s and the camera frame rate set to 200 FPS. The entire flight mission consisted of 5 swathes, but only one was used for the experiments in this work, which has been highlighted in red in Figure 11. From this swath, two smaller portions of 1024 hyperspectral frames were cut out for simulations. A closer view of these selected areas is also displayed in Figure 11. Once again, they are false RGB representations extracted from the acquired hyperspectral data. For more details about the flight campaigns, we encourage the reader to see [48].

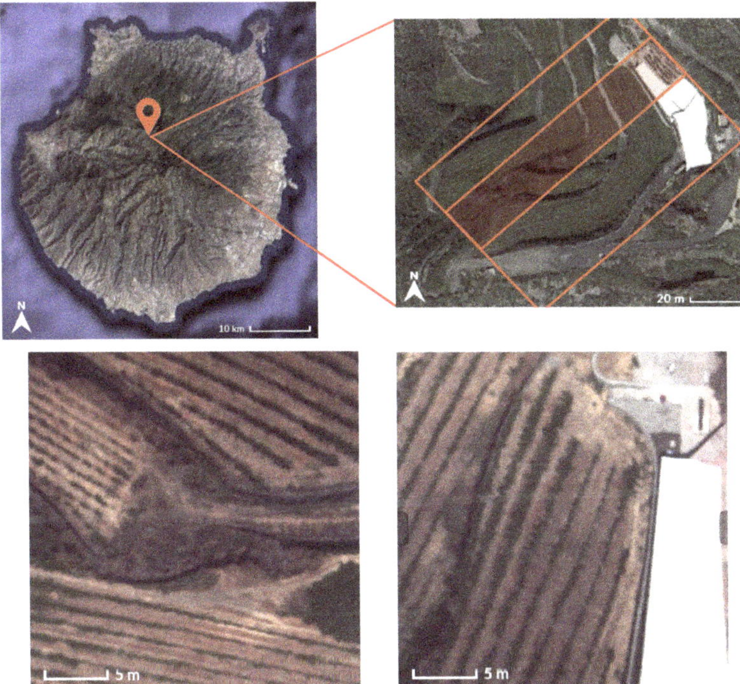

Figure 11. Google Maps pictures of the vineyard areas sensed during the second flight campaign. False RGB representations of the hyperspectral images employed in the experiments.

3. Results

3.1. Evaluation of the HyperLCA Compression Performance

The goodness of the I12 version of the HyperLCA algorithm proposed in this work has been evaluated and compared with previous I32 and I16 versions of the algorithm presented in [37] and the single precision floating-point (F32) implementation presented in [29]. For doing so, the hyperspectral imagery described in Section 2.3 has been compressed/decompressed using different settings of the HyperLCA compressor input parameters.

In this context, the information lost after the lossy compression process has been analyzed using three different quality metrics. Concretely, the *Signal-to-Noise Ratio (SNR)*, the *Root Mean Squared Error*

(*RMSE*) and the *Maximum Absolute Difference (MAD)*, which are shown in Equations (3)–(5), respectively. The *SNR* and the *RMSE* give an idea of the average information lost in the compression-decompression process. Bigger values of *SNR* are indicative of better compression performance. On the contrary, higher *RMSE* values mean that the lossy compression has introduced bigger data losses. The *MAD* assesses the amount of lost information for the worst reconstructed image value. For our targeted application, the dynamic range is $2^{12} = 4096$ and hence, the worst possible *MAD* value is 4095. For the sake of clarity, the aforementioned metrics have been calculated using the entire compressed–decompressed images, i.e., after the HyperLCA algorithm has finished to compress all image blocks, $\mathbf{M}_K$.

$$\text{SNR} = 10 \cdot \log_{10}\left(\frac{\sum_{i=1}^{nb}\sum_{j=1}^{np}(I_{i,j})^2}{\sum_{i=1}^{nb}\sum_{j=1}^{np}(I_{i,j} - Ic_{i,j})^2}\right) \tag{3}$$

$$\text{RMSE} = \frac{1}{np \cdot nb} \cdot \sqrt{\sum_{i=1}^{nb}\sum_{j=1}^{np}(I_{i,j} - Ic_{i,j})^2} \tag{4}$$

$$\text{MAD} = \max(I_{i,j} - Ic_{i,j}) \tag{5}$$

Table 2 shows the average results obtained for each configuration of the HyperLCA compressor input parameters using the data set described in Section 2.3. Different conclusions can be drawn from these results. First of all, it is confirmed that the I12 version offers significantly better-quality compression results than previous I16 version, employing the same hardware resources. These gaps are even wider for smaller compression ratios. This is because the losses introduced by the decrease in the data precision compare to the previous I16 version, as mentioned in Section 2.1.6, is disguised by the bigger losses introduced by the compression process itself for higher compression ratios.

Table 2. Comparison of the compression results for the four versions of the HyperLCA algorithm under study: I12, I16, I32 and F32.

Nbits	BS	CR	SNR				MAD				RMSE			
			I12	I16	I32	F32	I12	I16	I32	F32	I12	I16	I32	F32
12	1024	12	43.01	38.01	43.12	42.75	24.50	39.00	25.00	25.50	3.12	5.55	3.08	3.22
		16	42.27	38.57	42.27	41.97	32.25	41.50	32.25	32.50	3.40	5.21	3.40	3.52
		20	41.31	39.13	41.31	41.06	41.25	46.50	41.25	41.25	3.73	4.88	3.80	3.91
	512	12	42.99	38.95	43.03	42.68	26.25	34.50	25.00	25.00	3.13	4.98	3.12	3.24
		16	42.45	39.36	42.47	42.14	30.75	38.00	30.00	30.50	3.33	4.75	3.33	3.45
		20	41.50	39.92	41.48	41.24	39.50	43.75	39.50	40.00	3.72	4.46	3.72	3.83
	256	12	43.03	40.00	43.02	42.67	25.00	34.50	25.00	25.50	3.12	4.42	3.12	3.25
		16	42.33	40.51	42.32	42.02	34.75	37.25	34.75	35.00	3.38	4.16	3.38	3.50
		20	40.94	40.59	40.59	40.73	54.75	57.75	54.75	54.75	3.96	4.13	3.97	4.06
8	1024	12	41.80	36.91	42.19	41.68	23.25	41.50	22.00	22.25	3.62	6.32	3.47	3.69
		16	41.63	37.42	41.73	41.27	25.50	41.25	25.50	26.50	3.69	5.95	3.65	3.86
		20	41.20	37.89	41.20	40.79	32.25	42.00	32.50	32.25	3.87	5.64	3.87	4.07
	512	12	42.49	37.99	42.70	42.26	23.00	38.00	22.75	22.25	3.33	5.57	3.25	3.42
		16	42.07	38.64	42.09	41.69	27.50	36.00	26.50	26.75	3.49	5.17	3.48	3.65
		20	41.61	39.01	41.60	41.25	31.25	39.50	30.50	31.75	3.68	4.95	3.68	3.84
	256	12	42.79	39.35	42.82	42.39	24.75	35.50	25.00	25.00	3.20	4.76	3.19	3.36
		16	42.15	39.96	42.13	41.76	30.00	37.00	29.75	29.75	3.45	4.43	3.46	3.61
		20	41.80	40.21	41.78	41.44	35.75	37.00	35.75	36.00	3.59	4.31	3.60	3.74

Additionally, deviations in the values of the three quality metrics employed in this work between I32, I12 and F32 versions are almost negligible, with the advantage of halving the memory space required for storing **C**. Second, it can be also concluded that the HyperLCA lossy compressor is able to compress the hyperspectral data with high compression ratios and without introducing significant spectral information losses.

3.2. Evaluation of the HyperLCA Hardware Accelerator

Section 2.1 describes the FPGA-based implementation of the HyperLCA compressor as defined in Algorithm 1. The architecture of the HWacc is divided into two blocks, *Transform* and *Coder* that run in parallel following a producer-consumer approach. Therefore, for the performance analysis of the whole solution, the slowest block is the one determining the productivity of the proposed architecture.

The *HyperLCA Transform* block bears most of the complexity and computational burden of the compression process. For that reason, several optimizations (see Section 2.1.3) have been applied during its design in order to achieve a high degree of parallelism and, thus, reduce the latency. One of the most important improvements is the realization of the map-reduce programming model, to enable an architecture with multiple PEs (see Figure 8) working concurrently on several bands. The experiments carried out over the different alternatives for the *HyperLCA Transform* block are intended to evaluate how the performance and resource usage of the FPGA-based solution scales, as the number of PEs instantiated by the architecture grows up.

The configuration of the input parameters has been set as follows: *CR* parameter has been set to 12, 16 and 20; the *BS* parameter has been set to 1024, 512 and 256 and; N_{bits} parameter gets the values 12 and 8. The value of p_{max} is obtained at design time from these parameters following Equation (1), and are listed in the last column of Table 3. Concerning the sizing of the various memory elements present in the architecture, it is determined by two parameters: the number of PEs to instantiate (parallel processing of hyperspectral bands) and the size of the image block to be compressed. It is worth mentioning that in this version of the HWacc, the number of PEs must be a divisor of the number of hyperspectral bands in order to simplify the design of the datapath logic.

First, the data width of the architecture must be defined. Such parameter is obtained multiplying the number of PEs by the size of the data type used to represent a pixel band. In the version of the HWacc under evaluation, the I12 alternative has been selected, due to its good behavior (comparable to the I32 version as discussed in Section 2.1.6) and the resource savings it brings. For the I12 version, a band is represented with an *unsigned short int* which turns into 16-bit words in memory. On the contrary, choosing the I32 version, the demand for memory resources and internal buffers (such as the *SBuffer*) would double, because an *unsigned int* data type is used in the model definition. Thus, if the HWacc only instantiates a PE, the data width will be 16-bits, whereas if the HWacc instantiates 12 PEs (i.e., the HWacc performs 12 operations over the set of bands in parallel), the data width will be 192-bits. Second, the depth of the *SBuffer* must be calculated following Equation (6), where *BS* is the block size, *nb* the number of bands (in our case is fixed to 180), and N_{PEs} is the number of processing elements.

$$SBufferDepth_{min} = \frac{BS \cdot nb}{N_{PEs}} \tag{6}$$

Being optimal as to the use of BRAM blocks within the FPGA fabric is compulsory since this resource is highly demanded as the number of PEs increases (seen in Table 4). On top of that, keeping the use of resources under control brings along some benefits such as helping the synthesis tool to generate a better datapath and, therefore, to obtain an implementation with a shorter critical path or contain the consumption of energy.

Table 3. Design-time configurations parameters of the HyperLCA transform hardware block for hyperspectral images with 180 bands.

N_{bits}	BS	CR	Min SBuffer Depth						p_{max}
			1 PE	2 PEs	4 PEs	6 PEs	10 PEs	12 PEs	
12	1024	12							12
		16	184,320	92,160	46,080	30,720	18,432	15,360	9
		20							7
	512	12							10
		16	92,160	46,080	30,720	18,432	15,360	7680	8
		20							6
	256	12							8
		16	46,080	30,720	18,432	15,360	7680	3840	6
		20							4
8	1024	12							17
		16	184,320	92,160	46,080	30,720	18,432	15,360	13
		20							10
	512	12							14
		16	92,160	46,080	30,720	18,432	15,360	7680	10
		20							8
	256	12							10
		16	46,080	30,720	18,432	15,360	7680	3840	7
		20							6
Data Width (bits)			16	32	64	96	160	192	

Table 3 lists the memory requirements demanded by *SBuffer* for the 108 different configurations of the *HyperLCA Transform* block evaluated in this work. The *SBuffer* component is, by far, the largest memory instantiated by the architecture and it is implemented as a FIFO. *SBuffer* component has been generated with the FIFO generator tool provided by Xilinx which only allows depths that are power of two. Therefore, from a technical point of view, it is not possible to match the minimum required space of *SBuffer* with the obtained from the vendor tools. To mitigate the waste of memory space derived from such constraint of the tool, the *SBuffer* has been broken into two concatenated FIFOs (i.e., the output of the first FIFO is the input of the second) but keeping the facade of a single FIFO to the rest of the system. For example, the minimum depth of *SBuffer* for $PE = 1$ and $BS = 256$ is 46,080. With a single FIFO, the smallest depth with enough capacity generated by Xilinx tools would be 65,536. Therefore, the unused memory space represents approximately 30% of the overall resources for *SBuffer*. However, by using the double FIFO approach, one FIFO of 32,768 words plus another one of 16384 would be use. Only $\approx 6\%$ of the assigned resources to *SBuffer* would be misused.

To evaluate the *HyperLCA Transform* hardware accelerator, the proposed HWacc architecture has been implemented using the Vivado Design suite. This toolchain is provided by Xilinx and features a HLS tool (Vivado HLS) devoted to optimize the developing process of IP (*Intellectual Property*) components FPGA-based solutions for their own devices. The first implemented prototype instantiated one HWacc targeting the XC7Z020-CLG484 version of the Xilinx Zynq-7000 SoC. This FPGA has been selected because of its low-cost, low-weight and high flexibility, features that make it an interesting device to be integrated in aerial platforms, such as drones. The aim of this first prototype is to evaluate the capability of a mid-range reconfigurable FPGAs such as the XC7Z020 chip, for a specific application such as the HyperLCA compression algorithm. Hence, and due to the amount of resources available on the target device, the maximum possible number of PEs for the single-HWacc prototype is 12.

Table 4 summarizes the resources required, which have been extracted from post-synthesis reports, for each of the 108 versions of the HWacc that process different block sizes of 180-band hyperspectral images of the data set (see Section 2.3).

Table 4. Post-Synthesis results for the different versions of the HyperLCA Transform for a Xilinx Zynq-7020 programmable SoC and image block up to 180 bands.

BS	Num. PEs	BRAM18K		DSP48E		FlipFlops		LUTs	
	1	56	*(20.0%)*	9	*(4.09%)*	6942	*(6.52%)*	5194	*(9.76%)*
	2	59	*(21.07%)*	16	*(7.27%)*	9112	*(8.56%)*	8735	*(16.42%)*
256	4	71	*(25.36%)*	30	*(13.64%)*	20,171	*(18.96%)*	15,811	*(29.72%)*
	6	71	*(25.36%)*	62	*(28.18%)*	29,368	*(27.6%)*	23,530	*(44.23%)*
	10	84	*(30.0%)*	102	*(46.36%)*	47,350	*(44.5%)*	38,221	*(71.84%)*
	12	71	*(25.36%)*	122	*(55.45%)*	56,297	*(52.91%)*	45,867	*(86.22%)*
	1	101	*(36.07%)*	9	*(4.09%)*	6988	*(6.57%)*	5270	*(9.91%)*
	2	102	*(36.43%)*	16	*(7.27%)*	9159	*(8.61%)*	8820	*(16.58%)*
512	4	113	*(40.36%)*	30	*(13.64%)*	20,215	*(19.0%)*	16,040	*(30.15%)*
	6	113	*(40.36%)*	62	*(28.18%)*	29,406	*(27.64%)*	23,761	*(44.66%)*
	10	124	*(44.29%)*	102	*(46.36%)*	47,420	*(44.57%)*	38,352	*(72.09%)*
	12	113	*(40.36%)*	122	*(55.45%)*	56,342	*(52.95%)*	45,857	*(86.20%)*
	1	192	*(68.57%)*	9	*(4.09%)*	7114	*(6.69%)*	5468	*(10.28%)*
	2	190	*(67.86%)*	16	*(7.27%)*	9278	*(8.72%)*	8969	*(16.86%)*
1024	4	198	*(70.71%)*	30	*(13.64%)*	20,301	*(19.08%)*	16,210	*(30.47%)*
	6	199	*(71.07%)*	62	*(28.18%)*	29,524	*(27.75%)*	23,916	*(44.95%)*
	10	204	*(72.86%)*	102	*(46.36%)*	47,514	*(44.66%)*	38,485	*(72.34%)*
	12	199	*(71.07%)*	122	*(55.45%)*	56,463	*(53.07%)*	46,116	*(86.68%)*

Several conclusions can be derived from these figures. In first place, the amount of digital signal processors (DSPs), flipflops (FFs) and look-up-tables (LUTs) resources increases with the number of PEs but are similar for different values of the BS parameter. On the contrary, the demand of BRAMs depends directly on BS and increases slightly with PE for a given block size. The $PE = 10$ version needs a special remark. Such version represents an anomaly in the linear behavior of the resource demand. Even with the use of a double FIFO approach, as explained before, the total capacity of the BRAM used to instantiate *SBuffer* is clearly oversized for that datawidth to assure that a hyperspectral block and its transformations ($\mathbf{M}_k$ and $\mathbf{C}$, respectively) could be stored in-circuit. Second, in addition to the resources needed by the HWacc of the *HyperLCA Transform*, it is necessary to take into account those corresponding to the other components in the system such as the *Coder* or the DMA (*Direct Memory Access*) to move the hyperspectral data ($\mathbf{M}_k$) from/to DDR to/from the hardware accelerators. These extra components will make use of the remaining resources (specially LUTs, which is the most demanded as can be seen in Table 4), establishing a maximum of 12 for the number of PEs.

Table 5 shows the post-synthesis results for the *HyperLCA Coder* block. The resources demanded by the coder does not depend on the BS parameter or the number of PEs. It is important to mention that the majority of the BRAM, FFs and LUTs resources are assigned to the two AXI-Memory interfaces that the HLS tool generates for the *Pixel Reader* and the *Bitstream Packer* modules (see Figure 9).

Table 5. Post-Synthesis results for the HyperLCA Coder block for a Xilinx Zynq-7020 programmable SoC and pixel size up to 180 bands.

BS	BRAM18K		DSP48E		FlipFlops		LUTs	
2,565,121,024	7	*(2.5%)*	1	*(0.45%)*	3464	*(3.25%)*	4106	*(7.71%)*

Table 6 shows the throughput, expressed as the maximum frame rate, for a specific configuration of the HWacc using two clock frequencies: 100 MHz and 150 MHz (Table A1 extends this information by adding the number of cycles to compute a hyperspectral block). Columns labeled as *PE* denote the number of *processing elements* instantiated by the HWacc, which in combination with the input parameters (N_{bits}, BS, CR and p_{max}), show the average number of hyperspectral blocks that the HWacc is able to compress per second (FPS). The maximum frame rate has been normalized to 1024 hyperspectral pixels per block, which is the size of the frame delivered by the acquisition system.

Table 6. Maximum frame rate obtained using the FPGA-based solution on a Xilinx ZynQ-7020 programmable SoC for hyperspectral images with 180 bands.

N_{bits}	BS	CR	Max Frame Rate											
			1 PE		2 PEs		4 PEs		6 PEs		10 PEs		12 PEs	
			f_1	f_2	f_1	f_2	f_1	f_2	f_1	f_2	f_1	f_2	f_1	f_2
12	1024	12	37	56	72	108	131	196	178	266	248	370	275	411
		16	47	71	91	137	167	250	225	336	310	463	343	511
		20	58	87	112	168	204	305	273	408	373	555	410	611
	512	12	43	65	83	125	152	228	205	307	284	424	314	469
		16	52	78	100	151	183	273	245	366	336	501	370	552
		20	69	98	126	189	229	342	304	454	411	612	452	672
	256	12	52	78	100	149	181	270	242	361	330	493	364	543
		16	65	97	125	187	227	339	301	449	405	604	444	661
		20	87	130	167	251	303	453	397	591	523	778	570	847
8	1024	12	27	41	53	79	96	144	131	197	186	278	207	310
		16	34	52	67	100	122	183	166	248	232	347	258	386
		20	43	65	84	126	153	229	206	309	286	427	317	473
	512	12	32	49	62	94	114	170	155	232	217	324	241	360
		16	43	65	83	125	152	227	205	307	284	424	370	469
		20	52	78	100	151	183	273	245	366	336	501	370	552
	256	12	43	65	83	124	150	225	202	303	279	417	309	460
		16	57	86	111	166	201	301	268	401	364	543	400	596
		20	65	97	125	187	227	339	301	448	405	603	444	661

f_1 Clock frequency: 100 MHz. f_2 Clock frequency: 150 MHz.

It is worth noting that these results include the transform and coding steps, which are performed in parallel. Table 7 shows the coding times (in clock cycles) compared to the time needed by the transform step with the best configuration possible (i.e., $PE = 12$). The *HyperLCA Coder* takes roughly 50% less time on average and, since the relation between both hardware components is a dataflow architecture, the latency of the whole process corresponds to the maximum; that is, the delay of the *Transform* step.

One key factor is the minimum frame rate that must be supported for the targeted application. Ideally, such threshold would correspond to the maximum frame rate provided by the employed hyperspectral sensor (i.e., 330 FPS). However, the experimental validation of the camera set-up in the drone (Section 2.3), tells us that frame rates between 150 and 200 are enough to obtain hyperspectral images with the desired quality, given the speed and altitude of the flights. Therefore, a threshold value of 200 FPS is established as the minimum throughput to validate the viability of the HyperLCA hardware core. In Table 7, it has been highlighted (bold type-faced cells) the configurations that would be valid given this minimum. Thus, it can be observed that the $PE = 12$ version, for both clock frequencies, and the $PE = 10$ version at 150 MHz reach the minimum frame rate, even for the most demanding scenario ($N_{bits} = 8$, $BS = 1024$ and $CR = 12$).

In turn, by using lower values for N_{bits} the compression rate is reduced due to the higher number of **V** vectors extracted by the HyperLCA Transform. This means that more computations must be performed to compress a hyperspectral block. It must be mentioned that the $PE = 10$ version meets the FPS target in all the scenarios but the most demanding one when the clock frequency is set to 100 MHz. As for the $PE = 10$ version, the $PE = 6$ version does not reach the minimum FPS in a few scenarios (concentrated in $N_{bits} = 8$ and $BS = 1024$), but in can be a viable solution given the actual needs of the application set-up.

Table 7. Comparison of the computation effort made by the *Transform* and *Coding* stages.

| N_{bits} | BS | CR | Cycles per Block | | p_{max} |
			Transform Delay (12 PEs)	Coder Delay	
		12	364,594	184,999	12
	1024	16	293,101	137,379	9
		20	245,360	105,836	7
12		12	159,901	88,235	10
	512	16	135,691	70,468	8
		20	111,546	52,593	6
		12	69,030	44,095	8
	256	16	56,649	33,294	6
		20	44,265	22,587	4
		12	483,832	194,170	17
	1024	16	388,391	147,055	13
		20	316,731	111,919	10
8		12	208,129	94,584	14
	512	16	159,834	67,697	10
		20	135,748	54,037	8
		12	81,374	44,579	10
	256	16	62,850	31,424	7
		20	56,658	27,163	6

In addition to the performance results listed in Table 6, Figure 12 graphically shows the speed-up gained by the FPGA-based implementation as the number of PEs increases. The values have been normalized, using the average time for the $PE = 1$ version as the baseline. Several conclusions can be drawn from this figure. First, it is observable that the $PE = 12$ version of the HWacc performs $\times 7$ times ($N_{bits} = 12$) to $\times 7.6$ times ($N_{bits} = 8$) faster than $PE = 1$ version. This configuration is the one that guarantees the fastest compression results. Second, the speed-up gain is nearly linear for $PE = 2$ and $PE = 4$ versions, whereas the scalability of the accelerator drops as the number of PEs is higher (see Figure 12a,b). This behavior is seen for both $N_{bits} = 8$ and $N_{bits} = 12$ configurations (see Figure 12c,d). However, for higher values of BS and CR the shape of the curve shows a better trend though not reaching the desired linear speed-up.

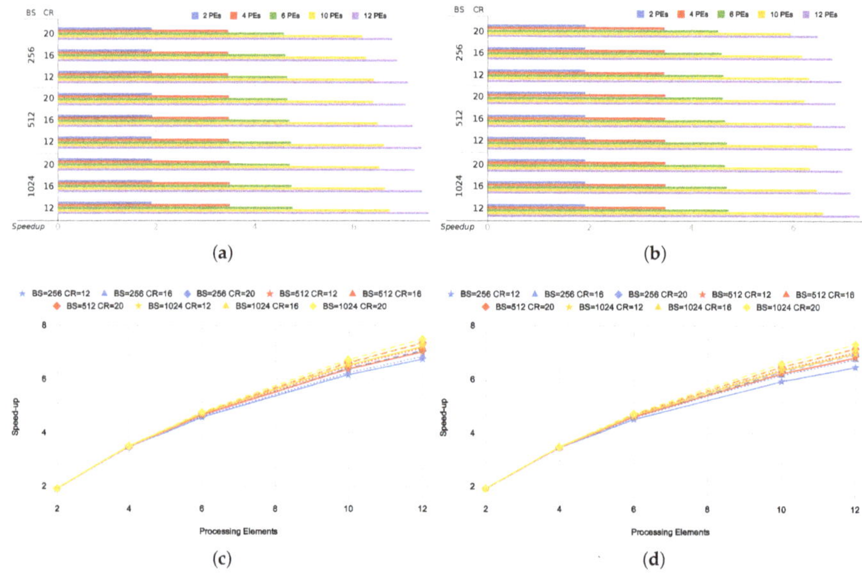

Figure 12. Speed-up obtained for multiple PEs compared to the 1 PE version of the HyperLCA HW compressor. (**a**) Speed-up $N_{bits} = 8$; (**b**) Speed-up $N_{bits} = 12$; (**c**) Speed-up curve $N_{bits} = 8$; (**d**) Speed-up curve $N_{bits} = 12$.

4. Discussion

In this paper, we present a detailed description of an FPGA-based implementation of the HyperLCA algorithm, a new lossy transform-based compressor. As fully discussed in Section 3.2, the proposed HWacc meets the requirements imposed by the targeted application in terms of the frame rate range employed to capture quality hyperspectral images. In this section, we would like to also provide a comprehensive analysis of the suitability of the FPGA-based HyperLCA compressor implementation and a comparison with the results obtained by an embedded System-on-Module (SoM) based on a GPU that has been recently published [29].

Concretely, María Díaz et al. introduce in [29] three implementation models of the HyperLCA compressor. In this previous work, the parallelism inherent to the HyperLCA algorithm was exploited beyond the thread-level concurrency of the GPU programming model taking advantage of the CUDA steams. In particular, the third approach, referred to as *Parallel Model 3* in [29], achieves the highest speed-up, especially for bigger image blocks ($BS = 1024$). This implementation model bets for pipelining the data transfers between the host and the device and the kernel executions for the different image blocks ($\mathbf{M}_k$). To do this, such proposal exploits the benefits of the concurrent kernel execution through the management of CUDA streams. Unlike the FPGA-based implementation model proposed in this paper, only the *HyperLCA Transform* stage is accelerated in the GPU. In this case, the codification stage is also pipelined with the *HyperLCA Transform* stage but executed on the Host using another parallel CPU process. Table 2 collects the quality results of the compression process issued by the aforementioned GPU-based implementation model in terms of SNR, MAD and RMSE (F32 version). For more details, we encourage the reader to see [29].

To compare the performance of the GPU-based implementation model and the FPGA solution presented here, we are going to use two assessment metrics: the maximum number of frames compressed in a second (FPS) and the power efficiency in terms of FPS per watt. The latter figure of merit is of great importance given the target application, since it is critical to maximize the battery life of the drone. Additionally, the GPU-based model introduced in [29] has been executed in three different NVIDIA Jetson embedded computing devices: Jetson Nano, Jetson TX2 and the most recent supercomputer Jetson Xavier NX.

These modules have been selected for the reasonable computational power provided at a relatively low power consumption. Table 8 summarizes the most relevant technical characteristics of these embedded computing boards. As can be seen, the Jetson Nano module integrates the less advanced, oldest generation of the three GPU architectures, instantiating the fewer execution units or CUDA cores as well. On the contrary, Jetson Xavier NX represents one of the latest NVIDIA power-efficient products, which offers more than 10X the performance of its widely adopted predecessor, Jetson TX2.

Table 8. Most relevant characteristics of the NVIDIA modules Jetson Nano, Jetson TX2 and Jetson Xavier NX.

	Jetson Nano
LPGPU	GPU NVIDIA Maxwell architecture with 128 NVIDIA CUDA cores
CPU	Quad-core ARM Cortex-A57 MPCore processor
Memory	4 GB LPDDR4, 1600MHz 25.6 GB/s
	Jetson TX2
LPGPU	GPU NVIDIA Pascal with 256 CUDA cores
CPU	HMP Dual Denver 2/2 MB L2 + Quad ARM A57/2 MB L2
Memory	8 GB LPDDR4, 128 bits bandwidth, 59.7 GB/s
	Jetson Xavier NX
LPGPU	GPU NVIDIA Volta with 384 NVIDIA CUDA cores and 48 Tensor cores
CPU	6-core NVIDIA Carmel ARM®v8.2 64-bit CPU 6 MB L2 + 4 MB L3
Memory	8 GB 128-bit LPDDR4x @ 51.2GB/s

Table 9 collects the performance results obtained for the three GPU-based implementations of the *HyperLCA* and the most powerful implementation of the HWacc in a Zynq-7020 SoC ($PE = 12$). For each implementation, it is specified the clock frequency and power budget. Several algorithm parameters have been tested over 180-band input images. In addition, Figure 13 displays the obtained FPS according to different configurations of the input image block size (BS). For the sake of simplicity, we only represent results for $N_{bits} = 8$ since the behavior is similar for $N_{bits} = 12$ as the reader can see in Table 9.

From the performance point of view, the most competitive FPGA results, compared to the fastest GPU implementations on Jetson Nano and Jetson TX2, are for the smallest input block size ($BS = 256$), resulting very similar to $BS = 512$ and even better compared to Jetson Nano. For the largest block size ($BS = 1024$) the performance of the solutions based on GPUs, is higher because of how the parallelism is inferred in both architectures. On the one hand, the FPGA architecture can process 12 bands at a time, overlapping the computation of several groups of pixels due to the internal pipeline architecture. Therefore, processing time linearly increases as the number of pixels in the image block is higher. On the other hand, GPUs can process all the pixels in an image block in parallel, regardless the size of the block. Thus, processing time is nearly constant, independently of the value of parameter BS. However, this assumption does not hold when it comes to reality, since it has to be taken in consideration the time required for memory transfers, kernel launches and data dependencies. In this context, the time spent to setting up and launching the instructions to execute a kernel or perform a memory transfer must be also taken into account. As analyzed in [29], the time used in transferring image blocks of 256 pixels is negligible in relation to the overhead of launching the memory transfers and the additional required logic. However, the time required for transferring image blocks of 1024 pixels is comparable with the overhead of initializing the copy. Consequently, the bigger the size of the block, the better performance obtained by GPU-based implementation. For this reason, the trend of the FPGA performance function (see blue line in Figure 13) decreases as BS increases, while the opposite behavior is shown for the GPU-based model regardless of the desired CR.

Table 9. Maximum frame rates obtained by the proposed FPGA implementation and the GPU-based model introduced in [29] for the NVIDIA boards Jetson Nano, Jetson TX2 and Jetson Xavier NX.

			Max Frame Rate				
			FPGA		Jetson Nano	GPU Jetson TX2	Jetson Xavier NX
N_{bits}	BS	CR	XC7Z020-CLG484 (MicroZed)				
			100 MHz 3.12 W	150 MHz 3.74 W	921.6 MHz 10 W	1.12 GHz 15 W	800 MHz 10 W
12	1024	12	275	411	533	558	2062
		16	343	511	638	762	2422
		20	410	611	729	923	2682
	512	12	315	469	351	506	1405
		16	371	553	408	599	1582
		20	452	672	487	748	1767
	256	12	365	543	225	372	909
		16	445	662	275	465	1052
		20	570	847	357	620	1251
8	1024	12	207	310	421	452	1730
		16	258	386	511	568	2020
		20	317	473	606	700	2294
	512	12	241	360	276	384	1149
		16	371	469	350	511	1409
		20	371	552	409	596	1574
	256	12	309	461	192	309	794
		16	401	662	249	414	973
		20	444	663	276	463	1053

This pattern is also present when analyzing the results for the Jetson Xavier NX. However, in this case, the GPU clearly outperforms the maximum number of FPS achieved by the FPGA for all algorithm settings. Nevertheless, it should be noticed by the reader that the Jetson Xavier NX represents one of the latest, most advanced NVIDIA single-board computers whereas the Xilinx Zynq-7020 SoC that mounts the ZedBoard (i.e., XC7Z020-CLG484) is a mid-range FPGA several technological generations behind the Jetson Xavier GPU. Despite the fact that there are more powerful FPGA devices currently on the market, one of the main objectives of this work is to assess the feasibility of the reconfigurable logic technology for high-performance embedded applications such as HyperLCA under real-file conditions. At the same time, it is also a goal of this work to explore the minimum requirements of an FPGA-based computing platform that is able to fulfil the performance demands and constraints of the hyperspectral application under study. Thus, the selected version of the Xilinx Zynq-7020 SoC meets the demand for all algorithm configurations, at a lower cost.

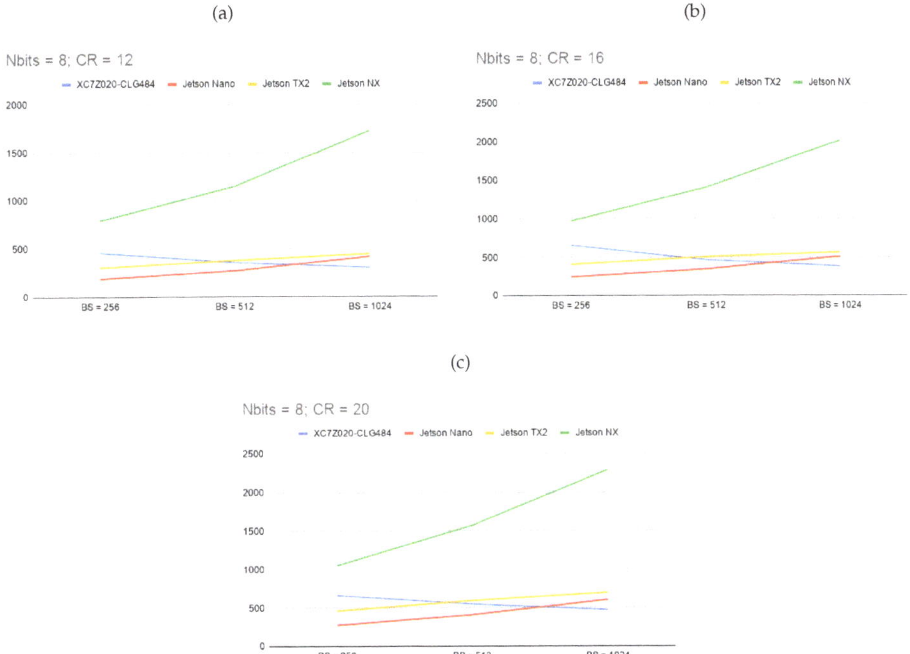

Figure 13. Comparison of the speed-up obtained in the compression process, in terms of FPS and the input parameter *BS*, reached by a Xilinx Zynq-7020 programmable SoC following the design-flow proposed in this work versus the GPU-based implementation model described in [29] performed onto some NVIDIA power-efficient embedded computing devices, such as Jetson Nano, Jetson TX2 and Jetson Xavier NX. (**a**) FPS $N_{bits} = 12$, $CR = 12$; (**b**) FPS $N_{bits} = 12$, $CR = 16$; (**c**) FPS $N_{bits} = 12$, $CR = 20$.

Going deeper in the analysis of the results, Figure 14 plots the power efficiency for each targeted device, measured as the FPS achieved divided by the average power budget. The picture shows how the efficiency varies in relation to the size of the input image blocks (*BS*). Jetson boards are designed with a high-efficient Power Management Integrated Circuit that handles voltage regulators, and a power tree to optimize power efficiency. According to [49–51], the typical power budgets of the selected boards amount to 10 W, 15 W and 10 W for the Jetson Nano, Jetson TX2 and Jetson Xavier NX modules, respectively. In the case of the XC7Z020-CLG484 FPGA, the estimated power consumption after *Place & Route* stage in Vivado toolchain goes up to 3.74 W at 150 MHz. Based on the trend lines shown in Figure 14, it can be concluded that the FPGA-based platform is by far more efficient in terms of power consumption that the Jetson Nano and TX2 NVIDIA boards, for all algorithm configurations. As in the case of the performance analysis (Figure 13), the power efficiency of the FPGA slightly decreases with higher *BS* values while GPU-based implementations present an opposite behavior. As a result, the FPGA-based solution remains a more power-efficient approach for the smallest image block size (*BS* = 256) and shows similar figures for *BS* = 512 and higher *CR*. Nevertheless, for *BS* = 1024, Jetson Xavier NX clearly outperforms the proposed FPGA-based solution.

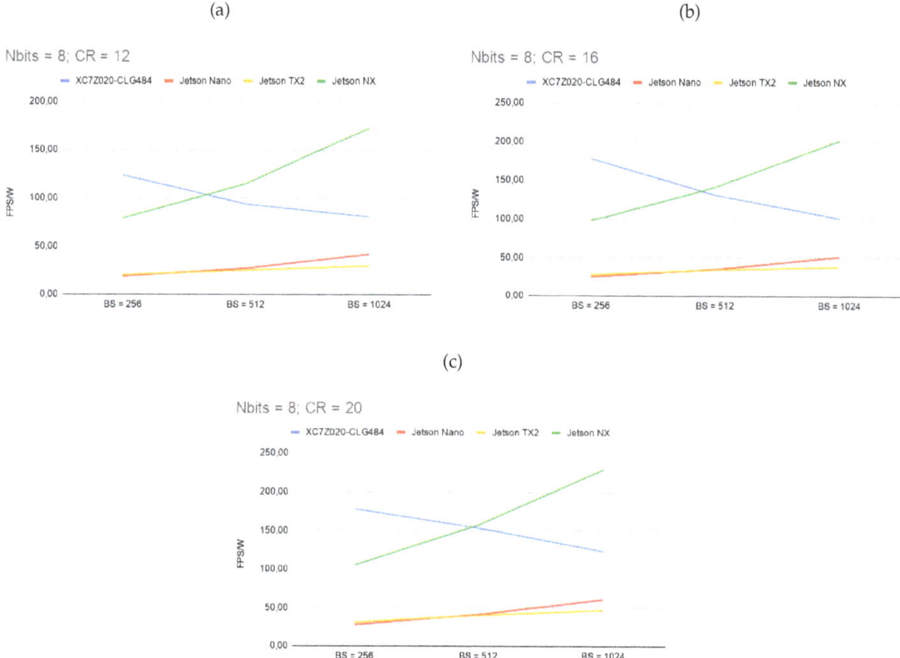

Figure 14. Comparison of the energy efficiency in the compression process, in terms of the ratio between obtained FPS and power consumption and the input parameter *BS*, reached by a Xilinx Zynq-7020 programmable SoC following the design-flow proposed in this work versus the GPU-based implementation model described in [29] performed onto some NVIDIA power-efficient embedded computing devices, such as Jetson Nano, Jetson TX2 and Jetson Xavier NX. (**a**) FPS $N_{bits} = 12$, $CR = 12$; (**b**) FPS $N_{bits} = 12$, $CR = 16$; (**c**) FPS $N_{bits} = 12$, $CR = 20$.

The reasons that explain this behavior root in the fact that GPU-embedded platforms have been able to significantly increase their performance while maintaining or even reducing the power demand. The combination of architectural improvements and better IC manufacturing processes have paved the way to an scenario where embedded-GPU platforms are gaining ground and can be seen as competitors of FPGAs concerning power efficiency.

Although the initial FPGA solution has proved to be sufficient, given the real-life requirements of the targeted application, we have ported the proposed design to a larger FPGA device. The objective is two-fold. First, to evaluate if it is possible to reach the same level of performance (in terms of FPS) than that obtained by the Jetson Xavier NX implementation with current FPGA technology. Second, to study how FPGA power efficiency evolves as the complexity of the design increases, and compare it to the results obtained by the Jetson Xavier NX.

Thus, a multi-HWacc version of the design was developed using the XC7Z100-FFV1156-1 FPGA, one of the biggest Xilinx Zynq-7000 SoCs, as the target platform. The new FPGA allows up to 5 instances of the *HyperLCA* component working in parallel. The selected baseline scenario is the configuration where the single-core FPGA design obtained the worst results compared to the Jetson Xavier NX (i.e., $BS = 1024$

and $Nbits = 8$ for all CR values). Synthesis and implementation was carried out by means of Vivado toolchain using the *Flow_AreaOptimized_high* and *Power_ExploreArea* strategies, respectively. As can be seen in Figure 15a, the performance of the multi-HWacc version grows almost linearly with the number of instances. There is a loss due to the concurrent access to memory in order to get the hyperspectral frames and the necessary synchronization of the process, which is the responsibility of the software. The new multi-core FPGA-based *HyperLCA* computing platform can reach the same level of performance for the maximum number of instances. As to the efficiency in terms of energy consumption per frame processed by the collection of HWaccs, with just three instances the FPGA is comparable to the GPU (above the Jetson Xavier NX results for $CR = 16$ and $CR = 20$) and better for four and five instances of the HWacc.

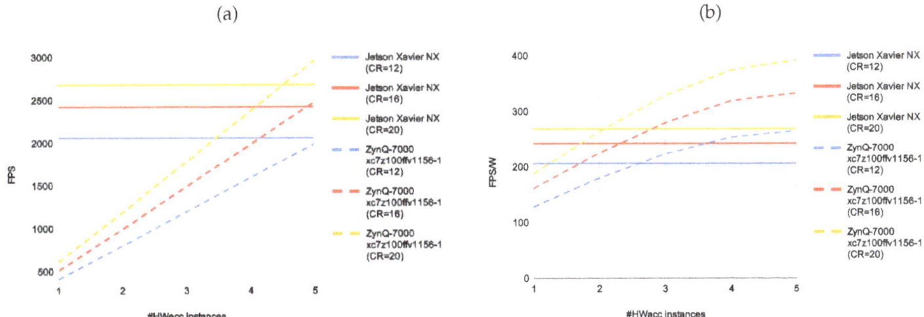

Figure 15. Evolution of the performance (**a**) and energy efficiency (**b**) of a multi-core version of the FPGA-based *HyperLCA* computing platform. Comparison with the GPU-based implementation model described in [29] performed onto NVIDIA Jetson Xavier NX ($N_{bits} = 8$, $BS = 1024$, using a Xilinx Zynq-7000 XC7Z100-FFV1156-1).

5. Conclusions

The suitability of the HyperLCA algorithm for being executed using integer arithmetic was already examined in further detail in previous state-of-the-art publications. Nonetheless, in this work we have contributed to its optimization providing a performance-enhancing alternative that has brought about a substantial performance improvement along with a significant reduction in hardware resources, especially aimed at overcoming the scarcity of in-chip memory, one of the weakness of FPGAs.

In this context, the aforementioned modified version of the HyperLCA lossy compressor has been implemented onto an heterogeneous Zynq-7000 SoC in pursuit of accelerating its performance and thus, complying with the requirements imposed by a UAV-based sensing platform that mounts a hyperspectral sensor, which is characterized by a high frame rate. The adopted solution combines modules using VHDL and synthesized HLS models, bringing to life an efficient dataflow that fulfils the real-life requirements of the targeted application. On this basis, the designed HWaccs are able to reach frame rates of compressed hyperspectral image blocks higher than 330 FPS, setting the baseline scenario in 200 FPS, using a small number of FPGA resources and low power consumption.

Remote Sens. **2020**, *12*, 3741

Additionally, we also provide a comprehensive comparison in terms of energy efficiency and performance between the FPGA-based implementation developed in this work and a state-of-the-art GPU-based model of the algorithm on three low-power NVIDIA computing boards, namely Jetson Nano, Jetson TX2 and Jetson Xavier NX. Conclusions drawn from the discussion show that although the FPGA-based platform is by far more efficient in terms of power consumption than the oldest-generation NVIDIA boards, such as the Jetson Nano and the Jetson TX2, the newest embedded-GPU platforms, such as the Jetson Xavier NX, are gaining ground and can be seen as competitors of FPGAs concerning power efficiency.

On account of that, we have also introduced a multi-HWacc version of the developed FPGA-based approach in order to analyze its evolution in terms of performance and power consumption when the number of accelerators increases in a larger FPGA. Results conclude that the new multi-core FPGA-based version can reach the same level of performance as the most efficient embedded GPU systems. Also, looking at the energy consumption, FPGA performance per watt is comparable from just three instances of the HWaccs.

Finally, we would like to conclude that although the work described in this manuscript has been focused on a UAV-based application, it can be easily extrapolated to other work in the space domain. In this regard, FPGAs have been established as the mainstream solution for on-board remote-sensing applications due to their smaller power consumption and above all, the accessibility to radiation-tolerant FPGAs [6]. That is why the FPGA-based model proposed in this manuscript efficiently implements all HyperLCA compression stages in the programmable logic (PL) of the SoC, that is the FPGA. Hence, it can easily be adapted to be performed on other space-grade certified FPGAs.

Author Contributions: Investigation, J.C. and J.B.; Methodology, M.D. and R.G.; Software, M.D. and R.G.; Supervision, J.B.; Validation, J.C.; Writing—original draft, J.C., M.D., J.B. and S.L.; Writing—review & editing, J.C., M.D., J.B., J.A.d.l.T. and S.L. All authors have read and agreed to the published version of the manuscript.

Funding: This research is partially funded by the Ministry of Economy and Competitiveness (MINECO) of the Spanish Government (PLATINO project, no. TEC2017-86722-C4, subprojects 1 and 4), the Agencia Canaria de Investigación, Innovación y Sociedad de la Información (ACIISI) of the Conserjería de Economía, Industria, Comercio y Conocimiento of the Gobierno de Canarias, jointly with the European Social Fund (FSE) (POC2014-2020, Eje 3 Tema Prioritario 74 (85%)), and the Regional Government of Castilla-La Mancha (SymbIoT project, no. SBPLY-17-180501-000334).

Conflicts of Interest: The authors declare no conflict of interest.

Appendix A

Table A1. Evaluation of the results obtained using the FPGA-based solution on a Xilinx Zynq-7020 programmable SoC for hyperspectral images with 180 bands.

N_{bits}	BS	CR	Cycles per Block (Max Frame Rate)												p_{max}
			1 PE		2 PEs		4 PEs		6 PEs		10 PEs		12 PEs		
			100 MHz	150 MHz	100 MHz	150 MHz	100 MHz	150 MHz	100 MHz	150 MHz	100 MHz	150 MHz	100 MHz	150 MHz	
12	1024	12	2,666,775		1,388,085		763,648		563,553		404,613		364,594		12
			37	56	72	108	131	196	178	266	248	370	275	411	
		16	2,093,523		1,088,779		599,742		445,886		323,820		293,101		9
			47	71	91	137	167	250	225	336	310	463	343	511	
		20	1,711,355		889,312		490,447		367,439		269,887		245,360		7
			58	87	112	168	204	305	273	408	373	555	410	611	
	512	12	1,145,407		596,547		328,928		244,157		176,829		159,901		10
			43	65	83	125	152	228	205	307	284	424	314	469	
		16	953,703		496,418		273,941		204,570		149,589		135,691		8
			52	78	100	151	183	273	245	366	336	501	370	552	
		20	761,999		396,125		218,877		165,082		122,367		111,546		6
			69	98	125	189	229	342	304	454	411	612	452	672	
	256	12	479,335		250,030		138,409		103,597		76,005		69,030		8
			52	78	100	149	181	270	242	361	330	493	364	543	
		16	382,863		199,499		110,479		83,400		62,056		56,649		6
			65	97	125	187	227	339	301	449	405	604	444	661	
		20	286,391		148,931		82,691		63,382		48,200		44,265		4
			87	130	167	251	303	453	397	591	523	778	570	847	

Table A1. *Cont.*

N_{bits}	BS	CR	1 PE		2 PEs		4 PEs		6 PEs		10 PEs		12 PEs		p_{max}
			100 MHz	150 MHz	100 MHz	150 MHz	100 MHz	150 MHz	100 MHz	150 MHz	100 MHz	150 MHz	100 MHz	150 MHz	
	1024	12	3,622,195		1,886,981		1,036,739		759,669		539,143		483,832		17
			27	41	53	79	96	144	131	197	186	**278**	**207**	**310**	
		16	2,857,859		1,487,981		818,201		602,802		431,461		388,391		13
			34	52	67	100	122	183	166	**248**	**232**	**347**	**258**	**386**	
		20	2,284,607		1,188,572		654,337		485,088		350,730		316,731		10
			43	65	84	126	153	**229**	**206**	**309**	**286**	**427**	**317**	**473**	
8	512	12	1,528,815		797,038		438,876		323,247		231,359		208,129		14
			32	49	62	94	114	170	155	**232**	**217**	**324**	**241**	**360**	
		16	1,145,407		596,571		328,949		244,100		176,809		159,834		10
			43	65	83	125	152	**227**	**205**	**307**	**284**	**424**	**370**	**469**	
		20	953,703		496,495		274,002		204,592		149,566		135,748		8
			52	78	100	151	183	**273**	**245**	**366**	**336**	**501**	**370**	**552**	
	256	12	575,808		300,575		166,183		123,677		89,908		81,374		10
			43	65	83	124	150	**225**	**202**	**303**	**279**	**417**	**309**	**460**	
		16	431,099		224,781		124,409		93,463		68,997		62,850		7
			57	86	111	166	**201**	**301**	**268**	**401**	**364**	**543**	**400**	**596**	
		20	382,863		199,486		110,540		83,520		62,093		56,658		6
			65	97	125	187	**227**	**339**	**301**	**448**	**405**	**603**	**444**	**661**	

References

1. Plaza, A.; Benediktsson, J.A.; Boardman, J.W.; Brazile, J.; Bruzzone, L.; Camps-Valls, G.; Chanussot, J.; Fauvel, M.; Gamba, P.; Gualtieri, A.; et al. Recent advances in techniques for hyperspectral image processing. *Remote. Sens. Environ.* **2009**, *113*, S110–S122. [CrossRef]
2. Noor, N.R.M.; Vladimirova, T. Integer KLT design space exploration for hyperspectral satellite image compression. In *International Conference on Hybrid Information Technology*; Springer: Berlin, Germany, **2011**, pp. 661–668.
3. Radosavljević, M.; Brkljač, B.; Lugonja, P.; Crnojević, V.; Trpovski, Ž.; Xiong, Z.; Vukobratović, D. Lossy Compression of Multispectral Satellite Images with Application to Crop Thematic Mapping: A HEVC Comparative Study. *Remote Sens.* **2020**, *12*, 1590. [CrossRef]
4. Villafranca, A.G.; Corbera, J.; Martín, F.; Marchán, J.F. Limitations of hyperspectral earth observation on small satellites. *J. Small Satell.* **2012**, *1*, 19–29.
5. Valentino, R.; Jung, W.S.; Ko, Y.B. A Design and Simulation of the Opportunistic Computation Offloading with Learning-Based Prediction for Unmanned Aerial Vehicle (UAV) Clustering Networks. *Sensors* **2018**, *18*, 3751. [CrossRef] [PubMed]
6. Lopez, S.; Vladimirova, T.; Gonzalez, C.; Resano, J.; Mozos, D.; Plaza, A. The promise of reconfigurable computing for hyperspectral imaging onboard systems: A review and trends. *Proc. IEEE* **2013**, *101*, 698–722. [CrossRef]
7. George, A.D.; Wilson, C.M. Onboard processing with hybrid and reconfigurable computing on small satellites. *Proc. IEEE* **2018**, *106*, 458–470. [CrossRef]
8. Fu, S.; Chang, R.; Couture, S.; Menarini, M.; Escobar, M.; Kuteifan, M.; Lubarda, M.; Gabay, D.; Lomakin, V. Micromagnetics on high-performance workstation and mobile computational platforms. *J. Appl. Phys.* **2015**, *117*, 17E517. [CrossRef]
9. Ortenberg, F.; Thenkabail, P.; Lyon, J.; Huete, A. Hyperspectral sensor characteristics: Airborne, spaceborne, hand-held, and truck-mounted; Integration of hyperspectral data with Lidar. *Hyperspectral Remote Sens. Veg.* **2011**, *4*, 39–68.
10. Board, N.S.; Council, N.R. *Autonomous Vehicles in Support of Naval Operations*; National Academies Press: Washington DC, USA, 2005.
11. Gómez, C.; Green, D.R. Small unmanned airborne systems to support oil and gas pipeline monitoring and mapping. *Arab. J. Geosci.* **2017**, *10*, 202. [CrossRef]
12. Bioucas-Dias, J.M.; Plaza, A.; Camps-Valls, G.; Scheunders, P.; Nasrabadi, N.; Chanussot, J. Hyperspectral remote sensing data analysis and future challenges. *IEEE Geosci. Remote Sens. Mag.* **2013**, *1*, 6–36. [CrossRef]
13. Keymeulen, D.; Aranki, N.; Hopson, B.; Kiely, A.; Klimesh, M.; Benkrid, K. GPU lossless hyperspectral data compression system for space applications. In Proceedings of the 2012 IEEE Aerospace Conference, Big Sky, MT, USA, 3–10 March 2012; pp. 1–9.
14. Huang, B. *Satellite Data Compression*; Springer Science & Business Media: New York, NY, USA, 2011.
15. Consultative Committee for Space Data Systems (CCSDS). Image Data Compression. CCSDS, Green Book 120.1-G-2. Available online: https://public.ccsds.org/Pubs/120x1g2.pdf (accessed on 10 July 2020).
16. Motta, G.; Rizzo, F.; Storer, J.A. *Hyperspectral Data Compression*; Springer Science & Business Media: New York, NY, USA, 2006.
17. Penna, B.; Tillo, T.; Magli, E.; Olmo, G. Transform coding techniques for lossy hyperspectral data compression. *IEEE Trans. Geosci. Remote Sens.* **2007**, *45*, 1408–1421. [CrossRef]
18. Marcellin, M.W.; Taubman, D.S. JPEG2000: image compression fundamentals, standards, and practice. In *International Series in Engineering and Computer Science, Secs 642*; Springer: New York, NY, USA, 2002.
19. Chang, L.; Cheng, C.M.; Chen, T.C. An efficient adaptive KLT for multispectral image compression. In Proceedings of the 4th IEEE Southwest Symposium on Image Analysis and Interpretation, Austin, TX, USA, 2–4 April 2000; pp. 252–255.

20. Hao, P.; Shi, Q. Reversible integer KLT for progressive-to-lossless compression of multiple component images. In Proceedings of the 2003 International Conference on Image Processing (Cat. No. 03CH37429), Barcelona, Spain, 14–17 September 2003; Volume 1, pp. 1–633.

21. Abrardo, A.; Barni, M.; Magli, E. Low-complexity predictive lossy compression of hyperspectral and ultraspectral images. In Proceedings of the 2011 IEEE International Conference on Acoustics, Speech and Signal Processing (ICASSP), Prague, Czech Republic, 22–27 May 2011; pp. 797–800.

22. Kiely, A.B.; Klimesh, M.; Blanes, I.; Ligo, J.; Magli, E.; Aranki, N.; Burl, M.; Camarero, R.; Cheng, M.; Dolinar, S.; et al. The new CCSDS standard for low-complexity lossless and near-lossless multispectral and hyperspectral image compression. In Proceedings of the 2018 Onboard Payload Data Compression Workshop, Matera, Italy, 20–21 September 2018; pp. 1–7.

23. Auge, E.; Santalo, J.; Blanes, I.; Serra-Sagrista, J.; Kiely, A. Review and implementation of the emerging CCSDS recommended standard for multispectral and hyperspectral lossless image coding. In Proceedings of the 2011 First International Conference on Data Compression, Communications and Processing, Palinuro, Italy, 21–24 June 2011; pp. 222–228.

24. Augé, E.; Sánchez, J.E.; Kiely, A.B.; Blanes, I.; Serra-Sagrista, J. Performance impact of parameter tuning on the CCSDS-123 lossless multi-and hyperspectral image compression standard. *J. Appl. Remote Sens.* **2013**, *7*, 074594. [CrossRef]

25. Santos, L.; Magli, E.; Vitulli, R.; López, J.F.; Sarmiento, R. Highly-parallel GPU architecture for lossy hyperspectral image compression. *IEEE J. Sel. Top. Appl. Earth Obs. Remote Sens.* **2013**, *6*, 670–681. [CrossRef]

26. Barrios, Y.; Sánchez, A.J.; Santos, L.; Sarmiento, R. SHyLoC 2.0: A Versatile Hardware Solution for On-Board Data and Hyperspectral Image Compression on Future Space Missions. *IEEE Access* **2020**, *8*, 54269–54287. [CrossRef]

27. Santos, L.; López, J.F.; Sarmiento, R.; Vitulli, R. FPGA implementation of a lossy compression algorithm for hyperspectral images with a high-level synthesis tool. In Proceedings of the 2013 NASA/ESA Conference on Adaptive Hardware and Systems (AHS-2013), Torino, Italy, 24–27 June 2013; pp. 107–114.

28. Guerra, R.; Barrios, Y.; Díaz, M.; Santos, L.; López, S.; Sarmiento, R. A New Algorithm for the On-Board Compression of Hyperspectral Images. *Remote Sens.* **2018**, *10*, 428. [CrossRef]

29. Díaz, M.; Guerra, R.; Horstrand, P.; Martel, E.; López, S.; López, J.F.; Roberto, S. Real-Time Hyperspectral Image Compression Onto Embedded GPUs. *IEEE J. Sel. Top. Appl. Earth Obs. Remote Sens.* **2019**, 1–18. [CrossRef]

30. Guerra, R.; Santos, L.; López, S.; Sarmiento, R. A new fast algorithm for linearly unmixing hyperspectral images. *IEEE Trans. Geosci. Remote Sens.* **2015**, *53*, 6752–6765. [CrossRef]

31. Díaz, M.; Guerra, R.; López, S.; Sarmiento, R. An algorithm for an accurate detection of anomalies in hyperspectral images with a low computational complexity. *IEEE Trans. Geosci. Remote Sens.* **2017**, *56*, 1159–1176. [CrossRef]

32. Díaz, M.; Guerra, R.; Horstrand, P.; López, S.; Sarmiento, R. A Line-by-Line Fast Anomaly Detector for Hyperspectral Imagery. *IEEE Trans. Geosci. Remote Sens.* **2019**, pp. 8968–8982. [CrossRef]

33. Diaz, M.; Guerra Hernández, R.; Lopez, S. A Novel Hyperspectral Target Detection Algorithm For Real-Time Applications With Push-Broom Scanners; In Proceedings of the 10th Workshop on Hyperspectral Imaging and Signal Processing: Evolution in Remote Sensing (WHISPERS), Amsterdam, The Netherlands, 24–26 September 2019; pp. 1–5. [CrossRef]

34. Díaz, M.; Guerra, R.; Horstrand, P.; López, S.; López, J.F.; Sarmiento, R. Towards the Concurrent Execution of Multiple Hyperspectral Imaging Applications by Means of Computationally Simple Operations. *Remote Sens.* **2020**, *12*, 1343. [CrossRef]

35. Consultative Committee for Space Data Systems (CCSDS). Blue Books: Recommended Standards. Available online: https://public.ccsds.org/Publications/BlueBooks.aspx (accessed on 11 March 2019).

36. Howard, P.G.; Vitter, J.S. Fast and Efficient Lossless Image Compression. In Proceedings of the DCC '93: Data Compression Conference, Snowbird, UT, USA, 30 March–2 April 1993; pp. 351–360.

37. Guerra Hernández, R.; Barrios, Y.; Diaz, M.; Baez, A.; Lopez, S.; Sarmiento, R. A Hardware-Friendly Hyperspectral Lossy Compressor for Next-Generation Space-Grade Field Programmable Gate Arrays. *IEEE J. Sel. Top. Appl. Earth Obs. Remote Sens.* **2019**, pp. 1–17. [CrossRef]

38. Oberstar, E.L. Fixed-point representation & fractional math. *Oberstar Consult.* **2007**, *9*. Available online: https://www.superkits.net/whitepapers/Fixed%20Point%20Representation%20&%20Fractional%20Math.pdf (accessed on 15 March 2019).

39. Hu, C.; Feng, L.; Lee, Z.; Davis, C.; Mannino, A.; Mcclain, C.; Franz, B. Dynamic range and sensitivity requirements of satellite ocean color sensors: Learning from the past. *Appl. Opt.* **2012**, *51*, 6045–6062. [CrossRef] [PubMed]

40. Kumar, A.; Mehta, S.; Paul, S.; Parmar, R.; Samudraiah, R. Dynamic Range Enhancement of Remote Sensing Electro-Optical Imaging Systems. In Proceeding of the Symposium at Indian Society of Remote Sensing, Bhopal, India, 2012.

41. Xilinx Inc. *UltraFast Vivado HLS Methodology Guide*; Technical Report; Xilinx Inc.: San Jose, CA, USA, 2020.

42. Cong, J.; Liu, B.; Neuendorffer, S.; Noguera, J.; Vissers, K.; Zhang, Z. High-Level Synthesis for FPGAs: From Prototyping to Deployment. *IEEE Trans. Comput.-Aided Design Integr. Circuits Syst.* **2011**, *30*, 473–491. [CrossRef]

43. Bailey, D.G. The Advantages and Limitations of High Level Synthesis for FPGA Based Image Processing. In Proceedings of the 9th International Conference on Distributed Smart Cameras, September 2015; pp. 134–139. [CrossRef]

44. de Fine Licht, J.; Meierhans, S.; Hoefler, T. Transformations of High-Level Synthesis Codes for High-Performance Computing. *arXiv* **2018**, arXiv:1805.08288

45. Dean, J.; Ghemawat, S. MapReduce: Simplified Data Processing on Large Clusters. *Commun. ACM* **2008**, *51*, 107–113. [CrossRef]

46. Caba, J.; Rincón, F.; Dondo, J.; Barba, J.; Abaldea, M.; López, J.C. Testing framework for on-board verification of HLS modules using grey-box technique and FPGA overlays. *Integration* **2019**, *68*, 129–138. [CrossRef]

47. Caba, J.; Cardoso, J.M.P.; Rincón, F.; Dondo, J.; López, J.C. Rapid Prototyping and Verification of Hardware Modules Generated Using HLS. In Proceedings of the Applied Reconfigurable Computing, Architectures, Tools, and Applications-14th International Symposium, ARC 2018, Santorini, Greece, 2–4 May 2018; Lecture Notes in Computer Science; Springer: Berlin, Germany, 2018; Volume 10824, pp. 446–458. [CrossRef]

48. Horstrand, P.; Guerra, R.; Rodríguez, A.; Díaz, M.; López, S.; López, J.F. A UAV platform based on a hyperspectral sensor for image capturing and on-board processing. *IEEE Access* **2019**, *7*, 66919–66938. [CrossRef]

49. NVIDIA Corporation. NVIDIA Jetson Linux Developer Guide 32.4.3 Release. Power Management for Jetson Nano and Jetson TX1 Devices. Available online: https://docs.nvidia.com/jetson/l4t/index.html#page/Tegra%20Linux%20Driver%20Package%20Development%20Guide/power_management_nano.html (accessed on 7 September 2019).

50. NVIDIA Corporation. NVIDIA Jetson Linux Driver Package Software Features Release 32.3. Power Management for Jetson TX2 Series Devices. Available online: https://docs.nvidia.com/jetson/archives/l4t-archived/l4t-3231/index.html#page/Tegra%20Linux%20Driver%20Package%20Development%20Guide/power_management_tx2_32.html (accessed on 7 September 2019).

51. NVIDIA Corporation. *NVIDIA Jetson Linux Developer Guide 32.4.3 Release. Power Management for Jetson Xavier NX and Jetson AGX Xavier Series Devices.* Available online: https://docs.nvidia.com/jetson/l4t/index.html#page/Tegra%20Linux%20Driver%20Package%20Development%20Guide/power_management_jetson_xavier.html (accessed on 7 September 2019).

Publisher's Note: MDPI stays neutral with regard to jurisdictional claims in published maps and institutional affiliations.

Article

An FPGA Accelerator for Real-Time Lossy Compression of Hyperspectral Images

Daniel Báscones [1,*], **Carlos González** [1] **and Daniel Mozos** [1,2]

[1] Departamento de Arquitectura de Computadores y Automática, Facultad de Informática, Universidad Complutense de Madrid, 28040 Madrid, Spain; carlosgo@ucm.es
[2] The Instituto de Tecnología del Conocimiento, University of Central Missouri, Warrensburg, MO 64093, USA; mozos@ucm.es
* Correspondence: danibasc@ucm.es; Tel.: +34-91-394-7581

Received: 10 July 2020; Accepted: 6 August 2020; Published: 9 August 2020

Abstract: Hyperspectral images offer great possibilities for remote studies, but can be difficult to manage due to their size. Compression helps with storage and transmission, and many efforts have been made towards standardizing compression algorithms, especially in the lossless and near-lossless domains. For long term storage, lossy compression is also of interest, but its complexity has kept it away from real-time performance. In this paper, JYPEC, a lossy hyperspectral compression algorithm that combines PCA and JPEG2000, is accelerated using an FPGA. A tier 1 coder (a key step and the most time-consuming in JPEG2000 compression) was implemented in a heavily pipelined fashion. Results showed a performance comparable to that of existing 0.18 μm CMOS implementations, all while keeping a small footprint on FPGA resources. This enabled the acceleration of the most complex step of JYPEC, bringing the total execution time below the real-time constraint.

Keywords: hyperspectral image; lossy compression; real time; FPGA; PCA; JPEG2000; EBCOT

---

1. Introduction

Remote sensing covers a broad range of techniques that are used to perform a variety of analyses remotely and with no close up interactions with the subjects of interest. Hyperspectral imaging is one of these techniques that has been growing since its inception.

It extends the concept of remote imaging by capturing information related not only to the visible part of the spectrum, but also in wavelengths that the human eye cannot see. A typical hyperspectral image will have from tens to hundreds of samples per pixel [1,2], each recording information of the light perceived at a specific wavelength.

The data collected at one wavelength are grouped in bands that span the whole image. The combination of multiple bands creates a spectral signature per pixel, providing information on a scale that helps with military applications such as target detection [3,4] and terrain trafficability [5], mineral identification [6,7], ground and water studies [8], vegetation and crop control [9,10], and many more.

It is the amount of information that is the bottleneck of hyperspectral imaging: storage and transmission are often limited in remote scenarios, and optimizing them is a must for uninterrupted image capture. Some of the most popular sensors such as EnMAP [11] reach speeds of 700 Mb/s and work uninterrupted for hours or days on a satellite.

Hyperspectral image compression has been explored in many ways, and has provided great results, especially in the lossless and near-lossless domains. Standards such as CCSDS [12] (a simple algorithm targeting on-board compression in real time) have emerged for these applications. However, for long-term storage, a more powerful lossy algorithm is also of interest.

The literature has shown a variety of approaches, generally inspired by traditional image compression techniques. Predictive models ([13] Ch.2) have been optimized—even reordering bands ([13] Ch. 3) for increased correlation. Vector quantization (VQ) has also been extensively used to exploit pixel similarities [14]. Block-based approaches [15] have been used to reduce algorithm complexity. Wavelet techniques [16] have been extended from the 2D domain to anisotropic decompositions [17]. The most promising ones combine spectral decorrelation followed by compression in the spatial domain for each decorrelated band, using the Karhunen–Loève Transform ([18] Ch.9), or principal component analysis (PCA) [19].

The JYPEC algorithm [20] follows this approach by employing PCA as the decorrelation step, followed by JPEG2000 [21] as the spatial domain compressor. Since PCA decorrelates the most information-heavy bands first, variable bit-depths are used, thereby allocating more bits to the first components and using less bits for the ones that convey less information according to PCA (i.e., less variance).

This process is computationally intensive, and not suited for real-time in general purpose processors. When analyzed with care, the most intensive part has been found to be the JPEG2000 compression step, and more specifically, its encoder.

JPEG2000's drawback is complexity, which is much higher [22] than that of the well-known and more extensively used JPEG [23]. The main reason is its block coding step, which takes up to 70% of the total execution time [24,25]. It includes simple operations but is heavy in conditional execution, making it unsuitable for traditional CPUs.

In order to level the playing field and bring lossy hyperspectral compression to real-time, acceleration techniques can be used. Field programmable gate arrays (FPGAs) offer a viable option with which to accomplish this task, and have been widely popular in accelerating block coding [24,26–32].

FPGAs each offer a reconfigurable fabric in which any circuit can be synthesized, and as a consequence have seen many applications [33,34]. They offer higher performance than a CPU or GPU, and the cost when compared to an ASIC is orders of magnitude smaller. They are also very efficient power-wise and radiation-hardened models [35] exist out of the box. All of these characteristics have made them very good candidates for remote sensing scenarios, wherein power is limited but performance and flexibility are still requirements. Great results have already been achieved for lossless and near-lossless hyperspectral image compression [36–38].

In [38] we present an FPGA implementation of the low complexity predictive lossy compression (LCPLC) algorithm. A highly pipelined architecture was designed which allows for real-time throughput while keeping FPGA usage low. However, when we want to obtain a very high compression ratio, the LCPLC algorithm does not obtain a distortion ratio as well as the JYPEC algorithm. Therefore in this study, JYPEC was accelerated using an FPGA. A tier 1 coder (a key step and the most time-consuming in JPEG2000 compression) was implemented in a heavily pipelined fashion. This enabled the acceleration of the most complex step of JYPEC, bringing the total execution time below the real-time constraint.

The rest of the paper is organized as follows: First, the JYPEC algorithm is looked at as a whole by detailing the tier 1 coder within JPEG2000, which was found to be the bottleneck for real-time execution. Secondly, an in-depth review of existing FPGA and ASIC implementations is presented, focusing on the different approaches for acceleration. Finally, a custom FPGA implementation of the tier 1 coder is presented based on the best techniques developed over the years, showing results that greatly improve on previous works, putting it in the context of real-time lossy hyperspectral image compression.

2. The JYPEC Algorithm

The JYPEC algorithm is a lossy algorithm aimed at hyperspectral image compression. It extends the concept presented in [19] of using PCA+JPEG2000 for compression. An optional vector quantization step is added prior to the dimensionality reduction, a variable bit-depth is used for each band, and the

JPEG2000 coding is optimized for the whole image instead of for progressive decoding afterwards. This improves compression ratios and quality. A general diagram is seen in Figure 1.

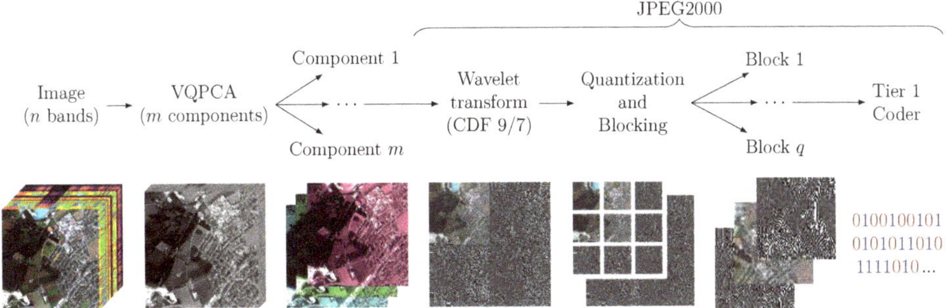

Figure 1. Diagram of the full JYPEC algorithm.

2.1. Dimensionality Reduction

Even though JYPEC supports multiple dimensionality reduction algorithms, for this approach only PCA was used. It has a very low complexity (being a candidate for real-time), and has also a very good distortion-ratio performance [20], only falling short when targeting very high qualities at low ratios (against vector quantization PCA (VQPCA)).

First, a preprocessing step selects random pixels in the image. These pixels are processed by PCA, generating a set of vectors that indicate the directions of maximum variance within the set. Since the selection is random and the sample size is big due to the size of hyperspectral images, it matches well the total variance even when using only 1% of the total pixels [39]. This process generates a projection matrix to reduce the data size, as well as a recovery matrix to go back from the reduced data to the original size. Each band of the reduced image is compressed by JPEG2000.

The process can be undone by first uncompressing each reduced band, and then using the recovery matrix to project back into the original space.

2.2. The JPEG2000 Algorithm

The JPEG2000 standard is a multi-step algorithm which takes advantage of different image characteristics for compression. As well as being used for its main purpose of image compression, it has also found applications in compressing electroencephalography [40], video [41], and hyperspectral images [20], among others.

It follows a series of simple steps (Figure 2) to compress an image:

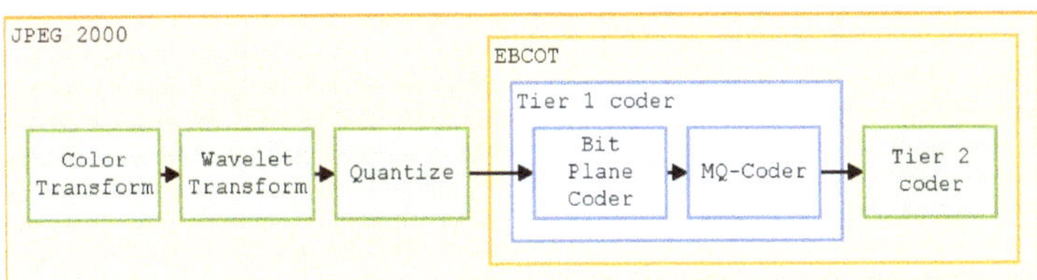

Figure 2. Diagram of the full JPEG2000 algorithm. In green, steps that deal with the whole image at once. In blue, steps that deal with small blocks of the image.

- A color transformation is done per pixel, converting the input color space (usually RGB) into a luminance (brightness) and chrominance (color) model, since human vision is more sensitive to brightness than color. The color channels can be down sampled with no perceivable loss in quality, reducing a large chunk of the data bits.

- Every channel is then subjected to a wavelet transform [42]. A wavelet transform consists of a high-pass and a low-pass [43] filter that are applied both horizontally and vertically to all rows and columns respectively. This can be done in a reversible (lossless) or irreversible (lossy) way, and in either case the result is a partitioned channel in which different zones present different patterns that can be compressed to a higher degree than the original data.
- After doing the wavelet transform, the resulting values are quantized to integer values; some information is lost when the lossy wavelet transform is used.
- Finally, the values are encoded. The image is split into blocks of up to 4096 samples; each one is encoded individually, thereby exploiting local redundancies and the patterns left by the wavelet transform.

The color transform is not needed for hyperspectral compression, since the spectral dimension is already decorrelated by the dimensionality reduction step. Only the wavelet transform and encoding steps are performed.

The wavelet transform is fast when compared to encoding, which takes up to 70% of the total execution time [24,25] of JPEG2000. Within JYPEC, it also dominates execution times [39]. This paper focuses on a FPGA implementation that greatly improves encoding, bringing the full algorithm execution time down as much as possible by accelerating the bottleneck.

In the following subsections, the encoder algorithm is detailed to give a better understanding of the implemented accelerator later.

2.2.1. Encoding

Encoding is done in a lossless way over blocks of up to 4096 samples (usually 64×64 squares, with a depth of 16 bits (15 magnitude + 1 sign). The encoding technique used in JPEG2000 is called embedded block coding with optimal truncation [44] (EBCOT). Two different coders lie within EBCOT: the tier 1 and tier 2 coders:

The **tier 1** coder compresses the original block. The resulting stream has the prefix property: any prefix of that stream, once decoded, gives an approximation of the block. Longer prefixes provide better reconstruction accuracy.

The **tier 2** coder splits the output streams from each block into sections, with each section refining the information decoded by the previous one. Sections from different blocks are interleaved by first storing the ones which better approximate the original image. This technique extends the concept of the prefix property from blocks to the full image. This part of the coder is not used here, since JYPEC targets full image compression and not progressive decoding.

2.2.2. Tier 1 Coder

Two phases lie within: First, the so called "bit plane coder" (BPC) pairs each bit with some context. This creates context—data pairs (CxD pairs). The distribution of all bits paired with the same context is highly skewed, making coding more efficient.

Second, CxD pairs are processed by the MQ-coder (it belongs to the family of arithmetic coders), generating the compressed output stream. The more skewed the input distributions are, the fewer bits the MQ-coder will output.

The **BPC** works by scanning the different bit planes of the block. First, the most significant bit of each sample is coded; then the second-most significant; and so on. This is the trick that later allows for progressive decoding.

The sign bit receives a special treatment, and is only coded when needed (i.e., when a sample is known to be nonzero, the sign bit for that sample is coded, but not before). This avoids coding sign bits for samples that are zero.

To exploit redundancies, bits are scanned in a zig-zag pattern in up to three passes per bit plane p (Figure 3). To keep track of what bit is scanned in what pass, a "significance" value $\sigma[j]$ is kept per sample $v[j]$, where j is the 2D position within the block/bit plane.

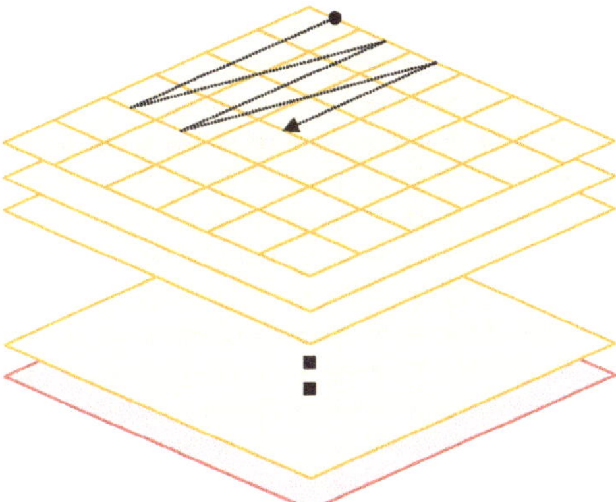

Figure 3. Bit plane coder. The zig-zag pattern can be seen (four rows visited column by column). This is repeated three times (three passes) until all bits from a plane have been visited. The sign bit plane (red) is coded separately.

This value indicates whether a sample is significant (i.e., at least one of its already coded bits is one) or insignificant (if all bits have been zero up until the current pass in previous bit planes). A significant sample is either positive or negative depending on its sign bit.

Three distinct passes exist: First, a significance propagation pass in which bits from samples that are expected to turn significant in the current plane are coded. Second, a refinement pass that codes bits from all samples that are already significant. Finally, a cleanup pass that codes the remaining bits.

The cleanup pass will mostly code zeros and the significance pass will mostly code ones, while the refinement pass is more random. These skewed distributions are the key for compression. To increase efficiency, each bit is paired with a context based on the significance state of neighboring samples. The context is used to predict the value of the bit, and if right, can save even more space in the compressed stream.

The **MQ-coder** starts by mapping each context to two values:

- A bit x which is the current prediction for the given context.
- A state (of which there are 47 different ones) indicating the probability p of the prediction x being right.

The basic idea of the MQ-coder is that of arithmetic coders: The input data will be compressed as a subinterval $I \subset [0,1)$. Starting with the interval $[0,1)$, the data are subdivided with each CxD pair.

Given the probability p, $I = [c, c+a)$ is divided in $I_1 = [c, c+ap)$ and $I_2 = [c+ap, c+a)$. If the predictive model is right, and the probability p high, I_1 (the bigger subinterval) will be kept. The bigger the final interval is, the less bits are needed to represent it.

In practice, infinite precision cannot be achieved, so 27 and 16 bits are allocated for c and a respectively as per JPEG2000 standard, calling them registers C and A.

Each state has an associated probability of finite precision $\bar{p} \in \{0, \ldots, 2^{16} - 1\}$ mapping to the $[0,1)$ interval when dividing by 2^{16}, indicating the probability of its prediction's success.

Since $\bar{p}$ is fixed for a given state, its change is accomplished with two transition functions: most probable symbol (MPS) and least probably symbol (LPS). The MPS or LPS functions will be used depending on whether the prediction turns out to be right or not, and will change the state to one where $\bar{p}$ is expected to better approximate the current bit input distribution. The prediction is also inverted when certain states are reached, doubling the possible states.

$\bar{p}$ is used to update the interval $[C, C + A)$ by either adding $\bar{p}$ to C or subtracting it from A. When both ends get close together, a shift left is performed to keep the length of the interval longer than the maximum value of $\bar{p}$.

A is kept under control by periodically resetting it, and C eventually overflows. The overflow is saved and forms the compressed output stream. To finish compression, C is flushed out. The process can then be reversed, and the original inputs recreated for decompression.

3. Existing Implementations

The objective of this study was to bring lossy compression to real time. For that, JYPEC was chosen as the target algorithm, and a tier 1 coder implementation of JPEG2000 (a step within JYPEC) is presented. It accelerates both the BPC and MQ-coder by making a single high-speed pipeline with both. To show that this effort was justified, a timing breakdown of compression of different images with JYPEC is shown in Figure 4.

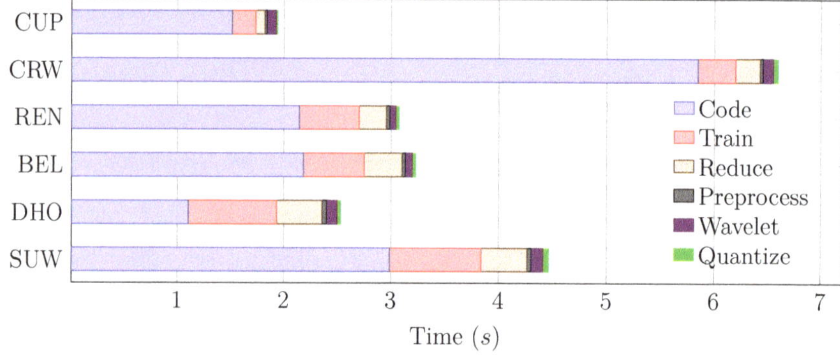

Figure 4. Timing breakdown when compressing a set of hyperspectral images with JYPEC. Steps are sorted by time. While training takes a fair bit of time, coding is clearly the bottleneck of the process.

It is clear that coding is the slowest part of the algorithm, and that any efforts to speed up the algorithm should be dedicated to it. Many implementations of the full tier 1 coder have been proposed, but more efforts have been focused towards accelerating the MQ-coder alone. In the following paragraphs, the literature is explored in this regard.

3.1. Bit Plane Coder

In [45], the authors designed a basic BPC which goes over the full block following the zig-zag pattern. Its controller is a 24 state machine which goes over the different passes bit by bit, producing at most one CxD pair per cycle.

Improving on that, in [25] the authors introduced the concept of skipping. They loaded full columns with four bits at once, and marked them with flags when they were no longer required in a certain pass. This way, the BPC could skip them when not needed, saving clock cycles. They also introduced flags for groups of columns and even full passes, allowing one to skip big chunks of idle cycles in some cases. Finally, since they loaded the full column at once, they also checked which bits needed to be encoded, and skipped the others within the column. In the end, savings of around 60% of clock cycles were achieved.

A different approach was reported in [27]. Instead of skipping, the authors processed whole columns at once, producing up to 10 CxD pairs per cycle. CxD generation is not independent, so a series of cascading dependencies were taken into account. Despite the added complexity, this idea doubles the throughput of the sample-skipping technique without the extra memory. In [46], the authors went even further by allowing multiple planes to be coded simultaneously by using non-default coding options. Despite the small loss in compression efficiency, the throughput grew by a factor of 8 in CMOS 0.35 μm technology when using gray-scale images. This, however, deviates from the standard implementation, since multiple MQ-coders would have had to be used for a single block in order to keep up with the BPC.

3.2. MQ-Coder

The MQ-coder has seen more optimizations [47] than the BPC, since traditionally the MQ-coder always was the bottleneck.

MQ-coder receives and processes CxD pairs from the bit plane coder, generating a compressed bit stream which can be further processed by the tier 2 coder. It is important to note that by design, the CxD pairs are processed serially, so no parallelization is possible at this stage.

Two main approaches have been used to accelerate its execution:

- **Pipelining**: As with many other designs, pipelining can be the key to improving performance. Distinct stages have been identified (mainly separating the update of A and C, and the output of coded byte(s)).
- **Dual symbol processing**: Some bit plane coders can produce two CxD pairs in one cycle. This has motivated the design of MQ coders with the capability of processing two pairs at the same time. Since this can not be done in parallel, these MQ-coders incorporate two cascaded processing units.

In [31], a three stage pipelined MQ coder is proposed. It performs all arithmetic operations in the first stage, A and C register shifts are done in a second stage, and a third stage emits bytes. The drawback of the authors' approach was that the second stage could stall the first if the number of shifts was greater than one, since the authors did not use barrel shifters. In the end, they worked around this limitation by having two clock domains increasing the speed of the shifting stage, ensuring that stalling occurred only 0.64% of the time, achieving a performance of around 145.9 MS/s on a Stratix EP1S10B672C6 board.

In [30], an implementation of the full tier 1 coder with no pipelining nor dual symbol processing is presented for the Virtex II Pro FG 456 board. They note that, at 112 MS/s, the arithmetic coder is the bottleneck, with the (context, bit) generation being five times faster. Speed was later increased by having up to four simultaneous instances of the tier 1 coder working in parallel.

In [48], two techniques were used to create a pipelined design that works at 413 MS/s: They first employed "traced pipelining" which consists of designing a pipeline for the most likely cases, and processing unlikely, more costly cases in a separate unit that stalls the pipeline if necessary. The second technique is based on eliminating cascading shifts (such as the ones from [29]) by looking ahead at the number of necessary shifts and performing them all at once. All of this is made possible by working on 0.18 μm CMOS technology.

Both pipelining and dual processing are employed in the approach from [32], in which improvements are made to the multiple approaches from [26]. Dual processing is solved by having four different units processing all four possible scenarios (taking the lower or upper interval two times in any combination). Pipelining was used to separate the A register update, C register update, and byte output procedures, achieving in the end performances of 96.6 MS/s on FPGA and 440.2 MS/s on 0.18 μm CMOS technology.

A different pipelined approach is presented in [49]. The A register update, C register update, and byte out procedures are kept in three distinct stages, and two more are added at the beginning by using two memory modules. The first one stores contextual information (namely state and predicted symbol) and the second one is a ROM which outputs state change information. If two consecutive contexts are equal, the second memory will be read with the updated state that is sent to the first one. This splits reading into a two-step process that accelerates the pipeline. The other notable technique is that shifting is limited to seven bits per cycle, reducing the critical path at the cost of one cycle stall in the unlikely case that the shift amount is greater than seven. In the end, a speed of 192.77 MS/s was achieved on 0.18 μm technology.

4. Implementation

The presented design includes both the BPC and MQ-coder, chained together to form the full tier 1 coder. The basic structure of the tier 1 coder is shown in Figure 5.

Figure 5. Tier 1 coder architecture.

The bit plane coder receives data from memory and generates CxD pairs. These are coded by the MQ-coder and the output stream is generated.

4.1. BPC

Internally, the BPC generates CxD pairs for four samples at a time (a full column of the zig-zag pattern), following the techniques of [27]. To generate the CxD pairs for a sample, flags from a 3×3 neighborhood around the current sample are taken into account. When dealing with columns, this neighborhood grows to 6×3. This is seen in Figure 6.

The main problem comes from the fact that the context generation modules are slow. They are based on lookup tables that require many LUT levels during FPGA implementation. They also need to be cascaded to generate the output significance that is required by the following bit strip. To avoid delays, a special module that predicts the next significance state in a faster way is used. It is shown in Figure 7.

The other modules are straightforward, with context generation being a ROM that outputs the context associated with the neighborhood by simple lookup. The cleanup predictor does a job similar to the significance predictor by looking at the first bit that is nonzero,and setting it and the following ones as significant if needed.

Up to 10 CxD pairs are generated per cycle. A serializer is used to order them sequentially before sending them to the MQ-coder (see Figure 8).

With a big enough vector queue, the problem of running into idle cycles when few CxD pairs are generated is avoided (e.g., the first refinement pass or the last significance and cleanup passes), since a big enough buffer exists to keep feeding the MQ coder. The serializer is designed to output one pair per cycle as long as the vector queue is not empty, so it can feed the MQ-coder without forcing a stall.

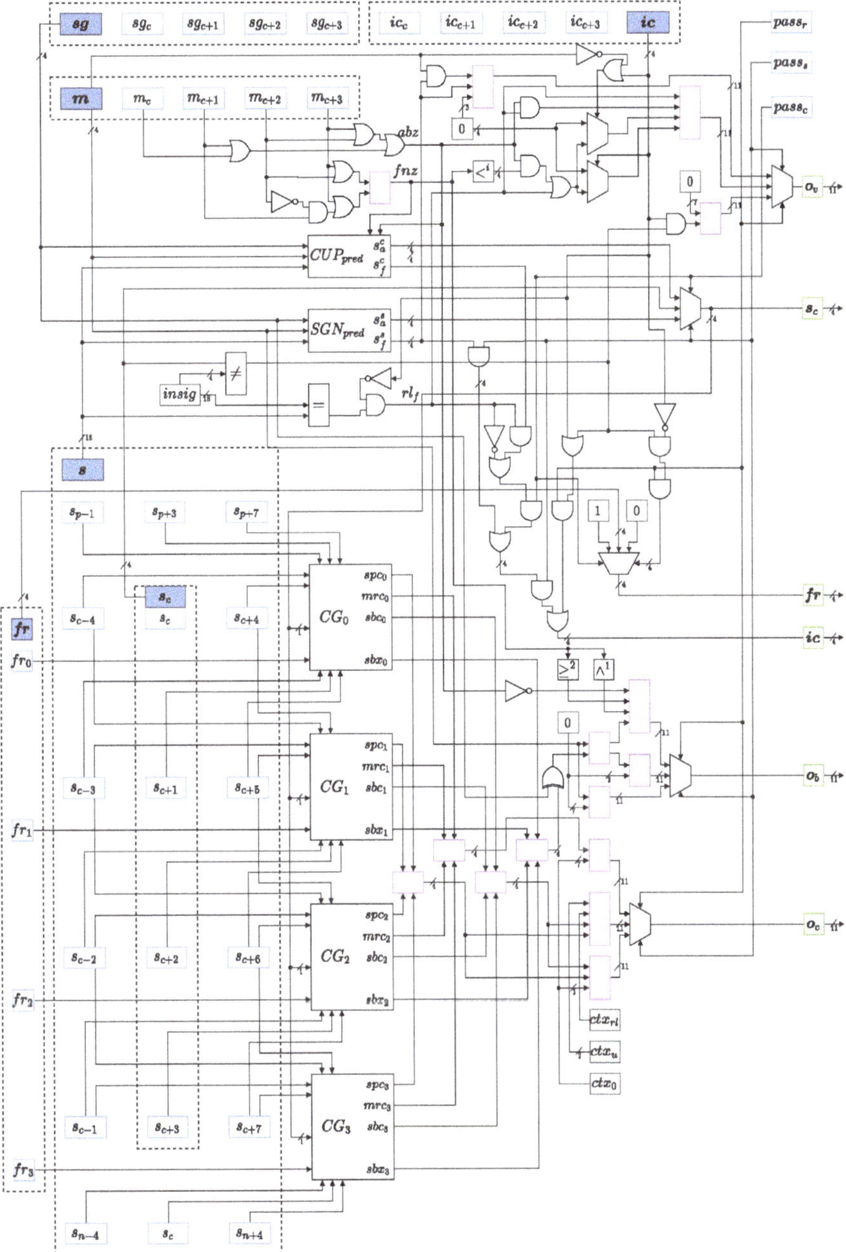

Figure 6. BPC structure. sg indicates sign bits for the current strip. m indicates magnitude bits for the current strip. ic is the coded flag for the strip, indicating whether each bit is coded. fr is yet another flag indicating, for each bit, whether it is being refined for the first time. s is the significance status for the neighborhood of the strip. $pass_x$ are flags indicating the current pass (significance, cleanup, refinement). Flags are updated for the next pass and output to memory. The contexts for each pass exit the context generation (CG) modules for all three passes (spc, mrc, sbc) along with the xor bit for the sign. This ends up being output as a vector of triplets o_c, o_b, o_v, containing the context–data (CxD) pairs as well as a valid bit.

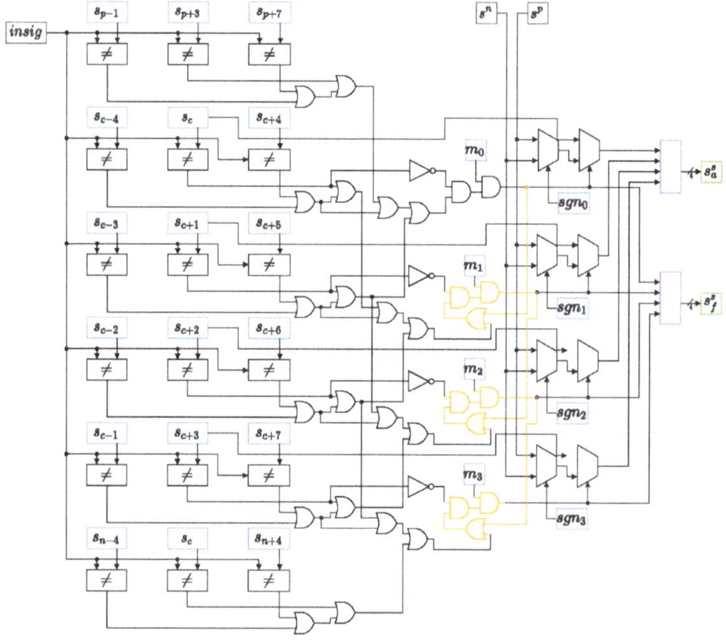

Figure 7. Significance predictor module. It cascades the possible significance conversion of all bits in the strip by using a small critical path, instead of waiting for the cascade of CG modules. The significances s, magnitude bits m, and sign bits sgn are used to produce the new significance s_f^s. It is output along with a flag s_a^s, indicating if it is a newly acquired significance this cycle.

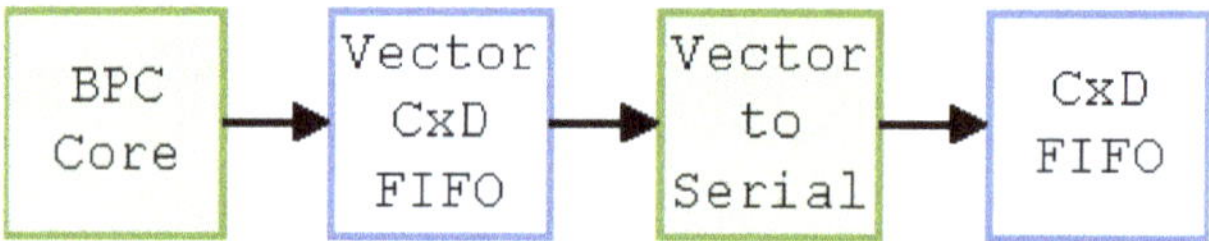

Figure 8. BPC architecture. The BPC core generates up to 10 CxD pairs per cycle, which are then serialized before sending them to the CxD FIFO.

4.2. MQ-Coder

The extensive work seen in Section 3.2 can be summed up in two different approaches: pipelining and dual-symbol processing. Since the target platform is an FPGA, pipelining is ideal to avoid a longer critical path which, on reconfigurable hardware, has a higher timing penalty than on fixed silicon. It will later be seen that a bottleneck arises in the CxD generation, so this area-efficient approach is the right choice since it can keep up with previous stages.

The only common step of all pipelining approaches is separating the byte out procedure in a last stage. Register updates are often split in two stages, treating A and C updates independently. Memory access is also split in some implementations. The point is, there are no obvious stages in the algorithm given the great dependency of the different stages. In fact, most designs implement the most likely execution path, having to stall in the event of an unexpected input.

Most implementations are offering a theoretical throughput that varies depending on the data being compressed. The goal with this paper was to design an implementation which can consistently process CxD pairs at a certain speed. By pipelining the design and inserting queues in between stages, any potential stall in one stage gets absorbed by the queues and does not affect the others. The result can be seen on Figure 9.

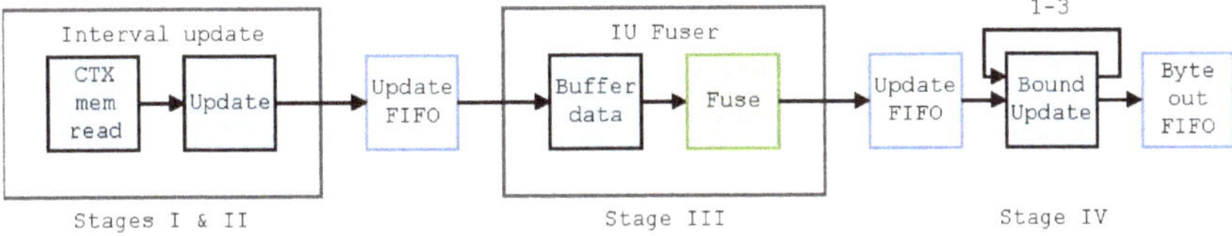

Figure 9. MQ-coder architecture. The interval updates are fused when possible, having less bound updates which could stall the pipeline.

Four main stages can be seen.

4.2.1. First and Second Stages

First, the context memory is accessed. This memory outputs "context information", which has the prediction for that context, as well as the MPS and LPS transitions, XOR bit, and probability.

This memory is written with the context from the third stage, so care is taken whenever the same context is encountered twice in a three context window span:

- If the same context is found in cycles n and $n + 2$, a write–read cycle is skipped and input data are directly multiplexed to the memory's output.
- To avoid stalling in the case where the same context appears in cycles n and $n + 1$, a second memory is present in the second stage, which outputs state information. The state is decided from the MPS and LPS transitions, and used to read this second memory. In this case, the context memory will be updated the next cycle. But those values are required in the current cycle, so a mux is used to bypass it from the second memory, avoiding a stall.

All in all, reading is segmented in two different stages, without the stalling that sometimes could happen in implementations such as [49].

The second stage is the most complex one, where the critical path lies. First, the context information to be used is decided, which comes from the context memory unless the same context appeared twice in a row, in which case it is fetched from the state memory and previous prediction.

State and prediction changes are fed to the state and context memories.

The prediction is adjusted, and the A register is updated. This can happen in one of four ways: either the A register is not shifted, it is shifted once or twice, or the contents are assigned from memory. In the last case, the number of shifts is calculated in advance.

The number of shifts $\bar{s}$, value $\bar{p}$ to add to the C register, and hit flag h (indicating C needs to be updated with $\bar{p}$) are sent to the next stage.

Both stages are seen in Figure 10.

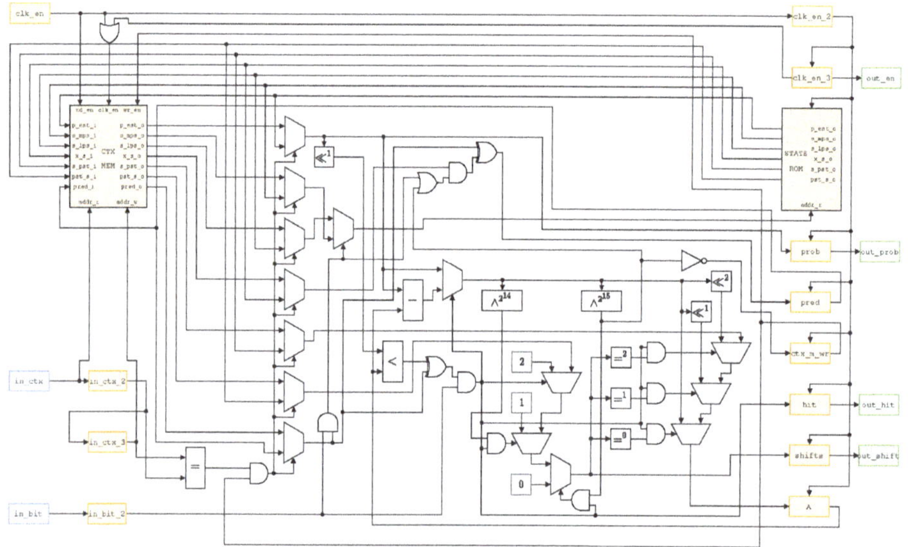

Figure 10. MQ-coder first and second stages. The context and bit from the BPC are input, and a probability, shift, and hit (correct prediction or not) are output.

4.2.2. Third Stage

The fourth stage can stall the pipeline. In order to minimize that risk, the inputs from multiple cycles are combined into just one update, so as to send the minimum number of updates ahead. This can be done under two scenarios:

- If $h_i = h_{i+1} = 0$ and $\bar{s}_i + \bar{s}_{i+1} \leq 15$, both updates can be merged by setting $h' = 0$, $\bar{s}' = \bar{s}_i + \bar{s}_{i+1}$, $\bar{p}' = 0$. This merges two consecutive shifts that are under the maximum shift length of 15.
- If $\bar{s}_i = 0$ and $\bar{p}_i + \bar{p}_{i+1} \leq 2^{16} - 1$, then both updates can be merged with $h' = h_i \oplus h_{i+1}$, $\bar{s}' = \bar{s}_{i+1}$, $\bar{p}' = \bar{p}_i + \bar{p}_{i+1}$. This is because the addition of $\bar{p}$ to C happens before the shift $\bar{s}'$. Both probabilities can be added at once because they are below the limit of 2^{16}.

Both these merging techniques can be done recursively.

4.2.3. Fourth Stage

The C register is updated by adding $\bar{p}$ and shifting it. Whenever it fills up, a byte is output. When the register overflows or a special byte $0xFF$ is output, padding needs to be added to avoid special markers used to indicate the end of stream. Up to three bytes might be output per update, and the control logic for all possibilities would make this stage too slow.

In order to avoid that problem, shifting is done byte by byte. If the shift amount is greater than one byte, the pipeline will stall for one cycle. Studies have shown [48] this problem to be negligible (<1% of the time). This stage is seen in Figure 11.

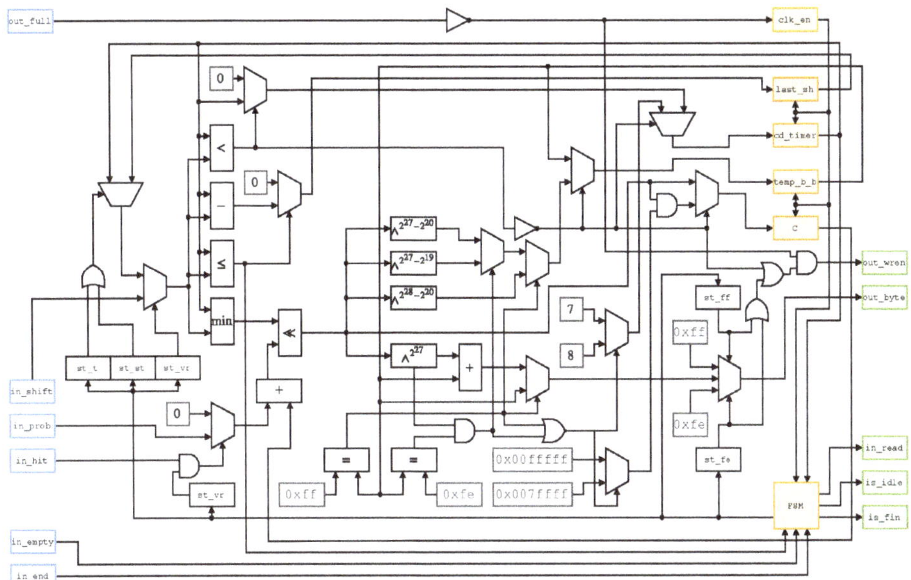

Figure 11. MQ-coder last stage. It interfaces with two FIFOs, reading updates from the previous stage and making sure space is available at the output to send bytes out.

4.3. The Full Tier 1 Coder

By chaining together both the BPC and MQ-coder, the tier 1 coder for JPEG2000 is formed. The basic segmentation is two stages for the BPC coder and four for the MQ-coder. Joining the different stages are multiple FIFOs. These help maintain a constant flow of data:

- When the MQ-coder stage IV stalls (because it has to output more than one byte), the fuser queue can hold updates until a fused one is sent (effectively canceling out the stalling).
- When the BPC-core is producing many CxD vectors, the vector queue avoids a stall from the BPC-core.
- Conversely, when the BPC-core does not produce vectors, the queue serves as a buffer to keep the next stages busy.

The full pipeline, taking into account the different queues, has a total of 15 stages, as seen in Figure 12. Despite the amount of stages, this has negligible impact in the final speed, since the coding of a full $64 \times 64 \times 16$ block takes a minimum of $1024 \cdot 3 \cdot 14 + 1024 = 44,032$ cycles, so filling the pipeline takes at most $15/44,032 \cdot 100 = 0.03\%$ of the total cycles.

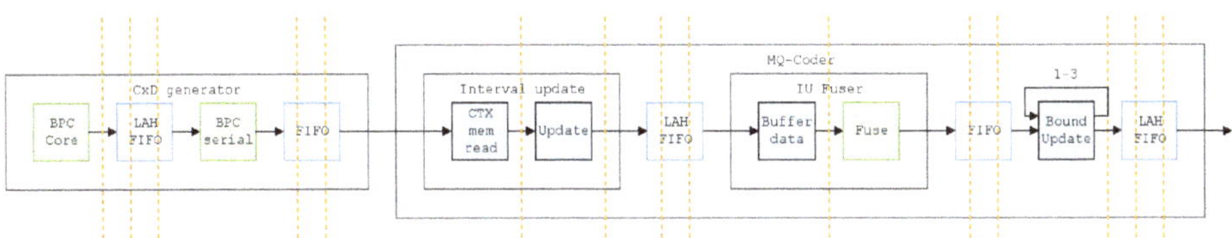

Figure 12. Detailed pipeline of the tier 1 coder. In dotted orange, the separation between stages. Each FIFO introduces two stages (read/write).

5. Results

The hardware architecture described in Section 4 has been implemented using VHDL language for the specification of the tier 1 coder. Moreover, we have used the 2020 Xilinx Vivado Design Suite environment to specify the complete system. The full system has been implemented on a VC709 board, a reconfigurable board with a single Virtex-7 XC7VX690T, two DDR3 SDRAM DIMM slots which hold up to 4 GB each, a RS232 port, and some additional components not used by our implementation. The HDLmodel has been verified via simulation and physical prototyping using a memory controller for input/output.

Table 1 shows the frequency and FPGA slice occupancy for the full tier 1 coder and its modules and submodules. More details are given in the following list:

- The BPC-core processes a full block of 65,536 bits in 44,032 cycles, working at a speed of 380 Mb/s.
- The BPC-serial can produce up to 390 MCxD pairs per second.
- The MQ-I/II stages processes 322 MCxD pairs per second, generating up to 322 M updates per second.
- The third stage is a bit faster, being able to merge 535 M updates per second.
- The last MQ stage processes up to 331 M updates per second.
- The intermediate FIFO queues have no problem at all keeping up with the speed requirements.

Table 1. Frequency and occupancy for the different modules that make the full tier 1 coder. Results are for the Virtex-7 XC7VX690T board with a depth of 32 set for all queues.

Module	Frequency (MHz)	Slices	BRAM
Tier 1 coder	255	2708	4
BPC-core	248	731	2
BPC-serial	390	142	0
MQ	321	1778	0
MQ-I/II	322	1326	0
MQ-III	535	47	0
MQ-IV	331	231	0
FIFOs	927	57	2

All in all, the full tier 1 coder is able to work at 255 MHz. At that speed, the bottleneck is the number of CxD pairs processed by the MQ-coder at 255 MCxD/s. By studying how many CxD pairs are produced, the input speed is calculated:

- The minimum number of updates for a $64 \times 64 \times 16$ block is seen when it is all zero, having successful run-length coding throughout the block. In this case, a total of $15 \times 1024 = 15,360$ updates are generated. That is, 0.234 per bit.
- Conversely, an upper limit for the number of updates is given by a cleanup pass with run-length interruptions at every position, followed by 14 refinement passes. In this case, the number of updates is $1024 \times 10 + 4096 \times 14 = 67,584$ updates. Exactly 1.03125 per bit.

Thus, the input rate to generate 255 MS/s would range from 1.01 Gb/s to 247.3 Mb/s. However, the BPC-core is only capable of processing 380 Mb/s, so in practice this range is limited to 247.3 to 380 Mb/s.

The exact value within this range of course depends on the redundancy of the data. For [31], they compressed five images of size $512 \times 512 \times 10$ and noted that the average p/b rate was 0.56. This means that, on average, the input rate for 255 MS/s would be 455 Mb/s. Thus it is safe to say that the tier 1 coder will consistently perform at its 380 Mb/s limit.

5.1. Comparison

A comparison with other implementations can be seen in Table 2. Only the best implementations found in the literature have been taken into account.

As seen, this implementation of the BPC works more than four times faster than other FPGA implementations, surpassing even ASIC designs in throughput.

With regard to the MQ-coder, this design doubles the performance of previous FPGA designs, only falling short of 0.18 μm CMOS. It was expected that porting the design to this technology would make it faster than the competition, since other implementations have experienced [32] a speedup of $4\times$ when doing so.

Table 2. Comparison with other implementations.

Coder	Ref.	Technology	Frequency	Speed	Slices	BRAM/b
BPC	[27]	APEX20KE FPGA	51.7 MHz	73.44 Mb/s	956	n/a **
	[28]	XCV600e-6BG432	52.0 MHz	94.4 Mb/s	n/a	n/a **
	[50]	Altera EP20K600EFC672–3	100.0 MHz	40.5 Mb/s	1850	0 **
	This	Virtex-7 FPGA	247.8 MHz	368.8 Mb/s	731	2
MQ	[51]	0.35 μm	90.0 MHz	180.0 MCxD/s	n/a	n/a
	[46]	0.35 μm	150.0 MHz	300.0 MCxD/s	n/a	n/a
	[26]	Stratix	48.8 MHz	97.7 MCxD/s	1596	8192 b
	[26]	0.18 μm	211.8 MHz	423.7 MCxD/s	n/a	n/a
	[50]	Altera EP20K600EFC672–3	26.3 MHz	52.6 MCxD/s	1811	n/a
	[29]	Stratix FPGA	153.0 MHz	137.7 MCxD/s	279	1344 b
	[48]	0.18 μm	413.0 MHz	413.0 MCxD/s	n/a	n/a
	[52]	Stratix FPGA	106.2 MHz	212.4 MCxD/s	1267	0
	[32]	XC4VLX80 FPGA	48.3 MHz	96.6 MCxD/s	6974	1509 b
	[32]	0.18 μm	220.0 MHz	440.0 MCxD/s	n/a	n/a
	[53]	Stratix EP1S10B672C6	136.9 MHz	136.9 MCxD/s	695	3301 b
	[31]	Stratix FPGA	146.0 MHz	146.0 MCxD/s	824	428 b
	[49]	0.18 μm	208.0 MHz	192.8 MCxD/s	n/a	n/a
	[54]	Stratix II FPGA	106.2 MHz	212.4 MCxD/s	1267	1321 b
	This	Virtex-7 FPGA	321.5 MHz	321.5 MCxD/s	1778	0
Tier 1	[25]	0.35 μm	50.0 MHz	36.5 Mb/s	n/a	n/a
	[24]	Virtex II XC2V1000	50.0 MHz	91.2 Mb/s	4420	3120 b **
	[30]	Virtex II Pro FG 456	112.0 MHz	181.6 Mb/s *	2504	28
	This	Virtex-7 FPGA	255.3 MHz	380.0 Mb/s	2708	4

* Although not specified, the architecture is similar to the one presented here so a similar relationship between frequency and speed was expected. ** Requires external memory for data and/or internal variables.

5.2. Acceleration of JYPEC

To see its impact on hyperspectral image compression under JYPEC, six images from two libraries have been compressed by JYPEC with and without acceleration. Four from the Spectir [55] library and two from the CCSDS 123 test data set [56]. The image characteristics are seen in Table 3 and a preview in Figure 13.

Table 3. Images used for testing.

Image	N_X	N_Y	N_Z	Bit Depth	Description
CUP [56]	350	350	188	16	Cuprite valley in Nevada
SUW [55]	320	1200	360	16	Lower Suwannee natural reserve
DHO [55]	320	1260	360	16	Deepwater Horizon oil spill
BEL [55]	320	600	360	16	Crop fields in Belstville
REN [55]	320	600	356	16	Urban and rural mixed area
CRW [56]	614	512	224	16	Cuprite valley full image

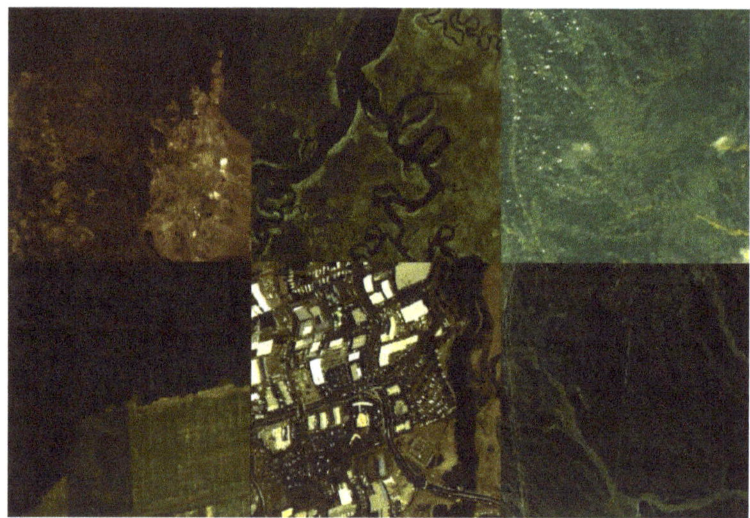

Figure 13. Small cutouts of the images from Table 3. In reading order: CUP, SUW, DHO, BEL, REN, CRW.

The results have been acquired in a DELL XPS 13 9360 computer, with an i7-7500U processor with a thermal design power (TDP) of 15 W, 8 GB of RAM running at 1866 MHz, and 256 GB of SSD PCIe storage. For the accelerated version, the time of coding in the processor was replaced with the time of coding in the FPGA itself. Memory transfer times ertr not taken into account, because the PCIe of the VC709 board works at 25 GB/s and the typical image size is 500 MB, so it was transferred in 20 ms, not impacting results.

The speedup attained is shown in Figure 14. The average speedup obtained was 3.6, ranging from 1.6 for the DHOimage to 7.5 for the CRWimage.

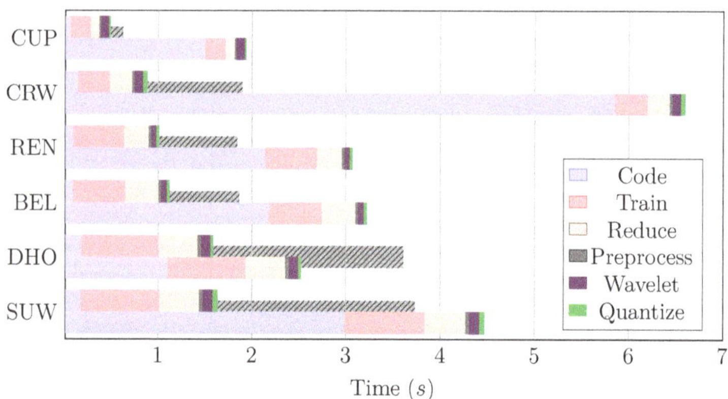

Figure 14. Speedup when using an FPGA as an accelerator. For each image, the top bars indicate the sped-up version, and the bottom bars are the non-sped-up one. A dashed bar indicates the real-time threshold, which without acceleration was only met by the DHO image.

The code for the software JYPEC implementation can be accessed in [57], and the accelerator code was uploaded in [58].

6. Conclusions

JYPEC is a complex algorithm that demands high-performing hardware for a fast execution in real time. The most costly part is the tier 1 coder within JPEG2000, since code with erratic branching is very hard to optimize for traditional processors.

Very simple arithmetic and logic operations, however, make this part of the algorithm ideal for execution on a FPGA. A very fast architecture for the full tier 1 coder within JPEG2000 has been developed based on two main ideas:

- First, the bit plane coder concurrently processes bits in groups of four, greatly accelerating execution. A system of FIFOs and buffers ensure that a constant stream of CxD pairs reach the MQ-coder.
- Second, the coder itself is highly optimized in a pipelined fashion. Stalling of the pipeline, a problem previous designs had, is avoided by fusing together multiple updates when possible.

The presented design doubles the speed of any previous design on FPGA, coming close in performance to 0.18 μm CMOS technology in single-core tests.

In the context of hyperspectral imaging, it brings complex lossy compression to real-time performance under the AVIRIS-ng sensor threshold (30–72 MS/s for a total of 491.52 Mb/s). This allows for very high data rates to be reduced for long-term storage on-the-fly, while keeping great quality for posterior analyses.

Author Contributions: D.B. designed the algorithm; C.G. and D.M. conceived and designed the experiments; D.B. performed the experiments; D.B., C.G. and D.M. analyzed the data; D.B., C.G. and D.M. wrote the paper. All authors have read and agreed to the published version of the manuscript.

Funding: This work has been supported by the by the Spanish MINECO projects TIN2013-40968-P and TIN2017-87237-P.

Acknowledgments: The authors would like to thank the anonymous reviewers. Their comments and suggestions greatly improved this paper.

References

1. DataBase, I. A Database for Remote Sensing Indices. Available online: https://www.indexdatabase.de/db/s.php (accessed on 24 February 2020).
2. eoPortal. Airborne Sensors. Available online: https://directory.eoportal.org/web/eoportal/airborne-sensors (accessed on 14 May 2020).
3. Briottet, X.; Boucher, Y.; Dimmeler, A.; Malaplate, A.; Cini, A.; Diani, M.; Bekman, H.; Schwering, P.; Skauli, T.; Kasen, I.; et al. Military applications of hyperspectral imagery. *Targets Backgr. XII Charact. Represent.* **2006**, *6239*, 62390B. [CrossRef]
4. Tiwari, K.C.; Arora, M.K.; Singh, D. An assessment of independent component analysis for detection of military targets from hyperspectral images. *Int. J. Appl. Earth Obs. Geoinf.* **2011**, *13*, 730–740. [CrossRef]
5. Slocum, K.; Surdu, J.; Sullivan, J.; Rudak, M.; Colvin, N.; Gates, C. Trafficability Analysis Engine. *Cross Talk J. Def. Softw. Eng.* **2003**, 28–30.
6. Murphy, R.J.; Monteiro, S.T.; Schneider, S. Evaluating classification techniques for mapping vertical geology using field-based hyperspectral sensors. *IEEE Trans. Geosci. Remote Sens.* **2012**, *50*, 3066–3080. [CrossRef]
7. Kurz, T.H.; Buckley, S.J.; Howell, J.A.; Schneider, D. Geological outcrop modelling and interpretation using ground based hyperspectral and laser scanning data fusion. *Int. Arch. Photogramm. Remote Sens. Spat. Inf. Sci.* **2008**, *37*, 1229–1234.
8. van der Meer, F.D.; van der Werff, H.M.; van Ruitenbeek, F.J.; Hecker, C.A.; Bakker, W.H.; Noomen, M.F.; van der Meijde, M.; Carranza, E.J.M.; de Smeth, J.B.; Woldai, T. Multi- and hyperspectral geologic remote sensing: A review. *Int. J. Appl. Earth Obs. Geoinf.* **2012**, *14*, 112–128. [CrossRef]
9. Thenkabail, P.; Lyon, J.; Huete, A. (Eds.) Hyperspectral Indices and Image Classifications for Agriculture and Vegetation. 2018. Available online: https://www.taylorfrancis.com/books/9781315159331/chapters/10.1201/9781315159331-1 (accessed on 16 July 2020).

10. Lu, J.; Yang, T.; Su, X.; Qi, H.; Yao, X.; Cheng, T.; Zhu, Y.; Cao, W.; Tian, Y. Monitoring leaf potassium content using hyperspectral vegetation indices in rice leaves. *Precis. Agric.* **2019**, *21*, 324–348. [CrossRef]

11. eoPortal. EnMAP (Environmental Monitoring and Analysis Program). Available online: https://directory.eoportal.org/web/eoportal/satellite-missions/e/enmap (accessed on 5 August 2020).

12. CCSDS. Low-Complexity Lossless and Near-Lossless Multispectral and Hyperspectral Image Compression. Available online: https://public.ccsds.org/Pubs/123x0b2c1.pdf (accessed on 8 August 2020).

13. Motta, G.; Rizzo, F.; Storer, J.A. *Hyperspectral Data Compression*; Springer Science & Business Media: Berlin, Germany, 2006; p. 417.

14. Ryan, M.J.; Arnold, J.F. The lossless compression of aviris images by vector quantization. *IEEE Trans. Geosci. Remote Sens.* **1997**, *35*, 546–550. [CrossRef]

15. Abrardo, A.; Barni, M.; Magli, E. Low-complexity predictive lossy compression of hyperspectral and ultraspectral images. In Proceedings of the 2011 IEEE International Conference on Acoustics, Speech and Signal Processing (ICASSP), Prague, Czech Republic, 22–27 May 2011; pp. 797–800. [CrossRef]

16. Fowler, J.; Rucker, J. 3D wavelet-based compression of hyperspectral imagery. In *Hyperspectral Data Exploitation: Theory and Applications*; John Wiley & Sons: Hoboken, NJ, USA, 2007; pp. 379–407.

17. Christophe, E.; Mailhes, C.; Duhamel, P. Hyperspectral image compression: Adapting SPIHT and EZW to anisotropic 3-D wavelet coding. *IEEE Trans. Image Process.* **2008**, *17*, 2334–2346. [CrossRef]

18. Huang, B. *Satellite Data Compression*; Springer Science & Business Media: Berlin, Germany, 2011.

19. Du, Q.; Fowler, J.E. Hyperspectral image compression using JPEG2000 and principal component analysis. *IEEE Geosci. Remote Sens. Lett.* **2007**, *4*, 201–205. [CrossRef]

20. Báscones, D.; González, C.; Mozos, D. Hyperspectral Image Compression Using Vector Quantization, PCA and JPEG2000. *Remote Sens.* **2018**, *10*, 907. [CrossRef]

21. Taubman, D.; Marcellin, M. *JPEG2000 Image Compression Fundamentals, Standards and Practice*; Springer Science & Business Media: Berlin, Germay, 2012; Volume 642, p. 773. [CrossRef]

22. Skodras, A.; Christopoulos, C.; Ebrahimi, T. The JPEG 2000 Still Image Compression Standard. *IEEE Signal Process. Mag.* **2001**, *18*, 36–58. [CrossRef]

23. Wallace, G.K. The JPEG still picture compression standard. *IEEE Trans. Consum. Electron.* **1992**. [CrossRef]

24. Gangadhar, M.; Bhatia, D. FPGA based EBCOT architecture for JPEG 2000. *Microprocess. Microsyst.* **2003**, *29*, 363–373. [CrossRef]

25. Lian, C.J.; Chen, K.F.; Chen, H.H.; Chen, L.G. Analysis and architecture design of block-coding engine for EBCOT in JPEG 2000. *IEEE Trans. Circuits Syst. Video Technol.* **2003**, *13*, 219–230. [CrossRef]

26. Dyer, M.; Taubman, D.; Nooshabadi, S.; Gupta, A.K. Concurrency techniques for arithmetic coding in JPEG2000. *IEEE Trans. Circuits Syst. I Regul. Pap.* **2006**, *53*, 1203–1213. [CrossRef]

27. Gupta, A.K.; Taubman, D.S.; Nooshabadi, S. High speed VLSI architecture for bit plane encoder of JPEG 2000. *IEEE Midwest Sympos. Circuits Syst.* **2004**, *2*, 233–236. [CrossRef]

28. Kai, L.; Chengke, W.; Yunsong, L. A high-performance VLSI arquitecture of EBCOT block coding in JPEG2000. *J. Electron.* **2006**, *23*, 1–5.

29. Sarawadekar, K.; Banerjee, S. Low-cost, high-performance VLSI design of an MQ coder for JPEG 2000. In Proceedings of the IEEE 10th International Conference on Signal Processing Proceedings, Beijing, China, 24–28 October 2010; Number D, pp. 397–400.

30. Saidani, T.; Atri, M.; Tourki, R. Implementation of JPEG 2000 MQ-coder. In Proceedings of the 2008 3rd International Conference on Design and Technology of Integrated Systems in Nanoscale Era, Tozeur, Tunisia, 25–27 March 2008; pp. 1–4. [CrossRef]

31. Sarawadekar, K.; Banerjee, S. VLSI design of memory-efficient, high-speed baseline MQ coder for JPEG 2000. *Integr. VLSI J.* **2012**, *45*, 1–8. [CrossRef]

32. Liu, K.; Zhou, Y.; Li, Y.S.; Ma, J.F.; Song Li, Y.; Ma, J.F. A high performance MQ encoder architecture in JPEG2000. *Integr. VLSI J.* **2010**, *43*, 305–317. [CrossRef]

33. Sulaiman, N.; Obaid, Z.A.; Marhaban, M.H.; Hamidon, M.N. Design and Implementation of FPGA-Based Systems—A Review. *Aust. J. Basic Appl. Sci.* **2009**, *3*, 3575–3596.

34. Trimberger, S.M. Three ages of FPGAs: A retrospective on the first thirty years of FPGA technology. *Proc. IEEE* **2015**, *103*, 318–331. [CrossRef]

35. Xilinx. Space-Grade Virtex-5QV FPGA. Available online: www.xilinx.com/products/silicon-devices/fpga/virtex-5qv.html (accessed on 8 August 2020).

36. Báscones, D.; González, C.; Mozos, D. Parallel Implementation of the CCSDS 1.2.3 Standard for Hyperspectral Lossless Compression. *Remote Sens.* **2017**, *9*, 973. [CrossRef]

37. Báscones, D.; Gonzalez, C.; Mozos, D. FPGA Implementation of the CCSDS 1.2.3 Standard for Real-Time Hyperspectral Lossless Compression. *IEEE J. Sel. Top. Appl. Earth Obs. Remote Sens.* **2017**, *11*, 1158–1165. [CrossRef]

38. Bascones, D.; Gonzalez, C.; Mozos, D. An Extremely Pipelined FPGA Implementation of a Lossy Hyperspectral Image Compression Algorithm. *IEEE Trans. Geosci. Remote Sens.* **2020**, 1–13. [CrossRef]

39. Báscones, D. Implementación Sobre FPGA de un Algoritmo de Compresión de Imágenes Hiperespectrales Basado en JPEG2000. Ph.D. Thesis, Universidad Complutense de Madrid, Madrid, Spain, 2018.

40. Higgins, G.; Faul, S.; McEvoy, R.P.; McGinley, B.; Glavin, M.; Marnane, W.P.; Jones, E. EEG compression using JPEG2000 how much loss is too much? In Proceedings of the 2010 Annual International Conference of the IEEE Engineering in Medicine and Biology Society, EMBC'10, Buenos Aires, Argentina, 31 August–4 September 2010; pp. 614–617. [CrossRef]

41. Marpe, D.; George, V.; Cycon, H.L.; Barthel, K.U. Performance evaluation of Motion-JPEG2000 in comparison with H . 264 / AVC operated in pure intra coding mode. *SPIE Proc.* **2003**, *5266*, 129–137. [CrossRef]

42. Van Fleet, P.J. *Discrete Wavelet Transformations: An Elementary Approach with Applications*; John Wiley & Sons: Hoboken, NJ, USA, 2011. [CrossRef]

43. Makandar, A.; Halalli, B. Image Enhancement Techniques using Highpass and Lowpass Filters. *Int. J. Comput. Appl.* **2015**, *109*, 12–15. [CrossRef]

44. Taubman, D. High performance scalable image compression with EBCOT. *IEEE Trans. Image Process.* **2000**, *9*, 1158–1170. [CrossRef]

45. Andra, K.; Acharya, T.; Chakrabarti, C. Efficient VLSI implementation of bit plane coder of JPEG2000. *Appl. Digit. Image Process. Xxiv* **2001**, *4472*, 246–257. [CrossRef]

46. Li, Y.; Bayoumi, M.A. A three level parallel high speed low power architecture for EBCOT of JPEG 2000. *IEEE Trans. Circuits Syst. Video Technol.* **2006**, *16*, 1153–1163. [CrossRef]

47. Jayavathi, S.D.; Shenbagavalli, A. FPGA Implementation of MQ Coder in JPEG 2000 Standard—A Review. **2016**, *28*, 76–83.

48. Rhu, M.; Park, I.C. Optimization of arithmetic coding for JPEG2000. *IEEE Trans. Circuits Syst. Video Technol.* **2010**, *20*, 446–451. [CrossRef]

49. Ahmadvand, M.; Ezhdehakosh, A. A New Pipelined Architecture for JPEG2000. *World Congress Eng. Comput. Sci.* **2012**, *2*, 24–26.

50. Mei, K.; Zheng, N.; Huang, C.; Liu, Y.; Zeng, Q. VLSI design of a high-speed and area-efficient JPEG2000 encoder. *IEEE Trans. Circuits Syst. Video Technol.* **2007**, *17*, 1065–1078. [CrossRef]

51. Chang, Y.W.; Fang, H.C.; Chen, L.G. High Performance Two-Symbol Arithmetic Encoder in JPEG 2000. 2000. pp. 4–7. Available online: https://video.ee.ntu.edu.tw/publication/paper/[C][2004][ICCE][Yu-Wei.Chang][1].pdf (accessed on 8 August 2020).

52. Kumar, N.R.; Xiang, W.; Wang, Y. An FPGA-based fast two-symbol processing architecture for JPEG 2000 arithmetic coding. In Proceedings of the 2010 IEEE International Conference on Acoustics, Speech and Signal Processing, Dallas, TX, USA, 14–19 March 2010; pp. 1282–1285. [CrossRef]

53. Sarawadekar, K.; Banerjee, S. An Efficient Pass-Parallel Architecture for Embedded Block Coder in JPEG 2000. *IEEE Trans. Circuits Syst.* **2011**, *21*, 825–836. [CrossRef]

54. Kumar, N.R.; Xiang, W.; Wang, Y. Two-symbol FPGA architecture for fast arithmetic encoding in JPEG 2000. *J. Signal Process. Syst.* **2012**, *69*, 213–224. [CrossRef]

55. Spectir. Free Data Samples. Available online: https://www.spectir.com/free-data-samples/ (accessed on 5 August 2020).

56. CCSDS. Collaborative Work Environment. Available online: https://cwe.ccsds.org/sls/default.aspx (accessed on 5 August 2020).

57. Báscones, D. Jypec. Available online: github.com/Daniel-BG/Jypec (accessed on 17 January 2018).
58. Báscones, D. Vypec. Available online: github.com/Daniel-BG/Vypec (accessed on 25 January 2018).

Article

Analysis of Variable-Length Codes for Integer Encoding in Hyperspectral Data Compression with the k^2-Raster Compact Data Structure

Kevin Chow *, Dion Eustathios Olivier Tzamarias, Miguel Hernández-Cabronero, Ian Blanes and Joan Serra-Sagristà

Department of Information and Communications Engineering, Universitat Autònoma de Barcelona, 08193 Cerdanyola del Vallès, Barcelona, Spain; dion.tzamarias@uab.cat (D.E.O.T.); miguel.hernandez@uab.cat (M.H.-C.); ian.blanes@uab.cat (I.B.); joan.serra@uab.cat (J.S.-S.)

* Correspondence: kevin.chow@uab.cat

Received: 19 May 2020; Accepted: 18 June 2020; Published: 20 June 2020

Abstract: This paper examines the various variable-length encoders that provide integer encoding to hyperspectral scene data within a k^2-raster compact data structure. This compact data structure leads to a compression ratio similar to that produced by some of the classical compression techniques. This compact data structure also provides direct access for query to its data elements without requiring any decompression. The selection of the integer encoder is critical for obtaining a competitive performance considering both the compression ratio and access time. In this research, we show experimental results of different integer encoders such as Rice, Simple9, Simple16, PForDelta codes, and DACs. Further, a method to determine an appropriate k value for building a k^2-raster compact data structure with competitive performance is discussed.

Keywords: compact data structure; k^2-raster; DACs; Elias codes; Simple9; Simple16; PForDelta; Rice codes; hyperspectral scenes

---

1. Introduction

Hyperspectral scenes [1–10] are data taken from the air by sensors such as AVIRIS (Airborne Visible/Infrared Imaging Spectrometer) or by satellite instruments such as Hyperion and IASI (Infrared Atmospheric Sounding Interferometer). These scenes are made up of multiple bands from across the electromagnetic spectrum, and data extracted from certain bands are helpful in finding objects such as oil fields [11] or minerals [12]. Other applications include weather prediction [13] and wildfire soil studies [14], to name a few. Due to their sizes, hyperspectral scenes are usually compressed to facilitate their transmission and reduce storage size.

Compact data structures [15] are a type of data structure where data are stored efficiently while at the same time providing real-time processing and compression of the data. They can be loaded into main memory and accessed directly by means of the rank and select functions [16] in the structures. Compressed data provide reduced space usage and query time, i.e., they allow more efficient transmission through limited communication channels, as well as faster data access. There is no need to decompress a large portion of the structure to access and query individual data as is the case with data compressed by classical compression algorithms such as gzip or bzip2 and by specialized algorithms such as CCSDS123.0-B-1 [17] or KLT+JPEG 2000 [18,19]. In this paper, we are interested in lossless compression of hyperspectral scenes through compact data structures. Therefore, reconstructed scenes should be identical to the originals before compression. Any deterministic analysis process will necessarily yield the same results. Figure 1 shows several images from our datasets.

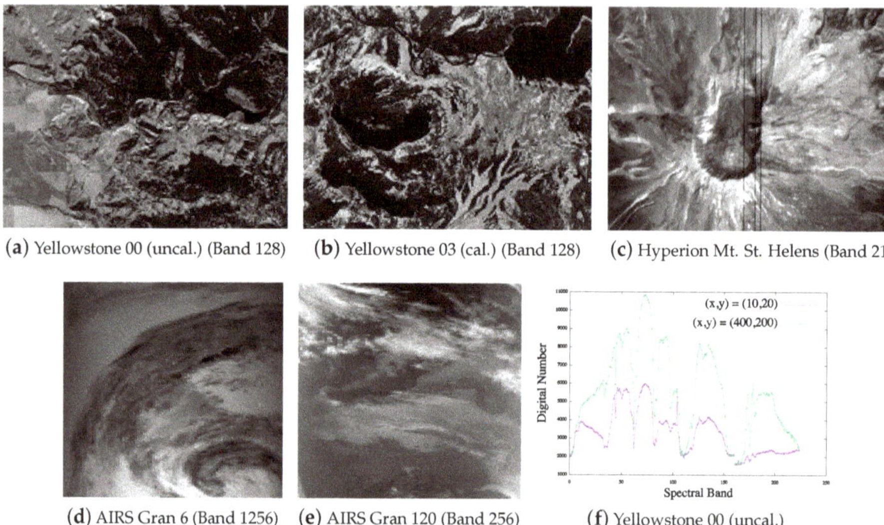

(**a**) Yellowstone 00 (uncal.) (Band 128)   (**b**) Yellowstone 03 (cal.) (Band 128)   (**c**) Hyperion Mt. St. Helens (Band 21)

(**d**) AIRS Gran 6 (Band 1256)   (**e**) AIRS Gran 120 (Band 256)   (**f**) Yellowstone 00 (uncal.)

Figure 1. Several hyperspectral scenes used in this paper. The original and the decompressed scenes discussed in this paper are numerically identical. Depicted also are two spectral signatures for AVIRIS Yellowstone 00 (uncal.). The AVIRIS images in this figure are courtesy of NASA/JPL-Caltech.

The compact data structure used in this paper is called k^2-raster. It is a tree structure developed from another compact data structure called k^2-tree. k^2-raster is built from a raster matrix with its pixel cells filled with integer values, while k^2-tree is from a bitmap matrix with zero and one values. During the construction of the k^2-raster tree, if the neighboring pixels have equal values such as clusters (spatial correlation), the number of nodes in the tree that need to be saved is reduced. If the values are similar, as discussed later in this paper, the values will be made even smaller. They are then compressed or packed in a more compact form by the integer encoders, and with these small integers, the compression results are even better. Moreover, when it comes to querying cells, a tree structure speeds up the search, saving access time. Another added advantage of some of the integer encoders is that they provide direct random access to the cells without any need for full decompression.

Currently, huge amounts of remote sensing data have been produced, transmitted, and archived, and we can foresee that in the future, the amount of larger datasets is expected to keep growing at a fast rate. The need for their compression is becoming more pressing and critical. In view of this trend, we take on the task of remote sensing compression and make it as one of our main objectives. In this research work, we reduce hyperspectral data sizes by using compact data structures to produce lossless compression. Early on, we began by examining the possibility of taking advantage of the spatial correlation and spectral correlation in the data. In our previous paper [20], we presented a predictive method and a differential method that made use of these correlations in hyperspectral data with favorable results. However, in this paper, we would like to focus on selecting a suitable integer encoder that is employed in the k^2-raster compact data structure, as that is also a major factor in providing competitive compression ratios.

Compression of integer data in the most effective and efficient way, in relation to compact data structures, has been the focus of many studies over the past several decades. Some include Elias [21–23], Rice [24–26], PForDelta [27–29], and Directly Addressable Codes (DACs) [30–32]. In our case, we need to store non-negative, typically small integers in the k^2-raster structure. This structure is a tree built in such a way that the nodes are not connected by pointers, but can still be reached with the use of a compact data structure linear rank function. When the data are saved, no pointers need to be stored, thus keeping the size of the structure small. Additionally, we use a fixed code ([15], §2.7) to help us save even more space. In what follows, we investigate the effectiveness of some of these integer encoders.

The rest of the paper is organized as follows: In Section 2, we describe the k^2-raster structure, followed by the various variable-length integer encoders such as Elias, Rice, PForDelta, and DACs. Section 3 presents experimental results for finding the best and optimal values for k and exploring the different integer encoders for k^2-raster. This is done in comparison with classical compression techniques. Lastly, some conclusions and final thoughts on future work are put forth in Section 4.

2. Materials and Methods

In this section, we describe k^2-raster [33] and the integer encoders Elias, Rice, Simple9, Simple16, PForDelta codes, and DACs. This is followed by a discussion on how to obtain the best value of k and two related works on raster compression: heuristic k^2-raster [33] and 3D-2D mapping [34].

2.1. K²-Raster

The k^2-tree structure was originally proposed by Ladra et al. [35] as a compact representation of the adjacency matrix of a directed graph. Its applications include web graphs and social networks. Based on k^2-tree, the same authors also proposed k^2-raster [33], which is specifically designed for raster data including images. A k^2-raster is built from a matrix of width w and height h. If the matrix can be partitioned into k^2 square subquadrants of equal size, it can be used directly. Otherwise, it is necessary to enlarge the matrix to size $s \times s$, where s is computed as:

$$s = k^{\lceil \log_k max(w,\, h) \rceil},$$ (1)

setting the new elements to 0. This extended matrix is then recursively partitioned into k^2 submatrices of identical size, referred to as quadrants. This process is repeated until all cells in a quadrant have the same value, or until the submatrix has size 1×1 and cannot be further subdivided. This partitioning induces a tree topology, which is represented in a bitmap T. Elements can then be accessed via a rank function. At each tree level, the maximum and minimum values of each quadrant are computed. These are then compared with the corresponding maximum and minimum values of the parent, and the differences are stored in the $Vmax$ and $Vmin$ arrays of each level. Saving the differences instead of the original values results in lower values for each node, which in turn allows a better compression with DACs or other integer encoders such as Simple9, PForDelta, etc. An example of a simple 8×8 matrix is given in Figure 2 to illustrate this process. A k^2-raster is constructed from this matrix with maximum and minimum values as given in Figure 3. Differences from the parents' extrema are then computed as explained above, resulting in the structure shown in Figure 4. Next, with the exception of the root node at the top level, the $Vmax$ and $Vmin$ arrays at all levels are concatenated to form $Lmax$ and $Lmin$, respectively. Both arrays are then compressed by an integer encoder such as DACs. The root's maximum ($rMax$) and minimum ($rMin$) values remain uncompressed. The resulting elements, which fully describe this k^2-raster structure, are given in Table 1.

2.2. Unary Codes and Notation

We denote x as a non-negative integer. The expression $|x|$ gives the minimum bit length needed to express x, i.e., $|x| = \lfloor \log_2 x \rfloor + 1$.

Unary codes are generally used for small integers. Unary codes have the following form:

$$u(x) = 0^x\, 1\,,$$ (2)

where the superscript x indicates the number of consecutive 0 bits in the code. For example, $u(1_d) = 0^1\, 1 = 01_b$, $u(6_d) = 0^6\, 1 = 0000001_b$, $u(9_d) = 0^9\, 1 = 0000000001_b$. Here, bits are denoted by a subscript b and decimal numbers by a subscript d. Furthermore, when codes are composed of two parts, they are spaced apart for readability purposes. In general, the notation used in [15] is adopted in this paper.

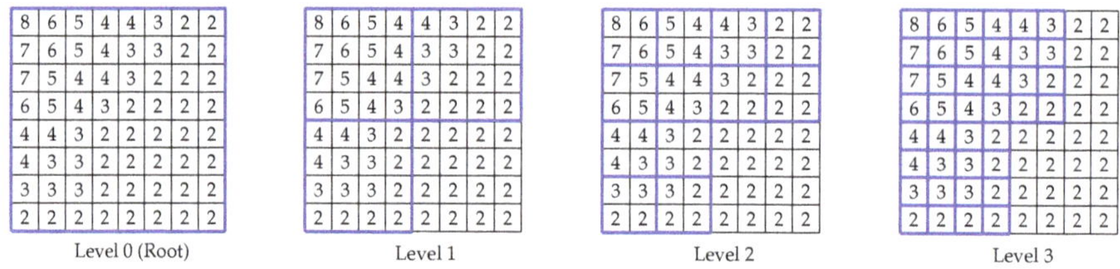

Level 0 (Root)   Level 1   Level 2   Level 3

Figure 2. Subdivision of an example 8×8 matrix for k^2-raster ($k = 2$).

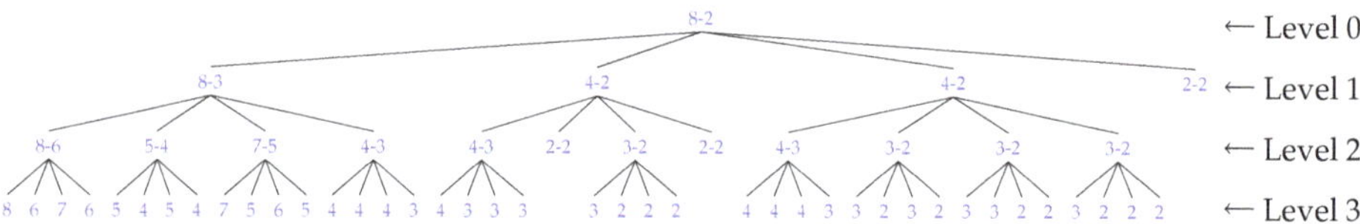

Figure 3. A k^2-raster ($k = 2$) tree storing the maximum and minimum values for each quadrant of every recursive subdivision of the matrix in Figure 2. Every node contains the maximum and minimum values of the subquadrant, separated by a dash. On the last level, only one value is shown as each subquadrant contains only one cell.

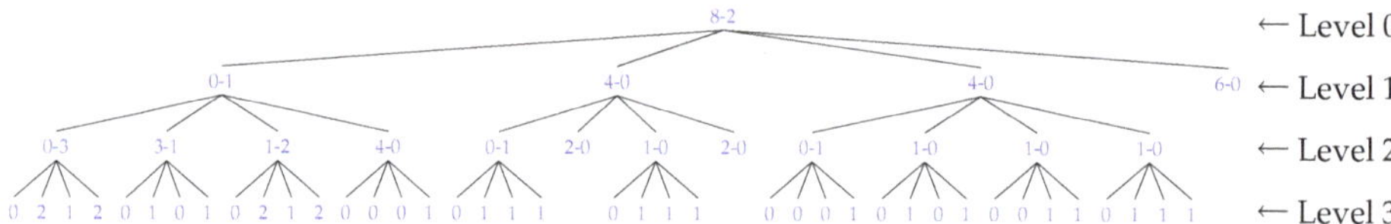

Figure 4. Based on the tree in Figure 3, the maximum value of each node is subtracted from that of its parent while the minimum value of the parent is subtracted from the node's minimum value. These differences then replace their corresponding values in the node. The maximum and minimum values of the root remain the same.

Table 1. An example of the elements of a k^2-raster based on Figures 2–4.

T Bitmap	binary		1110 1111 1010 1111
Lmax	decimal	Level 1	0446
		Level 2	0314 0212 0111
		Level 3	0212 0101 0212 0001 0111 0111 0001 0101 0011 0111
Lmin	decimal	Level 1	100
		Level 2	3120 10 1000
rMax	decimal		8
rMin	decimal		2

2.3. Elias Codes

Elias codes include Gamma (γ) codes and Delta (δ) codes. They were developed by Peter Elias [21] to encode natural numbers, and in general, they work well with sequences of small numbers.

Gamma codes have the following form:

$$\gamma(x) = 0^{|x|-1} [x]_{|x|} = u(|x| - 1) [x]_{|x|-1} , \tag{3}$$

where $[x]_l$ represents the l least significant bits of x. For example, $\gamma(1_d) = \gamma(1_b) = 1_b$, $\gamma(4_d) = \gamma(100_b)$ $= 001\ 00_b$, $\gamma(6_d) = \gamma(110_b) = 001\ 10_b$, $\gamma(9_d) = \gamma(1001_b) = 0001\ 001_b$, $\gamma(14_d) = \gamma(1110_b) = 0001\ 110_b$.
Delta codes have the following form:

$$\delta(x) = \gamma(\mid x \mid)\,[x]_{\mid x \mid -1}\,. \tag{4}$$

For values that are larger than 31, Delta codes produce shorter codewords than Gamma codes. This is due to the use of Gamma codes in forming the first part of their codes, which provides a shorter code length for Delta codes as the number becomes larger. Some examples are: $\delta(1_d) = \delta(1_b) = 1_b$, $\delta(6_d) = \delta(110_b) = 011\ 10_b$, $\delta(9_d) = \delta(1001_b) = 00100\ 001_b$, $\delta(14_d) = \delta(1110_b) = 00100\ 110_b$.

2.4. Rice Codes

Rice codes [25] are a special case of Golomb codes. Let x be an integer value in the sequence, and let $y = \lfloor x/2^l \rfloor$, where l is a non-negative integer parameter. The Rice codes for this parameter are defined as:

$$R_l(x) = u(y+1)\,[x]_l\,. \tag{5}$$

Some examples are shown for different values of l in Table 2.

Table 2. Some examples of Rice codes.

Value v		Rice Code $R_l(v)$			
Decimal	Binary	$l=1$	$l=2$	$l=3$	$l=4$
1_d	1_b	1_b	1_b	1_b	1_b
6_d	110_b	$0001\ 0_b$	$01\ 10_b$	110_b	110_b
9_d	1001_b	$00001\ 1_b$	$001\ 01_b$	$01\ 001_b$	1001_b
14_d	1110_b	$00000001\ 0_b$	$0001\ 10_b$	$01\ 110_b$	1110_b

To obtain optimal performance among Rice codes, l should be selected to be close to the expected value of the input integers. In general, Rice codes give better compression performance than Elias γ and δ codes.

2.5. Simple9, Simple16, and PForDelta

Apart from Elias codes and Rice codes, the codes in this section store the integers in single or multiple word-sized elements to achieve data compression. They have been shown to have good compression ratios [30].

Simple9 [36] assigns a maximum possible number of a certain bit length to a 28-bit segment or packing space of a 32-bit word. The other 4 bits contain a selector that has a value ranging from 0 to 8. Each selector has information that indicates how the integers are stored, and that includes the number of these integers and the maximum number of bits that each integer is allowed in this packing space. For example, Selector 0 tests to see if the first 28 integers in the data have a value of 0 or 1, i.e., a bit length of 1. If they do, then they are stored in this 28-bit segment. Otherwise, Selector 1 tests to see if it can pack 14 integers into the segment with a maximum bit length of 2 bits for each. If this still does not work, Selector 2 tests to see if 9 integers can each be packed into a maximum bit length of 3 bits. This testing goes on until the right number of data are found that can be stored in these 28 bits. Table 3 shows the 9 different ways of using 28 bits in a word of 32 bits in Simple9.

Simple16 [37] is a variant of Simple9 and uses all 16 combinations in the selector bits. Their values range from 0 to 15. Table 4 shows the 16 different ways of packing integers into the 28-bit segment in Simple16.

PForDelta [27] is also similar to both Simple9 and Simple16, but encodes a fixed group of numbers at a time. To do so, 128- or 256-bit words are used.

Due to its relative simplicity, Simple9 is used here as an example to illustrate how an integer sequence is stored in the encoders described in this section. This sequence $<3591\ 25\ 13\ 12\ 15\ 12\ 11\ 26\ 20\ 8\ 13\ 8\ 9\ 7\ 13\ 10\ 12\ 0\ 10>_d$ is taken from the *Lmax* array of one of our data scenes AG9, and the bit-packing is shown in Table 5. There are 19 integers in the sequence. Assuming the integer is 16 bits each, the sequence has a total size of 38 bytes. After packing into the array, the sequence occupies only 16 bytes.

Table 3. Nine different ways of encoding numbers in the 28-bit packing space in Simple9.

Selector	Number of Integers n	Width of Integers $\lfloor 28/n \rfloor$ (Bits)	Wasted Bits
0	28	1	0
1	14	2	0
2	9	3	1
3	7	4	0
4	5	5	3
5	4	7	0
6	3	9	1
7	2	14	0
8	1	28	0

Table 4. Sixteen different ways of encoding numbers in the 28-bit packing space in Simple16. There are no wasted bits in any of the selectors.

Selector	Number of Integers	Width of Integers (Bits)		
0	28			28×1 bit
1	21		7×2 bits,	14×1 bit
2	21	7×1 bit,	7×2 bits,	7×1 bit
3	21		14×1 bit,	7×2 bits
4	14			14×2 bits
5	9		1×4 bits,	8×3 bits
6	8	1×3 bits,	4×4 bits,	3×3 bits
7	7			7×4 bits
8	6		4×5 bits,	2×4 bits
9	6		2×4 bits,	4×5 bits
10	5		3×6 bits,	2×5 bits
11	5		2×5 bits,	3×6 bits
12	4			4×7 bits
13	3		1×10 bits,	2×9 bits
14	2			2×14 bits
15	1			1×28 bits

Table 5. Example to show how the integer sequence $<3591\ 25\ 13\ 12\ 15\ 12\ 11\ 26\ 20\ 8\ 13\ 8\ 9\ 7\ 13\ 10\ 12\ 0\ 10>_d$ is stored with Simple9.

Element	Selector	Number of Integers	Integers Stored (Decimal)	Integers Stored (Binary)
0	7_d (0111_b)	2	3591 25	00111000000111 00000000011001
1	4_d (0100_b)	5	13 12 15 12 11	01101 01100 01111 01100 01011
2	4_d (0100_b)	5	26 20 8 13 8	11010 10100 01000 01101 01000
3	3_d (0011_b)	7	9 7 13 10 12 0 10	1001 0111 1101 1010 1100 0000 1010

2.6. Directly Addressable Codes

Directly Addressable Codes (DACs) can be used to compress k^2-raster and provide access to variable-length codes. Based on the concept of compact data structures, DACs were proposed in the papers published by Brisaboa et al. in 2009 [30] and 2013 [31]. This structure is proven to yield good compression ratios for variable-length integer sequences. By means of the rank function, it gains fast direct access to any position of the sequence in a very compact space. The original authors also asserted that it was best suited for a sequence of integers with a skewed frequency distribution toward smaller integer values.

Different types of encoding are used for DACs, and the one that we are interested in for k^2-raster is called VBytecoding. Consider a sequence of integers x. Each integer x_i, which is represented by $\lfloor \log_2 x_i \rfloor + 1$ bits, is broken into chunks of bits of size C_S. Each chunk is stored in a block of size $C_S + 1$ with the additional bit used as a control bit. The chunk occupies the lower bits in the block and the control bit the highest bit. The block that holds the most significant bits of the integer has its control bit set to 0, while the others have it set to 1. For example, if we have an integer 41_d (101001_b), which is 6 bits long, and if the chunk size is $C_S = 3$, then we have 2 blocks: $\underline{0}101\ \underline{1}001_b$. The control bit in each block is shown underlined. To show how the blocks are organized and stored, we again illustrate it with an example. Given five integers of variable length: 7_d (111_b), 41_d (101001_b), 100_d (1100100_b), 63_d (111111_b), 427_d (110101011_b), and a chunk size of 3 (the block size is 4), their representations are listed in Table 6.

Table 6. Example of an integer sequence and the corresponding DACs blocks of the integers.

Decimal	Binary	DACs Blocks
7_d	$\underline{0}111_b$	$(B_{1,1}A_{1,1})$
41_d	$\underline{0}101\ \underline{1}001_b$	$(B_{2,2}A_{2,2}\ B_{2,1}A_{2,1})$
100_d	$\underline{0}001\ \underline{1}100\ \underline{1}100_b$	$(B_{3,3}A_{3,3}\ B_{3,2}A_{3,2}\ B_{3,1}A_{3,1})$
63_d	$\underline{0}111\ \underline{1}111_b$	$(B_{4,2}A_{4,2}\ B_{4,1}A_{4,1})$
427_d	$\underline{0}110\ \underline{1}101\ \underline{1}011_b$	$(B_{5,3}A_{5,3}\ B_{5,2}A_{5,2}\ B_{5,1}A_{5,1})$

We store them in three blocks of arrays A and control bitmaps B. This is depicted in Figure 5. To retrieve the values in the arrays A, we make use of the corresponding bitmaps B with the rank function. This function returns the number of bits, which are set to 1 from the beginning position to the one being queried in the control bitmap B_i. An example of how the function is used follows: If we want to access the third integer (100_d) in the sequence in Figure 5, we start looking for the third element in the array A_1 in Block$_1$ and find $A_{3,1}$ with its corresponding control bitmap $B_{3,1}$. The function rank($B_{3,1}$) then gives a result of 2, which means that the second element $A_{3,2}$ in the array A_2 in Block$_2$ contains the next block. With the control bit in $B_{3,2}$, we compute the function rank($B_{3,2}$) and obtain a result of 1. This means the next block in Block$_3$ can be found in the first element $A_{3,3}$. Since its corresponding control bitmap $B_{3,3}$ is set to 0, the search ends here. All the blocks found are finally concatenated to form the third integer in the sequence.

More information on DACs and the software code can be found in the papers [30,31] by Ladra et al.

Block$_1$	A_1	111 ($A_{1,1}$)	001 ($A_{2,1}$)	100 ($A_{3,1}$)	111 ($A_{4,1}$)	011 ($A_{5,1}$)
	B_1	0 ($B_{1,1}$)	1 ($B_{2,1}$)	1 ($B_{3,1}$)	1 ($B_{4,1}$)	1 ($B_{5,1}$)
Block$_2$	A_2	101 ($A_{2,2}$)	100 ($A_{3,2}$)	111 ($A_{4,2}$)	101 ($A_{5,2}$)	
	B_2	0 ($B_{2,2}$)	1 ($B_{3,2}$)	0 ($B_{4,2}$)	1 ($B_{5,2}$)	
Block$_3$	A_3	001 ($A_{3,3}$)	110 ($A_{5,3}$)			
	B_3	0 ($B_{3,3}$)	0 ($B_{5,3}$)			

Figure 5. Organization of 3 DACs blocks.

2.7. Selection of the k Value

Following the description of Subsection 2.1, using different k values leads to the creation of *Lmax* and *Lmin* arrays of different lengths. This, in turn, affects the final results of the size of k^2-raster. With this in mind, we present a heuristic approach that can be used to determine the best k value for obtaining the smallest storage size. First, we compute the sizes of the extended matrix for different values of k within a suitable range using Equation (1). Then, we find the k value that corresponds to the matrix with the smallest size, and the result can be considered as the best k value. Before the start of the k^2-raster building process, the program can find the best k value and use it as the default.

2.8. Heuristic k^2-Raster

In the k^2-raster paper by Ladra et al. [33], a variant of this structure was also proposed whereby the elements at the last level of the tree structure are stored by using an entropy-based heuristic approach. This is denoted by k_H^2-raster. For example, for $k = 2$, each set of the 4 nodes that are from the same parent forms a codeword. It is possible that at this same level of the tree, these codewords may be repeated, and their frequencies of occurrences can be computed. These sets of codewords and their frequencies are then compressed and saved. In effect, the more these codewords are repeated, the less storage space they take up. An example of codeword frequency based on the k^2-raster discussed in Section 2.1 is shown in Table 7. According to experiments conducted by the authors of [33], it saves space in the final representation.

Table 7. Codeword frequency in Level 3 of the *Lmax* bitmap in the k^2-raster structure in Figure 1.

Codeword	Frequency
0111	3
0212	2
0101	2
0001	2
0011	1

2.9. 3D-2D Mapping

A study on compact representation of raster images in a time-series was proposed by Cruces et al. in [34]. This method is based on the 3D to 2D mapping of a raster where 3D tuples $<x, y, z>$ are mapped into a 2D binary grid. That is, a raster of size $w \times h$ with values in a certain range, between 0 and v inclusive, has a binary matrix of $w \times h$ columns and $v+1$ rows. All the rasters are then concatenated into a 3D matrix and stored as a 3D-k^2-tree.

3. Experimental Results

In this section, we present an exhaustive comparison of the different integer encoders for use with k^2-raster. First, though, we report results from experiments for finding the best k value. Reported also are the experimental results to find out if the heuristic k^2-raster and 3D-2D mapping would give better storage sizes. All storage sizes in this section are expressed as bits per pixel per band (bpppb).

The hyperspectral scenes were captured by different sensors: Atmospheric Infrared Sounder (AIRS), AVIRIS, Compact Reconnaissance Imaging Spectrometer for Mars (CRISM), Hyperion, and IASI. Except for IASI, all of them are publicly available for download (http://cwe.ccsds.org/sls/docs/sls-dc/123.0-B-Info/TestData). The hyperspectral scenes used are listed in Table 8.

The implementations for k^2-raster and k_H^2-raster were based on the algorithms presented in the paper by Ladra et al. [33]. The sdsl-lite implementation of k^2-tree by Simon Gog [38] (https://github.com/simongog/sdsl-lite/blob/master/include/sdsl/k2_tree.hpp) was used for testing 3D-2D mapping described in the paper by Cruces et al. [34]. The DACs software was downloaded from a package called "DACs, optimization with no further restrictions" at the Universidade da Coruña's Database Laboratory website (http://lbd.udc.es/research/DACS/). The programming code for the Rice, PForDelta, Simple9, and Simple16 codes was written by the programmers Diego Caro, Michael Dipperstein, and Christopher Hoobin and was downloaded from these authors' GitHub web pages. Slight modifications to the code were made to meet our requirements to perform the experiments. All programs for this paper were written in C and C++ and compiled with gnu g++ 5.4.0 20160609 with -Ofast optimization. The experiments were carried out on an Intel Core 2 Duo

CPU E7400 @2.80GHz with 3072KB of cache and 3GB of RAM. The operating system was Ubuntu 16.04.5 LTS with kernel 4.15.0-47-generic (64 bits). The software code is available at http://gici.uab.cat/GiciWebPage/downloads.php.

Table 8. Hyperspectral scenes used in our experiments. Also shown are the bit rate and bit rate reduction using k^2-raster. x is the scene width, y the scene height, and z the number of spectral bands. bpppb, bits per pixel per band; CRISM, Compact Reconnaissance Imaging Spectrometer for Mars; IASI, Infrared Atmospheric Sounding Interferometer.

Sensor	Name	C/U *	Acronym	Original Dimensions ($x \times y \times z$)	Bit Depth (bpppb)	Best k Value	k^2-raster Bit Rate (bpppb)	k^2-raster Bit-Rate Reduction (%)
AIRS	9	U	AG9	$90 \times 135 \times 1501$	12	6	9.49	21%
	16	U	AG16	$90 \times 135 \times 1501$	12	6	9.12	24%
	60	U	AG60	$90 \times 135 \times 1501$	12	15	9.72	19%
	126	U	AG126	$90 \times 135 \times 1501$	12	6	9.61	20%
	129	U	AG129	$90 \times 135 \times 1501$	12	6	8.65	28%
	151	U	AG151	$90 \times 135 \times 1501$	12	6	9.53	21%
	182	U	AG182	$90 \times 135 \times 1501$	12	6	9.68	19%
	193	U	AG193	$90 \times 135 \times 1501$	12	15	9.30	23%
AVIRIS	Yellowstone sc. 00	C	ACY00	$677 \times 512 \times 224$	16	6	9.61	40%
	Yellowstone sc. 03	C	ACY03	$677 \times 512 \times 224$	16	6	9.42	41%
	Yellowstone sc. 10	C	ACY10	$677 \times 512 \times 224$	16	6	7.62	52%
	Yellowstone sc. 11	C	ACY11	$677 \times 512 \times 224$	16	6	8.81	45%
	Yellowstone sc. 18	C	ACY18	$677 \times 512 \times 224$	16	6	9.78	39%
	Yellowstone sc. 00	U	AUY00	$680 \times 512 \times 224$	16	9	11.92	25%
	Yellowstone sc. 03	U	AUY03	$680 \times 512 \times 224$	16	9	11.74	27%
	Yellowstone sc. 10	U	AUY10	$680 \times 512 \times 224$	16	9	9.99	38%
	Yellowstone sc. 11	U	AUY11	$680 \times 512 \times 224$	16	9	11.27	30%
	Yellowstone sc. 18	U	AUY18	$680 \times 512 \times 224$	16	9	12.15	24%
CRISM	frt000065e6_07_sc164	U	C164	$640 \times 420 \times 545$	12	6	10.08	16%
	frt00008849_07_sc165	U	C165	$640 \times 450 \times 545$	12	6	10.37	14%
	frt0001077d_07_sc166	U	C166	$640 \times 480 \times 545$	12	6	11.05	8%
	hrl00004f38_07_sc181	U	C181	$320 \times 420 \times 545$	12	5	9.97	17%
	hrl0000648f_07_sc182	U	C182	$320 \times 450 \times 545$	12	5	10.11	16%
	hrl0000ba9c_07_sc183	U	C183	$320 \times 480 \times 545$	12	5	10.65	11%
Hyperion	Agricultural	C	HCA	$256 \times 3129 \times 242$	12	16	8.52	29%
	Coral Reef	C	HCC	$256 \times 3127 \times 242$	12	8	7.62	36%
	Urban	C	HCU	$256 \times 2905 \times 242$	12	16	8.85	26%
	Erta Ale	U	HUEA	$256 \times 3187 \times 242$	12	8	7.76	35%
	Lake Monona	U	HULM	$256 \times 3176 \times 242$	12	8	7.82	35%
	Mt. St. Helena	U	HUMS	$256 \times 3242 \times 242$	12	8	7.91	34%
IASI	Level 0 1	U	I01	$60 \times 1528 \times 8359$	12	12	6.32	47%
	Level 0 2	U	I02	$60 \times 1528 \times 8359$	12	12	6.38	47%
	Level 0 3	U	I03	$60 \times 1528 \times 8359$	12	12	6.31	47%
	Level 0 4	U	I04	$60 \times 1528 \times 8359$	12	12	6.43	46%

*: Calibrated (C) or Uncalibrated (U).

3.1. Best k Value Selection

From our previous research [20], the selection of the k value when building a k^2-raster was shown to have a great effect on the resulting size of the structure, as well as the access time to query its elements. In order to further investigate this idea, we extended our research to finding ways of choosing the best k value. One way was to build the k^2-raster structure with different k values for scene data from each sensor to see how the matrix size affected the choice of the k value. Additionally, we measured the time it took to build the k^2-raster and the size of the structure. The results are shown in Table 9. For most tested data, the k value leading to the smallest extended matrix size (attribute S in the table) usually provided the fastest build time and the smallest storage size. With these results, we could say that, in general, when $k = 2$, the compressed data size was large, sometimes even larger than the size of the original scene. As the value of k became larger, beginning with $k = 3$, the compressed data size was reduced. As far as the compressed size was concerned, the best value was in the range from three to 10 for matrices with a small raster size (i.e., if both the original width and original height were less than 1000) such as the ones for the AIRS Granule or AVIRIS Yellowstone scenes. If at least one dimension was larger than 1000 such as Hyperion calibrated or uncalibrated scenes, a larger range, typically between three and 20, needed to be considered.

Table 9. Results for different k values using the scene data from each sensor for the following attributes: (S) the extended matrix Size (pixels), (C) the k^2-raster Compressed storage data rate (bpppb), and (B) the time to Build the k^2-raster (seconds). The original scene width and height are shown in the first column. The best results are highlighted in blue.

Scene Data (w × h) *		k=2	3	4	5	6	7	8	9	10	11	12	13	14	15	16	17	18	19	20
AG9 (90 × 135)	S	256	243	256	625	216	343	512	729	1000	1331	144	169	196	225	256	289	324	361	400
	C	13.06	10.11	10.03	10.47	9.49	9.98	10.68	9.89	10.65	12.98	11.23	10.33	11.29	9.53	11.57	11.72	10.78	12.52	12.13
	B	5.3	3.2	4.1	10.9	4.2	10.9	12.6	10.7	17.5	29.6	2.9	4.1	3.0	4.3	4.6	6.6	6.8	4.9	7.3
ACY00 (677 × 512)	S	1024	729	1024	3125	1296	2401	4096	729	1000	1331	1728	2197	2744	3375	4096	4913	5832	6859	8000
	C	12.34	10.20	9.76	10.70	9.61	9.91	10.26	9.69	9.83	9.87	9.95	10.24	10.20	10.51	10.24	10.55	10.61	10.49	10.73
	B	19.5	10.7	10.8	30.7	10.5	19.3	42.0	8.8	9.3	11.5	13.2	17.1	23.2	29.1	45.5	55.8	72.6	101.1	131.0
AUY00 (680 × 512)	S	1024	729	1024	3125	1296	2401	4096	729	1000	1331	1728	2197	2744	3375	4096	4913	5832	6859	8000
	C	15.31	12.93	12.20	13.06	12.08	12.35	12.47	11.92	12.11	12.13	12.17	12.52	12.43	12.84	12.44	12.83	12.87	12.69	12.96
	B	18.4	10.7	10.1	30.7	11.4	20.9	41.4	7.7	8.5	10.9	12.7	17.1	22.9	29.3	44.3	55.8	73.0	101.0	130.6
C164 (640 × 420)	S	1024	729	1024	3125	1296	2401	4096	729	1000	1331	1728	2197	2744	3375	4096	4913	5832	6859	8000
	C	12.60	10.42	10.17	11.35	10.08	10.46	11.12	10.34	10.20	10.76	10.48	10.96	10.66	10.77	11.19	11.18	11.55	11.80	11.30
	B	47.1	28.8	27.9	74.3	27.7	47.3	98.6	19.3	21.1	24.8	29.7	38.7	49.4	69.6	96.2	133.3	179.3	231.1	314.9
HCA (256 × 3129)	S	4096	6561	4096	15625	7776	16807	4096	6561	10000	14641	20736	28561	38416	3375	4096	4913	5832	6859	8000
	C	17.2	15.64	9.79	-	10.47	-	8.54	9.13	9.7	-	-	-	-	8.65	8.52	8.75	9.16	9.07	8.92
	B	121.9	183.6	68.7	-	186.6	-	55.2	115.6	238.9	-	-	-	-	44.3	56.7	70.6	91.9	121.8	156.6
HUEA (256 × 3187)	S	4096	6561	4096	15625	7776	16807	4096	6561	10000	14641	20736	28561	38416	3375	4096	4913	5832	6859	8000
	C	16.02	14.63	8.89	-	9.68	-	7.76	8.46	9.00	-	-	-	-	8.50	7.80	8.69	8.68	8.46	8.27
	B	131.1	189.0	74.4	-	172.3	-	60.3	120.8	245.3	-	-	-	-	49.5	60.4	75.7	95.3	123.3	159.4
I01 (60 × 1528)	S	2048	2187	4096	3125	7776	2401	4096	6561	10000	14641	1728	2197	2744	3375	4096	4913	5832	6859	8000
	C	21.99	12.59	17.98	10.25	24.60	7.71	9.33	12.23	16.97	26.49	6.32	7.28	8.28	6.80	7.64	8.54	9.44	10.43	8.25
	B	1021.1	780.5	1728.7	986.5	5167.5	635.6	1426.7	3938.6	7870.7	17973.9	339.7	474.3	658.9	944.0	1498.1	1826.6	2789.8	3543.2	4810.3

*: w = scene width; h = scene height.

The above experiments were repeated to compare the access time for the different k values. For each scene, the average time over 100,000 consecutive queries is reported. Results are shown in Table 10, and Figure 6 shows how the access time and the size varied depending on the k value. As can be observed, access time became smaller and smaller as the value of k became larger. The plotted data suggested that there was a trade-off between access time and size with respect to the k value. We considered the optimal k value to be the one that created a relatively small size with a minimal access time. For example in AG9, when comparing the results between $k = 6$ and $k = 15$, the difference in bits per pixel per band for storage size was not very significant, but the reduction in access time was. Therefore, for this scene, $k = 15$ was considered an optimal value.

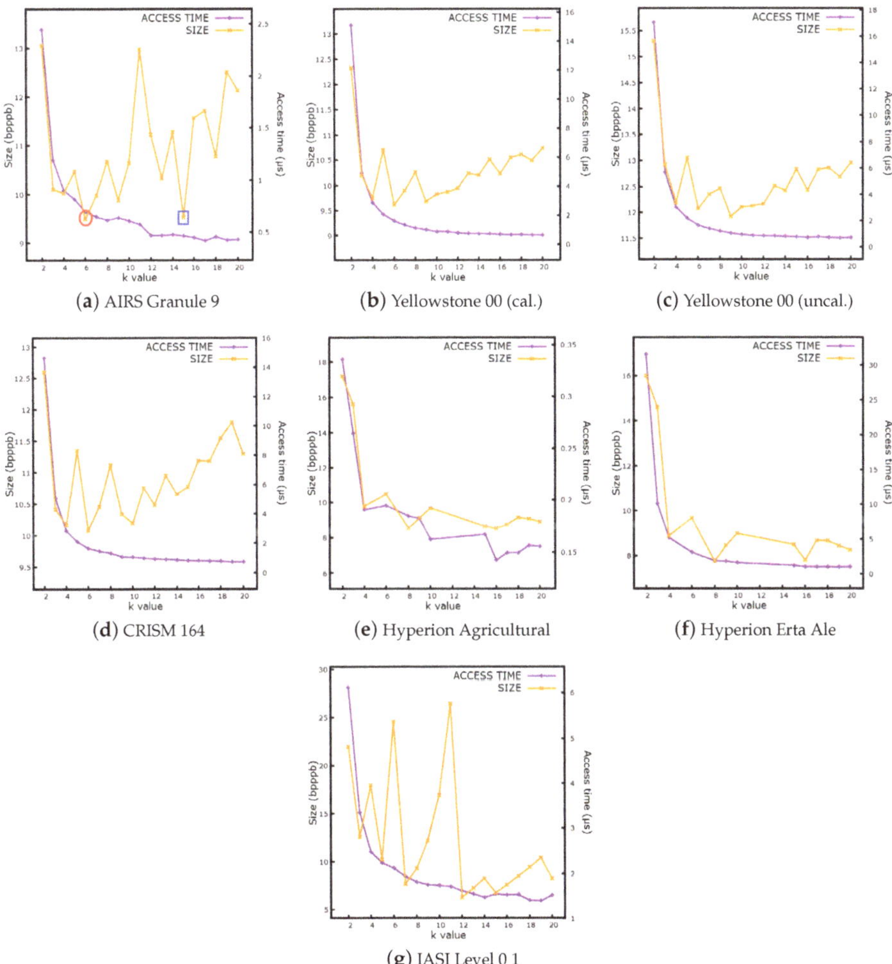

(**a**) AIRS Granule 9

(**b**) Yellowstone 00 (cal.)

(**c**) Yellowstone 00 (uncal.)

(**d**) CRISM 164

(**e**) Hyperion Agricultural

(**f**) Hyperion Erta Ale

(**g**) IASI Level 0 1

Figure 6. A comparison of the storage size (bpppb) and access time (μs) for different k values of k^2-raster built from scenes in our datasets. Access time is the average time of 100,000 consecutive queries. For AIRS Granule 9, the best value is marked with a red circle, and the optimal value is marked with a blue square.

Table 10. Access time (μs) for a random cell query with different kvalues. Each result is the average time over 100,000 consecutive queries. The best results are highlighted in blue.

Scene Data (w $\times$ h) *	$k = 2$	3	4	5	6	7	8	9	10	11	12	13	14	15	16	17	18	19	20
AG9 (90 $\times$ 135)	2.45	1.20	0.91	0.82	0.70	0.65	0.61	0.64	0.61	0.57	0.47	0.47	0.48	0.47	0.45	0.42	0.46	0.43	0.43
ACY00 (677 $\times$ 512)	15.10	4.95	2.89	2.07	1.60	1.33	1.13	1.01	0.89	0.88	0.79	0.76	0.74	0.73	0.69	0.67	0.68	0.66	0.64
AUY00 (680 $\times$ 512)	17.07	5.70	3.03	2.19	1.66	1.40	1.20	1.05	0.94	0.88	0.85	0.83	0.80	0.77	0.72	0.77	0.72	0.70	0.72
C164 (640 $\times$ 420)	14.66	5.09	2.84	2.12	1.67	1.48	1.34	1.10	1.08	1.01	0.95	0.93	0.88	0.83	0.81	0.81	0.79	0.74	0.74
HCA (256 $\times$ 3129)	0.34	0.26	0.19	-	0.19	-	0.18	0.18	0.16	-	-	-	-	0.17	0.14	0.15	0.15	0.16	0.16
HUEA (256 $\times$ 3187)	31.59	10.09	5.24	-	3.11	-	1.87	1.81	1.60	-	-	-	-	1.22	1.02	1.01	1.01	1.00	1.01
I01 (60 $\times$ 1528)	6.13	3.35	2.48	2.24	2.13	1.94	1.82	1.76	1.74	1.72	1.62	1.56	1.48	1.55	1.54	1.54	1.41	1.40	1.53

*: w = scene width; h = scene height.

3.2. Heuristic k^2-Raster

In this section, we present the results of the experiments using the heuristic k^2-raster proposed by Ladra et al. [33] on some of our datasets. Table 11 reports results for two hyperspectral scene datasets: AIRS Granule and AVIRIS Uncalibrated Yellowstone. In the experiments, we found that only when $k = 2$ would there be enough repeated sets of codewords in the last level of nodes to help us save space. When $k \geq 3$, there were no repeated sets of codewords. From the table, it can be seen that there was not much size reduction with k_H^2-raster in most cases. However, if we built a k^2-raster using the best or optimal k value, the size was considerably smaller. Therefore, we can see that k_H^2-raster structure did not produce a better size.

Table 11. Comparison of the structure size (bpppb) built from k^2-raster and k_H^2-raster where $k = 2$. The sizes for k^2-raster using the best k value and the optimal k value are also shown. The best results are highlighted in blue.

AIRS Granule	k^2-Raster ($k = 6$) (Best)	k^2-Raster ($k = 15$) (Optimal)	k^2-Raster ($k = 2$)	k_H^2-Raster ($k = 2$)
AG9	9.49	9.53	13.06	13.22
AG16	9.12	9.17	12.72	12.85
AG60	9.81	9.72	13.65	13.86
AG126	9.61	9.72	13.42	13.59
AG129	8.65	8.72	11.98	11.95
AG151	9.53	9.56	13.19	13.35
AG182	9.68	9.71	13.32	13.47
AG193	9.44	9.30	13.29	13.43

AVIRIS Uncalibrated	k^2-Raster ($k = 9$) (Best)	k^2-Raster ($k = 9$) (Optimal)	k^2-Raster ($k = 2$)	k_H^2-Raster ($k = 2$)
AUY00	11.92	11.92	15.31	15.19
AUY03	11.74	11.74	15.03	14.74
AUY10	9.99	9.99	12.85	11.86
AUY11	11.27	11.27	14.27	14.08
AUY18	12.15	12.15	15.36	15.25

3.3. 3D-2D Mapping

As discussed earlier, Cruces et al. [34] proposed a 3D to 2D mapping of raster images using k^2-tree as an alternative to achieve a better compression ratio. We used the k^2-tree implementation in sdsl-lite software to obtain the sizes for one of our datasets (AG9) from $k = 2$ to $k = 4$. Note that similar to k^2-raster, if the 2D binary matrix cannot be partitioned into square subquadrants of equal size, it needs to be expanded using Equation (1), and the extra elements are set to zero. The results are presented in Table 12. The sizes for a range of bands from 1481 to 1500 of the scene are also given for comparison.

From the results for AG9, we can see that the 3D-2D mapping did not make the size smaller. Instead, it became larger when the k value increased, and therefore, the method did not produce competitive results.

3.4. Comparison of Integer Encoders for k^2-Raster

Experiments were conducted to determine whether other variable-length encoders of integers might serve as a better substitute for DACs, which were the original choice in the k^2-raster structure initially proposed by Ladra et al. [33]. The performance of DACs was compared to that of other encoders such as Rice, Simple9, PForDelta, Simple16 codes, and gzip. In these experiments, the *Lmax* and *Lmin* arrays were encoded using these codes, and the results are shown in Table 13. For Rice codes, the l value, as explained in Section 2.4, produced different results depending on the mean of the raster's elements, and only the ones with the best l value are shown.

Table 12. The sizes of AIRS Granule (AG9) produced by 3D to 2D mapping from $k = 2$ to $k = 4$. The individual band sizes ranging from 1481 to 1500 are also shown. Sizes for individual bands are in bits per pixel (bpp), while the ones for all bands are in bits per pixel per band (bpppb).

Band	Original Size	k^2-Tree $k = 2$	k^2-Tree $k = 3$	k^2-Tree $k = 4$
All bands	16	16.53	20.57	26.57
1481	16	17.56	22.00	28.45
1482	16	17.27	21.54	27.84
1483	16	17.19	21.47	27.67
1484	16	17.45	21.81	28.18
1485	16	16.93	21.10	27.29
1486	16	17.09	21.27	27.50
1487	16	16.82	21.06	27.02
1488	16	17.01	21.21	27.34
1489	16	17.23	21.51	27.78
1490	16	16.94	21.10	27.20
1491	16	16.80	20.86	26.96
1492	16	16.56	20.64	26.51
1493	16	16.80	20.91	26.89
1494	16	16.84	20.93	26.98
1495	16	16.69	20.88	26.72
1496	16	16.66	20.75	26.66
1497	16	16.70	20.87	26.73
1498	16	16.61	20.70	26.58
1499	16	16.67	20.73	26.78
1500	16	16.39	20.40	26.18

Table 13. A comparison of the storage size (in bpppb) using different integer encoders on *Lmax* and *Lmin* from the k^2-raster built from our datasets. The combined entropies for *Lmax* and *Lmin* are listed as a reference. The *l* value that was used in Rice codes is enclosed in brackets. The best and optimal *k* values for DACs are also enclosed in brackets. Except for the entropy, the best rates for each scene's data are highlighted in blue.

Hyperspectral Scene	Entropy (*Lmax* + *Lmin*)	Rice (*l* Value)	Simple9	PForDelta	Simple16	DACs (Best *k*)	DACs (Optimal *k*)	gzip
AG9	8.29	10.10 (7)	10.06	9.88	9.69	9.49 (6)	9.53 (15)	12.45
AG16	7.92	9.88 (7)	9.64	9.55	9.30	9.12 (6)	9.17 (15)	11.96
AG60	8.58	10.31 (7)	10.50	10.19	10.12	9.72 (15)	9.81 (6)	12.79
AG126	8.42	10.34 (7)	10.25	9.98	9.81	9.61 (6)	9.72 (15)	12.55
AG129	7.47	9.66 (7)	9.01	9.01	8.61	8.65 (6)	8.72 (15)	11.21
AG151	8.36	10.39 (7)	9.99	9.79	9.54	9.53 (6)	9.56 (15)	12.39
AG182	8.44	10.58 (7)	10.44	10.09	10.01	9.68 (6)	9.71 (15)	12.71
AG193	8.25	10.26 (7)	10.06	9.93	9.65	9.30 (15)	9.44 (6)	12.33
ACY00	8.81	9.89 (7)	10.37	9.80	10.11	9.61 (6)	9.69 (9)	12.56
ACY03	8.48	9.70 (7)	9.80	9.40	9.57	9.42 (6)	9.50 (9)	11.98
ACY10	6.88	9.18 (7)	7.34	7.43	7.18	7.62 (6)	7.74 (9)	9.32
ACY11	8.12	9.45 (7)	9.32	9.02	9.09	8.81 (6)	9.00 (9)	11.61
ACY18	8.96	10.58 (7)	10.52	9.84	10.28	9.78 (6)	9.88 (9)	12.66
AUY00	11.16	17.59 (7)	14.01	11.93	13.79	11.92 (9)	11.92 (9)	15.13
AUY03	10.83	16.59 (7)	13.54	11.56	13.29	11.74 (9)	11.74 (9)	14.59
AUY10	9.26	12.87 (7)	10.90	9.61	10.54	9.99 (9)	9.99 (9)	12.29
AUY11	10.60	15.16 (7)	13.12	11.24	12.89	11.27 (9)	11.27 (9)	14.47
AUY18	11.38	20.70 (7)	14.19	12.10	14.01	12.15 (9)	12.15 (9)	15.53
C164	9.18	10.33 (7)	11.35	10.44	11.14	10.08 (6)	10.08 (6)	12.85
C165	9.48	10.91 (7)	11.78	10.69	11.57	10.37 (6)	10.37 (6)	13.17
C166	10.02	12.83 (7)	12.99	11.41	12.74	11.05 (6)	11.05 (6)	13.61
C181	9.16	9.96 (7)	10.93	10.53	10.72	9.97 (5)	9.97 (5)	13.37
C182	9.27	10.17 (7)	11.24	10.67	10.99	10.11 (5)	10.11 (5)	13.26
C183	9.60	11.15 (7)	12.33	11.21	12.05	10.65 (5)	10.65 (5)	13.32

Table 13. *Cont.*

Hyperspectral Scene	Entropy ($Lmax + Lmin$)	Rice (l Value)	Simple9	PForDelta	Simple16	DACs (Best k)	DACs (Optimal k)	gzip
HCA	7.59	8.94 (7)	9.79	8.80	9.56	8.52 (16)	8.54 (8)	11.20
HCC	6.75	8.20 (7)	8.28	7.60	7.93	7.62 (8)	7.71 (16)	9.51
HCU	7.87	9.78 (7)	10.30	8.91	10.04	8.85 (16)	8.86 (8)	11.35
HUEA	6.66	7.67 (5)	8.30	7.99	8.00	7.76 (8)	7.80 (16)	9.85
HULM	6.71	7.66 (5)	8.38	8.11	8.10	7.82 (8)	7.88 (16)	10.13
HUMS	6.77	7.90 (5)	8.48	8.14	8.20	7.91 (8)	7.94 (16)	10.12
I01	5.39	6.51 (4)	6.26	6.54	5.94	6.32 (12)	6.80 (15)	7.46
I02	5.46	6.56 (4)	6.27	6.55	5.96	6.38 (12)	6.84 (15)	7.51
I03	5.42	6.51 (4)	6.19	6.48	5.89	6.31 (12)	6.79 (15)	7.39
I04	5.51	6.62 (4)	6.37	6.65	6.04	6.43 (12)	6.90 (15)	7.63

The results showed that, in most cases, DACs still provided the best storage size compared to other encoders for our datasets. They also had the added advantage of direct random access to individual elements of the matrix whilst the other encoders would need to decompress each raster in order to retrieve the element, thus requiring much longer access time. When DACs did not yield the best performance, DACs results were usually only less than 0.1 bpppb worse. In the worst cases, DACs results lagged behind by, at most, 0.4 bpppb.

4. Conclusions

In this research, we examined the possibility of using different integer coding methods for k^2-raster and concluded that this compact data structure worked best when it was used in tandem with DACs encoding. The other variable-length encoders, though having competitive compression ratios, lacked the ability to provide users with direct access to the data. We also studied a method whereby we could obtain a k value that gave a competitive storage size and, in most cases, also a suitable access time.

For future work, we are interested in investigating the feasibility of modifying elements in a k^2-raster structure, facilitating data replacements without having to go through cycles of decompression and compression for the entire compact data structure.

Author Contributions: Conceptualization, K.C., D.E.O.T., M.H.-C., I.B., and J.S.-S.; methodology, K.C., D.E.O.T., M.H.-C., I.B., and J.S.-S.; software, K.C.; validation, K.C., I.B., and J.S.-S.; formal analysis, K.C., D.E.O.T., M.H., I.B., and J.S.-S.; investigation, K.C., D.E.O.T., M.H.-C., I.B., and J.S.-S.; resources, K.C., D.E.O.T., M.H.-C., I.B., and J.S.-S.; data curation, K.C., I.B., and J.S.-S.; writing, original draft preparation, K.C., I.B., and J.S.-S.; writing, review and editing, K.C., M.H.-C., I.B., and J.S.-S.; visualization, K.C., I.B., and J.S.-S.; supervision, I.B. and J.S.-S.; project administration, I.B. and J.S.-S.; funding acquisition, M.H.-C., I.B., and J.S.-S. All authors read and agreed to the published version of the manuscript.

Funding: This research was funded by the Spanish Ministry of Economy and Competitiveness and the European Regional Development Fund under Grants RTI2018-095287-B-I00 and TIN2015-71126-R (MINECO/FEDER, UE) and BES-2016-078369 (Programa Formación de Personal Investigador), by the Catalan Government under Grant 2017SGR-463, by the postdoctoral fellowship program Beatriu de Pinós, Reference 2018-BP-00008, funded by the Secretary of Universities and Research (Government of Catalonia), and by the Horizon 2020 program of research and innovation of the European Union under the Marie Skłodowska-Curie Grant Agreement #801370.

Conflicts of Interest: The authors declare no conflict of interest.

References

1. Clark, R.N.; Roush, T.L. Reflectance spectroscopy: Quantitative analysis techniques for remote sensing applications. *J. Geophys. Res. Solid Earth* **1984**, *89*, 6329–6340. [CrossRef]

2. Goetz, A.F.; Vane, G.; Solomon, J.E.; Rock, B.N. Imaging spectrometry for earth remote sensing. *Science* **1985**, *228*, 1147–1153. [CrossRef] [PubMed]

3. Kruse, F.A.; Lefkoff, A.; Boardman, J.; Heidebrecht, K.; Shapiro, A.; Barloon, P.; Goetz, A. The spectral image processing system (SIPS)-interactive visualization and analysis of imaging spectrometer data. *AIP Conf. Proc.* **1993**, *283*, 192–201.

4. Joseph, W. Automated spectral analysis: A geologic example using AVIRIS data, north Grapevine Mountains, Nevada. In Proceedings of the Tenth Thematic Conference on Geologic Remote Sensing, Environmental Research Institute of Michigan, San Antonio, TX, USA, 9–12 May 1994; pp. 1407–1418.

5. Asner, G.P. Biophysical and biochemical sources of variability in canopy reflectance. *Remote Sens. Environ.* **1998**, *64*, 234–253. [CrossRef]

6. Parente, M.; Kerekes, J.; Heylen, R. A Special Issue on Hyperspectral Imaging [From the Guest Editors]. *IEEE Geosci. Remote Sens. Mag.* **2019**, *7*, 6–7. [CrossRef]

7. Ientilucci, E.J.; Adler-Golden, S. Atmospheric Compensation of Hyperspectral Data: An Overview and Review of In-Scene and Physics-Based Approaches. *IEEE Geosci. Remote Sens. Mag.* **2019**, *7*, 31–50. [CrossRef]

8. Khan, M.J.; Khan, H.S.; Yousaf, A.; Khurshid, K.; Abbas, A. Modern trends in hyperspectral image analysis: A review. *IEEE Access* **2018**, *6*, 14118–14129. [CrossRef]

9. Theiler, J.; Ziemann, A.; Matteoli, S.; Diani, M. Spectral Variability of Remotely Sensed Target Materials: Causes, Models, and Strategies for Mitigation and Robust Exploitation. *IEEE Geosci. Remote Sens. Mag.* **2019**, *7*, 8–30. [CrossRef]

10. Sun, W.; Du, Q. Hyperspectral band selection: A review. *IEEE Geosci. Remote Sens. Mag.* **2019**, *7*, 118–139. [CrossRef]

11. Scafutto, R.D.M.; de Souza Filho, C.R.; de Oliveira, W.J. Hyperspectral remote sensing detection of petroleum hydrocarbons in mixtures with mineral substrates: Implications for onshore exploration and monitoring. *ISPRS J. Photogramm. Remote Sens.* **2017**, *128*, 146–157. [CrossRef]

12. Bishop, C.A.; Liu, J.G.; Mason, P.J. Hyperspectral remote sensing for mineral exploration in Pulang, Yunnan Province, China. *Int. J. Remote Sens.* **2011**, *32*, 2409–2426. [CrossRef]

13. Le Marshall, J.; Jung, J.; Zapotocny, T.; Derber, J.; Treadon, R.; Lord, S.; Goldberg, M.; Wolf, W. The application of AIRS radiances in numerical weather prediction. *Aust. Meteorol. Mag.* **2006**, *55*, 213–217.

14. Robichaud, P.R.; Lewis, S.A.; Laes, D.Y.; Hudak, A.T.; Kokaly, R.F.; Zamudio, J.A. Postfire soil burn severity mapping with hyperspectral image unmixing. *Remote Sens. Environ.* **2007**, *108*, 467–480. [CrossRef]

15. Navarro, G. *Compact Data Structures: A Practical Approach*; Cambridge University Press: Cambridge, UK, 2016.

16. Jacobson, G. Space-efficient static trees and graphs. In Proceedings of the 30th Annual Symposium on Foundations of Computer Science, Research Triangle Park, NC, USA, 30 October–1 November 1989; pp. 549–554.

17. Consultative Committee for Space Data Systems (CCSDS). *Image Data Compression CCSDS 123.0-B-1*; Blue Book; CCSDS: Washington, DC, USA, 2012.

18. Jolliffe, I.T. *Principal Component Analysis*; Springer: Berlin, Germany, 2002; p. 487.

19. Taubman, D.S.; Marcellin, M.W. *JPEG 2000: Image Compression Fundamentals, Standards and Practice*; Kluwer Academic Publishers: Boston, MA, USA, 2001.

20. Chow, K.; Tzamarias, D.E.O.; Blanes, I.; Serra-Sagristà, J. Using Predictive and Differential Methods with K2-Raster Compact Data Structure for Hyperspectral Image Lossless Compression. *Remote Sens.* **2019**, *11*, 2461. [CrossRef]

21. Elias, P. Efficient storage and retrieval by content and address of static files. *J. ACM (JACM)* **1974**, *21*, 246–260. [CrossRef]

22. Ottaviano, G.; Venturini, R. Partitioned Elias-Fano indexes. In Proceedings of the 37th International ACM SIGIR Conference on Research & Development in Information Retrieval, Gold Coast, Australia, 6–11 July 2014; pp. 273–282.

23. Pibiri, G.E. Dynamic Elias-Fano Encoding. Master's Thesis, University of Pisa, Pisa, Italy, 2014.

24. Sugiura, R.; Kamamoto, Y.; Harada, N.; Moriya, T. Optimal Golomb-Rice Code Extension for Lossless Coding of Low-Entropy Exponentially Distributed Sources. *IEEE Trans. Inf. Theory* **2018**, *64*, 3153–3161. [CrossRef]

25. Rice, R.; Plaunt, J. Adaptive variable-length coding for efficient compression of spacecraft television data. *IEEE Trans. Commun. Technol.* **1971**, *19*, 889–897. [CrossRef]

26. Rojals, J.S.; Karczewicz, M.; Joshi, R.L. Rice Parameter Update for Coefficient Level Coding in Video Coding Process. U.S. Patent 9,936,200, 3 April 2018.

27. Zukowski, M.; Heman, S.; Nes, N.; Boncz, P. Super-scalar RAM-CPU cache compression. In Proceedings of the 22nd International Conference on Data Engineering (ICDE'06), Atlanta, GA, USA, 3–7 April 2006; pp. 59–59.

28. Silva-Coira, F. Compact Data Structures for Large and Complex Datasets. Ph.D. Thesis, Universidade da Coruña, A Coruña, Spain, 2017.

29. Al Hasib, A.; Cebrian, J.M.; Natvig, L. V-PFORDelta: Data compression for energy efficient computation of time series. In Proceedings of the 2015 IEEE 22nd International Conference on High Performance Computing (HiPC), Bengaluru, India, 16–19 December 2015; pp. 416–425.

30. Brisaboa, N.R.; Ladra, S.; Navarro, G. DACs: Bringing direct access to variable-length codes. *Inf. Process. Manag.* **2013**, *49*, 392–404. [CrossRef]

31. Brisaboa, N.R.; Ladra, S.; Navarro, G. Directly addressable variable-length codes. In Proceedings of the International Symposium on String Processing and Information Retrieval, Saariselkä, Finland, 25–27 August 2009; pp. 122–130.

32. Baruch, G.; Klein, S.T.; Shapira, D. A space efficient direct access data structure. *J. Discret. Algorithms* **2017**, *43*, 26–37. [CrossRef]

33. Ladra, S.; Paramá, J.R.; Silva-Coira, F. Scalable and queryable compressed storage structure for raster data. *Inf. Syst.* **2017**, *72*, 179–204. [CrossRef]

34. Cruces, N.; Seco, D.; Gutiérrez, G. A compact representation of raster time series. In Proceedings of the Data Compression Conference (DCC), Snowbird, UT, USA, 26–29 March 2019; pp. 103–111.

35. Brisaboa, N.R.; Ladra, S.; Navarro, G. k 2-trees for compact web graph representation. In Proceedings of the International Symposium on String Processing and Information Retrieval, Saariselkä, Finland, 25–27 August 2009; pp. 18–30.

36. Anh, V.N.; Moffat, A. Inverted index compression using word-aligned binary codes. *Inf. Retr.* **2005**, *8*, 151–166. [CrossRef]

37. Zhang, J.; Long, X.; Suel, T. Performance of compressed inverted list caching in search engines. In Proceedings of the 17th International Conference on World Wide Web, Beijing, China, 21–25 April 2008; pp. 387–396.

38. Gog, S.; Beller, T.; Moffat, A.; Petri, M. From theory to practice: Plug and play with succinct data structures. In Proceedings of the International Symposium on Experimental Algorithms, Copenhagen, Denmark, 29 June–1 July 2014; pp. 326–337.

Article

Lossy Compression of Multispectral Satellite Images with Application to Crop Thematic Mapping: A HEVC Comparative Study

Miloš Radosavljević [1,†], Branko Brkljač [1,*,†], Predrag Lugonja [2], Vladimir Crnojević [2], Željen Trpovski [1], Zixiang Xiong [3] and Dejan Vukobratović [1]

[1] Department of Power, Electronic and Communication Engineering, Faculty of Technical Sciences, University of Novi Sad, Trg Dositeja Obradovića 6, 21000 Novi Sad, Serbia; milos.r@uns.ac.rs (M.R.); zeljen@uns.ac.rs (Ž.T.); dejanv@uns.ac.rs (D.V.)

[2] BioSense Institute, Zorana Djindjića 1, 21000 Novi Sad, Serbia; lugonjap@uns.ac.rs (P.L.); crnojevic@uns.ac.rs (V.C.)

[3] Department of Electrical and Computer Engineering, Texas A&M University, College Station, TX 77843, USA; zx@ece.tamu.edu

* Correspondence: brkljacb@uns.ac.rs; Tel.: +381-21-485-2517

† These authors contributed equally to this work.

Received: 12 March 2020; Accepted: 25 April 2020 ; Published: 16 May 2020

Abstract: Remote sensing applications have gained in popularity in recent years, which has resulted in vast amounts of data being produced on a daily basis. Managing and delivering large sets of data becomes extremely difficult and resource demanding for the data vendors, but even more for individual users and third party stakeholders. Hence, research in the field of efficient remote sensing data handling and manipulation has become a very active research topic (from both storage and communication perspectives). Driven by the rapid growth in the volume of optical satellite measurements, in this work we explore the lossy compression technique for multispectral satellite images. We give a comprehensive analysis of the High Efficiency Video Coding (HEVC) still-image intra coding part applied to the multispectral image data. Thereafter, we analyze the impact of the distortions introduced by the HEVC's intra compression in the general case, as well as in the specific context of crop classification application. Results show that HEVC's intra coding achieves better trade-off between compression gain and image quality, as compared to standard JPEG 2000 solution. On the other hand, this also reflects in the better performance of the designed pixel-based classifier in the analyzed crop classification task. We show that HEVC can obtain up to 150:1 compression ratio, when observing compression in the context of specific application, without significantly losing on classification performance compared to classifier trained and applied on raw data. In comparison, in order to maintain the same performance, JPEG 2000 allows compression ratio up to 70:1.

Keywords: HEVC; intra coding; JPEG 2000; high bit-depth compression; multispectral satellite images; crop classification; Landsat-8; Sentinel-2

---

1. Introduction and Motivation

Satellite imaging has experienced proliferation in the recent period and there are more possibilities for realization of its applications than ever before. However, data burden is expected to become even greater in the future with the advancement of sensors with higher spatial resolutions and more measurement bands. Therefore, there is a need to revisit and possibly complement some previous findings that were related to efficient satellite image compression, like ones presented in [1], or more recently in [2], and offer users more insights into effects of different choices in such case. Of particular interest is the field of *lossy image compression* where high compression ratios can be easily achieved, but at the cost of inevitable information loss. However, the question remains what level of image quality degradation could be acceptable for some remote sensing application, and even more importantly, how sensitive such application would be on the particular choice of an image compression strategy. This paper tries to offer some answers to such questions and tracks possibilities of using one of the latest, industry confirmed, general purpose coding standard, which could be expected to be easily accessible to broad community of users, and easy to implement in custom, application dependant, processing frameworks. More exactly, we present a comparative study in which the effectiveness of the *intra coding* part of the High Efficiency Video Coding (HEVC) standard [3,4] is analyzed against the well established JPEG 2000 lossy compression [5,6], under two selected application scenarios:

(a) HEVC as a general purpose lossy compression solution for medium and high resolution multispectral satellite images acquired by the Landsat-8's OLI [7,8], and Sentinel-2's multispectral instrument (MSI) [9,10], which are considered as frequently used and freely available image sources; and

(b) HEVC as an application oriented lossy compression solution for thematic mapping, i.e., pixel-based crop classifications like the ones presented in [11,12]—however with an important difference that multispectral time series in our study consist of lossy compressed reflectance measurements as compared to original (uncompressed or lossless) measurements that were used in the referred studies.

The choice of the second application scenario has been influenced by the fact that the corresponding classifier design, and the subsequent production phases, where the trained model is going to be deployed, require large volumes of data. Therefore, such tasks could easily benefit from the effective lossy image compression, resulting in the more efficient data handling. They could also benefit from the possible side effects, like the noise cancelling, leading to the better classifier performance, as in [13,14]. Naturally, this could be only possible if such lossy compression classification approach, usually denoted as compress-then-analyze, would prove to be not highly sensitive to the expected information loss. Thus, the aim of this work is to offer an application oriented insight into specific compression vs. classification performance trade-off in the pixel-based crop classification tasks, through estimation of the corresponding classification performance sensitivity, similarly as in [15]. Also, since we advocate the use of the standardized compression solutions, instead of custom-made codecs that cannot benefit to such extent from the large community support (optimized hardware implementations and continuously maintained software libraries), HEVC effectiveness is tested against the JPEG 2000, which has been often utilized in Earth Observation (EO) data distribution systems.

Besides the possible advantages arising from utilizing the baseline (still-image) HEVC intra mode, we also point out that the particular thematic mapping application that has been chosen in this study represents a sort of typical remote sensing tasks. However, it has not been frequently analyzed from the application perspective of the general data compression community. Moreover, currently there is not so much research on this topic, and a comprehensive analysis of the compression effects on remote sensing applications has been relatively scarce. Therefore, the EO community could benefit from more interest of the compression community for the specific application oriented compression scenarios, as well as from the comprehension of how some finely tuned and well established (off-the-shelf) data compression technique could effect different bands of acquired multispectral images, and to what extent this choice could be expected to have influence on some particular remote sensing application.

Results presented in this paper confirm that HEVC, as a hybrid and feature rich method based on transform and predictive coding techniques, outperforms transform based JPEG 2000 compression approach that has been currently used in some ground-based data distribution systems and has been considered as a general purpose solution for lossless or lossy satellite image compression. The presented results confirm initial assumptions regarding HEVC's applicability and effectiveness in lossy compression of multispectral images, opening possibly new research directions towards its application in hyperspectral image compression. However, even more importantly, presented results give a new perspective on the possibilities of exploiting general purpose lossy compression solutions, like HEVC, in contemporary ground segment data distribution operations, oriented towards users interested in different levels of quality of service, like in the standard streaming application scenarios.

The rest of the paper is organized as follows. In Section 2, a brief discussion on the current state and future trends in the field of satellite image compression has been given, specifically placing in a given context HEVC as one of the possible compression solution for the future EO data distribution systems. Review of other related coding techniques has been given in the same section. In the Section 3, the underlying data, used as part of experiments, has been briefly described and motivated. Experimental setup and methods used in this study are explained in Section 4. It includes the design characteristics of the crop classification method, that is used as an example of application oriented benchmark in this work, as well as settings of the codecs used for the compression purpose. The results, and additional discussions of the outcomes of this study, are emphasized in Section 5. Presented results in this work include both quantitative and qualitative assessments of compression effects on the image quality, as well as on the overall accuracy of the considered crop thematic maps. The paper is concluded in Section 6, with a short reference to the possible future research directions.

2. Provisioning of EO Data Through Lossy Compression Schemes in the Ground Segment

Broadly speaking, utility of an image compression scheme is mostly affected by the type of data and planned application. Essentially it goes down to the amount of useful information inherent in the data, as compared to the overall data quantity [16], sometimes requiring a global, mission specific approach in a quest for an optimal compression trade-off [17].

2.1. Downlink Related Work

Downlink systems, Figure 1, usually prefer lossless or near-lossless image coding designed in such a way to ensure high fidelity of possible applications. Due to a limited on-board power capacity, low complexity compression schemes, e.g., [18–24], have been the main choice towards the practical implementation. For example, in such restricted environment, even JPEG 2000 has been considered a too complex method. Consequently, the Consultative Committee for Space Data Systems (CCSDS) has been constituted to research on a family of recommended compression schemes specifically designed for space data systems. Works like these [25–27] developed by CCSDS have emerged, following a general paradigm of low complexity requirements in order to facilitate applications such as on-board compression. Aforementioned approaches are just some of the wide range of different coding choices that meet requirements imposed by power-constrained EO missions, with a limited hardware resources specifically designed to function in orbit. However, those techniques are often used with compression ratios that could be considered as moderate [28], sometimes also due to the the nature of the lossless compression itself. Nevertheless, limited bandwidths and ever-growing data volumes make these methods inevitable in today's EO missions. As a consequence of the huge increment in the data-rate of the new generation sensors (despite that the lossless or near-lossless compression algorithms have been traditionally preferred for on-board compression scenarios), a research effort targeting lossy on-board compression has been reported in [29,30]. Also, lossy extensions of CCSDS standards have been proposed [26,31], however still bearing in mind available on-board technologies related with tolerance to solar radiation and low power consumption, e.g., see [32]. As a matter of fact, it is generally accepted that the lossy coding schemes can be (if required) derived from the corresponding lossless modes through introduction of a suitable quantization (or coefficient truncation) scheme [31,33].

2.2. Focusing towards the Users Side Compression

As opposed to the constraints related to the on-board processing and downlink communication requirements, end-user oriented EO applications, that are placed at the other end of the data distribution chain, offer much broader range of design possibilities. Liberating from the tendency to suggest low-power codec solutions, and approaching closer to end-users, different compression options may appear, especially considering theoretically unlimited resources in the cloud where complexity is no longer a critical issue during the encoding step. Thus, low-complexity requirement imposed over on-board solutions can be significantly relaxed, which can be particularly suited for a more complex compression algorithms that are trying to achieve better trade-off between image quality and compression ratio.

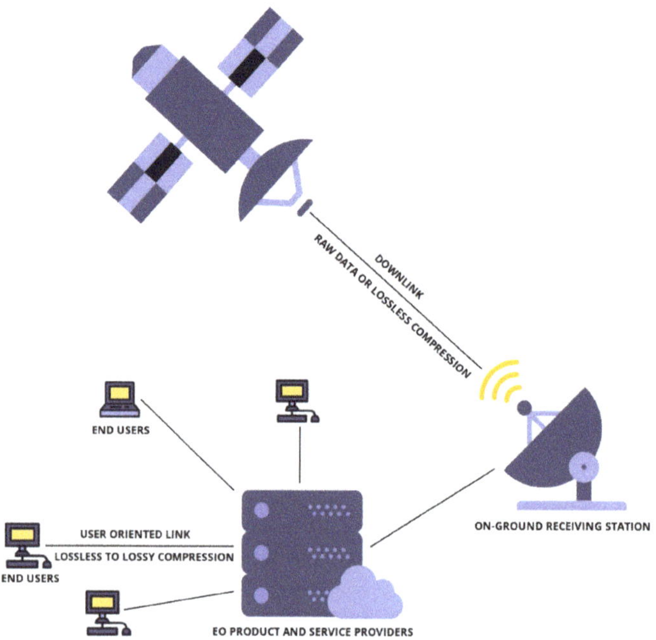

Figure 1. General scheme of Earth Observation (EO) communication and data distribution system.

Traditionally, those methods are mainly based on various decorrelating transforms [34,35], leading to wide range of different coding approaches, where usual precondition for efficiently going below the nominal entropy rate of the source is to decorrelate acquired measurements in spatial and/or spectral domain. In particular, general purpose coding schemes, originally designed to be applicable in broader progressive lossless-to-lossy use-case scenarios, such as EZW [36], SPIHT [37–39] or SPECK [40,41], proved to be useful in achieving higher compression ratio requirements, however bringing in higher processing and memory requirements when compared to less complex on-board solutions (see Section 2.1). Aforementioned research has established that, among transform coding architectures, many systems usually follow a general coding pipeline in which, at the first stage, a suitable transform is applied in the spectral domain (e.g., wavelets or some covariance-based decorrelation method like Karhunen-Loève transform (KLT) [42] or its low-complexity approximation named Pairwise Orthogonal Transform [43,44]), which is afterwards followed by some of the standard coding schemes (EZW, SPIHT, SPECK). Similarly, JPEG 2000 can be deployed in order to compress each of the previously computed, spectrally decorrelated, image components (usually by implicitly performing an additional 2D spatial transform on each of resulting channels from the first step, followed by rate or quality allocator, and entropy coder), leading to three-dimensional (3-D) decorrelation [13,35,45,46]. In particular, a combination of spectral decorrelator such as KLT in conjunction with JPEG 2000 proved to provide state-of-the-art performance according to [13,43]. Other noteworthy research efforts are also reported in [47–50].

In addition, authors note that very few works utilize standard video coding schemes, or any similar hybrid-based approaches (combination of predictive coding and transform coding), as a compression choices for high-dimensional satellite data. More precisely, we found out only work presented in [51] worth mentioning. However it does not use HEVC as a base codec setting.

2.3. Applications Oriented Lossy Compression—Related Work

Although higher compression gains may be achieved using lossy compression, there is always a question to what extent some artifacts introduced by the specific compression algorithm affect a certain type of application. Some previous works, that provide insights into this important question, and which could be considered as closely related in the context of the presented study, are given in [15,52–55]. In [15], as well as [56], it was shown that a small amount of lossy compression can have beneficial effects for the overall performance of crop thematic mapping. This was also independently confirmed by the experimental results of our study. However, in comparison to [15], we have considered different types of compression methods and application scenarios. On the other hand, regarding the specific task of crop thematic mapping, in comparison to [15] our study considers only a multitemporal classification scenario, which can be considered as a more realistic one, as opposed to a single-date classification approach, since it is well known that the plant's phenology is captured much better by independent observations made throughout the vegetation season. Also, according to [15], multitemporal information allows one to utilize much higher compression ratios in comparison to single-date classification, and still achieve almost the same classification performance as in the case of original data. One of the main conclusions in [52] was that the level of scene fragmentation determines the acceptable level of lossy compression, which is consistent with the general rule of thumb that image's scene complexity determines the overall information content (entropy of the image), and thus level of required spatial details for successful, discriminative classification. Related to the mentioned fragmentation aspect, scene content complexity was also considered by the first set of experiments in our study, which involves algorithms benchmarking against satellite images with the varying scene content, captured at the approximately same time by the image sensors with varying spatial and spectral characteristics. In addition, influence of lossy compression on the quality of different vegetation indices, which also represent potentially valuable features for crop classification, was analyzed in [53,54]. Favoring image channels that contain more task specific information was also proposed in [54], where the group of channels characteristic for vegetation was encoded at a higher bit rate than the remaining one.

Besides the above mentioned methods, there were also many others which have evaluated compression efficiency against the specific remote sensing tasks like classification [35,43,55,57–60], image segmentation [61], atmospheric parameter retrieval [14,62], spectral unmixing [63], endmember extraction [64], vegetation [53,54] and spectral [65] indices computation, and biophysical variables estimation [66]. In [59] it was concluded that in the context of the general applications it is hard to define optimal level of image quality that should be preserved by the lossy compression. As a possible answer to this open question, authors in [59] have proposed a general, unsupervised classification-accuracy based method for image quality assessment, which could provide some indicators regarding effectiveness of different elements of the considered compression procedures. In [59] it was also pointed out that the spectral decorrelation usually removes peaks from spectral signatures of hyperspectral measurements, even at high reconstructed image qualities. On the other hand, Ref. [60] proposed a specific classification oriented lossy compression approach, inspired by optimal signal representation in the spectral domain. This approach was later extended in [67]. An overview of the contemporary approaches for measuring how classification accuracy is affected by the compression can also be found in a recent study in [68]. In addition, they establish a prediction model between the features extracted from compressed images and the corresponding classification accuracy.

2.4. HEVC as a Matter of Choice

Generally speaking, role of an efficient compression scheme would be to enable and facilitate deployment of other technologies that are expected to be highly dependant on the efficient and timely distribution of image data, and which would possible benefit from the achieved reduction in the required storage, energy consumption, data access and processing time. This could prove to be extremely important for data vendors, but even more so for individual users, on-demand service providers, or third party stakeholders in general. At the same time, positioning at the ground segment could open the access to the larger compression community that could also be crucial for the implementation of new forms of compression techniques that at the time being have not been considered in the RS applications, e.g., HEVC. In that sense, this work aims to position itself on the link between the EO data service providers and the end-users, see Figure 1. Also, in the context of this work, we consider classification task just as one of the possible user specific applications that could benefit from the reduced data burden that is easily achieved by the lossy compression of multispectral satellite images. In addition, as the remote sensing technologies advance, new applications require constant improvement of existing coding schemes, allowing higher compression gains, which usually (however not necessarily) comes at the price of higher complexity.

Eventually, it is important to note that widely accepted industry standards for image and video compression could ensure interoperability and can be adopted by the broader community of EO practitioners. At the same time, it is more likely to be implemented in a custom, application dependant, processing framework, since they usually come with the open source reference software, e.g., [69]. This implies that efficient, fast and reliable implementation of the standardized codecs may be realized with much less effort than implementing custom based codec from the ground. By doing so, schemes such as HEVC may be deployed efficiently, making available to process large amount of data utilizing computationally complex features like those incorporated in HEVC. Hence, building compression processing unit, on cloud or locally on a workstation (or even personal computer), is significantly facilitated since available computing power comes at reasonable cost. Efficient widely used and industry accepted compression techniques are therefore easier to be integrated in the existing infrastructure, or implemented using existing equipment. In addition, even though HEVC has high computational complexity, it is highly parallelizable and implementation friendly scheme [70,71] (it uses integer logic with shifts and additions). For example, work that implements fast and efficient HEVC encoder was described in [72].

Emerging from the above, in our research, we consider the use of standardized and ubiquitous, industry accepted solution such as HEVC. Technical specifications for the data distribution and management of the most spaceborne programs has identified JPEG 2000 as a straightforward compression solution, e.g., [73]. In this work we want to show that more recent, state-of-the-art standardized compression schemes may provide performance improvement compared to the JPEG 2000. Given the current computing technologies, besides HEVC's higher computational complexity, it is practically feasible to use such scheme in a ground-based satellite data distribution chain. Hence, we have done comparative study in which we analyze effectiveness of the HEVC standard. We show that using intra-band (still-image) coding part of HEVC can improve compression gain over the JPEG 2000 coding scheme which is specified as a baseline (and mostly used) compression method for EO data distribution at the ground segment. Indeed, HEVC has been reported to perform better than any other international standardized coding scheme currently on the market. It was reported in [74] that HEVC intra coding reduced the average bit rate by 17% compared to H.264/AVC, 23% compared to JPEG-2000, 30% compared to JPEG-XR, and 44% compared to classic JPEG. However, the previous work has been focused on the HEVC's performance evaluation by using representative test set of standard photographic still images. Furthermore, with the completion of its second version that supports up to 16 bits per pixel (b/p) [75], it became available to other specialized applications besides standard consumer data, such as remote sensing applications among others [76]. Thus, in this work, we quantify HEVC's intra coding performance, however this time applied on a set of multidimensional satellite data.

3. Datasets Preparation and Design

3.1. General Compression Performance Evaluation Dataset

The set of test images, which have been used by the first group of experiments, consists of the selected image scenes presented in Table 1. Similar set was originally proposed in [77] with the aim of having a representative collection of the scene content variations in order to derive the optimal estimates of the tasselled cap transformation parameters, specifically adapted to the corresponding Landsat-8 TOA reflectances. Therefore, in this paper, it was also considered as a suitable image collection for the general analysis of the compression effects in the case of the proposed HEVC solution. In line with the above mentioned, the set of images includes variety of scenes consisting of different land use and land cover classes (LULC), corresponding to urban areas, crop fields, mountainous forests, water bodies of different size, bare soil and various urban features, Table 1.

Table 1. Characteristics of scenes and images used in general compression performance comparisons.

Scene No.	Location	Description	Landsat-8 Path/Row	Landsat-8 Acq. Date	Sentinel-2A/B Granule ID	Sentinel-2A/B Acq. Date
1	Jiangxi, China	Poyang lake, urban and vegetation areas	121/40	10 April	T50RMT/R132	7 April
2	Jiangxi, China	Poyang lake, urban and vegetation areas	121/40	31 July	T50RMT/R132	31 July
3	Northern Nebraska, South Dakota	Variety of crop fields, urban areas, airport and bare soil	31/30	8 July	T14TLN/R055	10 July
4	Northern Nebraska, South Dakota	Variety of crop fields, urban areas, airport and bare soil	31/30	9 August	T14TLN/R055	9 August
5	West Virginia	Mountainous forests, water bodies and urban features	16/34	12 May	T17SPB/R097	14 May
6	West Virginia	Mountainous forests, water bodies and urban features	16/34	29 June	T17SPB/R097	18 June
7	North Carolina	Agricultural holdings, forest and water bodies	16/35	12 May	T17SPV/R097	14 May
8	North Carolina	Agricultural holdings, forest and water bodies	16/35	29 June	T17SPV/R097	18 June
9	Idaho	Large water bodies and crop fields	39/31	14 June	T12TUM/R127	10 June
10	Idaho	Large water bodies and crop fields	39/31	30 June	T12TUM/R127	25 June
11	Indus River Basin	Indus River, extensive crop fields	152/41	8 July	T42RVR/R021	7 August
12	Indus River Basin	Indus River, extensive crop fields	152/41	25 August	T42RVR/R021	27 August

However, all images were always used and analyzed as a whole, and the corresponding image quality metrics were aggregated at the scene level, which makes the class specific analysis of the corresponding compression effects hard to follow. Nevertheless, some of the selected scenes are mutually discriminative enough, thus allowing a generalization of results and analysis to a wide range of image contents. In addition to Landsat-8 multispectral images, corresponding to the specific scenes in Table 1, an additional subset of the benchmark dataset was designed. It consists of Sentinel-2 images of the same ground areas that were closest in space and time to the specified Landsat-8 scenes (images listed in the last two columns of Table 1). Thus, the proposed dataset has been designed to conduct the first group of experiments aim to objectively characterize HEVC's general compression performance in comparison to the baseline JPEG 2000 solution. Visual preview of selected images is given in Figure 2.

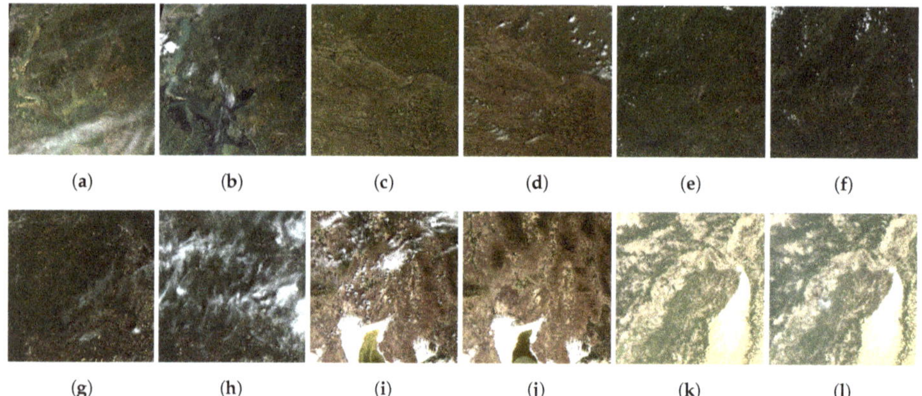

Figure 2. Visual overview of the scenes 1–12 in Table 1. True color illustrations (a–l) correspond to 60 m Sentinel-2 image channels. Some contain clouds and haze due to the adopted selection criteria which favours the Sentinel-2 images that were the ones closest in time to the original Landsat-8 counterparts.

Since Sentinel-2A/B were still not operational at the time when the original scenes in [77] were proposed, we have selected measurements from 2018 when both satellites were available, with the acquisition dates that were closest to the original ones made by Landsat-8 in [77]. It should be noted that in [77], for some of the location listed in Table 1, only one acquisition was selected, while for some others there were up to two acquisitions relatively close in time. Therefore, for all of the scenes of the dataset used in this study, we have selected two close acquisitions for each Landsat-8 and Sentinel-2 in order to have richer test environment with more samples of representative LULC categories in the given scenes.

All selected images are Level-2 atmospherically corrected image products, containing surface level reflectances, stored as **16 b/p** integers. Selection of multispectral images of the same test sites, but originating from two different satellite instruments, was motivated by the fact that Sentinel-2 and Landsat-8 imagers have different spatial resolutions, which could enable comparison of compression effects under different spatial resolution scenarios. In addition, individual channels of Sentinel-2 images are also of varying spatial resolutions, which was taken into account in results presentation, where in addition to performance measures aggregated at the scene level over all channels, the results corresponding to the groups of image channels with different spatial resolutions are also presented.

In case of Landsat-8 images, experiments were performed on all OLI based bands except panchromatic channel 8 and cirrus channel 9 [8], e.g., on all atmospherically corrected Level-2 image products of spatial resolution of 30 m [78]. These include 4 VNIR (visual and near-infrared) and 2 SWIR (shortwave infrared) narrowband channels, and one coastal/aerosol channel 1.

In case of Sentinel-2 images [9], those originating from both Sentinel-2A, as well as -2B, were used interchangeably in order to find an image with the acquisition time closest to the corresponding Landsat-8 image of the same area. Although Sentinel-2 images were selected to match with Landsat-8 acquisitions (path and row in the corresponding world reference system), it should be noted that, due to higher spatial resolution, Sentinel-2 images cover significantly lower area, of approximately 100 km by 100 km, as compared to Landsat-8's 185 km wide swath and 170 km long range. Sentinel-2 MSI channels [10], that were used in this study, include all channels that were available after atmospheric correction, at different spatial resolutions (10, 20 and 60 m). Thus, at 10 m spatial resolution, in total

4 VNIR channels were used, consisting of three B02-B04 narrowbands, and somewhat wider B08; at 20 m resolution 7 VNIR channels were used (B02-B07, B8a) and additional 2 SWIR channels (B11 and B12), thus in total nine channels; while at 60 m all channels except B08 and B10 were utilized, i.e., in total 11 channels of which 9 VNRI and 2 SWIR. As specified in [73] and illustrated in [10], at 20 m resolution only 6 channels B05-B8a and B11-B12 are originally acquired by the MSI, while the rest of 5 channels are synthesized from the original higher resolution 10 m measurements. Similar also holds for 60 m image channels, where out of 11 utilized channels only the image channels B01 and B09 are acquired at original, nominal spatial resolution, while the rest of nine 60 m image channels, that are delivered as part of Level-2 atmospherically corrected Sentinel-2 image product, are synthesized through adequate postprocessing.

Intrinsic spatial resolutions of Sentinel-2 image bands are: $10,980 \times 10,980$, 5490×5490, 1830×1830 pixels, corresponding to spatial resolutions of 10, 20, and 60 m respectively. On the other hand, intrinsic spatial resolution of Landsat-8 OLI images is not fixed and varies around $\approx 7800 \times 7900$ pixels. In addition, substantial part of Landsat-8 images is always occupied by no-data regions (along the corners of the scene), which is not always the case with Sentinel-2 images. To be more precise, it turns out that in our dataset Sentinel-2 images are barely occupied with such no data regions. From the compression point of view, such regions have high information redundancy, and therefore are easy to compress, which could be considered as a source of the potential positive bias in the overall evaluation of the compressed image quality. However, the same influence is present in the case of all analyzed compression scenarios, and therefore if present it could be considered as a relative one, which is a more favourable option for performance comparison of different methods. Also, according to corresponding product specifications, like [78], original data are always distributed in the predefined format that is adopted by the end-user community, which makes the use of standard, of-the-shelf codec implementations, more suitable, as opposed to some optimal, custom solutions, which would have a special treatment of such no-data regions.

Most of the above presented information related to characteristics of multispectral image products, that were utilized in the first type of experiments, also hold for the data that were used in the second type of experiments (pixel-based classification) described in Section 4.2. Therefore, in the next subsection, only the specific differences arising from the characteristics of the chosen application scenario will be additionally discussed and highlighted.

3.2. Crop Thematic Mapping Dataset

The second group of experiments *relies on time series* of multispectral images corresponding to the specific ground test site (area) where additional field level data were collected. Selected satellite images were used as the source of reflectance data for the design of pixel-based crop classification models, which are mutually compared and benchmarked against different lossy compression scenarios, utilizing the proposed HEVC and the chosen baseline JPEG 2000 solutions. As the result, constructed classification models differ only in the type of the training data that were used for their estimation. Thus, they differ in the degree to which different compression types influence the performance sensitivity of the chosen type of classification model (random forest), which belongs to the widely accepted and utilized family of learning machines that are present in many contemporary pixel-based thematic mapping systems.

Due to the varying cloud cover and limited vegetation season, only a relatively small subset of all available images for the chosen test site was utilized. In total, two different time series of Sentinel-2A images were chosen, corresponding to two adjacent and spatially overlapping relative orbits R036 and R136, respectively. Two orbits cover the main agricultural region in the Republic of Serbia, known as Vojvodina. Acquisition dates and the percent of cloud coverage for the two described Sentinel-2A multispectral time series are shown in Figure 3.

Orbits from Figure 3 have some spatial overlapping, as illustrated in Figure 4, however they were used separately and independently in all experiments. This was done, since the field level data were distributed uniformly over the whole region of interest, but also since images corresponding to one relative orbit have significant percent of no-data region pixels, and therefore were expected to achieve higher compression ratios in comparison to images from the other one. Also, having two classification experiments with almost the same ground-truth data, but different image sources, enables one to additionally assess stability of the obtained results and their significance.

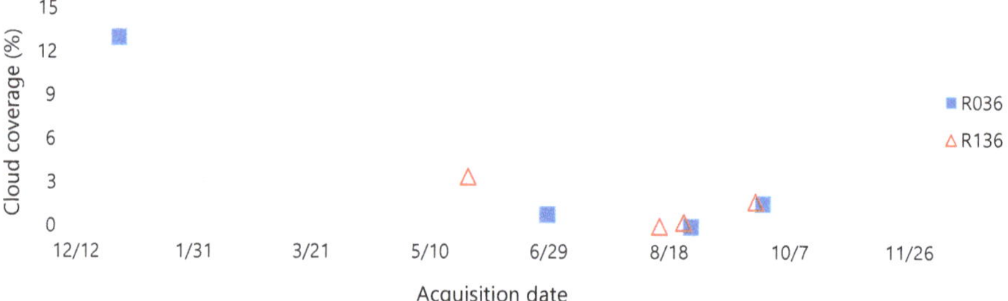

Figure 3. Acquisition dates and the percentage of cloud coverage (CC) for the selected Sentinel-2A images in the crop thematic mapping. Data correspond to images with small CC from two adjacent relative orbits of Sentinel-2A in 2016, i.e., time series R036 and R136. Since the winter wheat harvest season in Serbia usually starts in the beginning of June, and in R036 were no cloud free acquisitions before the end of the June, in order to improve winter wheat classification R036 also includes one snow-free image from the winter period.

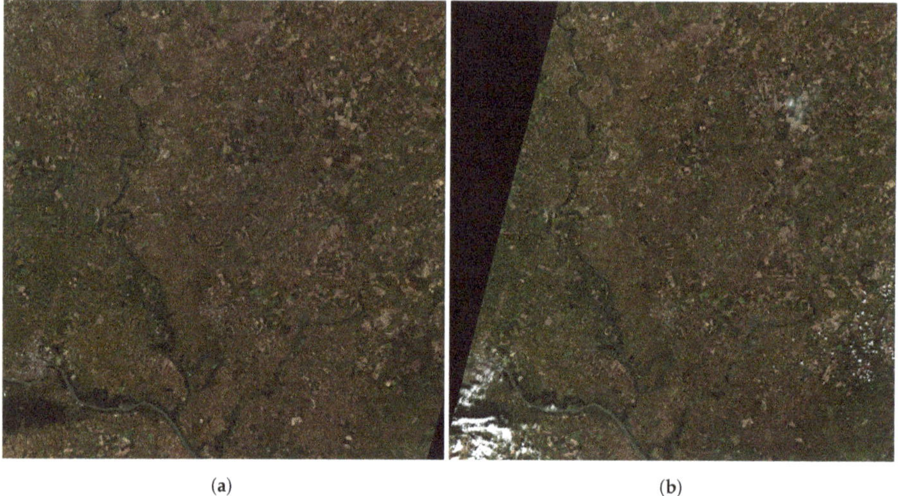

(a)   (b)

Figure 4. Degree of spatial overlap between two adjacent Sentinel-2A relative orbits, (**a**) R036 and (**b**) R136. Shown 20 m images were acquired on 28th and 25th August 2016, as parts R036 and R136.

As the result, two distinct classification datasets were designed and prepared by labeling individual pixels from the images in multispectral time series, i.e., by using ground-truth crop type labels extracted from the corresponding shapefiles that were labeled at the level of individual agricultural plots/parcels.

Characteristics of utilized Level-2 image products were already described in Section 3.1. However, it should be mentioned that the second type of experiments does not include Landsat-8 images, due to relatively small size of agricultural parcels in Vojvodina, which in the case of Landsat-8 can be only a few pixels wide, see [11]. Therefore, in order to eliminate possible influence of the spectral mixing effects onto performed compression sensitivity analysis, we have decided to exclude Landsat-8 from thematic mapping experiments, since there was no possibility to only exclude pixels laying in the crop field boundaries, due to the highly narrow shape of agricultural plots and the general lack of pixels within the fields in the case of Landsat-8.

Field level data consist of manually labeled plots corresponding to five predefined crop types (wheat, maize, sugar beet, sunflower, soybean) and an additional class corresponding to other land cover types. Each of the labeled plots was planted with only one crop type, and there was no double cropping. Labels were assigned to individual pixels inside the parcel, and their distribution per each of the categories, as well as distribution of number of corresponding agricultural parcels per each of the categories, are shown in the lower part of Tables 2 and 3. For example, note the table entries in the rows denoted by the symbols "R036" and "R136", corresponding to the total number of the parcels falling into paths of the two distinct orbits of interest, respectively.

Table 2. Number of agricultural fields per each of the categories in 10-fold cross-validation of mapping experiments. Partition subsets, denoted by $N^{\underline{o}}$, consist of manually labeled crop fields from two different relative orbits in Figure 4, R036 and R136. Abbreviations: Ma-Maize, Wh-Wheat, So-Soybean, Sb-Sugar beet, Sn-Sunflower, Ot-Other.

$N^{\underline{o}}$	Ma	Wh	So	Sb	Sn	Ot	Σ	$N^{\underline{o}}$	Ma	Wh	So	Sb	Sn	Ot	Σ
1	26	28	14	4	11	8	91	1	23	25	11	2	11	8	80
2	26	28	14	4	11	8	91	2	23	25	11	2	11	8	80
3	26	28	14	4	11	8	91	3	23	25	11	2	10	8	79
4	26	27	14	4	11	8	90	4	23	25	11	2	10	8	79
5	26	27	14	4	11	8	90	5	23	25	11	2	10	8	79
6	26	27	14	4	10	8	89	6	22	25	11	2	10	8	78
7	26	27	14	4	10	8	89	7	22	25	11	2	10	8	78
8	26	27	14	3	10	8	88	8	22	24	11	1	10	7	75
9	26	27	14	3	10	8	88	9	22	24	11	1	10	7	75
10	25	27	14	3	10	8	87	10	22	24	11	1	10	7	75
R036	259	273	140	37	105	80	894	R136	225	247	110	17	102	77	778

Table 3. Number of labeled pixels in 10^3 per each of the categories of random partition in Table 2.

$N^{\underline{o}}$	Ma	Wh	So	Sb	Sn	Ot	Σ	$N^{\underline{o}}$	Ma	Wh	So	Sb	Sn	Ot	Σ
1	9.5	16.8	19.0	5.3	3.5	1.7	55.9	1	21.0	9.2	3.6	1.8	10.1	1.6	47.2
2	22.3	13.5	4.2	3.3	9.0	4.0	56.1	2	5.2	26.2	4.1	6.2	2.2	2.0	46.1
3	21.1	14.5	10.7	5.4	2.0	2.5	56.1	3	22.3	6.4	2.9	1.0	7.3	6.5	46.5
4	24.4	7.8	4.4	2.4	10.0	6.1	55.2	4	14.7	17.5	5.2	0.4	2.5	6.4	46.8
5	9.5	20.5	4.1	13.8	2.9	4.1	55.0	5	5.2	11.4	9.9	3.5	7.7	8.9	46.6
6	12.3	23.5	3.1	5.6	3.9	7.3	55.7	6	14.9	17.6	5.1	2.3	2.4	4.0	46.3
7	9.5	26.1	4.4	5.0	2.6	8.0	55.6	7	7.7	16.0	14.4	0.8	2.7	4.3	45.9
8	15.1	15.7	7.0	7.3	9.1	2.4	56.7	8	25.9	9.4	4.3	0.4	3.7	2.3	46.0
9	12.6	25.4	10.8	1.2	2.5	2.5	55.1	9	11.8	9.5	9.4	3.5	5.7	7.0	46.9
10	15.4	19.5	9.3	3.0	1.9	7.1	56.3	10	10.0	24.8	6.5	0.6	2.4	2.4	46.8
R036	151.9	183.5	76.8	52.4	47.3	45.8	557.9	R136	138.7	147.9	65.3	20.9	46.9	45.5	465.1

Final benchmark dataset of the labeled pixels, corresponding to the multispectral time series R036 and R136, which was utilized in all classification (thematic mapping) experiments, was obtained by random partitioning of original labeled dataset into ten disjunctive subsets. Partitioning was performed at the plot level in order to avoid that some pixels from the same field enter into another partition subset. Consequently, distribution of pixels corresponding to different crop types is not completely uniform over the 10 different, randomly generated subsets of the partition, Table 3. However, distribution of number of plots that are corresponding to different crop types over the 10 different subsets of the partition is approximately uniform, as illustrated by the entries in different rows of the Table 2.

The same subsets of the random partition shown in Table 3 were used in all performed experiments for training and testing of classification models, and the average results obtained by the corresponding 10-fold cross-validation performance evaluation procedure were reported in all cases.

4. Experimental Setup and Methods

As previously stated, experiments are divided into two groups, or experimental setups, illustrated in Figure 5. In the following we give a detailed descriptions of all steps and explore the motivations for conducted numerical analyses presented in Section 5.

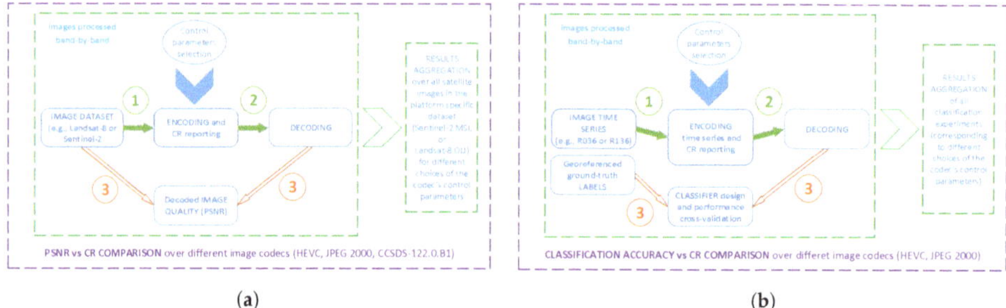

(a)   (b)

Figure 5. Processing workflows for comparison of High Efficiency Video Coding (HEVC)'s intra coding against widely accepted lossy compression solutions: (**a**) general performance comparison and (**b**) application oriented comparison.

4.1. First Set of Experiments—General Coding Performance Comparison

This experimental setup was designed in order to perform the general rate-distortion performance comparison of the HEVC's still-image intra-band coding, relative to the standard band-per-band JPEG 2000 lossy compression. Objective compression performance is measured by the means of Peak Signal-to-Noise Ratio (PSNR) vs. compression ratio curves. PSNR formulation is given with $10 log_{10} \frac{L^2}{MSE}$, MSE is mean squared error, and L represents the dynamic range of pixel intensities, e.g., $2^{16} - 1$ for 16 b/p input data. Besides aiming at the objective assessment of lossy compression effect, we have also analyzed compression artefacts introduced by different coding approaches. A performance analysis utilizes multispectral satellite image dataset described in Section 3.1.

In the case of JPEG 2000, as the baseline lossy compression solution, a reference JPEG 2000 implementation made by [79], version 2.3.0, was used with default set of encoder parameters. Such setup includes Reversible Color Transformation (RCT) and 5/3 wavelet filter, and number of wavelet resolutions (discrete wavelet transform (DWT) decomposition level) is set to 6. Wavelet transform is applied to the whole image, thus single tile/segment is used, and the resulting wavelet transform coefficients are grouped into code-blocks (rectangular regions in the sub-band wavelet domain) of size 64×64. In addition, quality control parameter Q, which corresponds to the desired PSNR level, is set to some of the discrete values in the range between $[30, 136]$ [dB], with the step size of 2 [dB] in-between different experiments. This gives 54 unique lossy compression quality levels obtained by JPEG 2000 per each multispectral image, per each of image bands. It is important to note that the input quality parameter is targeted value, however exact reconstruction quality level should be precisely calculated after image reconstruction by performing decoding process, e.g., see Figure 5a.

For comparison purpose only, in our general coding performance experiments we also included CCSDS-122.0.B1 image compression standard (we use CCSDS to refer to the given standard). It is based on similar coding scheme as JPEG 2000 (both are wavelet-based codecs), however with several features specifically adopted to reduce its complexity [28]. Implementation of the standard, used in our experiments, is Bit Plane Encoder (BPE) [80]. In contrast to JPEG 2000 wavelet transform, CCSDS uses integer 9/7 wavelet filter with maximum number of decomposition levels set to 3 (as defined by standard). As a quality regulation parameter, compression ratio in terms of b/p is used. In total, 55 different quality levels were obtained during experiments. CCSDS has been used in frame-based setup (with only one transform segment consisting of all image blocks), which is the most similar to computation of DWT in JPEG 2000. However, coding blocks in CCSDS are limited to 8×8 pixels. We have also performed some experiments in strip-based mode (with transform segments in form of strips or lines consisting of image blocks collected in raster order). However, those results were noticeably lower in terms of achieved PSNR vs. compression ratio performance.

The implementation of the HEVC's encoder and decoder according to the standardized bit-stream is represented by the HM-15.0+RExt-8.1 reference software [69]. The set of tools is restricted to intra coding only by using *encoder_intra_main_rext.cfg* configuration setup, a.k.a. all intra (AI), which is the preferred configuration of the HM codec for still images. In our experiments, we use default coding settings with the respect of the given intra configuration and HM reference software. Tiles are not used in HM experiments. Essentially, the standardized version of HEVC [3] accepts input data up to 16 b/p, which makes it suitable for the applications that use high-dimensional and high bit-depth satellite data. Thus, it is also very important not to downsample satellite images prior to compression since some sensitive visual information can be lost, and this is the main reason why we have to use the range extension (RExt) version of HEVC [75]. Therefore, code is build with RExt_HIGH_BIT_DEPTH_SUPPORT and with *ExtendedPrecision* flag set to 1, in order to accommodate 16 b/p extension as suggested in [81]. Quality control has been given by the means of the input quantization parameter Q_p, where we vary its value in the range of $[-48, 51]$ with step size of 2 in-between experiments. Note that RExt version of HEVC extends Q_p for high bit-depth inputs to negative values by performing additional internal offset calculations before mapping Q_p to actual quantization step size. Thus, we obtain 50 different quality levels in our experiments, which resulted in variety of compression ratios, from near lossless to high compression ratios. Although, within HEVC standardization process, common test conditions for HEVC range extensions [81] have been specified in order to have unified experimental setup used by multiple groups, we have not complied strictly with this process, especially as we turn to the specific application which is not covered under the given specification. In addition, it only defines two tiers of quality estimates (each tier uses only 4 Q_p values taken in-between -37 and -16), which in our use-case is not enough to cover wider range of compression ratios and to precisely obtain application's sensitivity to the introduced compression distortion.

Also, used default configuration of the reference codecs resulted in the best coding performance for both JPEG 2000, CCSDS, and HEVC. It should be mentioned that JPEG 2000 and CCSDS in all settings were used without an additional spectral decorrelation step before actual band-by-band image compression, which also holds for the proposed HEVC. Although there are many examples of custom encoders from the literature that suggest such spectral decorrelation step, it is usually in the context of hyperspectral images. In such case, there is much more redundancy present in the spectral domain than in the case of multispectral images. Also, since baseline JPEG 2000 performs only spatial decorrelation, HEVC analyses were focused only on the still-image coding and no spectral decorrelation was considered, e.g., utilization of inter coding mode. That part of the codec is left for our future research.

Therefore, in our setup we use *per-band* coding, meaning that each band is compressed individually as a separate file. Considering HEVC as video based codec, we could also re-arrange all bands of an image, resembling standard video sequence, and thereafter code all bands jointly as a sequence of bands resulting in one compressed file (although still by using intra mode only to code all bands). However, we observed that overhead information introduced by band-by-band compression is negligible when compared to the sequence coding, and almost identical performance has been obtained. Thus we chose to code bands individually to be consistent with current EO systems which are organized in a manner to perform per-band distribution. Similarly can be said for the video based Motion-JPEG 2000 (MJP2) [82], where MJP2 only defines file format for sequences of JPEG 2000 images, and where each image is coded as an independent JPEG 2000 codestream. Thus, results are obtained on per-band basis, and aggregated over image.

Regarding the PSNR vs. compression ratio performance diagrams that will be presented in Section 5.1, it should be emphasized that the presented PSNR values are averages, computed over all satellite images in Table 1. However, averaging in the case of the Landsat-8 and the Sentinel-2A/B images is slightly different, due to the fact that analyzed Sentinel-2A/B image products have varying number of image channels per each of the supported spatial resolutions of 10, 20, and 60 m (4, 9, and 11 channels, respectively). Thus, in the case of Sentinele-2A/B, in order to have a more representative average, PSNRs of individual channels where first averaged over the same spatial resolution, and then three distinct PSNR values were aggregated in the final PSNR of the same image.

Since in the most of the currently operational ground based image distribution systems user specified image orders are usually provided in the form of bulk downloads, in Section 5 compression ratio (CR) is computed over all lossy compressed images, without averaging. Such CR values correspond exactly to the possible use case scenario, like the one considered in this paper, where due to the specific characteristics of the crop thematic mapping, time series were required to be downloaded as a whole for the purpose of developing analyzed classifier instances.

4.2. Second Set of Experiments—Crop Thematic Mapping

The other set of experiments is oriented towards the analysis of the compression induced distortion effects on the specific thematic mapping application. Experiments are conducted on the proposed Sentinel-2A benchmark dataset given in Section 3.2. Since presented analysis also aims to offer application oriented comparison of HEVC's performance, as the foundation for the second set of experiments, a classic problem of pixel-based crop classification in multispectral satellite images using time series of georeferrenced surface reflectance measurements has been chosen. Such data capture the time varying phenology of the plants and make required category discrimination much more effective. In order to simplify the experimental setup, and enable a fair comparison in terms of sensitivity of classification performance to various effects caused by different lossy compression scenarios, instead of object-based classification approach we have decided to rely on pixel-based classifier. As opposed to object-based approach, it performs feature space partitioning by using measurements aggregated only at the level of individual pixel, which eliminates possible improvements that could arise from incorporation of pixel's neighbourhood, or enhancements resulting from the postprocessing of pixel-based decisions at the object level. However, such pixel-based approach is widely accepted solution, which is used in many operational thematic mapping systems.

Thus, classification performance, which is intended to be used as an indicator of the thematic mapping's sensitivity to compression effects, is influenced only by the quality of measurements at the level of individual pixels, and not by the spatial resolution of the sensor nor the structure of pixel's neighbourhood. However, through comparison of classifier's performance with the same type of decision system utilizing original, uncompressed data, such pixel-based classifier is still capable of highlighting and reflecting the level of preserved quality of original measurements after compression, relative to the uncompressed baseline. Potentially, this could be very important in some other applications besides thematic mapping, which could be oriented more towards computation of plants' biophysical parameters, or that would rely on multispectral satellite images with coarser spatial resolution than Sentinel-2. On the other hand, experimentally measured performance of such pixel-based classification approach does not reflect the effects of compression on simple image primitives like edges and textures, which could be important in the case of higher resolution sensors and predefined or learned object-based image descriptors. However, such effects are possible to estimate in other ways, including visual inspection of higher resolution lossy compressed images and their comparison with their coarser resolution counterparts, like Landsat-8 images of the same scene. Therefore, this was particularly considered in the set of the conducted experiments of the first type, described in Section 4.1, where the compression performance was analyzed from the general, and not an application oriented perspective.

As the type of an adequate classification model for the specified thematic mapping task was used random forest [83], which possesses high generalization capability due to its intrinsic ensemble structure and the fact that decision trees [84] are particularly suitable for partitioning of feature space based on simple, yet very informative features corresponding to time varying spectral signatures of the Earth's surface.

In order to remove any classification performance bias coming from the specific classifier design or feature extraction pipeline, in all conducted experiments was used the same classification setting as proposed in [11]. It consists of feature level fusion that extracts surface reflectance measurements corresponding to the same image pixel from all bands of all images in the corresponding time series. This step is followed by an ensemble type random forest classifier with 100 trees of variable depth (adaptive splitting until all leaves are pure or contain only two samples), and splitting criterion based on the Gini index and sampling of $\sqrt{d}$ random features in each node, where d is the total dimensionality of the feature space. Classification was implemented in the Python programming language, using library [85].

However, since individual bands of Sentinel-2 images have varying spatial resolutions (see Section 3.2), before final feature extraction all multispectral images were: (1) compressed at their nominal spatial resolution (except in the case of classification model based on uncompressed image data), and afterwards (2) in the case of 20 m and 60 m Sentinel-2 image channels additionally upsampled to spatial resolution of 10 m using nearest neighbour interpolation method. The second step is necessary since we are interested in exploiting as much of available spatial information during the decision process, e.g., fine signal variations present in 10 m image channels are valuable in the presence of small crop fields. We would also like to add that the practical implementation of this step does not require allocation of new memory, since 20 m and 60 m image pixels are just indexed several times during the feature extraction process, in accordance with their spatial position in the physical coordinate system. Thus, classifier decisions were made at the highest level of detail, i.e., at the level of pixels with 10 m of spatial resolution.

Each of the experiments that are introduced in this subsection correspond to the performance analysis of the selected random forest classifier that was trained and tested using the labeled benchmark dataset described in Section 3.2, which was initially randomly partitioned into 10 subsets in order to enable classifier performance evaluation based on 10-fold cross-validation procedure. It should be noted that each of the experiments is utilizing different type of source data, i.e., the same Sentinel-2A multispectral images (time series), but with different *type* and *level* of *lossy compression*. Thus, the performance of the baseline classification model is evaluated using uncompressed image data, while the rest of experiments are conducted using lossy compressed image time series, obtained with different sets of encoding control parameters in each experiment, i.e., by varying HEVC and JPEG 2000 configurations. These configurations were varied in the same manner as in the case of the first set of general compression experiments that were described in Section 4.1. The only difference is that instead of both Landsat-8 and Sentinel-2A/B data, in classification experiments only Sentinel-2A data were utilized.

Since for the selected area we have collected two separate image time series, denoted as R036 and R136, each of the above mentioned experiments was conducted twice, once for each of the two Sentinel-2A relative orbits. Alternative option would be to combine all these measurements together. However, as already mentioned before, using data from R036 and R136 separately gives possibility to mutually compare results of two individual classification analyses under similar experimental setting. Also, combining data from R036 and R136 would slightly reduce the overall number of training and test samples per each class in the benchmark dataset. Therefore, such alternative dataset design strategy was avoided.

5. Results Analysis and Discussion

All results are presented according to the plan of experiments developed and presented in Section 4 with level of presentation details adapted to the large number of numerical simulations.

5.1. General Compression Performance Comparison

From the perspective of the general compression performance, results from the first set of experiments, Figure 5a, confirm initial hypothesis that HEVC still-image intra band coding can be considered as an adequate standardized solution for effective lossy compression of multispectral satellite images.

5.1.1. PSNR vs. CR—Results

Figure 6 and Table 4 neatly summarize the codecs general performance over the two proposed image datasets corresponding to—Landsat-8 and Sentinel-2A/B, which were described in detail in Figure 2 and Table 1. Presented results also highlight variations in compression performance due to differences in the type of studied image sensors.

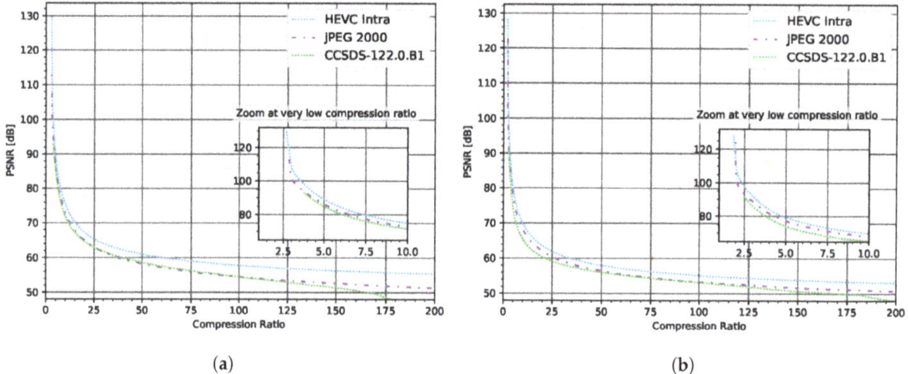

(a)   (b)

Figure 6. Average Peak Signal-to-Noise Ratio (PSNR) vs. compression ratio performance of the proposed High Efficiency Video Coding (HEVC), baseline JPEG 2000, and the Consultative Committee for Space Data Systems (CCSDS) 122.0.B1 standard, over: (**a**) Landsat-8 and (**b**) Sentinel-2A/B part of the dataset described in Figure 2 and Table 1.

Table 4. Achieved compression ratios (CR) at the same level of PSNR. Performance of HEVC is expressed relative to JPEG 2000 and CCSDS 122.0.B1 CRs over two parts of the proposed dataset.

	Landsat-8 CRs				
PSNR	**HEVC**	**JPEG 2000**	**Rel. Diff. [%]**	**CCSDS**	**Rel. Diff. [%]**
55	229.84	89.04	**158.13**	93.19	**146.63**
60	60.66	37.14	**63.33**	39.78	**52.48**
65	26.29	19.68	**33.59**	19.30	**36.22**
70	15.04	12.36	**21.68**	11.56	**30.10**
	Sentinel-2A/B CRs				
PSNR	**HEVC**	**JPEG 2000**	**Rel. Diff. [%]**	**CCSDS**	**Rel. Diff. [%]**
55	103.94	66.17	**57.08**	65.02	**59.85**
60	34.11	26.69	**27.80**	21.46	**58.94**
65	16.57	13.69	**21.04**	10.48	**58.11**
70	9.84	8.34	**17.99**	6.56	**50.00**

5.1.2. PSNR vs. CR—Analysis and Discussion

Regarding the performance comparison between JPEG 2000 and CCSDS image coding standards, from Figure 6, we can observe that CCSDS codec perform on par with JPEG 2000 over the majority of the CRs. The CCSDS results were close, or sometimes even lower, to the results obtained by the JPEG 2000. Therefore, this codec was left out from the further discussion, e.g., classification sensitivity analysis. In addition, at higher CRs we note significant deterioration in performance for CCSDS codec. However, it is expected since CCSDS was not designed to operate on such high CRs.

As measured by preserved image quality, HEVC has shown greater resilience to information loss, i.e., ability to achieve higher CRs at approximately the same image quality levels as compared to JPEG 2000. Thus, it can be seen that Landsat-8 images, Figure 6a, tend to be compressed more efficiently, e.g., have higher compression ratios for the same level of PSNR, when compared to their Sentinel-2 counterparts, Figure 6b. Therefore, relative difference between HEVC and JPEG 2000 performance is higher when benchmarked exclusively over Landsat-8 part of the adopted general purpose dataset, which is also summarized in Table 4. This could be explained by the presence of no-data regions, which are always found in Landsat-8 images (dark/blank areas in the corners of the scene, present due to orientation of satellite orbits with respect to the image reference system). In contrast, this is not always the case for Sentinel-2 images (especially in the proposed Sentinel-2A/B part of our general performance comparison dataset, which turns out to have a very low proportion of no-data regions). Such blank, uniform regions are naturally easier to be compressed by using block based approach (characteristic for HEVC), as opposed to wavelet decomposition that is performed on an image as a whole (in the selected baseline JPEG 2000).

We also remark that HEVC brings substantial savings in the required rate in comparison to JPEG 2000, as it can be seen from Figure 6 and Table 4. In particular, for mid compression ratios (high PSNR range, between 65 [dB] and 70 [dB]), HEVC brings from 17.99% up to 33.59% savings in the required rate with respect to JPEG 2000 (as measured by the relative difference in achieved compression ratios over Sentinel-2A/B and Landsat-8 parts of the proposed benchmark dataset, respectively, Table 4). Furthermore, HEVC achieves up to 63.33% reduction in rate over JPEG 2000, for high compression ratios (low PSNRs) represented by the 60 [dB] entry in Table 4 corresponding to the results obtained over Landsat-8 part of the proposed dataset. As it can be seen from Figure 6 and Table 4, savings are even higher at 55 [dB], but this range of image quality is already below the one desirable for crop thematic mapping applications (see Section 5.2). Thus, for very high compression ratios HEVC can achieve even up to 158.13% of rate improvement over JPEG 2000, as measured over Landsat-8 images, which is lower for Sentinel-2 images and goes to 57.08%, due to already mentioned lower presence of no-data regions that are characteristic for Landsat-8 images.

Additionally, in Figure 6 we can see that the performance on very low compression ratios (around 2–3) does not show significant difference between the two benchmarked codecs. Moreover, JPEG 2000 was able to slightly outperform HEVC, as shown in the zoomed part of the diagrams in Figure 6. However, performance for very low compression ratios should be additionally investigated in greater detail, e.g., by benchmarking obtained lossy compression results with the lossless versions of the same codecs, which should be considered as a more appropriate design choice for the described, specific operational range. Our assumption is that the same compression ratios could be achieved by lossless modes, leading to a perfect reconstruction. Also, lossless modes usually posses less computational complexity. However, such analysis is left out from this study and could be part of potential future work.

5.1.3. Resolution Effects—Results

Since Sentinel-2 image products contain image channels with varying spatial resolutions, this specific property was exploited to additionally investigate effects of the pixel size on the compression performance. These results are summarized in Figure 7, where for each spatial resolution of 10, 20, and 60 m, computed average PSNRs are depicted against the achieved compression ratios.

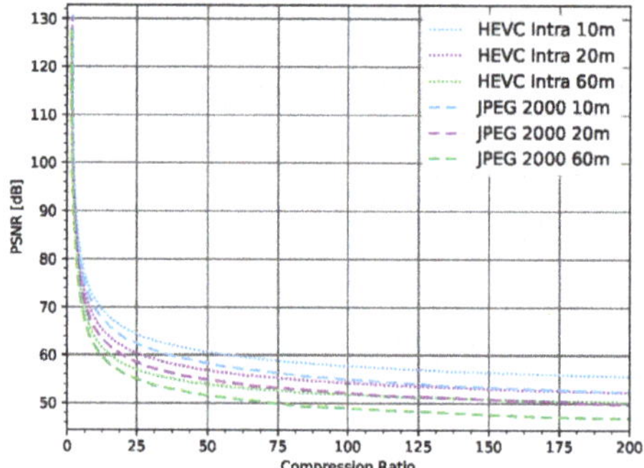

Figure 7. Compression performance comparison over satellite images with different spatial resolutions. Pixel resolutions of 10, 20, and 60 m correspond to channels of Sentinel-2A/B images listed in Table 1.

5.1.4. Resolution Effects—Analysis and Discussion

We can see that both codecs perform better on higher resolutions (blue lines). Moreover, HEVC (represented by dotted lines) always shows higher performance margin in comparison to JPEG 2000 (dashed lines), at all spatial resolutions (which was already expected based on the results presented in Figure 6). However, HEVC's advantage is particularly exaggerated at the highest level of details, corresponding to 10 m spatial resolution, where, e.g., at 60 [dB] we can see significantly higher margin between HEVC and JPEG 2000's compression ratios, as compared to other spatial resolutions. This was natural to expect, since images with higher spatial resolution have higher presence of uniform image regions, and these regions are also expected to have larger size (image extent in pixels) in comparison to images with coarser spatial resolution of 20 m, or 60 m per pixel.

5.1.5. Visual Comparison—Results

As an adequate testbed for qualitative assessment of the benchmarked codecs we have selected an image region with the complex scene content, corresponding to agricultural holdings, forest and water bodies in North Carolina, entry № 7 form Table 1. Visual illustrations of the obtained results are presented in Figure 8, where true color composites of original images are ordered according to Sentinel-2 image spatial resolutions (10, 20, and 60 m) and the selected, representative levels of PSNR (56, 60 and 70 [dB]). True color composites of lossy compressed 30 m Landsat-8 image of the same region are also presented in Figure 9.

5.1.6. Visual Comparison—Analysis and Discussion

As expected, visual inspection of results in Figures 8 and 9 reveals that in the case of both codecs the presence of compression artefacts is becoming more significant after the mid-range of 60 [dB] (the last two columns). Also, compression effects are dependant on spatial resolution, and the perceived rate of image degradation is higher towards the lower resolutions of 20, and 60 m, even at the same level of PSNR (e.g., Figure 8e,h or Figure 8f,i).

However, it is particularly interesting to note that, based on visual comparison, HEVC outperforms JPEG 2000 even at the same level of PSNR, consistently over all spatial resolutions. This is easiest to see at 60 [dB], even at the highest spatial resolution of 10 m, Figure 8b. In addition, savings introduced by the HEVC's higher compression ratio over JPEG 2000 at the same PSNR level of 60 [dB], were already shown to be 27.80%, Table 4. Thus, besides the significant gain in achieved compression ratios in the case of the proposed HEVC solution, visual inspection confirms that the object boundaries are also better preserved, i.e., object boundaries in the image are much sharper and have a more precise shape with less visual distortions (blurring effect). For example, lake and field boundaries in the upper part of Figure 8b, and objects at smaller spatial scale. For the best comparison, 10 m images in Figure 8b should be analyzed enlarged, side by side, at the computer screen.

Figure 8. Visual comparison of HEVC and JPEG 2000 at the same level of PSNR. **The first column** corresponds to Sentinel 2 images of 10, 20, and 60 m generated at high PSNR of 70 [dB]. In the **second column** are images with 60 [dB], while in the **last column** are images with approximate lower borderline quality of 56 [dB]. Each pair of rows (3 in total) represents satellite images at exclusively one of the following spatial resolutions: 10 m (**a–c**), 20 m (**d–f**), and 60 m (**g–i**), respectively. The **first row** in each pair corresponds to obtained HEVC results, e.g., the first row of (**a–c**), while the **second row** in each pair is reserved for JPEG 2000 results with the same level of PSNR and spatial resolution (e.g., 10 m for (**a–c**)). Presented image detail was taken from the test scene № 7 in Table 1, Figure 2g.

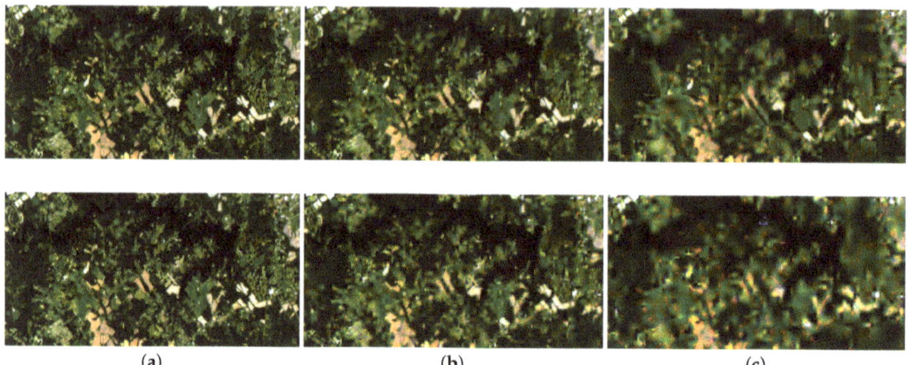

Figure 9. HEVC (**upper row**) vs. JPEG 2000 (**lower row**) on the example of an 30 m Landsat-8 image detail corresponding to the same area as shown in Figure 8. Results are the most comparable to those shown in Figure 8d–f. Columns (**a–c**) correspond to PSNR of 70, 60, and 56 [dB], respectively.

Differences in compression effects among HEVC and JPEG 2000 at high PSNR of 70 [dB] are definitely more subtle (the first column in Figure 8). However, based on objective assessment there is still significant relative difference in achieved compression ratios between HEVC and JPEG 2000 at this high PSNR range (from 17.99% up to 21.68% according to Table 4).

Regarding comparison over 30 m Landsat-8 image, Figure 9, it is worth mentioning that by visual inspection at mid and low PSNR range it is easier to become aware of blocking artifacts introduced by HEVC, in comparison to 20 m or even 60 m Sentinel-2 images with same image quality in Figure 8 (e.g., compare Figure 8h and Figure 9b). Latter is rather strange, since 30 m image contains four times more pixels than a 60 m one. However, it can be explained by the scale of objects present in the scene and the fact that a 60 m image channel inherently already has significant amount of spectral mixing (less features of the scene geometry), i.e., only large scale scene details are recorded in the uncompressed original image.

Again, HEVC shows better ability to preserve image details (e.g., lake boundaries), however it is less obvious than in the case of 10 m and 20 m Sentinel-2 images due to lower spatial resolution (especially at lower PSNRs, the last two columns). As expected, results also reveal that at lower spatial resolutions and PSNR levels more blocking effects are present. E.g., when the upper image in Figure 9b with a 60 [dB] is enlarged next to the upper image in Figure 9a (which at 70 [dB] is very close to the 30 m uncompressed original).

5.2. Application Oriented Compression Performance Comparison

In addition to the general compression performance analysis, it is also tempting to validate previously presented results from the standpoint of some practical remote sensing application. Thus, in the following are the analyses measuring sensitivity of the specific remote sensing classification task in the presence of lossy compressed input data. The same set of experiments, illustrated in Figure 5b and described in Section 4.2, was conducted twice, independently over two different subsets R036 and R136 of the prepared ground-truth dataset introduced in Section 3.2.

5.2.1. Classification Sensitivity—Results

Performed analyses consist of benchmarking classifiers' performance against the lossy compressed images generated by the HEVC and the JPEG 2000. This means that for the each of the codecs' configurations (driven by different values of the corresponding quality control parameters), datasets R036 and R136 were lossy compressed each time, and the new instance of the *same* learning model type was trained and tested.

Thus, in Figure 10, for each experiment (classifier instance), a pair of values corresponding to the overall classification accuracy (OA) and the achieved compression ratio (CR) was plotted. For the sake of completeness, in Table 5, are also provided results of the two classifier instances based on the original, uncompressed image data, while their OAs are plotted by the solid red lines in Figure 10.

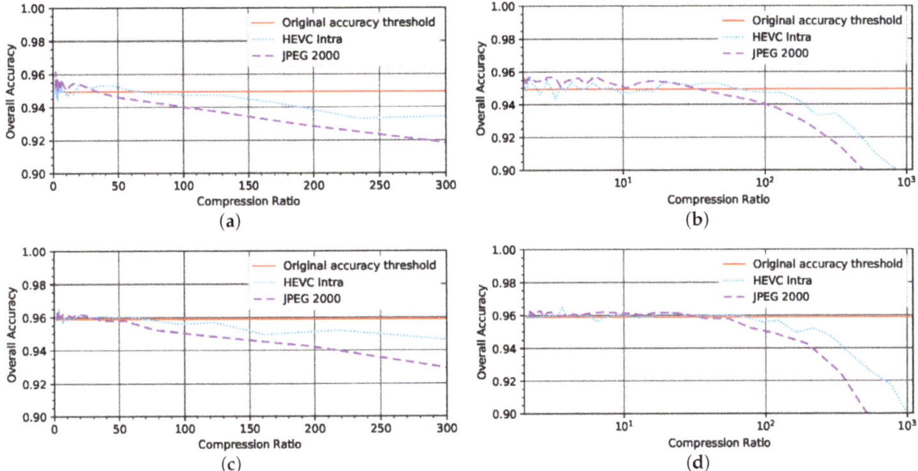

Figure 10. Sensitivity of classification performance, as illustrated by the overall classification accuracies (OAs) against the compression ratios (CRs) achieved by the HEVC intra and the JPEG 2000 image codecs. Diagrams (**a**,**b**) and (**c**,**d**) correspond to experiments based on time series R036 and R136, while the solid red lines denote OAs of the "original" classifier instances reported in Table 5. Higher CRs are given in log scale in (**b**,**d**).

Table 5. Accuracy assessment matrices for two classifier instances based on the original, uncompressed time series data, R036 and R136. Symbols are the same as in Table 2. Values represent number of pixels in 10^3. Column labels denote classifier decisions, while the row labels represent true crop categories. Overall accuracy (OA), Cohen's Kappa statistic (KP) and the KP confidence intervals (CI) are given at the bottom of the table.

R036	Ma	Wh	So	Sb	Sn	Ot	R136	Ma	Wh	So	Sb	Sn	Ot
Ma	148.6	0.4	1.5	0.2	0.9	0.1	Ma	137.0	0.3	0.5	0.0	0.6	0.1
Wh	0.5	181.6	0.6	0.0	0.3	0.4	Wh	0.3	146.7	0.0	0.0	0.2	0.7
So	4.7	0.3	67.0	1.4	3.2	0.2	So	0.9	1.4	60.6	0.1	2.0	0.3
Sb	0.6	0.1	0.5	51.0	0.2	0.0	Sb	0.0	0.0	0.2	20.5	0.1	0.0
Sn	1.3	0.4	0.4	0.3	44.7	0.1	Sn	1.0	0.5	0.1	0.2	45.0	0.1
Ot	4.9	3.1	0.3	0.1	0.8	36.6	Ot	0.5	8.4	0.2	0.1	0.0	36.2
OA = 94.95% KP = 0.94 CI = [0.9339 0.9354]							OA = 95.90% KP = 0.95 CI = [0.9471 0.9456]						

5.2.2. Classification Sensitivity—Analysis and Discussion

Based on Figure 10, it can be said that the initial assumption regarding the possibility of lossy compression having a positive impact on the classification performance at low compression ratios proved to be valid, since the compression side effects (e.g., noise canceling) led to better classifier performance at lower CRs, in the case of both benchmarked codecs and for the both R036 and R136 test subsets. Also, in Figure 10 we can see that the JPEG 2000 performs quite well for the CRs of up to **35:1** (dashed lines), mostly without any loss in performance when compared to the original classifiers, which were trained and tested on the uncompressed data (red lines). At the same time, HEVC (dotted lines) can achieve the same (or even better) classification accuracy in comparison to the original classifiers (OA's represented by the red lines), even at compression ratios of up to **70:1**, leading to a nearly **100%** relative improvement in achieved CR (potential savings) over JPEG 2000. The range of compression ratios leading to such OAs, which are close to the ones achieved by the original classifiers, could be regarded as the first operational range, when considered in the context of the selection of an optimal compression strategy. Therefore, points towards the end of the described range should be regarded as the effective codec configuration choices for efficient distribution of the lossy image data applied in crop thematic mapping. Moreover, since high temporal resolution signatures of the vegetation pixels require dense image time series, any potential savings in storage and communication requirements could be significant in the case of the constrained time or memory resources.

On the other hand, JPEG 2000 is able to provide sufficiently high classification accuracies when data are compressed with CRs of up to **60:1** for R036, or **70:1** for R136, maintaining OA within the acceptable *loss margin* of **0.5%**. However, HEVC has demonstrated ability to maintain approximately the same classification performance (losing no more than 0.5% of accuracy) for CRs of up to **150:1**, in the case of R036, and up to **140:1**, for R136 test subset, as in Figure 10. It is a limit after which distortions begin to have significant influence on the overall classification performance. For example, in comparison to HEVC, at same CR JPEG 2000 loses on OA are around **1.5%** and **1.2%** for R136 and R036, respectively.

Therefore, breaking points of HEVC and JPEG 2000 curves in Figure 10 could be considered as the examples of the indicators designed to measure codec's expected performance in application oriented lossy compression scenarios. Implications of the presented sensitivity analyses, conducted over both R036 and R136, confirm that at the relatively high CRs (which are determined by the application specific, performance loss margin) the proposed HEVC based solution is able to better preserve features important for more accurate classification, as compared to the JPEG 2000, when the same CR has been accounted for—leading to the higher savings in storage and communication requirements of the ground based data distribution segment.

Although it may not be considered to be of significant practical importance for the considered application task, codecs' behavior for the higher compression ratios (with loss margin larger than the 0.5% of the overall accuracy of the original classifier) is also shown in Figure 10b,d.

In the context of the crop thematic mapping, spatial dependant relationships may be an important part to analyze in more detail. However, significance of the particular, individual image channels and the impact of the level of distortion introduced during compression at different spatial resolutions on the overall classification accuracy will be the subject of our future work, especially from the standpoint of adaptive control parameter selection for optimal compression performance and classifier design.

5.2.3. Crop Thematic Maps—Results

Besides the comparative quantitative analyses presented in Figure 10, in order to further discuss the performance of HEVC and JPEG 2000 based solutions, we also provide a type of qualitative assessment of the achieved classification performance. It is based on the visual inspection of the obtained crop thematic maps and their differences, see Figure 11.

As the result of HEVC's higher effectiveness, as shown in Section 5.1, it was expected that HEVC based lossy compressed images should also generally provide better application oriented performance (higher quality crop thematic maps). In particular, it was expected that even at low CRs, where there is less difference between HEVC's and JPEG 2000's performance, visual inspection of the corresponding thematic maps should also reveal better visual quality of classification maps produced using HEVC encoded images. This was the main motivation for illustrations presented in Figure 11, which have confirmed described initial assumptions, and gave additional insights into complex relationship between lossy compressed image quality and classification performance.

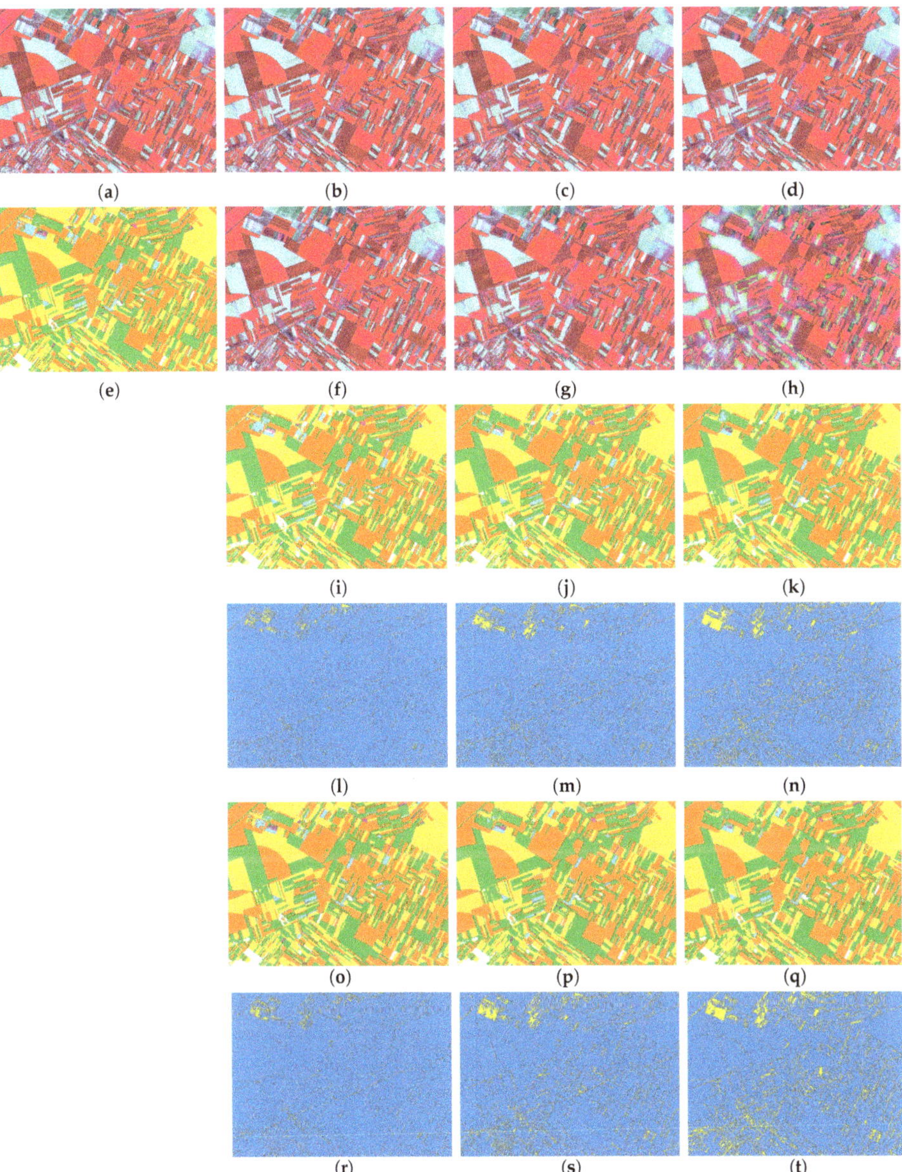

Figure 11. Illustration of the compression effects on the quality of generated crop thematic maps. Each of the columns 2–4 gives visual insights into sensitivity at achieved image compression ratios (CRs) of approximately **25:1**, **75:1**, and **200:1**, respectively. Classification map (**e**) in the first column is based on the the original, uncompressed image time series, shown in (**a**). Images (**b–d**) and (**f–h**) are the NIR-red-green false color composites of the same scene detail produced by the HEVC and the JPEG 2000 for the considered CRs, respectively. Similarly, group of images (**i–k**) are the crop thematic maps made by classifier instances tbased on the proposed HEVC solution, while the maps (**o–q**) are the results of applying JPEG 2000. Differences between HEVC based classification maps (**i–k**), as well as JPEG 2000 based maps (**o–q**), in comparison to the original map (**e**) are illustrated by the yellow pixels in (**l–n**) for the HEVC, and (**r–t**) for the JPEG 2000.

5.2.4. Visual Comparison of Classification Maps—Analysis and Discussion

Examples in Figure 11 were selected in order to indicate an important fact - that the classification maps obtained by the HEVC and the JPEG 2000 with approximately the same CRs (around **25:1**, **75:1**, and **200:1**) result in the crop thematic maps that are visibly more accurate in the case of the HEVC compression. This can be confirmed by the visual comparison of the corresponding difference images, computed between the map produced by original classifier and each of the maps produced by HEVC and JPEG 2000 based solutions (obtained at the same CR), i.e., by comparison of: Figure 11l vs. Figure 11r; Figure 11m vs. Figure 11s; and Figure 11n vs. Figure 11t. Difference images reveal changes in decision maps made by the original classifier (relying on uncompressed data), Figure 11e, and the maps produced by the three selected lossy compression instances of the same classification model, illustrated in Figure 11i–k,o–q.

Thus, by inspection of the given pairs of figures, it is clear that at all considered CRs (corresponding to different columns of Figure 11), there are always more yellow pixels (dissimilarities) in the case of JPEG 2000, as opposed to HEVC based solution. It is interesting to see that differences are also visible in the case of low compression ratio, e.g., 25:1, where HEVC produces visually better map, compare Figure 11l against Figure 11r, even that classifiers' OA is approximately the same at the given CR, e.g., see Figure 10. The same trend is even more exaggerated at the higher CR's (column 3–4), where according to the previously presented results HEVC achieves even better level of the corresponding quality-compression trade-off against JPEG 2000. However, it should also be noted that in both cases, at low CRs (second column) classification maps are the most sensitive to compression effects affecting fine image details, like the boundaries between fields or the lines corresponding to roads. This is understandable, since at the boundaries of different objects in the scene there is naturally more spectral mixing and it is harder to delineate different categories in the corresponding feature space, which makes them the most sensitive to any signal variations. On the other hand, as expected, higher CRs result in noisier maps, or even complete misclassification. Again, smaller fields, which are harder to delineate, are more suspectable to errors. In addition to the above discussion, we would also like to note that the presented map differences at lower CRs (higher PSNR) are not necessarily corresponding to classification errors, since the map in Figure 11e does not represent the ground-truth, but the decisions of the original classifier instance (which is also prone to errors), while the previously presented quantitative analyses revealed that the small amount of lossy compression can have a noise canceling effect and bring additional improvements in the overall classification performance.

5.3. Classification Results from the Statistical Significance Perspective

Since presented sensitivity analyses in Section 5.2 were based on extensive numerical simulations, there is a question of statistical significance of difference between results corresponding to different classifier instances, i.e., lossy compression experiments driven by different values of corresponding HEVC and JPEG 2000 control parameters, Figure 5b.

In total, for each R036 and R136 dataset there were 10^4 unique numerical experiments, 50 corresponding to HEVC and 54 to JPEG 2000, covering approximately the same range of image quality levels. Each of the reported outcomes, plotted in Figure 10, represents the statistically cross-validated classification performance measure, i.e., in the given case overall accuracy (OA). However, for the purpose of measuring statistical significance of difference between each pair of conducted experiments, we have decided to analyse another aggregate performance measure, the Cohen's Kappa coefficient [86], which compactly integrates both "type I" and "type II" classification errors into one representative performance measure, i.e., efficiently aggregates the off-diagonal elements present in the corresponding accuracy assessment matrix of each experiment. Thus, for each pair of the experiments, represented by the estimated Kappa coefficients, the null hypothesis of two Kappa values being the same was tested with the level of 5% of statistical significance. Results of all statistical tests are graphically presented by the upper triangular matrix in Figure 12, with circles denoting the positive outcomes (rejection of null hypothesis).

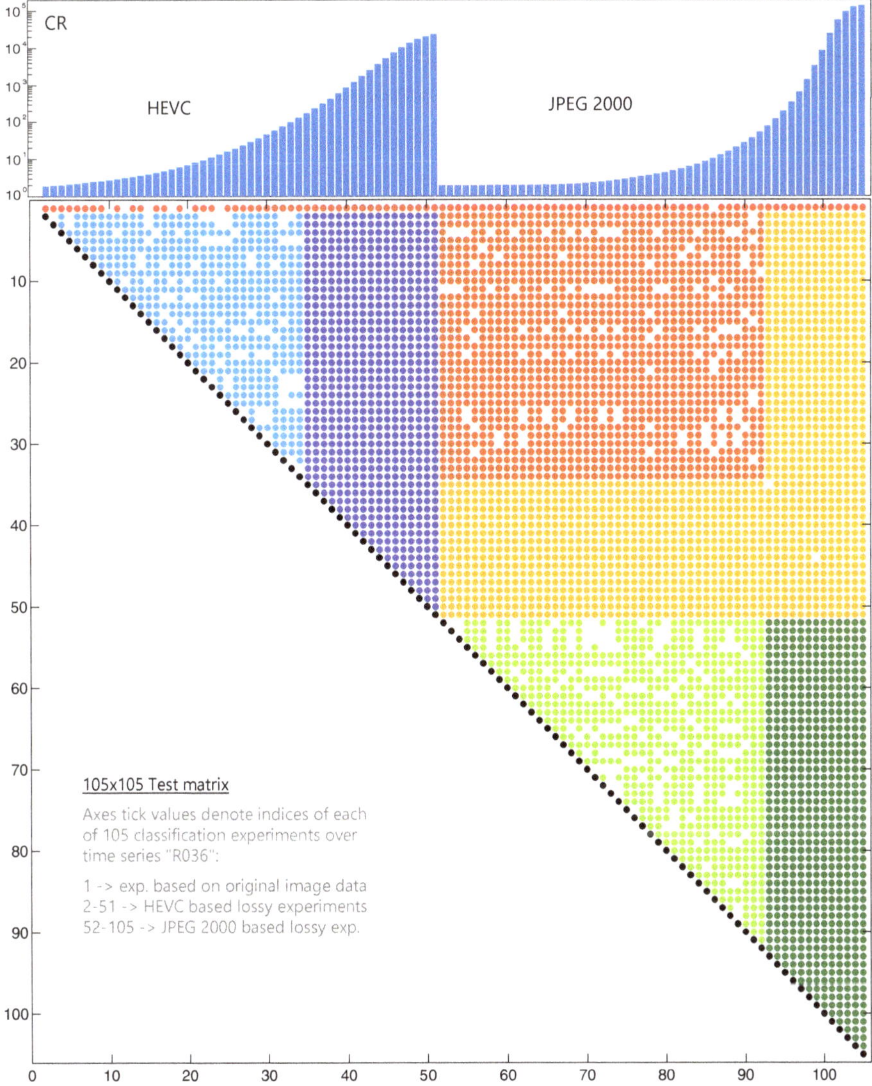

Figure 12. Results of the tests of statistical significance of difference among the classification results produced by the crop thematic mappers based on the HEVC and the JPEG 2000. Missing entries indicate pairs of experiments with non-significant differences. Coloured circles above the main diagonal indicate rejections of the null hypothesis, and denote the pairs of experiments with significantly different classification performance. Tests were based on the Cohen's Kappa statistic and the significance level of 5%. **Black** circles on the main diagonal correspond to the tests of all 104 experiments against the classification results expected to be produced by chance, while the circles in *red* correspond to the tests against original classifier based on the uncompressed image data in the time series R036. Other regions are coloured according to the type of test (exclusively within the HEVC based results, *light* and *dark blue*; or exclusively within the JPEG 2000, the *green* ones). *Orange* and *yellow* regions are the crosswise tests between the HEVC and the JPEG 2000. Aligned above the test matrix is the bar diagram representing compression ratios (CRs) achieved by the each of the experiments. Interpretations of regions illustrated by different colours and discussion of tests implications are given in Section 5.3.1.

Thus, when present, circles *above* the main diagonal denote that the experiments' outcomes (Kappa coefficients) were significantly different, while the elements on the main diagonal (except the first one) denote that the each of the associated experiments resulted in classification that was significantly different from the one expected by chance. Finally, indices of the first element in the $10^5 \times 10^5$ test matrix in Figure 12 denote the original classifier, based on the uncompressed image data from the time series R036.

5.3.1. Analysis of Observed Patterns and Discussion of Their Implications

Figure 12 shows that in the most cases experiments based on different (*lossy* compressed) input data resulted in different classification performance. Presented illustration also reveals that for some choices of codecs' configuration parameters (varied in the defined range of values by the constant step size) there appear small "clusters" of empty circles, i.e., experiments resulting in classification performance that could be regarded as the same in the statistical sense. However, the most interesting insight is that observed clusters can be visually grouped into different "compression comparison zones", or *regions*, illustrated by different colours in Figure 12. Moreover, the regions seem to be not arbitrary, but aligned with the experiments corresponding to different working regimes (compression ratios) of the compared HEVC and JPEG 2000 image codecs. Therefore, besides the test matrix, upper part of Figure 12 also includes the graphical illustration of the computed CR for each experiment (the bar diagram that is aligned with the matrix columns and shown in the log scale). The bar diagram also has an additional purpose, to delineate matrix columns (i.e., rows) corresponding to experiments considering HEVC and JPEG 2000. Therefore, with the help of the bar diagram, it is easy to distinguish between the two, by considering the two distinctive parts of the diagram—the first one on the left, representing the HEVC related CRs, and the second one on the right, depicting considered JPEG 2000 configurations.

In Figure 12 we have identified several regions, which is also reflected in the choice of the selected colours. Thus, the *light blue* and the *dark blue* regions correspond to the tests of statistical significance among the experiments exclusively utilizing the HEVC based data with different CRs, while the *light green* and the *dark green* correspond to the experiments exclusively based on the JPEG 2000 input data. The remaining regions, the *orange* one and the "L" shaped *yellow* one, correspond to *crosswise* tests among experiments based on different configurations of the HEVC and the JPEG 2000.

Last, but not the least important, is the group of tests illustrated by the *red* colour, the first row of the test matrix in Figure 12. It illustrates the significance of difference among all 104 "lossy classifiers" and the "original" classifier (based on the uncompressed image data and represented by the the first element of the test matrix, i.e., by the first "missing" circle in the first row). The *red* region, i.e., the first row, reveals that there are more empty spaces (missing circles) among the set of columns in the first row that are representing the HEVC solution, as opposed to the one associated with the JPEG 2000. This is another confirmation that among the HEVC based classifiers operating at the relatively moderate compression ratios (up to **16:1**), there are more classifiers that can achieve the classification performance that is (in the statistical sense) the same as as the one achieved by the original classifier—as compared to the JPEG 2000, which has only one empty circle in the *red* region of the given example in Figure 12 (at CR of **13:1**). Moreover, when compared by using described Kappa based tests of statistical significance, it is also clear that the bar values (CRs) corresponding to the missing *red* circles in Figure 12 grow faster in the case of the HEVC, as compared to the CR entries preceding the missing *red* circle in the case of the the JPEG 2000. This is in full accordance with the results and implications of the previously discussed sensitivity analysis in Figure 10a, where the HEVC was able to preserve the OAs close to the original one much longer, in comparison to the JPEG 2000 (up to significantly higher CRs).

We would also like to note that in the Figure 10a the breaking points of the HEVC and the JPEG 2000 performance with respect to the OA of the original classifier (red line), were at CRs of **70:1** and **35:1**, respectively, Figure 10a. In both cases, these values are significantly higher than the above mentioned Kappa based CRs with similar interpretation, in Figure 12. This indicates that establishing equivalence between lossy compression based solutions and original classifier, besides being application dependant also requires careful consideration of the adopted metrics. However, we should also emphasize that the *red* circles in the first row of Figure 12 do not indicate to *what extent* an experiment with the OA close to the one achieved by the original classifier *is different* (like in the case of the *red* circle in *column 32* of the test matrix, closely corresponding to the HEVC's CR limit of **70:1**). However, in the given case, based on Figure 10a, we can say that their (OA) performance is almost the same.

In the column direction, the triangular *light blue* HEVC region in Figure 12 ends at CR of $\approx$ **130:1** (*column 34* of the test matrix), while the *light green* JPEG 2000 region ends at CR of $\approx$ **53:1** (*column 92*). The shared characteristic of both *triangular* regions is that they correspond to relatively low CRs, as well as to the statistical tests performed exclusively among the pairs of experiments based on the *same* type of compression algorithm (HEVC or JPEG 2000).

The fact that in the *light blue* and the *light green* matrix regions there is a significant number of missing circles is also interesting and indicates that many of the classifier instances could be regarded as mutually equivalent at relatively low CRs. In that sense, the following hypothesis could also be established. Let us consider the HEVC based classifier with CR closest to the previously discussed performance limit of **70:1** (highest CR in Figure 10a at which OA is still the same as the one of the original classifier). The closest experiment to this point in Figure 12 is the *red* circle in *column 32* of the test matrix (to be precise, this circle has CR of $\approx$ **75:1**). According to previous discussion, this experiment results in classification that is statistically different from the original one (due to the presence of the *red* circle). However, the test matrix in Figure 12 also shows that in the *second row* of the same *column 32* there is a missing *light blue* circle, which indicates that (in the *statistical sense*) the classification performance of the HEVC classifier with CR of **75:1** in *column 32* could be the same as the performance of the HEVC based mapper with very low CR of less than **2:1** in the *column 2* (i.e., the one denoted with the **black** circle in the *row 2*). Since CR of **2:1** should be expected to produce an image with very high PSNR and very close to the original (uncompressed) one, it could be said that based on Figure 12 the performance of the HEVC based mappper at CR of **70:1** is very close to the original classifier. Moreover, the same also holds for the *column 33* in Figure 12 with CR of **99:1**, where the similar type of the equivalence with the original classifier could be established.

Finally, the limit of the *light blue* region at CR of $\approx$**130:1** is very close to the HEVC CR limit of **150:1**, which according to Figure 10a corresponds to the *lower bound* of the **0.5%** OA performance *loss margin* in Figure 10a. On the other hand, the *dark blue* HEVC region in Figure 12, characterized by no missing circles, starts at the *column 35* of the test matrix, and corresponds to the CR of $\approx$ **170:1**. Since 150 is an exact average of the two values (130 and 170), results of the tests in Figure 12 indirectly confirm that the HEVC's CR limit of **150:1** (determined from the HEVC curve in Figure 10a by the 0.5% accuracy loss margin) corresponds exactly to the HEVC experiment that would be positioned on the border between the *light blue* and the *dark blue* region in Figure 12. This result is also interesting since it provides an empirical answer to the question what should be considered as an acceptable level of the performance loss (in the terms of the overall classification accuracy) in the case of the thematic mapping based on lossy compression in comparison to the original classifier. Since the loss margin level of 0.5% has been previously proposed, presented results could be regarded as an empirical confirmation of these findings.

In the same line with the above mentioned, based on Figure 10a, CR limit of the JPEG 2000 is estimated to be around **60:1** (corresponding to the 0.5% drop in performance with respect to the original classifier), which is almost exactly the value that is obtained as the average of the CR values corresponding to the last column in the *light green* region of the test matrix in Figure 12 (*column 92*, CR **53:1**), and the first column in the *dark green* region (*column 93*, CR **78:1**) of the same matrix.

Regarding the rest of the entries in Figure 12, it is noticeable that in the case of the both HEVC and the JPEG 2000, when mutually compared at relatively low CRs (experiments determined by the row and column indices of the rectangular, *orange* region) there is a larger number of statistically similar codec configurations. However, it is also interesting that the empty circles are mostly clustered along the vertical or horizontal lines, which implies that at this high PSNR range there are some classifier instances that have almost the same performance as several classifiers based on the other type of encoder. Nevertheless, it is clear that the CR bars corresponding to row indices representing HEVC in this *orange* region are significantly higher than the bars representing CRs of the compared JPEG 2000 solutions (corresponding to column indices).

The *dark blue* and the *dark green* matrix regions confirm that the both HEVC and the JPEG 2000 mappers at *high* CRs produce classification results that are statistically significantly different among themselves, as well as in comparison to the original classifier. The same conclusion holds in the case of the HEVC vs. JPEG 2000 *crosswise* tests, represented by the "L" shaped *yellow* region in Figure 12, where there were confirmed only two non-significant entries, which could be regarded as possible outliers.

6. Conclusions and Future Work

In this work we gave comprehensive analysis of the HEVC still-image intra coding part applied to the multispectral image data. Obtained results were compared with standard JPEG 2000 solutions, and significantly better compression performance has been observed. In addition, specific crop classification task has been considered as an additional measure of HEVC's performance against JPEG 2000. We showed that HEVC is able to outperform JPEG 2000 with regard to the preservation of features important for accurate classification. According to the obtained results, HEVC has demonstrated ability to maintain approximately the same classification accuracy for CR up to **150:1**, which for JPEG 2000 on the same test set and the same accuracy has been shown to be up to **70:1** of CR. Results of the qualitative analysis based on visual inspection have also proved that HEVC can be regarded as a better compression solution. Even at the same level of PSNR (which also means higher CR in the case of HEVC), HEVC results were able to visually demonstrate higher level of the preserved image details in comparison to the JPEG 2000.

In future research, we plan to analyze in details the impact of the particular (individual) image channels and the level of distortion introduced during compression at different resolutions on classification accuracy. In particular, our research efforts will be oriented towards adaptive parameter selection for optimal compression performance considering different image scales with different level of useful information. Also, we will explore additional HEVC's features, and we will try to investigate different strategies on how to improve coding gain (including inter-based prediction). In that way, our plan will be to compare some custom coding approach based on the transform-based spectral decorrelation (which is considered as a state-of-the-art approach), with spectral decorrelation based on the standardized block-based prediction achieved by utilizing HEVC inter coding mode. Part of our focus will also be on HEVC's complexity reduction from the algorithmic standpoint of view. As we have successfully showed that the certain level of image quality degradation can be acceptable, we will also continue to investigate the effects of the HEVC compression on other applications in remote sensing. In that sense we believe that this work will contribute to the introduction of the lossy compression based solutions in the ground segment of the currently operational satellite image distribution systems.

Author Contributions: M.R. and B.B. contributed to paper's conceptualization and methodology; M.R., P.L. and B.B. designed software and performed investigation, formal analysis, data curation, and visualizations; B.B. and M.R. performed writing and editing; V.C. concept validation; Ž.T., D.V. and Z.X. contributed through supervision, validation, review and editing. All authors have read and agreed to the published version of the manuscript.

Funding: This work was partially supported by the Ministry of Education, Science and Technological Development of the Republic of Serbia, as part of the Faculty of Technical Sciences project 01-209/21-1, and by the European Union's Horizon 2020 research and innovation programme under Grant Agreement number 856697 (INCOMING).

Remote Sens. **2020**, *12*, 1590

Acknowledgments: Landsat-8 products are courtesy of the NASA Goddard Space Flight Center and the U.S. Geological Survey. Sentinel-2 images are courtesy of the European Space Agency (ESA), and were provided through Copernicus EO programme. The second author would also like to acknowledge the Era.Net ST2017-099 (HARMONIC) project, H2020 872614 (Smart4All), and NVIDIA corporation for their previous support.

Conflicts of Interest: The authors declare no conflict of interest.

References

1. Huang, B. Guest editorial: Satellite Data Compression. *J. Appl. Remote Sens.* **2010**, *4*, 1–4. [CrossRef]
2. Dusselaar, R.; Paul, M. Hyperspectral image compression approaches: Opportunities, challenges, and future directions: discussion. *J. Opt. Soc. Am. A* **2017**, *34*, 2170–2180. [CrossRef] [PubMed]
3. Sullivan, G.J.; Ohm, J.R.; Han, W.J.; Wiegand, T. Overview of the high efficiency video coding (HEVC) standard. *IEEE Trans. Circ. Syst. Video Technol.* **2012**, *22*, 1649–1668. [CrossRef]
4. Lainema, J.; Bossen, F.; Han, W.J.; Min, J.; Ugur, K. Intra coding of the HEVC standard. *IEEE Trans. Circ. Syst. Video Technol.* **2012**, *22*, 1792–1801. [CrossRef]
5. Rabbani, M. JPEG2000: Image compression fundamentals, standards and practice. *J. Electron. Imaging* **2002**, *11*, 286.
6. Skodras, A.; Christopoulos, C.; Ebrahimi, T. The JPEG 2000 still image compression standard. *IEEE Signal Process. Mag.* **2001**, *18*, 36–58. [CrossRef]
7. Roy, D.P.; Wulder, M.A.; Loveland, T.R.; Woodcock, C.E.; Allen, R.G.; Anderson, M.C.; Helder, D.; Irons, J.R.; Johnson, D.M.; Kennedy, R.; et al. Landsat-8: Science and product vision for terrestrial global change research. *Remote Sens. Environ.* **2014**, *145*, 154–172. [CrossRef]
8. Barsi, J.A.; Lee, K.; Kvaran, G.; Markham, B.L.; Pedelty, J.A. The spectral response of the Landsat-8 operational land imager. *Remote Sens.* **2014**, *6*, 10232–10251. [CrossRef]
9. Drusch, M.; Del Bello, U.; Carlier, S.; Colin, O.; Fernandez, V.; Gascon, F.; Hoersch, B.; Isola, C.; Laberinti, P.; Martimort, P.; et al. Sentinel-2: ESA's optical high-resolution mission for GMES operational services. *Remote Sens. Environ.* **2012**, *120*, 25–36. [CrossRef]
10. Gascon, F.; Bouzinac, C.; Thépaut, O.; Jung, M.; Francesconi, B.; Louis, J.; Lonjou, V.; Lafrance, B.; Massera, S.; Gaudel-Vacaresse, A.; et al. Copernicus Sentinel-2A calibration and products validation status. *Remote Sens.* **2017**, *9*, 584. [CrossRef]
11. Crnojević, V.; Lugonja, P.; Brkljač, B.N.; Brunet, B. Classification of small agricultural fields using combined Landsat-8 and RapidEye imagery: case study of northern Serbia. *J. Appl. Remote Sens.* **2014**, *8*, 083512. [CrossRef]
12. Matton, N.; Canto, G.; Waldner, F.; Valero, S.; Morin, D.; Inglada, J.; Arias, M.; Bontemps, S.; Koetz, B.; Defourny, P. An automated method for annual cropland mapping along the season for various globally-distributed agrosystems using high spatial and temporal resolution time series. *Remote Sens.* **2015**, *7*, 13208–13232. [CrossRef]
13. Du, Q.; Fowler, J.E. Hyperspectral image compression using JPEG2000 and principal component analysis. *IEEE Geosci. Remote Sens. Lett.* **2007**, *4*, 201–205. [CrossRef]
14. García-Sobrino, J.; Laparra, V.; Serra-Sagristà, J.; Calbet, X.; Camps-Valls, G. Improved Statistically Based Retrievals via Spatial-Spectral Data Compression for IASI Data. *IEEE Trans. Geosci. Remote Sens.* **2019**, *57*, 5651–5668. [CrossRef]
15. Zabala, A.; Pons, X. Impact of lossy compression on mapping crop areas from remote sensing. *Int. J. Remote Sens.* **2013**, *34*, 2796–2813. [CrossRef]
16. Blanes, I.; Magli, E.; Serra-Sagristà, J. A tutorial on image compression for optical space imaging systems. *IEEE Geosci. Remote Sens. Mag.* **2014**, *2*, 8–26. [CrossRef]
17. Christophe, E. Hyperspectral data compression tradeoff. In *Optical Remote Sensing: Advances in Signal Processing and Exploitation Techniques*; Prasad, S., Bruce, L., Chanussot, J., Eds.; Springer: Berlin, Germany, 2011; Chapter 2, pp. 9–30.
18. Magli, E.; Olmo, G.; Quacchio, E. Optimized onboard lossless and near-lossless compression of hyperspectral data using CALIC. *IEEE Geosci. Remote Sens. Lett.* **2004**, *1*, 21–25. [CrossRef]
19. Wang, H.; Babacan, S.D.; Sayood, K. Lossless hyperspectral-image compression using context-based conditional average. *IEEE Trans. Geosci. Remote Sens.* **2007**, *45*, 4187–4193. [CrossRef]

20. Mielikainen, J.; Toivanen, P. Lossless compression of hyperspectral images using a quantized index to lookup tables. *IEEE Geosci. Remote Sens. Lett.* **2008**, *5*, 474–478. [CrossRef]

21. Lin, C.C.; Hwang, Y.T. An efficient lossless compression scheme for hyperspectral images using two-stage prediction. *IEEE Geosci. Remote Sens. Lett.* **2010**, *7*, 558–562. [CrossRef]

22. Li, N.; Li, B. Tensor completion for on-board compression of hyperspectral images. In Proceedings of the 2010 IEEE International Conference on Image Processing, Hong Kong, China, 26–29 September 2010; pp. 517–520.

23. Zhang, J.; Li, H.; Chen, C.W. Distributed lossless coding techniques for hyperspectral images. *IEEE J. Sel. Top. Signal Process.* **2015**, *9*, 977–989. [CrossRef]

24. Conoscenti, M.; Coppola, R.; Magli, E. Constant SNR, rate control, and entropy coding for predictive lossy hyperspectral image compression. *IEEE Trans. Geosci. Remote Sens.* **2016**, *54*, 7431–7441. [CrossRef]

25. Lossless Data Compression. CCSDS, Blue Book 121.0-B-2. May 2012. Available online: https://public.ccsds.org/Pubs/121x0b2ec1.pdf (accessed on 27 April 2020).

26. Image Data Compression. CCSDS, Blue Book 122.0-B-2. September 2017. Available online: https://public.ccsds.org/Pubs/122x0b2.pdf (accessed on 27 April 2020).

27. Low-Complexity Lossless and Near-Lossless Multispectral and Hyperspectral Image Compression. CCSDS, Blue book 123.0-B-2. February 2019. Available online: https://public.ccsds.org/Pubs/123x0b2c1.pdf (accessed on 27 April 2020).

28. Image Data Compression. CCSDS, Green Book 120.1-G-2. February 2015. Available online: https://public.ccsds.org/Pubs/120x1g2.pdf (accessed on 27 April 2020).

29. Abrardo, A.; Barni, M.; Magli, E. Low-complexity predictive lossy compression of hyperspectral and ultraspectral images. In Proceedings of the 2011 IEEE International Conference on Acoustics, Speech and Signal Processing (ICASSP), Prague, Czech Republic, 22–27 May 2011; pp. 797–800.

30. Guerra, R.; Barrios, Y.; Díaz, M.; Santos, L.; López, S.; Sarmiento, R. A new algorithm for the on-board compression of hyperspectral images. *Remote Sens.* **2018**, *10*, 428. [CrossRef]

31. Valsesia, D.; Magli, E. A novel rate control algorithm for onboard predictive coding of multispectral and hyperspectral images. *IEEE Trans. Geosci. Remote Sens.* **2014**, *52*, 6341–6355. [CrossRef]

32. Santos, L.; Berrojo, L.; Moreno, J.; López, J.F.; Sarmiento, R. Multispectral and hyperspectral lossless compressor for space applications (HyLoC): A low-complexity FPGA implementation of the CCSDS 123 standard. *IEEE J. Sel. Top. Appl. Earth Obs. Remote Sens.* **2016**, *9*, 757–770. [CrossRef]

33. Wu, X.; Memon, N. Context-based lossless interband compression-extending CALIC. *IEEE Trans. Image Process.* **2000**, *9*, 994–1001.

34. Kaarna, A.; Parkkinen, J. Transform based lossy compression of multispectral images. *Pattern Anal. Appl.* **2001**, *4*, 39–50. [CrossRef]

35. Penna, B.; Tillo, T.; Magli, E.; Olmo, G. Transform coding techniques for lossy hyperspectral data compression. *IEEE Trans. Geosci. Remote Sens.* **2007**, *45*, 1408–1421. [CrossRef]

36. Cheng, K.J.; Dill, J. Lossless to lossy dual-tree BEZW compression for hyperspectral images. *IEEE Trans. Geosci. Remote Sens.* **2014**, *52*, 5765–5770. [CrossRef]

37. Dragotti, P.L.; Poggi, G.; Ragozini, A. Compression of multispectral images by three-dimensional SPIHT algorithm. *IEEE Trans. Geosci. Remote Sens.* **2000**, *38*, 416–428. [CrossRef]

38. Khelifi, F.; Bouridane, A.; Kurugollu, F. Joined spectral trees for scalable SPIHT-based multispectral image compression. *IEEE Trans. Multimedia* **2008**, *10*, 316–329. [CrossRef]

39. Christophe, E.; Mailhes, C.; Duhamel, P. Hyperspectral image compression: adapting SPIHT and EZW to anisotropic 3-D wavelet coding. *IEEE Trans. Image Process.* **2008**, *17*, 2334–2346. [CrossRef] [PubMed]

40. Khelifi, F.; Kurugollu, F.; Bouridane, A. SPECK-based lossless multispectral image coding. *IEEE Signal Process. Lett.* **2008**, *15*, 69–72. [CrossRef]

41. Tang, X.; Pearlman, W. Three-dimensional wavelet-based compression of hyperspectral images. In *Hyperspectral Data Compression*; Springer: Boston, MA, USA, **2006**; pp. 273–308.

42. Blanes, I.; Serra-Sagristà, J.; Marcellin, M.; Bartrina-Rapesta, J. Divide-and-conquer strategies for hyperspectral image processing: A review of their benefits and advantages. *IEEE Signal Process. Mag.* **2012**, *29*, 71–81. [CrossRef]

43. Blanes, I.; Serra-Sagristà, J. Pairwise orthogonal transform for spectral image coding. *IEEE Trans. Geosci. Remote Sens.* **2011**, *49*, 961–972. [CrossRef]

44. Blanes, I.; Hernández-Cabronero, M.; Aulí-Llinas, F.; Serra-Sagristà, J.; Marcellin, M. Isorange pairwise orthogonal transform. *IEEE Trans. Geosci. Remote Sens.* **2015**, *53*, 3361–3372. [CrossRef]
45. Penna, B.; Tillo, T.; Magli, E.; Olmo, G. Progressive 3-D coding of hyperspectral images based on JPEG 2000. *IEEE Geosci. Remote Sens. Lett.* **2006**, *3*, 125–129. [CrossRef]
46. Báscones, D.; González, C.; Mozos, D. Hyperspectral image compression using vector quantization, PCA and JPEG2000. *Remote Sens.* **2018**, *10*, 907. [CrossRef]
47. Delaunay, X.; Chabert, M.; Charvillat, V.; Morin, G. Satellite image compression by post-transforms in the wavelet domain. *Signal Process.* **2010**, *90*, 599–610. [CrossRef]
48. Amrani, N.; Serra-Sagristà, J.; Laparra, V.; Marcellin, M.W.; Malo, J. Regression wavelet analysis for lossless coding of remote-sensing data. *IEEE Trans. Geosci. Remote Sens.* **2016**, *54*, 5616–5627. [CrossRef]
49. Kozhemiakin, R.; Abramov, S.; Lukin, V.; Djurović, B.; Djurović, I.; Vozel, B. Lossy compression of Landsat multispectral images. In Proceedings of the 2016 5th Mediterranean Conference on Embedded Computing (MECO), Bar, Montenegro, 12–16 June 2016; pp. 104–107.
50. Álvarez-Cortés, S.; Amrani, N.; Hernández-Cabronero, M.; Serra-Sagristà, J. Progressive lossy-to-lossless coding of hyperspectral images through regression wavelet analysis. *Int. J. Remote Sens.* **2018**, *39*, 2001–2021. [CrossRef]
51. Santos, L.; Lopez, S.; Callico, G.M.; Lopez, J.F.; Sarmiento, R. Performance evaluation of the H.264/AVC video coding standard for lossy hyperspectral image compression. *IEEE J. Sel. Top. Appl. Earth Obs. Remote Sens.* **2012**, *5*, 451–461. [CrossRef]
52. Zabala, A.; Pons, X. Effects of lossy compression on remote sensing image classification of forest areas. *Int. J. Appl. Earth Obs. Geoinf.* **2011**, *13*, 43–51. [CrossRef]
53. Hagag, A.; Fan, X.; El-Samie, F.E.A. The effect of lossy compression on feature extraction applied to satellite Landsat ETM+ images. In Proceedings of the Eighth International Conference on Digital Image Processing (ICDIP 2016), Chengu, China, 20–22 May 2016; Volume 100333H, pp. 1–8.
54. Hagag, A.; Fan, X.; El-Samie, F. Lossy compression of satellite images with low impact on vegetation features. *Multidim. Syst. Signal Process.* **2017**, *28*, 1717–1736. [CrossRef]
55. Qiao, T.; Ren, J.; Sun, M.; Zheng, J.; Marshall, S. Effective compression of hyperspectral imagery using an improved 3D DCT approach for land-cover analysis in remote-sensing applications. *Int. J. Remote Sens.* **2014**, *35*, 7316–7337. [CrossRef]
56. Du, Q.; Ly, N.; Fowler, J.E. An operational approach to PCA+JPEG2000 compression of hyperspectral imagery. *IEEE J. Sel. Top. Appl. Earth Obs. Remote Sens.* **2014**, *7*, 2237–2245. [CrossRef]
57. Zabala, A.; Gonzalez-Conejero, J.; Serra-Sagristà, J.; Pons, X. JPEG2000 encoding of images with NODATA regions for remote sensing applications. *J. Appl. Remote Sens.* **2010**, *4*, 1–18.
58. Blanes, I.; Zabala, A.; Moré, G.; Pons, X.; Serra-Sagristà, J. Classification of hyperspectral images compressed through 3D-JPEG2000. In Proceedings of the International Conference on Knowledge-Based and Intelligent Information and Engineering Systems, Kaiserslautern, Germany, 12–14 September 2011; Springer: Berlin/Heidelberg, Germany, 2008; pp. 416–423.
59. Kaarna, A.; Toivanen, P.; Keränen, P. Compression and classification methods for hyperspectral images. *Pattern Recognit. Image Anals.* **2006**, *16*, 413–424. [CrossRef]
60. Lee, C.; Choi, E.; Jeong, T.; Lee, S.; Lee, J. Compression of hyperspectral images with discriminant features enhanced. *J. Appl. Remote Sens.* **2010**, *4*, 1–27. [CrossRef]
61. García-Sobrino, J.; Pinho, A.J.; Serra-Sagristà, J. Competitive segmentation performance on near-lossless and lossy compressed remote sensing images. *IEEE Trans. Geosci. Remote Sens.* **2019**, *17*, 834–838. [CrossRef]
62. García-Sobrino, J.; Serra-Sagristà, J.; Laparra, V.; Calbet, X.; Camps-Valls, G. Statistical atmospheric parameter retrieval largely benefits from spatial–spectral image compression. *IEEE Trans. Geosci. Remote Sens.* **2017**, *55*, 2213–2224. [CrossRef]
63. Garcia-Vilchez, F.; Munoz-Mari, J.; Zortea, M.; Blanes, I.; Gonzalez-Ruiz, V.; Camps-Valls, G.; Plaza, A.; Serra-Sagristà, J. On the impact of lossy compression on hyperspectral image classification and unmixing. *IEEE Geosci. Remote Sens. Lett.* **2011**, *8*, 253–257. [CrossRef]
64. Martin, G.; Gonzalez, V.R.; Plaza, A.; Ortiz, J.P.; Garcia, I. Impact of JPEG2000 compression on endmember extraction and unmixing of remotely sensed hyperspectral data. *J. Appl. Remote Sens.* **2010**, *4*, 41796.

65. Qian, S.E.; Hollinger, A.; Dutkiewicz, M.; Tsang, H.; Zwick, H.; Freemantle, J. Effect of lossy vector quantization hyperspectral data compression on retrieval of red-edge indices. *IEEE Trans. Geosci. Remote Sens.* **2001**, *39*, 1459–1470. [CrossRef]

66. Hu, B.; Qian, S.E.; Haboudane, D.; Miller, J.R.; Hollinger, A.B.; Tremblay, N.; Pattey, E. Retrieval of crop chlorophyll content and leaf area index from decompressed hyperspectral data: The effects of data compression. *Remote Sens. Environ.* **2004**, *92*, 139–152. [CrossRef]

67. Lee, C.; Youn, S.; Jeong, T.; Lee, E.; Serra-Sagristà, J. Hybrid compression of hyperspectral images based on PCA with pre-encoding discriminant information. *IEEE Geosci. Remote Sens. Lett.* **2015**, *12*, 1491–1495.

68. Chen, Z.; Hu, Y.; Zhang, Y. Effects of compression on remote sensing image classification based on fractal analysis. *IEEE Trans. Geosci. Remote Sens.* **2019**, *57*, 4577–4590. [CrossRef]

69. HEVC Reference Software HM. Available online: https://hevc.hhi.fraunhofer.de/trac/hevc/browser#tags (accessed on 27 February 2020).

70. Bossen, F.; Bross, B.; Suhring, K.; Flynn, D. HEVC complexity and implementation analysis. *IEEE Trans. Circuits Syst. Video Technol.* **2012**, *22*, 1685–1696. [CrossRef]

71. Chi, C.; Alvarez-Mesa, M.; Juurlink, B.; Clare, G.; Henry, F.; Pateux, S.; Schierl, T. Parallel scalability and efficiency of HEVC parallelization approaches. *IEEE Trans. Circuits Syst. Video Technol.* **2012**, *22*, 1827–1838.

72. Lemmetti, A.; Koivula, A.; Viitanen, M.; Vanne, J.; Hämäläinen, T.D. AVX2-optimized Kvazaar HEVC intra encoder. In Proceedings of the 2016 IEEE International Conference on Image Processing (ICIP), Phoenix, AZ, USA, 25–28 September 2016; pp. 549–553.

73. Gatti, A.; Bertolini, A. Sentinel-2 Products Specification Document, Issue 14.5. March 2018. Available online: https://sentinel.esa.int/documents/247904/685211/Sentinel-2-Products-Specification-Document (accessed on 27 February 2020).

74. Nguyen, T.; Marpe, D. Objective performance evaluation of the HEVC main still picture profile. *IEEE Trans. Circ. Syst. Video Technol.* **2014**, *25*, 790–797. [CrossRef]

75. Flynn, D.; Marpe, D.; Naccari, M.; Nguyen, T.; Rosewarne, C.; Sharman, K.; Sole, J.; Xu, J. Overview of the range extensions for the HEVC standard: Tools, profiles, and performance. *IEEE Trans. Circ. Syst. Video Technol.* **2016**, *26*, 4–19. [CrossRef]

76. Radosavljević, M.; Adamović, M.; Brkljač, B.; Trpovski, Ž.; Xiong, Z.; Vukobratović, D. Satellite image compression based on High Efficiency Video Coding standard—An experimental comparison with JPEG 2000. In Proceedings of the Conferenceon Big Data from Space. ESA, DLR, Munich, Germany, 19–21 February 2019; pp. 257–260.

77. Baig, M.; Zhang, L.; Shuai, T.; Tong, Q. Derivation of a tasselled cap transformation based on Landsat 8 at-satellite reflectance. *Remote Sens. Lett.* **2014**, *5*, 423–431. [CrossRef]

78. Zanter, K. Landsat 8 (L8) Data Users Handbook, Version 3.0. October 2018. Available online: https://prd-wret.s3-us-west-2.amazonaws.com/assets/palladium/production/s3fs-public/atoms/files/LSDS-1574_L8_Data_Users_Handbook.pdf (accessed on 27 February 2020).

79. OpenJPEG—JPEG 2000 Reference Implementation Written in C. Image and Signal Processing Group, Université Catholique de Louvain. Available online: https://github.com/uclouvain/openjpeg/ (accessed on 27 February 2020).

80. Bit Plane Encoder (BPE). CCSDS-122-0-B1 Recommended Standard Codec Implementation by the University of Nebraska-Lincoln 2011. Available online: http://hyperspectral.unl.edu/ (accessed on 20 April 2020).

81. Rosewarne, C.; Sharman, K.; Flynn, D. Common test conditions and software reference configurations for HEVC range extensions. In Proceedings of the 16th JCT-VC Meeting, San Jose, CA, USA, 9–17 January 2014; Document P1006.

82. Fukuhara, T.; Katoh, K.; Kimura, S.; Hosaka, K.; Leung, A. Motion-JPEG2000 standardization and target market. In Proceedings of the 2000 International Conference on Image Processing (Cat. No.00CH37101), Vancouver, BC, Canada, 10–13 September 2000; Volume 2, pp. 57–60.

83. Breiman, L. Random forests. *Mach. Learn.* **2001**, *45*, 5–32. [CrossRef]

84. Breiman, L.; Friedman, J.H.; Olshen, R.A.; Stone, C.J. *Classification and Regression Trees*; Wadsworth statistics/Probability Series; Chapman and Hall/CRC: Boca Raton, FL, USA, 1984.

85. Buitinck, L.; Louppe, G.; Blondel, M.; Pedregosa, F.; Mueller, A.; Grisel, O.; Niculae, V.; Prettenhofer, P.; Gramfort, A.; Grobler, J.; et al. API design for machine learning software: experiences from the scikit-learn project. ECML PKDD Workshop: Languages for Data Mining and ML. *arXiv* **2013**, arXiv:1309.0238.
86. Congalton, R.G.; Green, K. *Assessing the Accuracy of Remotely Sensed Data: Principles and Practices*; CRC Press: Boca Raton, FL, USA, 2019.

Article

Spectral Imagery Tensor Decomposition for Semantic Segmentation of Remote Sensing Data through Fully Convolutional Networks

Josué López [1,*], Deni Torres [1], Stewart Santos [2] and Clement Atzberger [3]

[1] Center for Research and Advanced Studies of the National Polytechnic Institute, Telecommunications Group, Av del Bosque 1145, Zapopan 45017, Mexico; dtorres@gdl.cinvestav.mx
[2] University of Guadalajara, Center of Exact Sciences and Engineering, Blvd. Gral. Marcelino García Barragán 1421, Guadalajara 44430, Mexico; stewart.santos@academicos.udg.mx
[3] University of Natural Resources and Life Science, Institute of Geomatics, Peter Jordan 82, Vienna 1180, Austria; clement.atzberger@boku.ac.at
* Correspondence: jalopez@gdl.cinvestav.mx

Received: 22 November 2019; Accepted: 11 January 2020; Published: 5 February 2020

Abstract: This work aims at addressing two issues simultaneously: data compression at input space and semantic segmentation. Semantic segmentation of remotely sensed multi- or hyperspectral images through deep learning (DL) artificial neural networks (ANN) delivers as output the corresponding matrix of pixels classified elementwise, achieving competitive performance metrics. With technological progress, current remote sensing (RS) sensors have more spectral bands and higher spatial resolution than before, which means a greater number of pixels in the same area. Nevertheless, the more spectral bands and the greater number of pixels, the higher the computational complexity and the longer the processing times. Therefore, without dimensionality reduction, the classification task is challenging, particularly if large areas have to be processed. To solve this problem, our approach maps an RS-image or third-order tensor into a core tensor, representative of our input image, with the same spatial domain but with a lower number of new tensor bands using a Tucker decomposition (TKD). Then, a new input space with reduced dimensionality is built. To find the core tensor, the higher-order orthogonal iteration (HOOI) algorithm is used. A fully convolutional network (FCN) is employed afterwards to classify at the pixel domain, each core tensor. The whole framework, called here HOOI-FCN, achieves high performance metrics competitive with some RS-multispectral images (MSI) semantic segmentation state-of-the-art methods, while significantly reducing computational complexity, and thereby, processing time. We used a Sentinel-2 image data set from Central Europe as a case study, for which our framework outperformed other methods (included the FCN itself) with average pixel accuracy (PA) of 90% (computational time $\sim$90s) and nine spectral bands, achieving a higher average PA of 91.97% (computational time $\sim$36.5s), and average PA of 91.56% (computational time $\sim$9.5s) for seven and five new tensor bands, respectively.

Keywords: fully convolutional network; semantic segmentation; spectral image; tensor decomposition

---

1. Introduction

Remote sensing RS images are of great use in many earth observation applications, such as agriculture, forest monitoring, disaster prevention, security affairs, and others [1]. The recent and upcoming availability of multispectral and hyperspectral satellites alleviates specific tasks, such as detection, classification, and semantic segmentation. In semantic segmentation, also called pixel-wise classification, each pixel in an RS image is assigned to one class [1]. This classification becomes easier when higher dimensional spectral information is acquired [1]. Spectral systems split, by physical filters, the incoming radiance, and provide a vector with spectral reflectance values called spectral signatures. The remotely sensed spectral signatures enable a precise interpretation and recognition of different elements of interest covering the earth surface [2].

Supervised and unsupervised classification of RS images is a very active research area in spectral analysis [3]. To reduce the data dimensionality, and to concentrate the information into a fewer number of features, a once widely used approach was to define various indices to facilitate the classification of diverse land cover [4]. For instance, normalized difference vegetation index (NDVI) [5] and normalized difference water index (NDWI) [6] use a combination of visible to near infrared (NIR) spectral reflectance respectively, to assess land cover, vegetation vitality, and water status [4]. Additionally, supervised machine learning techniques such as random forest [7], support vector machine (SVM) [8,9], decision trees [10], and ANN [11] have been used for RS spectral image classification and have achieved very high accuracy rates [12]. More recently, CNN has been used for semantic segmentation of multispectral images (MSI), promising to be an alternative for solving semantic segmentation issues [13].

The high spectral redundancy of spectral images produces a huge unnecessary number of computations in classification/segmentation algorithms. It is therefore advisable to implement these algorithms together with a dimensionality reduction preprocessing [14]. Spectral data are stored as three-dimensional arrays, so it seems possible to use tensor decomposition (TD) methods [15] for preprocessing, to reduce high redundancy while avoiding information loss [14]. Different to matrix-based decomposition algorithms, such as principal components analysis (PCA) [16] and SVD, TD approach allows to treat spectral data as third-order tensor preserving the spatial information, which sustains the pixel-wise classification task.

In this work we aim addressing two main issues: data compression at input space, and semantic segmentation; i.e., pixel-wise classification of RS imagery. We introduce a spectral data preprocessing that preserves tensor structure and reduces information loss through tensor algebra [17], with the ultimate aim of reducing processing time while keeping high accuracy in further semantic segmentation CNNs. This will produce MSI compression, preserving the spatial domain while reducing the spectral domain, decomposing the original tensor into a core tensor with same order but much lower dimensionality multiplied by a matrix in each mode in the context of tensor algebra [17]. The core tensor, with lower rank than the original data, is used as the input data to the semantic segmentation ANN instead of the MSIs, decreasing the number of computations and in turn the execution time. Previous experimental results demonstrate high performance in semantic segmentation with circa $10\times$ speed up in execution time [18].

The proposed framework can be applied to multispectral, hyperspectral, and even multitemporal datasets. As a particular case, in this study we performed experiments using RS multispectral dataset from the european space agency (ESA) program Sentinel-2 [19] with five classes (soil, water, vegetation, cloud, and shadow).

1.1. Related Work

In recent years, spectral data for earth surface classification has been a very active research area. Methods proposed by Kemker et al. [11,20], Hamida et al. [21], and López et al. [18] use CNNs for RS-CNNMSI pixel-wise classification. Nevertheless, processing raw spectral data with deep learning (DL) algorithms is computationally very expensive. Wang et al. [22] introduced a salient band selection method for HSIs by manifold ranking, and Li et al. [23] proposed a band selection method from the perspective of spectral shape similarity analysis of RS-HSIs to obtain less computational complexity. However, some surface materials differentiate from each other in specific bands, so cutting off spectral bands negatively affected further classification tasks.

More recently, the use of tensor approach for spectral images compression has been introduced; see Zhang et al. [24]. Many authors adopted dimensionality reduction algorithms, such as PCA [16] and singular value decomposition (SVD), for spectral image compression. Other authors have made efforts to reduce the computational cost in CNNs for image classification by using TD algorithms [25,26]. Astrid et al. in [25] proposed a CNN compression method based on CPD and the tensor power method where they achieved significant reduction in memory and computational cost. Chien et al. in [26] presents a tensor-factorized ANN, which integrates TD and ANNs for multi-way feature extraction and classification. Nevertheless, although the idea is to compress data in order to reduce computational cost and processing time, these works compress or decompose the data of the hyper-parameters within the network, which causes the training of the semantic segmentation or classification network to be slower due to the change of the weights in the tensor decomposition.

Recently, three works close to our research [27–29] were published. In [27] An et al. proposed an unsupervised tensor-based multiscale low rank decomposition (T-MLRD) method for hyperspectral image dimensionality reduction, and Li et al. in [28] proposed a low-complexity compression approach for multispectral images based on convolution neural networks CNNs with nonnegative Tucker decomposition (NTD). Nevertheless, these methods reduce the tensor in every dimension, which is self-defeating for a segmentation CNN. Besides, the non-negative decomposed tensor proposed in [28] causes slower convergence in DL algorithms. In [29] An et al. proposed a tensor discriminant analysis (TDA) model via compact feature representation, wherein the traditional linear discriminant analysis was extended to tensor space to make the resulting feature representation more discriminant. However, this approach still leads to a degradation of the spatial resolution, which disturbed the CNN performance. See Table 1 for a summary of the related works.

Table 1. Related work in spectral imagery semantic segmentation.

Reference	Input	Decomposition	Reduction	Classifier
Li, S. et al. [23] (2014)	HSI	-	Band selection	SVM
Zhang, L. et al. [24] (2015)	HSI	TKD	Spatial-Spectral	-
Wan, Q. et al. [22] (2016)	HSI	-	Band selection	SVM/kNN/CART
Kemke, R. et al. [11] (2017)	MSI	-	-	CNN
Hamida, A. et al. [21] (2017)	MSI	-	-	CNN
Li, J. et al. [28] (2019)	MSI	NTD-CNN	Spatial-spectral	-
An, J. et al. [27] (2019)	HSI	T-MLRD	Spatial-spectral	SVM/1NN
An, J. et al. [29] (2019)	HSI	TDA	Spatial-spectral	SVM/1NN
Our framework (2019)	MSI	HOOI	Spectral	FCN

1.2. Contribution

The contribution of this work is summarized into three main points:

1. RS-CNNMSI or -HSI, or third order tensors are compressed in the spectral domain through TKD preprocessing, preserving the pixel spatial structure and obtaining a core tensor representative of the original. These core tensors, with less new tensor bands, which belong to subspaces of the original space, build the new input space for any supervised classifier at pixel level, which delivers the corresponding prediction matrix of pixels classified element-wise. This approach achieves high or competitive performance metrics but with less computational complexity, and consequently, lower computational time.

2. This approach outperforms other methods in normalized difference indexes, PCA, particularly the same FCN with original data. Each core tensor is calculated using the HOOI algorithm, which achieves high orthogonality degree for the core tensor (all-orthogonality) and for its factor matrices (column-wise orthogonal); besides, it converges faster than others, such as TUCKALS3 [17].

3. The efficiency of this approach can be measured by one or more performance metrics, e.g., pixel accuracy (PA), as a function of the number of new tensor bands, orthogonality degree of the factor matrices and the core tensor, reconstruction error of the original tensor, and execution time. These results are shown in Section 6: Experimental Results.

The remainder of this work is organized as follows. Section 2 introduces tensor algebra notation and basic concepts to familiarize the reader with the symbology used in this paper. Section 3 presents the problem statement of this work and the mathematical definition. In Section 4, CNN theory is described for classification and semantic segmentation. Section 5 presents the framework proposed for compression and semantic segmentation of spectral images. Experimental results are presented in Section 6. Finally, Sections 7 and 8 present a discussion and conclusions based on the results obtained in the experiments.

2. Tensor Algebra Basic Concepts

For this work we used the conventional tensor algebra notation [15]. Hence, scalars or zero order tensors are represented by italic lowercase letters; e.g., a. Vectors or first order tensor are denoted by boldface lowercase letters; e.g., $\mathbf{a}$. Matrices or tensor of order two are denoted by boldface capital letters, e.g., $\mathbf{A}$, and three or higher order tensors by boldface Euler script letters, e.g., $\boldsymbol{\mathcal{A}}$. In a N-order tensor $\boldsymbol{\mathcal{A}} \in \mathbb{R}^{I_1 \times \cdots \times I_N}$, where $\mathbb{R}$ represents the set of real numbers, I_n indicates the size of the tensor in each mode $n = \{1, \ldots, N\}$. An element of $\boldsymbol{\mathcal{A}}$ is denoted with indices in lowercase letters, e.g., $a_{i_1 \ldots i_N}$ where i_n denotes the n-mode of $\boldsymbol{\mathcal{A}}$ [17]. A fiber is a vector, the result of fixing every index of a tensor but one, and it is denoted by $\mathbf{a}_{:i_2 i_3}$, $\mathbf{a}_{i_1 : i_3}$, and $\mathbf{a}_{i_1 i_2}$—for column, row, and tube fibers respectively for a third order tensor instance. A slice is a matrix, the result of fixing every index of a tensor but two, and it is denoted by $\mathbf{A}_{i_1::}$, $\mathbf{A}_{:i_2:}$, and $\mathbf{A}_{::i_3}$, or more compactly, $\mathbf{A}_{i_1}$, $\mathbf{A}_{i_2}$, and $\mathbf{A}_{i_3}$ for horizontal, lateral, and frontal slices respectively for a third order tensor instance. Finally, $\mathbf{A}^{(n)}$ denotes a matrix element from a sequence of matrices [17].

It is also necessary to introduce some tensor algebra operations and basic concepts used in later explanations. These notations were taken textually from [17].

2.1. Matricization

The mode-n matricization is the process of reordering the elements of a tensor into a matrix along axis n and it is denoted as $\mathbf{A}_{(n)} \in \mathbb{R}^{I_n \times \prod_{m \neq n} I_m}$.

2.2. Outer Product

The outer product of N vectors $\boldsymbol{\mathcal{X}} = \mathbf{a}^{(1)} \circ \cdots \circ \mathbf{a}^{(N)}$ produces a tensor $\boldsymbol{\mathcal{X}} \in \mathbb{R}^{I_1 \times \cdots \times I_N}$ where $\circ$ denotes the outer product and $\mathbf{a}^{(n)}$ denotes a vector in a sequence of N vectors and each element of the tensor is the product of the corresponding vector elements; i.e., $x_{i_1 i_2 \ldots i_N} = a_{i_1}^{(1)} \ldots a_{i_N}^{(N)}$.

2.3. Inner Product

The inner product of two tensors $\boldsymbol{\mathcal{A}}, \boldsymbol{\mathcal{B}} \in \mathbb{R}^{I_1 \times \cdots \times I_N}$ is the sum of the products of their entries; i.e., $\langle \boldsymbol{\mathcal{A}}, \boldsymbol{\mathcal{B}} \rangle = \sum_{i_1=1}^{I_1} \cdots \sum_{i_N=1}^{I_N} a_{i_1 \ldots i_N} b_{i_1 \ldots i_N}$.

2.4. N-Mode Product

It means the multiplication of a tensor $\boldsymbol{\mathcal{A}} \in \mathbb{R}^{I_1 \times \cdots \times I_N}$ by a matrix $\mathbf{U} \in \mathbb{R}^{J \times I_n}$ or vector $\mathbf{u} \in \mathbb{R}^{I_n}$ in mode n; i.e., along axis n. It is represented by $\boldsymbol{\mathcal{B}} = \boldsymbol{\mathcal{A}} \times_n \mathbf{U}$, where $\boldsymbol{\mathcal{B}} \in \mathbb{R}^{I_1 \times \cdots \times I_{n-1} \times J \times I_{n+1} \times \cdots \times I_N}$ [17].

2.5. Rank-One Tensor

A tensor $\boldsymbol{\mathcal{X}} \in \mathbb{R}^{I_1 \times \cdots \times I_N}$ is rank one if it can be written as the outer product of N vectors; i.e., $\boldsymbol{\mathcal{X}} = \mathbf{a}^{(1)} \circ \cdots \circ \mathbf{a}^{(N)}$.

2.6. Rank-R Tensor

The rank of a tensor rank($\mathcal{X}$) is the smallest number of components in a CPD; i.e., the smallest number of rank-one tensors that generate $\mathcal{X}$ as their sum [17].

2.7. N-Rank

The n-rank of a tensor $\mathcal{X} \in \mathbb{R}^{I_1 \times \cdots \times I_N}$ denoted $\text{rank}_n(\mathcal{X})$, is the column rank of $\mathbf{X}_{(n)}$; i.e., the dimension of the vector space spanned by the mode-n fibers. Hence, if $R_n \equiv \text{rank}_n(\mathcal{X})$ for $n = 1, \ldots, N$, we can say that $\mathcal{X}$ has a $\text{rank} - (R_1, \ldots, R_N)$ tensor.

All the tensor algebra notation presented until this point is summarized in Table 2 for simpler regarding.

Table 2. Tensor algebra notation summary

$\mathcal{A}, \mathbf{A}, \mathbf{a}, a$	Tensor, matrix, vector and scalar respectively
$\mathcal{A} \in \mathbb{R}^{I_1 \times \cdots \times I_N}$	N-order tensor of size $I_1 \times \cdots \times I_N$.
$a_{i_1 \ldots i_N}$	An element of a tensor
$\mathbf{a}_{:i_2 i_3}, \mathbf{a}_{i_1 :i_3},$ and $\mathbf{a}_{i_1 i_2 :}$	Column, row and tube fibers of a third order tensor
$\mathbf{A}_{i_1 ::}, \mathbf{A}_{:i_2 :}, \mathbf{A}_{::i_3}$	Horizontal, lateral and frontal slices for a third order tensor
$\mathbf{A}^{(n)}, \mathbf{a}^{(n)}$	A matrix/vector element from a sequence of matrices/vectors
$\mathbf{A}_{(n)}$	Mode-n matricization of a tensor. $\mathbf{A}_{(n)} \in \mathbb{R}^{I_n \times \prod_{m \neq n} I_m}$
$\mathcal{X} = \mathbf{a}^{(1)} \circ \cdots \circ \mathbf{a}^{(\mathbf{N})}$	Outer product of N vectors, where $x_{i_1 i_2 \ldots i_N} = a_{i_1}^{(1)} \ldots a_{i_N}^{(N)}$
$\langle \mathcal{A}, \mathcal{B} \rangle$	Inner product of two tensors.
$\mathcal{B} = \mathcal{A} \times_n \mathbf{U}$	n-mode product of tensor $\mathcal{A} \in \mathbb{R}^{I_1 \times \cdots \times I_N}$ by a matrix $\mathbf{U} \in \mathbb{R}^{J \times I_n}$ along axis n.

2.8. Tucker Decomposition (Tkd)

The TKD can be seen as a form of higher-order PCA [17]. This method decomposes a tensor $\mathcal{X} \in \mathbb{R}^{I_1 \times \cdots \times I_N}$ into a core tensor $\mathcal{G} \in \mathbb{R}^{J_1 \times \cdots \times J_N}$ multiplied by a matrix along each mode $n = 1, \ldots, N$ as

$$\mathcal{X} \approx \mathcal{G} \times_1 \mathbf{U}^{(1)} \cdots \times_N \mathbf{U}^{(N)} \tag{1}$$

where the core tensor preserves the level of interaction for each factor or projection matrix $\mathbf{U}^{(n)} \in \mathbb{R}^{I_n \times J_n}$. These matrices are usually, but not necessarily, orthogonal, and can be thought of as the principal components in each mode [17] (see Figure 1). J_n represents the number of components in the decomposition; i.e., the $\text{rank} - (R_1, \ldots, R_N)$. We compute $\text{rank} - (R_1, \ldots, R_N)$, where $\text{rank}_n(\mathcal{X}) = R_n$ for every n-mode, which generally does not exactly reproduce $\mathcal{X}$. Starting from (1), the reconstruction of an approximated tensor can be given by where $\hat{\mathcal{X}}$ is the reconstructed tensor. Then, we can acquire the core tensor $\mathcal{G}$ by the multilinear projection

$$\mathcal{G} = \mathcal{X} \times_1 \mathbf{U}^{(1)\mathrm{T}} \cdots \times_N \mathbf{U}^{(N)\mathrm{T}}, \tag{2}$$

where $\mathbf{U}^{(n)\mathrm{T}}$ denotes the transpose matrix of $\mathbf{U}^{(n)}$ for $n = 1, \ldots, N$. The reconstruction error ξ can be computed as

$$\xi(\hat{\mathcal{X}}) = ||\mathcal{X} - \hat{\mathcal{X}}||_F^2, \tag{3}$$

where $|| \cdot ||_F$ represents the Frobenius norm. To effectively compress data, the reconstructed lower-rank tensor $\hat{\mathcal{X}}$ should be close to the original tensor $\mathcal{X}$; this can be reached by an algorithm as HOOI, which is iterative, and it is described in Section 5.1.

$$\hat{\mathcal{X}} = \mathcal{G} \times_1 \mathbf{U}^{(1)} \cdots \times_N \mathbf{U}^{(N)}, \tag{4}$$

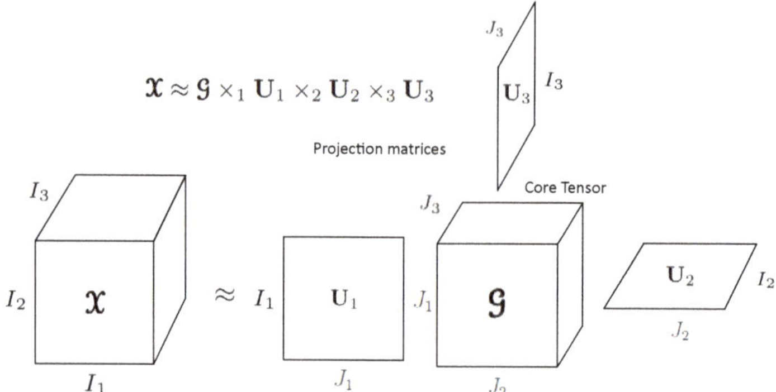

Figure 1. Tucker decomposition for a third-order tensor.

3. Problem Statement and Mathematical Definition

Spectral images are third-order arrays, which provide not only spatial, but also spectral features from RS scenes of interest. These properties aid CNNs to easily find features to characterize the behaviors of different materials over the earth's surface. However, the large amount of spectral data causes huge computational load, and therefore, large processing time using machine learning algorithms.

It is important to preserve the three-dimensional array structure of the RS spectral input image, in order to effectively classify each pixel of the image. In RS multi- or hyperspectral images, the spectral bands are highly correlated, and contain lot of redundancy. Therefore, we propose a TKD-based method as a preprocessing step to provide a better suited input for the semantic segmentation based on CNN. This will also considerably reduce high number of parameters, and in turn, processing time during training and testing. Our problem statement for RS spectral images can be described as follows.

3.1. Problem Statement

Given a pair $(\mathcal{X}, \mathbf{Y})$, where tensor $\mathcal{X} \in \mathbb{R}^{I_1 \times I_2 \times I_3}$ denotes a CNNMSI or HSI, and $\mathbf{Y} \in \mathbb{R}^{I_1 \times I_2}$ its corresponding ground truth matrix for a specific number of classes C, find another pair $(\mathcal{G}, \hat{\mathbf{Y}})$, where the tensor $\mathcal{G} \in \mathbb{R}^{J_1 \times J_2 \times J_3}$, used for classification, is representative of $\mathcal{X}$, and $\hat{\mathbf{Y}}$ is its associated matrix of predicted classes; preserving the spatial-domain $J_1 = I_1, J_2 = I_2$ but with fewer new tensor bands, i.e., $J_3 < I_3$, achieving higher or competitive performance metrics for pixel-wise classification, reducing the dimensionality, and therefore, decreasing computational complexity in the classification task.

3.2. Mathematical Definition

We can describe the problem stated in previous subsection mathematically as the following optimization problem

$$\min_{\mathcal{G}, \mathbf{U}^{(1)}, \mathbf{U}^{(2)}, \mathbf{U}^{(3)}} ||\mathcal{X} - \mathcal{G} \times_1 \mathbf{U}^{(1)} \times_2 \mathbf{U}^{(2)} \times_3 \mathbf{U}^{(3)}||_F^2$$

$$\text{subject to} \quad \mathbf{U}^{(n)} \in St_{I_n \times J_n} \quad \text{and} \quad St_{I_n \times J_n} \equiv \{\mathbf{U}^{(n)} \in \mathbb{R}^{I_n \times J_n} \mid \mathbf{U}^{(n)\mathrm{T}}\mathbf{U}^{(n)} = \mathbf{I}^{(n)}\},$$

$$J_1 = I_1, J_2 = I_2 \qquad \text{preserving the pixel domain,}$$

$$J_3 < I_3 \qquad \text{reducing spectral dimensionality}$$

$$\xi(\hat{\mathcal{X}}) \leq \psi \qquad \text{mesaure of how representative of } \hat{\mathcal{X}} \; \mathcal{G} \text{ is}$$

$$\tag{5}$$

where ψ denotes an error threshold defined depending on the accuracy or performance metrics required for each application and $St_{I_n \times J_n}$ represents the Stiefel manifold [30]. Embedding $\mathcal{G}$ into the objective function, as Lathhauwer proved in [31] Theorems 3.1, 4.1, and 4.2, (5), can be written by the equivalent under the same constraints as (5).

$$\max_{\mathbf{U}^{(1)},\mathbf{U}^{(2)},\mathbf{U}^{(3)}} ||\mathcal{X} \times_1 \mathbf{U}^{(1)\mathrm{T}} \times_2 \mathbf{U}^{(2)\mathrm{T}} \times_3 \mathbf{U}^{(3)\mathrm{T}}||_F^2 \tag{6a}$$

$$\text{where} \quad \mathcal{G} = \mathcal{X} \times_1 \mathbf{U}^{(1)\mathrm{T}} \times_2 \mathbf{U}^{(2)\mathrm{T}} \times_3 \mathbf{U}^{(3)\mathrm{T}} \tag{6b}$$

The subtensors $\mathcal{G}_{i_n}$ of the core tensor $\mathcal{G}$ satisfy the all-orthogonality property [32], which establishes that two subtensors $\mathcal{G}_{i_n=\alpha}$ and $\mathcal{G}_{i_n=\beta}$ are all-orthogonal

$$\langle \mathcal{G}_{i_n=\alpha}, \mathcal{G}_{i_n=\beta} \rangle = 0 \tag{7}$$

for all possible values of n, α, and β subject to $\alpha \neq \beta$, and the ordering property:

$$||\mathcal{G}_{i_n=1}||_F \geq ||\mathcal{G}_{i_n=2}||_F \geq \cdots \geq ||\mathcal{G}_{i_n=I_N}||_F. \tag{8}$$

Our optimization problem can be solved by several algorithms. In this work, the HOOI algorithm was selected (described in Section 5.1), due to its convergence and orthogonality performance. Once a tensor $\mathcal{G}$ is obtained, a classifier f that belongs to the hypothesis space H maps input data $\mathcal{G}$ into output data $\hat{Y}$; that is

$$\hat{Y} = f(\mathcal{G}) \tag{9}$$

where f is a pixel-wise classifier. In this paper, a FCN for semantic segmentation was used as classifier due to the need of classify each pixel of the input image and to its performance in pixel accuracy. The FCN used in this work is described in Section 4.

4. Convolutional Neural Networks (CNNs)

CNNs are supervised feed-forward DL-ANNs for computer vision. The idea of applying a sort of convolution of the synaptic weights of a neural network through the input data yields to a preservation of spatial features, which alleviates the hard task of classification and in turn semantic segmentation. This type of ANN works under the same linear regression model as every machine learning (ML) algorithm. Since images are three dimensional arrays, we can use tensor algebra notation to describe the input of CNNs as a tensor $\mathcal{A} \in \mathbb{R}^{I_1 \times I_2 \times I_3}$, where I_1, I_2, and I_3 represent height, width, and depth of the third order array respectively; i.e., the spatial and spectral domain of an image. We can write generally the linear regression model used for ANNs as

$$\hat{y} = \sigma\left(\mathbf{W}g + \mathbf{b}\right) \tag{10}$$

where $\hat{y}$ represents the output prediction of the network; σ denotes an activation function; g is the input dataset; $\mathbf{W}$ and $\mathbf{b}$ are the matrix of synaptic weights and the bias vector, respectively. These parameters are adjustable; i.e., their values are modified every iteration looking for convergence to minimize the loss in the prediction through optimization algorithms [33]. For simplicity, the bias vector can be ignored, assuming that matrix $\mathbf{W}$ will update until convergence independently of another parameter [33]. Considering that the input dataset to a CNN is a multidimensional array, we can represent (9) and (10) using tensor algebra notation as

$$\hat{\mathcal{y}} = \sigma\left(\mathcal{W}\mathcal{G}\right) \tag{11}$$

where $\hat{\mathcal{y}}$ represents the prediction output tensor of the ANN (in our case, a second order tensor or matrix $\hat{Y}$), $\mathcal{G}$ is the input dataset, and $\mathcal{W}$ is a $K_1 \times K_2 \times F_1$ tensor called filter or kernel with the adaptable synaptic weights. Different to conventional ANN, in CNNs, $\mathcal{W}$ is a shiftable square tensor is much smaller in height and width than the input data, i.e., $K_1 = K_2$ and $K_s << I_s$ for $s = 1, 2$; F_1 denotes the number of input channels; i.e., $F_1 = I_3$. For hidden layers, instead of the prediction tensor $\hat{\mathbf{y}}$, the output is a matrix called activation map $\mathbf{M} \in \mathbb{R}^{I_1 \times I_2}$, which preserves features from the original data in each domain. Actually, it is necessary to use much kernels $\mathbf{W}^{(f_2)}$ as activation maps, with different initialization values to preserve diverse features of the image. Hence, we can also define activation maps as a tensor $\mathbf{M} \in \mathbb{R}^{I_1 \times I_2 \times F_2}$ where F_2 denotes the number of activation maps produced by each filter (see Figure 2). Kernels are displaced through the whole input image as a discrete convolution operation. Then, each element of the output activation map $m_{i_1 i_2 f_2}$ is computed by the summary of the Hadamard product of kernel $\mathbf{W}^{(f_2)}$ and a subtensor from the input tensor $\mathcal{G}$ centered in position (i, j) and with same dimensions of $\mathbf{W}$, as follows

$$m_{i_1 i_2 f_2} = \sigma \left[\sum_{k_1=1}^{K_1} \sum_{k_2=1}^{K_2} \sum_{f_1=1}^{F_1} w_{k_1,k_2,f_1} \mathcal{G}_{i_1+k_1-o_1, i_2+k_2-o_2, f_1} \right] \tag{12}$$

where $m_{i_1 i_2 f_2}$ denotes the value of the output activation map f_2 at position i_1, i_2; σ represents the activation function; and o_1 and o_2 are offsets in spatial dimensions which depend on the kernel size, and equal $\frac{K_1+1}{2}$ and $\frac{K_2+1}{2}$ respectively (see Figure 2).

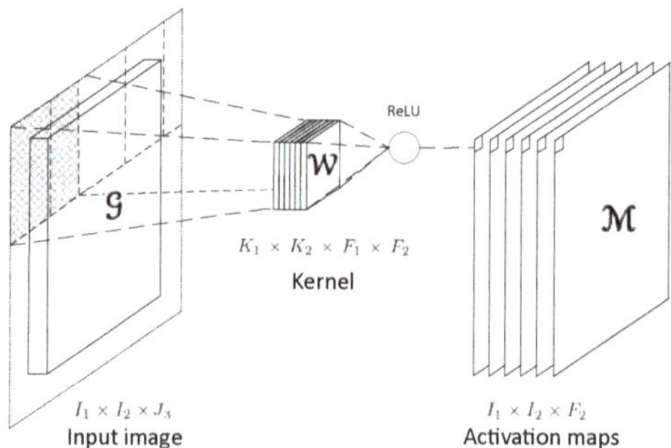

Figure 2. Convolutional layer with a $K_1 \times K_2 \times F_1 \times F_2$ kernel. Input channels F_1 must equal the spectral bands I_3. To preserve original dimensions at the output, zero padding is needed [18]. Output dimensions also depend on stride $S = 1$ to consider every piece of pixel information and to preserve original dimensions.

An ANN is trained by using iterative gradient-based optimizers, such as Stochastic gradient descent, Momentum, RMSprop, and Adam [33]. This drive the cost function $L(\mathbf{W})$ to a very low value by updating the synaptic weights $\mathbf{W}$. We can compute the cost function by any function that measures the difference between the training data and the prediction, such as Euclidean distance or cross-entropy [10]. Besides, the same function is used to measure the performance of the model during testing and validation. In order to avoid overfitting [33], the total cost function used to train an ANN combines one of the cost functions mentioned before, plus a regularization term.

$$J(\mathbf{W}) = L(\mathbf{W}) + R(\mathbf{W}), \tag{13}$$

where $J(\mathbf{W})$ denotes the total cost function and $R(\mathbf{W})$ represents a regularization function. Then, we can decrease $J(\mathbf{W})$ by updating the synaptic weights in the direction of the negative gradient. This is known as the method of steepest descent or gradient descent.

$$\mathbf{W}' = \mathbf{W} - \alpha \nabla_{\mathbf{W}} J(\mathbf{W}), \tag{14}$$

where $\mathcal{W}'$ represents the synaptic weights tensor in next iteration during training, α denotes the learning rate parameter, and $\nabla_{\mathcal{W}} J(\mathcal{W})$ the cost function gradient. Gradient descent converges when every element of the gradient is zero, or in practice, very close to zero [10].

CNNs has been successfully used in many image classification frameworks. This variation in architecture from other typical ANN models yields the network to learn spatial and spectral features, which are highly profitable for image classification. Besides, FCNs, constructed with only convolutional layers are able to classify each element of the input image; i.e., they yield pixel-wise classification, or in other words, semantic segmentation.

5. Hooi-Fcn Framework

In this work we propose a TKD-CNN-based framework called HOOI-FCN, which maps the original high-correlated spectral image into a low-rank core tensor, preserving enough statistical information to alleviate image pixel-wise classification. The aim is to improve performance while reducing processing time in semantic segmentation ANNs by compressing CNNMSI third-order tensors. Applying TD methods, relevant information is preserved, mainly acquired from the spectral domain, convenient for the classification FCN. This novel framework is in summary, a two step structure composed by an HOOI TD and a FCN for semantic segmentation described below (see Figure 3).

5.1. Higher Order Orthogonal Iteration (HOOI) for Spectral Image Compression

Quoting Kolda, "The truncated higher order singular value decomposition (HOSVD) is not optimal in terms of giving the best fit as measured by the norm of the difference, but it is a good starting point for an iterative alternating least square algorithm" [17]. HOOI is an iterative algorithm to compute a rank-$(R_1, \ldots, R_N)$ TKD. Let $\mathcal{X} \in \mathbb{R}^{I_1 \times \cdots \times I_N}$ be an N-th order tensor and $R_1, \ldots, R_N$ be a set of integers satisfying $1 \leq R_n \leq I_n$, for $n = 1, \ldots, N$; the rank $- (R_1, \ldots, R_N)$ approximation problem is to find a set of $I_n \times R_n$ matrices $\mathbf{U}^{(n)}$ column-wise orthogonal and a $R_1 \times \cdots \times R_N$ core tensor $\mathcal{G}$ by computing

$$\min_{\mathcal{G}, \mathbf{U}^{(1)}, \ldots, \mathbf{U}^{(N)}} ||\mathcal{X} - \mathcal{G} \times_1 \mathbf{U}^{(1)} \cdots \times_N \mathbf{U}^{(N)}||^2, \tag{15}$$

and from matrices $\mathbf{U}^{(n)}$, where $\mathbf{U}^{(n)\mathrm{T}} \mathbf{U}^{(n)} = \mathbf{I}^{(n)}$, the core tensor $\mathcal{G}$ is found to satisfy (2) [34]. For a third-order tensor decomposition, we can rewrite (4) as

$$\hat{\mathcal{X}} = \mathcal{G} \times_1 \mathbf{U}^{(1)} \times_2 \mathbf{U}^{(2)} \times_3 \mathbf{U}^{(3)} \tag{16}$$

where $\hat{\mathcal{X}}$ denotes the reconstruction approximation of the input spectral image $\mathcal{X}$, $\mathcal{G}$ is the $J_1 \times J_2 \times J_3$ core tensor, and $\mathbf{U}^{(1)} \in \mathbb{R}^{I_1 \times J_1}$, $\mathbf{U}^{(2)} \in \mathbb{R}^{I_2 \times J_2}$ and $\mathbf{U}^{(3)} \in \mathbb{R}^{I_3 \times J_3}$ are the projection matrices. Algorithm 1 shows HOOI for a third order tensor decomposition, but the extension to higher order tensors is straightforward. Thus, with Algorithm 1 we compute the tensor $\mathcal{G}$ with rank-(J_1, J_2, J_3) for each spectral image as third-order tensor.

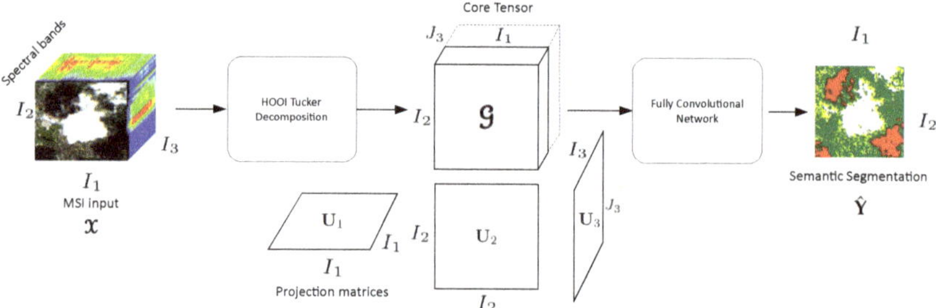

Figure 3. The big picture of the fast semantic segmentation framework proposed, with a fully convolutional network encoder-decoder architecture and a preprocessing HOOI tucker decomposition stage.

Algorithm 1: HOOI for MSI. ALS algorithm to compute the core tensor $\mathcal{G}$.

Function HOOI ($\mathcal{X}$, R_1, R_2, R_3):

    initialize $\mathbf{U}^{(n)} \in \mathbb{R}^{I_n \times R_n}$ for $n = 1, 2, 3$ using HOSVD;

    repeat

        for $n = 1, 2, 3$ **do**

            $\mathcal{D} \leftarrow \mathcal{X} \times_1 \mathbf{U}^{(1)\mathrm{T}} \times_2 \mathbf{U}^{(2)\mathrm{T}} \times_3 \mathbf{U}^{(3)\mathrm{T}}$

            $\mathbf{U}^{(n)} \leftarrow R_n$ leading left singular vectors of $\mathbf{D}_{(n)}$

        end

    until *fit ceases to improve or maximum iterations exhausted*;

    $\mathcal{G} \leftarrow \mathcal{X} \times_1 \mathbf{U}^{(1)\mathrm{T}} \times_2 \mathbf{U}^{(2)\mathrm{T}} \times_3 \mathbf{U}^{(3)\mathrm{T}}$

Output: $\mathcal{G}$, $\mathbf{U}^{(1)}$, $\mathbf{U}^{(2)}$, $\mathbf{U}^{(3)}$

5.2. Fcn for Semantic Segmentation of Spectral Images

We use a FCN model for semantic segmentation based on the proposed by Badrinarayanan et al. in [35] called Segnet. Each core tensor $\mathcal{G}$ obtained after decomposition, is the input to the SegNet for training and testing the network. Hence, the feature activation maps $\mathcal{M} \in \mathbb{R}^{I_1 \times I_2 \times F_2}$ for each hidden layer of the SegNet encoder-decoder FCN are computed by displacing the filters $\mathcal{W}$ through the whole input core tensor in strides $S = 1$. It is worth noting that kernel $\mathcal{W}$ is a four-order tensor $\mathcal{W} \in \mathbb{R}^{K_1 \times K_2 \times F_1 \times F_2}$, where K_1 and K_2 represent its spatial dimensions height and width; F_1 its depth, i.e., the spectral domain; and F_2 denotes the number of filters used to produce F_2 activation maps (Figure 2). We express this convolution operation as

$$\mathbf{M}^{(f_2)} = \sigma\left(\mathcal{W} \odot \mathcal{G}\right), \tag{17}$$

where $\mathbf{M}^{(f_2)}$ represents each activation map for $f_2 = 1, \ldots, F_2$, and each value $m_{i_1 i_2 f_2}$ is computed as in (12). σ denotes the rectified linear unit (ReLU) [33] function; i.e., $\sigma(z) = \max\{0, z\}$. Symbol $\odot$ is used in this paper to represent the convolution; i.e., the whole operation applied in convolutional layers (see Figure 2). These activation maps are the input for the subsequent layer in the SegNet FCN.

The last layer is used the softmax activation function [33] to produce a distribution probability, and so, predict values relating each pixel to one of the C classes of interest. Hence, for the last layer we rewrite (17) as

$$\hat{\mathbf{Y}} = \delta\left(\mathcal{W} \odot \mathcal{M}\right), \tag{18}$$

where $\hat{\mathbf{Y}}$ represents the output prediction, $\mathcal{M}$ is the feature activation maps at previous layer, δ the softmax activation function, and $\mathcal{W}$ the filter or kernel tensor with the adaptable synaptic weights.

The output of the FCN is a matrix $\hat{\mathbf{Y}}$ with the same spatial dimensions as the input, with a value of the most likely class for each pixels. Figure 4 shows the architecture of the SegNet model used in this work. Experiments present the behavior of this FCN with and without data compression in the spectral domain.

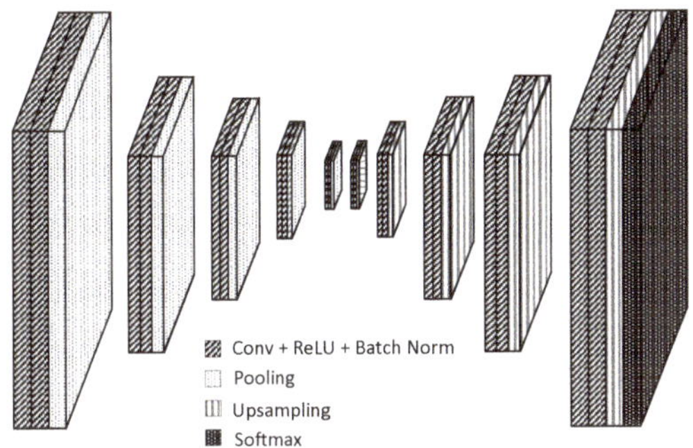

Figure 4. SegNet FCN. Encoder-decoder architecture with convolutional, pooling, and upsampling layers with their corresponding activation functions and batch normalization [33].

6. Experimental Results

6.1. Our Data

As case study, a CNNMSI dataset with 100 RS images was used for training and 10 for testing, all of them from central Europe with 128×128 pixels. These images are partitions of the original Sentinel-2 images without modification and all semi-manually labeled, and with abundant presence of the elements of interest. In Table 3 the 10 scenarios correspond to our 10 images for testing. We used only nine from the 13 available spectral bands from visible, NIR to SWIR wavelengths. Bands 2, 3, 4, and 8 have 10 m resolution, and bands 5, 6, 7, 11, and 12 have 20 m (oversampled to 10 m [18]). These bands provide decisive information for discrimination of different classes. Bands 1, 9, and 10 were dismissed because of their lower spatial resolution of 60 m. Band 8A, also with 20 m spatial resolution, was dismissed due to wavelength overlapping with band 8. It is worth mentioning that the framework proposed in this work can be applied to any kind of spectral image and multitemporal datasets [36].

6.1.1. The Training Space

For training, the input data ws a tensor $\mathcal{X} \in \mathbb{R}^{128 \times 128 \times 9 \times 100}$, where 128×128 is the spatial dimensions, 9 is the number of spectral bands, and 100 is the number of images used for training. Although the number of images seems low, taking into account that we work at pixel-domain, the real number of training points or vectors is high. Indeed, our FCN for semantic segmentation was trained with $128 \times 128 \times 100 = 1638400$ samples or vectors. To test whether the size of the data for training was sufficiently high, a smaller subtensor of $\mathcal{X}$, $\mathcal{X}_p \in \mathbb{R}^{128 \times 128 \times 9 \times 80}$, equivalent to 1310720 points or vectors, was used for a second training obtaining, for the same test set, an average PA of 91.48%; i.e., only 0.08% less than with 100 images, 91.56%. We also tested these results by a third training with an extended dataset of 120 images, $\mathcal{X}_q \in \mathbb{R}^{128 \times 128 \times 9 \times 120}$ equivalent to 1966080 vectors, and we found only a slight variation of +0.01% in the PA (91.57%), while the execution time for the training increased significantly.

6.1.2. The Labels

Our labels were acquired using the scene classification algorithm developed by the ESA [19], and subsequently modified, semi-manually, misclassified pixels.

6.1.3. The Testing Space

For testing, our input data were a $128 \times 128 \times 9 \times 10$ tensor; i.e., 10 different scenarios for pixel-wise classification, whose results are shown in Table 3. That is, the framework classifies $128 \times 128 \times 10 = 163,840$ pixels.

6.1.4. Downloading Data

Due to the big size of the data, format npy was used. Data are available in the link Dataset.

- The training dataset is in the file S2_TrainingData.npy.
- Labels of the training dataset are in the file S2_TrainingLabels.npy.
- A true color representation of the training dataset can be found in S2_Trainingtruecolor.npy.
- The testing dataset and the corresponding labels are in the file S2_TestData.npy.
- Labels of the test dataset are in the file S2_TestLabels.npy.
- Last, a true color representation of the test data can be found in S2_Testtruecolor.npy.

Code will be delivered by the corresponding author upon request for research purposes only.

6.2. Classes

The CNNMSI dataset has been semi-manually labeled for supervised semantic segmentation of $C = 5$ classes; vegetation, water, cloud, cloud shadow, and soil. These classes were selected according to their impact in RS research areas such as agriculture, forest monitoring, population growth analysis, and disaster prevention. It is worth mentioning that the detection of clouds and cloud shadows is an important prerequisite for almost all RS applications.

6.3. Metrics

6.3.1. Pixel Accuracy (PA)

We used the PA metric to compute a ratio between the amount of correctly classified pixels and the total number of pixels as

$$PA = \frac{\sum_{c=0}^{C} p_{cc}}{\sum_{c=0}^{C} \sum_{d=0}^{D} p_{cd}} \tag{19}$$

where we have a total of C classes and p_{ii} is the amount of pixels of class c correctly assigned to class c (true positives), and p_{cd} is the amount of pixels of class c inferred to belong to class d (false positives). We can see in Table 3 the PA values for our proposed framework in comparison with other state-of-the-art methods. From Table 3, we can see that:

- Indexes NDI are important references for pixel-wise classification but they show one of the lowest PAs and the highest computational time.
- Classic PCA with five components shows the lowest PA, although the computational time is similar to HOOI-FCN with five tensor bands.
- Due to the poor results of NDI and classical PCA, FCN (with raw data and nine components) is a good reference in terms of performance and computational time, and HOOI-FCN with seven and five tensor bands achieves the highest PA and the lowest computational time.

The PA and the computational times for FCN and HOOI-FCN with different numbers of tensor bands are shown in Figure 5.

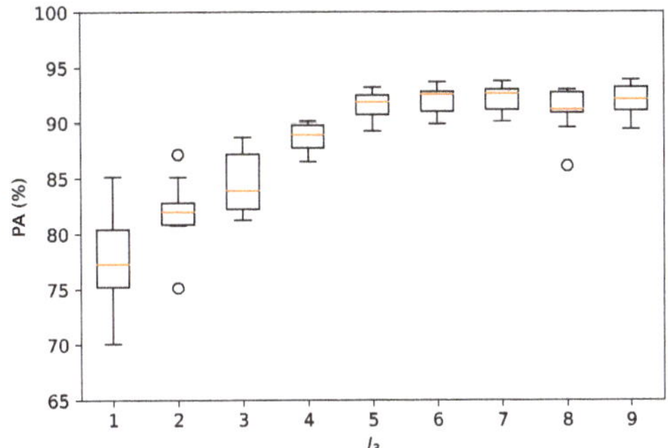

Figure 5. Box and whiskers plot of the pixel accuracy (PA) for the 10 testing scenarios shown in Table 3.

6.3.2. Relative Mean Square Error (rMSE)

In order to compute the reconstruction error of the tensor $\mathcal{X}$ for the implementation of HOOI, the rMSE was used:

$$rMSE\left(\hat{\mathcal{X}}\right) = \frac{1}{Q} \sum_{q=1}^{Q} \frac{\left\|\hat{\mathcal{X}}_q - \mathcal{X}_q\right\|_F^2}{\left\|\mathcal{X}_q\right\|_F^2}, \tag{20}$$

where $\mathcal{X}_q$ represents the q-th CNNMSI from our dataset with Q MSIs and $\hat{\mathcal{X}}_q$ its corresponding reconstruction computed by (4).

Figure 6a shows the behavior of the reconstruction rMSE for our 100 training images for $J_3 = 1, \ldots, I_3$. With this metric we can quantify how good the decomposition represents the input data. The rMSE is also one of the decisive parameters to set the value of the $rank_3(\mathcal{X}) = J_3$. To preserve a high performance in the pixel-wise classification task, we set the threshold ψ to a value for which the rMSE error is less than or equal to 0.05%, since deeper decomposition decrease the PA to less than 90%, as we can see in Figure 5. For a rank decomposition $(128, 128, 5)$ our rMSE is 0.04%, which means that we reduce the dimensionality of our input data to almost half with a very low loss in performance. Besides, comparing this error with matrix based methods as PCA, we can see that our tensor-based decomposition produces lower rMSE for every value of J_3 except for the first one.

6.3.3. Orthogonality Degree of Factor Matrices and Tensor Bands

A way to analyze the algorithm HOOI efficiency is computing the orthogonality degree of the core tensor $\mathcal{G}$ and the projection matrices $\mathbf{U}^{(n)}$. As we mentioned in Section 3, we use the all-orthogonality property proposed in [32] and described in (7) and (8) to evaluate the orthogonality degree of our core tensors. Table 4 shows the results of the inner products between each tensor band with the others from one of our training images. We can see that these values are practically zero, which means that our bands are orthogonal. Furthermore, we can see in Figure 6b that (8) is fulfilled.

It is also important to know the orthogonality degree in our projection matrices. From Theorem 2 in [32] we start from the condition $\mathbf{U}^{(n)\mathrm{T}}\mathbf{U}^{(n)} = \mathbf{I}^{(n)}$; then, we create a vector $\hat{\mathbf{o}}$ where the components are the trace of each resulting matrix, i.e., $\mathrm{tr}(\mathbf{I}^{(n)})$, and compute the MSE with respect to a vector rank $\mathbf{o} = (J_1, J_2, J_3)$ as

$$MSE(\hat{\mathbf{o}}) = \sum_{q=1}^{3} \left\|o_q - \hat{o}_q\right\|_F^2. \tag{21}$$

Using this orthogonality analysis, we obtain MSE values very close to zero, e.g., in order of 10^{-20}, which means that projection matrices present a high orthogonality degree.

Table 3. Quantitative results for 10 test MSIs running in a NVIDIA GeForce GTX 1050 Ti graphics processing unit (GPU), Intel core i7 processor, 8 Gb RAM, SSD 128 Gb, and HDD 1 Tb. Values in blue and red represent the highest PA and the lowest time, respectively.

Scenarios	NDI		FCN$_9$		PCA-FCN$_5$		HOOI-FCN$_7$		HOOI-FCN$_5$	
	PA (%)	Time (s)	PA (%)	Time (s)	PA (%)	Time (s)	PA (%)	Time (s)	PA (%)	Time (s)
1	88.20	363.03	91.05	101.21	85.12	9.85	91.12	37.84	90.63	9.13
2	84.75	412.89	92.21	87.54	84.60	9.83	90.12	36.54	89.23	9.06
3	92.34	307.56	93.67	93.45	88.32	10.00	93.75	36.02	93.22	9.03
4	90.08	382.31	91.72	98.92	86.08	9.73	92.85	36.79	92.18	8.93
5	87.14	400.12	89.91	103.57	86.36	9.12	92.13	35.88	91.84	9.67
6	89.75	312.15	90.95	95.21	87.65	10.15	92.95	37.23	92.71	10.09
7	85.73	373.84	89.92	107.13	88.47	9.63	93.06	35.56	92.59	9.55
8	91.49	308.00	90.17	95.45	85.78	9.76	90.23	36.34	90.12	9.14
9	89.38	397.92	90.74	80.33	87.91	10.26	92.50	37.09	92.18	10.11
10	90.01	352.66	88.52	112.85	84.32	9.88	91.17	35.53	90.97	9.85
Average	**88.87**	**361.04**	**90.88**	**97.56**	**86.46**	**9.82**	**91.97**	**36.48**	**91.56**	**9.45**

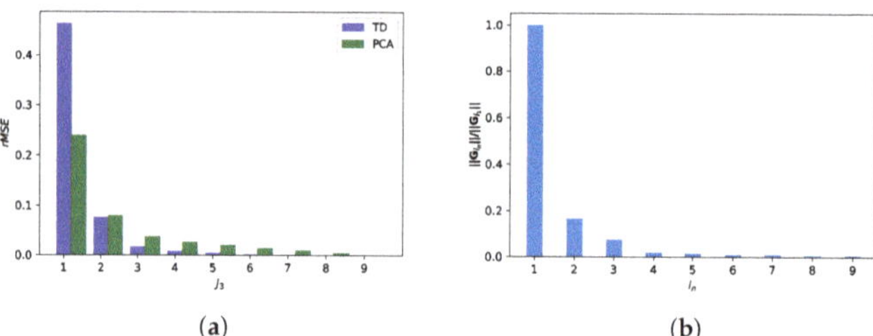

Figure 6. TD metrics (**a**) Reconstruction error computed by the relative mean square error (rMSE) for $J_3 = 1, ..., I_3$ and (**b**) norm of each subtensor $\mathcal{G}_{i_n}$, relative to the norm of the first tensor band $\mathcal{G}_{i_1}$.

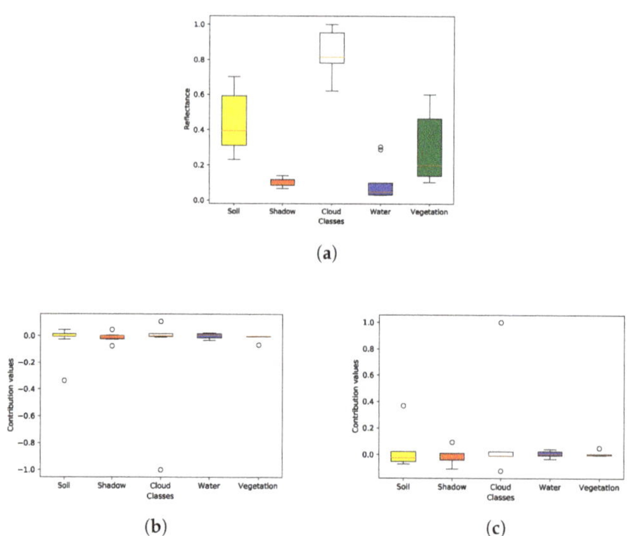

Figure 7. Box and whiskers plots of the behavior of five classes of interest: (**a**) in the original spectral domain, (**b**) the tensor band domain after decomposition for nine bands, and (**c**) the new tensor band domain for five bands.

6.4. Fcn Specifications

We used hyperparameter search [33] to set the learning rate to 1×10^{-3}. The model was run 100 epochs introducing 100 CNNMSI from our dataset. We used the Adam optimizer as our optimization algorithm. Xavier initialization was used for setting the initial values of the weights in the model. The Segnet FCN was used as the base model, since it achieves very high performance metrics in semantic segmentation [35].

6.5. Hardware/Software Specifications

Our framework was implemented using Python 3.7 with Tensorflow-GPU version 1.13. Experiments were run with a NVIDIA GeForce GTX 1050 Ti GPU. The processor used was an Intel core i7 with 8GB RAM, 128 GB SSD, and 1 TB HDD.

Table 4. Inner products of each tensor band with the others from one image of our dataset decomposed by HOOI.

Tensor Band	1	2	3	4	5	6	7	8	9
1	-	2.7×10^{-4}	8.0×10^{-5}	7.0×10^{-5}	4.1×10^{-5}	9.7×10^{-6}	2.0×10^{-5}	2.6×10^{-5}	8.6×10^{-5}
2	-	-	3.1×10^{-7}	8.5×10^{-6}	4.9×10^{-6}	3.2×10^{-6}	3.6×10^{-6}	6.0×10^{-6}	4.8×10^{-6}
3	-	-	-	8.4×10^{-7}	3.9×10^{-7}	4.4×10^{-7}	4.1×10^{-7}	1.8×10^{-9}	1.0×10^{-6}
4	-	-	-	-	5.0×10^{-8}	2.6×10^{-7}	1.2×10^{-8}	5.3×10^{-8}	1.2×10^{-7}
5	-	-	-	-	-	3.7×10^{-9}	8.3×10^{-9}	2.6×10^{-8}	8.9×10^{-9}
6	-	-	-	-	-	-	1.4×10^{-8}	7.2×10^{-8}	2.1×10^{-7}
7	-	-	-	-	-	-	-	1.2×10^{-8}	1.3×10^{-9}
8	-	-	-	-	-	-	-	-	1.6×10^{-7}

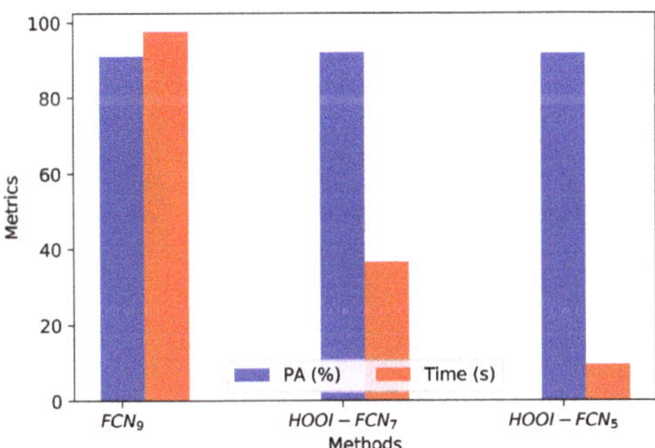

Figure 8. Comparison of the PA and the computational time of FCN with the proposed HOOI-FCN (seven and five bands) for semantic segmentation. See Table 3.

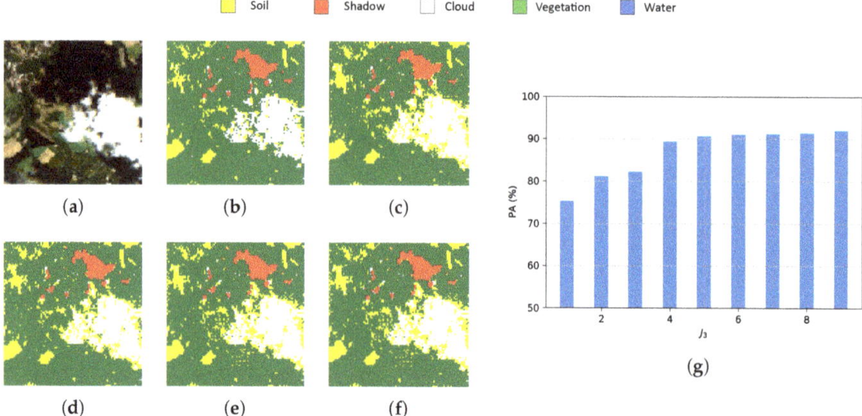

Figure 9. Qualitative results testing a scene of interest with abundant vegetation, and presence of shadows and clouds. (**a**) Original true color scenario of 128 × 128 pixels, in Central Europe: (**b**) five classes semi-manually labeled ground truth of the MSIs, (**c**) classification with an unsupervised normalized difference index (NDI) fusion algorithm, and (**d**) output prediction after 100 epochs in the FCN used for this work without data compression. (**e**) PCA-FCN framework output; (**f**) prediction of the whole framework HOOI-FCN proposed in this work; and (**g**) PA behavior of the HOOI-FCN versus number of tensor bands.

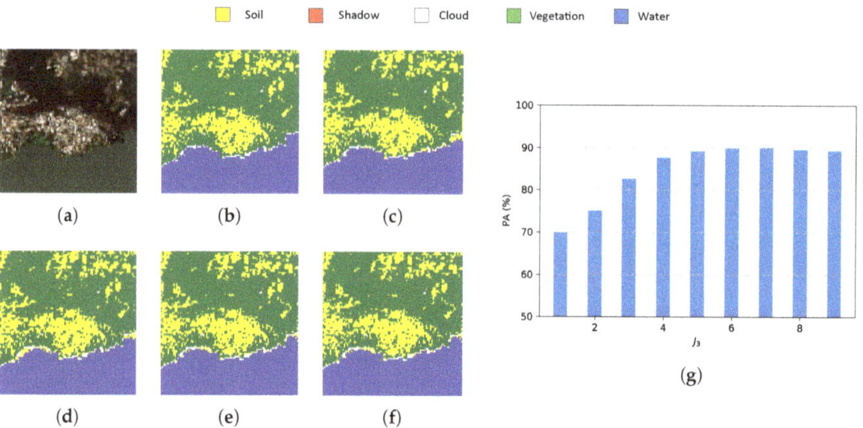

Figure 10. Qualitative results testing a scene of interest with abundant vegetation, and presence of shadows and clouds. (**a**) Original true color scenario of 128 × 128 pixels, in Central Europe: (**b**) five classes semi-manually labeled ground truth of the MSIs, (**c**) classification with an unsupervised normalized difference index (NDI) fusion algorithm, and (**d**) output prediction after 100 epochs in the FCN used for this work without data compression. (**e**) PCA-FCN framework output; (**f**) prediction of the whole framework HOOI-FCN proposed in this work; and (**g**) PA behavior of the HOOI-FCN versus number of tensor bands.

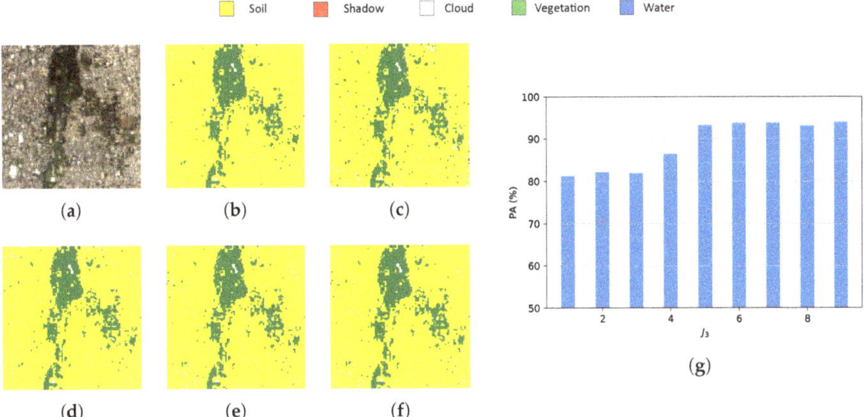

Figure 11. Qualitative results testing a scene of interest with abundant presence of soil. (**a**) Original true color scenario of 128 × 128 pixels, in Central Europe: (**b**) five classes semi-manually labeled ground truth of the MSIs, (**c**) classification with an unsupervised normalized difference index (NDI) fusion algorithm, and (**d**) output prediction after 100 epochs in the FCN used for this work without data compression. (**e**) PCA-FCN framework output; (**f**) prediction of the whole framework HOOI-FCN proposed in this work; and (**g**) PA behavior of the HOOI-FCN versus number of tensor bands.

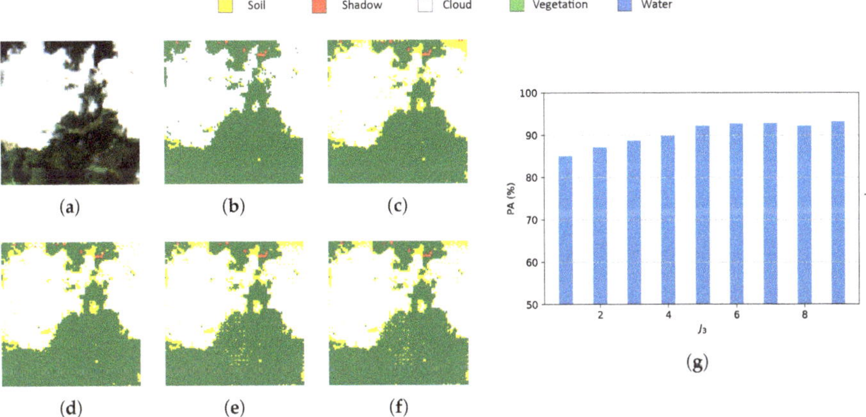

Figure 12. Qualitative results testing a scene of interest with abundant presence of clouds. (**a**) Original true color scenario of 128 × 128 pixels, in Central Europe: (**b**) five classes semi-manually labeled ground truth of the MSIs, (**c**) classification with an unsupervised normalized difference index (NDI) fusion algorithm, and (**d**) output prediction after 100 epochs in the FCN used for this work without data compression. (**e**) PCA-FCN framework output; (**f**) prediction of the whole framework HOOI-FCN proposed in this work; and (**g**) PA behavior of the HOOI-FCN versus number of tensor bands.

7. Discussion and Comparison with Other Methods

Original spectral bands (Figure 7a) were transformed or mapped into new tensor bands (Figure 7b,c) which preserved features of our classes of interest within the first tensor bands, avoiding the use of all the original spectral bands, thereby reducing computational load in further applications.

From Figure 7b,c, we can see that, for the classes of interest in this case study, the error margin selected ψ is indeed a good parameter to restrict the rank in the third mode, since the spectral information for differentiation of these five classes is a greater proportion than the first elements of the spectral domain. Nevertheless, if a smaller value for J_3 were used, there would be a trade off in the performance of the semantic segmentation.

Quantitative results in Figures 8–12 and Table 3 present a comparison of the processing time and PA from our proposed framework with a model without any preprocessing data decomposition algorithm and with a normalized differentiation index based method in different scenarios. The accuracy values obtained by the proposed HOOI-FCN framework are better in overall than those obtained by the other methods under same conditions and scenarios, but with a quite significant decrease of the processing time, in the order of 10 times. It is worth noting that our HOOI-FCN framework with seven and five tensor bands outperforms in PA to the same FCN with the original nine bands. This means that the decomposition produces better features for the classification ANN.

In the confusion matrix presented in Figure 13, we can see the accuracy of the framework proposed HOOI-FCN for each class and the overall accuracy. Rows correspond to the output class or prediction and the columns to the truth class. Diagonal cells show the correctly classified pixels. Off-diagonal cells show where the errors come from. The rightmost column shows the accuracy for each predicted class, while the bottom row shows the accuracy for each true class. It is important to note that vegetation and cloud classes are close to 95% accuracy, while for water and cloud shadows have less than 90% accuracy. The latter can be caused by the lack of samples with a greater contribution of these elements in the training dataset as well as the similarity of these elements to others in the scenes.

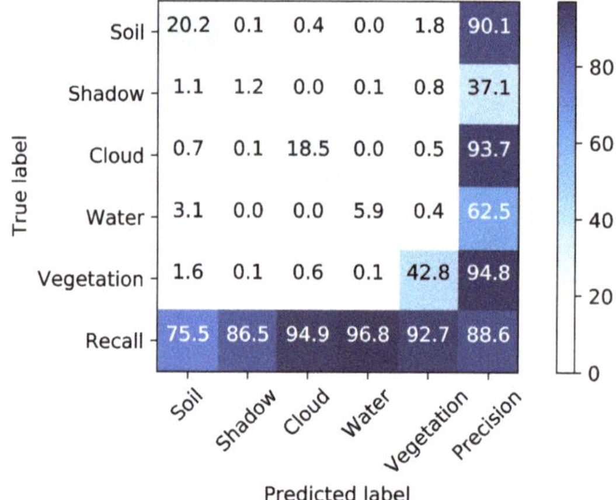

Figure 13. Confusion matrix of the proposed framework. The main diagonal indicates the pixel accuracy for each class in % for the ten selected scenarios.

8. Conclusions

Any RS-MSI or -HSI or third-order tensor image is mapped by the TKD to another tensor, called core tensor representative of the original, preserving its spatial structure, but with fewer tensor bands. In other words, a new subspace embedded in the original space was found and it was be used as the new input space for the task of pixel-level classification or semantic segmentation. Due to the success of DL for image processing, our approach employs an FCN network as the classifier, which delivers the corresponding prediction matrix of pixels classified element-wise.

The efficiency of the proposed higher order orthogonal iteration (HOOI)-FCN framework is measured by metrics such as pixel accuracy (PA) or recall as a function of the number of new tensor bands, which is defined by the reconstruction error computed by the rMSE. Another important parameter in the TKD is the orthogonality degree of each component, i.e., the core tensor and the factor matrices, computed by the inner products of each band with the others.

Our experimental results for a case study show that the proposed HOOI-FCN framework for CNNMSI semantic segmentation reduced the number of spectral bands from nine to seven or five tensor bands, for which PA values converge or are very close to the maximum.

State-of-the-art methods, such as normalized difference indexes, PCA with five principal components, and the same FCN network with nine original bands, with an average pixel accuracy 90% (computational time ∼90s), were outperformed by the HOOI-FCN framework, which achieved a higher average pixel accuracy of 91.97% (and computational time ∼36.5s), and average PA of 91.56% (computational time 9.5s) for seven and five new tensor bands respectively.

These results are very promising in RS, since the use of other algorithms for the calculation of core tensors and a deeper data analysis of weights and initialization of the convolutional neural network (CNN) can increase performance metrics of the segmentation for RS spectral data. Some limitations for a better validation of this approach are: denoising is not included; there is a need for new cases to enhance the input space; use of a greater number of classifiers is needed.

Finally, this research allows us to emphasize two main, relevant points. (1) RS images are characterized by a large number of bands, high correlation between neighbor bands, and high data redundancy; (2) besides, they are corrupted by several noises. Some issues related to our approach remain open.

Open Issues

- Compression affects not only the input data, but also the CNN network to reduce overall complexity and/or create new ANN architectures for specific RS-CNNMSI or HSI image applications.
- Instead of the HOOI algorithm, use greedy HOOI and other algorithms that determine the core tensor for a broad comparison.
- For classification purposes, use other machine learning algorithms, such as a SVM or random forest.
- Increase the input data with more scenarios and their corresponding ground truth to a deeper study of the behaviors of several classifiers, including those based on ANN, and the scope of the TD methods.
- Denoise the original input data for an improvement of the new subspace of reduced dimensionality.

Author Contributions: Conceptualization, S.S.; formal analysis, C.A.; investigation, J.L.; methodology, J.L., D.T., and S.S.; resources, C.A.; software, J.L.; supervision, D.T. and C.A.; validation, S.S. and C.A.; writing—original draft, J.L. and D.T. All authors have read and agreed to the published version of the manuscript.

Funding: This work was supported by the National Council of Science and Technology CONACYT of Mexico under grants 280975 and 253955.

Acknowledgments: We would like to thank the student E. Padilla and his advisor A. Méndez for their help in producing the results for the PCA decomposition and facilitate the comparative analysis between TD and PCA; and F. Hermosillo for useful observations about this work, all of them being from CINVESTAV, Guadalajara. We also would like to dedicate this work to the memory of Y. Shkvarko, who was an important mentor for the realization of this research.

Conflicts of Interest: The authors declare no conflict of interest.

Abbreviations

The following abbreviations are used in this manuscript:

ANN	artificial neural network
CNN	convolutional neural network
CPD	canonical polyadic decomposition
ESA	european space agency
DL	deep learning
FCN	fully convolutional network
GPU	graphics processing unit

HSI	hyperspectral image
HOOI	higher order orthogonal iteration
HOSVD	higher order singular value decomposition
MSE	mean square error
ML	machine learning
MSI	multispectral image
NIR	near-infrared
NTD	nonnegative Tucker decomposition
NDVI	normalized difference vegetation index
NDWI	normalized difference water index
PA	pixel accuracy
PCA	principal components analysis
ReLU	rectified linear unit
rMSE	relative mean square error
RS	remote sensing
SVD	singular value decomposition
SWIR	short wave infrared
SVM	support vector machine
T-MLRD	tensor-based multiscale low rank decomposition
TD	tensor decomposition
TDA	tensor discriminant analysis
TKD	tucker decomposition

References

1. Tempfli, K.; Huurneman, G.; Bakker, W.; Janssen, L.; Feringa, W.; Gieske, A.; Grabmaier, K.; Hecker, C.; Horn, J.; Kerle, N.; et al. *Principles of Remote Sensing: An Introductory Textbook*, 4th ed.; ITC: Geneva, Switzerland, 2009.

2. He, Z.; Hu, J.; Wang, Y. Low-rank tensor learning for classification of hyperspectral image with limited labeled sample. *IEEE Signal Process.* **2017**, *145*, 12–25. [CrossRef]

3. Richards, A.; Xiuping, J.J. Band selection in sentinel-2 satellite for agriculture applications. In *Remote Sensing Digital Image Analysis*, 4th ed.; Springer-Verlag: Berlin, Germany, 2006.

4. Zhang, T.; Su, J.; Liu, C.; Chen, W.; Liu, H.; Liu, G. Band selection in sentinel-2 satellite for agriculture applications. In Proceedings of the 23rd International Conference on Automation & Computing, University of Huddersfield, Huddersfield, UK, 7–8 September 2017.

5. Xie, Y.; Zhao, X.; Li, L.; Wang, H. Calculating NDVI for Landsat7-ETM data after atmospheric correction using 6S model: A case study in Zhangye city, China. In Proceedings of the 18th International Conference on Geoinformatics, Beijing, China, 18–20 June 2010.

6. Gao, B. NDWI—A normalized difference water index for remote sensing of vegetation liquid water from space. *Remote Sens. Environ.* **1996**, *58*, 1–6. [CrossRef]

7. Ham, J.; Chen, Y.; Crawford, M.; Ghosh, J. Investigation of the random forest framework for classification of hyperspectral data. *IEEE Trans. Geosci. Remote Sens.* **2005**, *43*, 492–501. [CrossRef]

8. Hearst, Marti A. Support Vector Machines. *IEEE Intell. Syst. J.* **1998**, *13*, 18–28. [CrossRef]

9. Huang, X.; Zhang, L. An SVM Ensemble Approach Combining Spectral, Structural, and Semantic Features for the Classification of High-Resolution Remotely Sensed Imagery. *IEEE Trans. Geosci. Remote Sens.* **2013**, *51*, 257–272. [CrossRef]

10. Delalieux, S.; Somers, B.; Haest, B.; Spanhove, T.; Vanden Borre, J.; Mucher, S. Heathland conservation status mapping through integration of hyperspectral mixture analysis and decision tree classifiers. *Remote Sens. Environ.* **2012**, *126*, 222—231. [CrossRef]

11. Kemker, R.; Salvaggio, C.; Kanan, C. Algorithms for semantic segmentation of multispectral remote sensing imagery using deep learning. *ISPRS J. Photogramm. Remote Sens.* **2018**, *145*, 60–77. [CrossRef]

12. Pirotti, F.; Sunar, F.; Piragnolo, M. Benchmark of machine learning methods for classification of a sentinel-2 image. In Proceedings of the XXIII ISPRS Congress, Prague, Czech Republic, 12–19 July 2016.

13. Mateo-García, G.; Gómez-Chova, L.; Camps-Valls, G. Convolutional neural networks for multispectral image cloud masking. In Proceedings of the IGARSS, Fort Worth, TX, USA, 23–28 July 2017.

14. Guo, X.; Huang, X.; Zhang, L.; Zhang, L.; Plaza, A.; Benediktsson, J. A. Support Tensor Machines for Classification of Hyperspectral Remote Sensing Imagery. *IEEE Trans. Geosci. Remote Sens.* **2016**, *54*, 3248–3264. [CrossRef]

15. Cichocki, A.; Mandic, D.; De Lathauwer, L.; Zhou, G.; Zhao, Q.; Caiafa, C.; Phan, H. Tensor Decompositions for Signal Processing Applications: From two-way to multiway component analysis. *IEEE Signal Process. Mag.* **2015**, *32*, 145–163. [CrossRef]

16. Jolliffe, I.T. *Principal Component Analysis*, 2nd ed.; Springer Verlag: New York, NY, USA, 2002.

17. Kolda, T.; Bader, B. Tensor Decompositions and Applications. *SIAM Rev.* **2009**, *51*, 455–500. [CrossRef]

18. Lopez, J.; Santos, S.; Torres, D.; Atzberger, C. Convolutional Neural Networks for Semantic Segmentation of Multispectral Remote Sensing Images. In Proceedings of the LATINCOM, Guadalajara, Mexico, 14–16 November 2018.

19. European Space Agency. Available online: https://sentinel.esa.int/web/sentinel/missions/sentinel-2 (accessed on 15 July 2019).

20. Kemker, R.; Kanan, C. Deep Neural Networks for Semantic Segmentation of Multispectral Remote Sensing Imagery. *arXiv* **2017**, arXiv:abs/1703.06452.

21. Hamida, A.; Benoît, A.; Lambert, P.; Klein, L.; Amar, C.; Audebert, N.; Lefèvre, S. Deep learning for semantic segmentation of remote sensing images with rich spectral content. In Proceedings of the IGARSS, Fort Worth, TX, USA, 23–28 July 2017.

22. Wang, Q.; Lin, J.; Yuan, Y. Salient Band Selection for Hyperspectral Image Classification via Manifold Ranking. *IEEE Trans. Neural Netw. Learn. Syst.* **2016**, *27*, 1279–1289. [CrossRef]

23. Li, S.; Qiu, J.; Yang, X.; Liu, H.; Wan, D.; Zhu. Y. A novel approach to hyperspectral band selection based on spectral shape similarity analysis and fast branch and bound search. *Eng. Appl. Artif. Intell.* **2014**, *27*, 241–250. [CrossRef]

24. Zhang. L.; Zhang, L.; Tao, D.; Huang, X.; Du, B. Compression of hyperspectral remote sensing images by tensor approach. *Neurocomputing* **2015**, *147*, 358–363. [CrossRef]

25. Astrid, M.; Lee, Seung-Ik. CP-decomposition with Tensor Power Method for Convolutional Neural Networks compression. In Proceedings of the BigComp, Jeju, Korea, 13–16 February 2017.

26. Chien, J.; Bao, Y. Tensor-factorized neural networks. *IEEE Trans. Neural Netw. Learn. Syst.* **2018**, *29*, 1998–2011. [CrossRef] [PubMed]

27. An, J.; Lei, J.; Song, Y.; Zhang, X.; Guo J. Tensor Based Multiscale Low Rank Decomposition for Hyperspectral Images Dimensionality Reductio. *Remote Sens.* **2019**, *11*, 1485. [CrossRef]

28. Li, J.; Liu, Z. Multispectral Transforms Using Convolution Neural Networks for Remote Sensing Multispectral Image Compression. *Remote Sens.* **2019**, *11*, 759. [CrossRef]

29. An, J.; Song, Y.; Guo, Y.; Ma, X.; Zhang, X. Tensor Discriminant Analysis via Compact Feature Representation for Hyperspectral Images Dimensionality Reduction. *Remote Sens.* **2019**, *11*, 1822. [CrossRef]

30. Absil, P.-A.; Mahony, R.; Sepulchre, R. *Optimization Algorithms on Matrix Manifolds*, 1st ed.; Princeton University Press: Princeton, NJ, USA, 2007.

31. De Lathauwer, L.; De Moor, B.; Vandewalle, J. On the best rank-1 and rank-(R 1, R 2, ···, R N) approximation of higher-order tensors. *SIAM J. Matrix Anal. Appl.* **2000**, *21*, 1324–1342. [CrossRef]

32. De Lathauwer, L.; De Moor, B.; Vandewalle, J. A Multilinear Singular Value Decomposition. *SIAM J. Matrix Anal. Appl.* **2000**, *21*, 1253–1278. [CrossRef]

33. Goodfellow, I.; Bengio, Y.; Courville, A. *Deep Learning*, 1st ed.; MIT Press, 2016.

34. Sheehan, B. N.; Saad, Y. Higher Order Orthogonal Iteration of Tensors (HOOI) and its Relation to PCA and GLRAM. In Proceedings of the 7th SIAM International Conference on Data Mining, Minneapolis, MN, USA, 26–28 April 2007.

35. Badrinarayanan, V.; Kendall, A.; Cipolla, R. SegNet: A Deep Convolutional Encoder-Decoder Architecture for Image Segmentation. *IEEE Trans. Pattern Anal. Mach. Intell.* **2017**, *39*, 2481–2495. [CrossRef] [PubMed]

36. Rodes, I.; Inglada, J.; Hagolle, O.; Dejoux, J.; Dedieu, G. Sampling strategies for unsupervised classification of multitemporal high resolution optical images over very large areas. In Proceedings of the 2012 IEEE International Geoscience and Remote Sensing Symposium, Munich, Germany, 22–27 July 2012.

 *remote sensing* 

Article

Using Predictive and Differential Methods with K²-Raster Compact Data Structure for Hyperspectral Image Lossless Compression †

Kevin Chow *, Dion Eustathios Olivier Tzamarias, Ian Blanes and Joan Serra-Sagristà

Department of Information and Communications Engineering, Universitat Autònoma de Barcelona, Cerdanyola del Vallès, 08193 Barcelona, Spain; dion.tzamarias@uab.cat (D.E.O.T.); ian.blanes@uab.cat (I.B.); joan.serra@uab.cat (J.S.-S.)

* Correspondence: kevin.chow@uab.cat

† This paper is an extended version of our paper published in the 6th ESA/CNES International Workshop on On-Board Payload Data Compression Proceedings.

Received: 31 August 2019; Accepted: 17 October 2019; Published: 23 October 2019

Abstract: This paper proposes a lossless coder for real-time processing and compression of hyperspectral images. After applying either a predictor or a differential encoder to reduce the bit rate of an image by exploiting the close similarity in pixels between neighboring bands, it uses a compact data structure called k^2-raster to further reduce the bit rate. The advantage of using such a data structure is its compactness, with a size that is comparable to that produced by some classical compression algorithms and yet still providing direct access to its content for query without any need for full decompression. Experiments show that using k^2-raster alone already achieves much lower rates (up to 55% reduction), and with preprocessing, the rates are further reduced up to 64%. Finally, we provide experimental results that show that the predictor is able to produce higher rates reduction than differential encoding.

Keywords: compact data structure; quadtree; k^2-tree; k^2-raster; DACs; 3D-CALIC; M-CALIC; hyperspectral images

1. Introduction

Compact data structures [1] are examined in this paper as they can provide real-time processing and compression of remote sensing images. These structures are stored in reduced space in a compact form. Functions can be used to access and query each datum or groups of data directly in an efficient manner without an initial full decompression. This compact data should also have a size which is close to the information-theoretic minimum. The idea was explored and examined by Guy Jacobson in his doctoral thesis in 1988 [2] and in a paper published by him a year later [3]. Prior to this, works had been done to express similar ideas. However, Jacobson's paper is often considered the starting point of this topic. Since then it has gained more attention and a number of research papers have been published. Research on algorithms such as FM-index [4,5] and Burrows-Wheeler transform [6] were proposed and applications were released, notable examples of which include bzip2 (https://linux.die.net/man/1/bzip2), Bowtie [7] and SOAP2 [8]. One of the advantages of using compact data structures is that the compressed data form can be loaded into main memory and accessed directly. The smaller compressed size also helps data move through communication channels faster. The other advantage is that there is no need to compress and decompress the data as is the case with data compressed by a classical compression algorithm such as gzip or bzip2, or by a specialized algorithm such as CCSDS 123.0-B-1 [9] or KLT+JPEG 2000 [10,11]. The resulting image will have the same quality as the original.

Hyperspectral images are image data that contain a multiple number of bands from across the electromagnetic spectrum. They are usually taken by hyperspectral satellite and airborne sensors. Data are extracted from certain bands in the spectrum to help us find the objects that we are specifically looking for, such as oil fields and minerals. However, due to their large sizes and the huge amount of data that have been collected, hyperspectral images are normally compressed by lossy and lossless algorithms to save space. In the past several decades, a lot of research studies have gone into keeping the storage sizes to a minimum. However, to retrieve the data, it is still necessary to decompress all the data. With our approach using compact data structures, we can query the data without fully decompressing them in the first place, and this is the main motivation for this work.

Prediction is one of the schemes used in lossless compression. CALIC (Context Adaptive Lossless Image Compression) [12,13] and 3D-CALIC [14] belong to this class of scheme. In 1994, Wu et al. introduced CALIC, which uses both context and prediction of the pixel values. In 2000, the same authors proposed a related scheme called 3D-CALIC in which the predictor was extended to the pixels between bands. Later in 2004, Magli et al. [15] proposed M-CALIC whose algorithm is related to 3D-CALIC. All these methods take advantage of the fact that in a hyperspectral image, neighboring pixels in the same band (spatial correlation) are usually close to each other and even more so for neighboring pixels of two neighboring bands (spectral correlation).

Differential encoding is another way of encoding an image by taking the difference between neighboring pixels and in this work, it is a special case of the predictive method. It only takes advantage of the spectral correlation. However, this correlation between the pixels in the bands will become smaller as the distance between the bands are further apart and therefore, its effectiveness is expected to decrease when the bands are far from each other.

The latest studies on hyperspectral image compression, both lossy and lossless, are focused on CCSDS 123.0, vector quantization, Principal Component Analysis (PCA), JPEG2000, and Lossy Compression Algorithm for Hyperspectral Image Systems (HyperLCA), among many others. Some of these research works are listed in [16–19]. In this work, however, we investigate lossless compression of hyperspectral images through the proposed k^2-raster for 3D images, which is a compact data structure that can provide bit-rate reduction as well as direct access to the data without full decompression. We also explore the use of a predictor and a differential encoder as preprocessing on the compact data structure to see if it can provide us with further bit-rate reduction. The predictive method and the differential method are also compared. The flow chart shown in Figure 1 depicts how the encoding/decoding of this proposal works.

This paper is organized as follows: In Section 2, we present the k^2-raster and discuss it in detail, beginning with quadtree, followed by k^2-tree and k^2-raster. Later in the same section, details of the predictive method and the differential method are discussed. Section 3 shows the experimental results on how the two methods fare using k^2-raster on hyperspectral images, and more results on how some other factors such as using different k-values can affect the bit rates. Finally, we present our conclusions in Section 4.

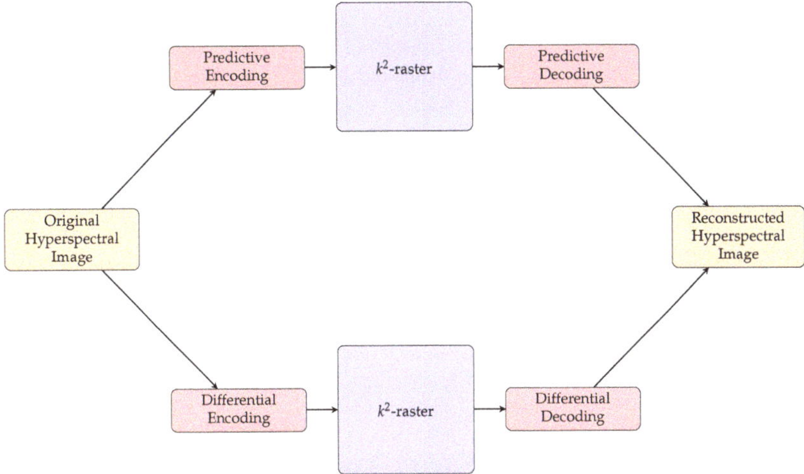

Figure 1. A flow chart showing the encoding and decoding of this coder.

2. Materials and Methods

One way to build a structure that is small and compact is to make use of a tree structure and do it without using pointers. Pointers usually take up a large amount of space, with each one having a size in the order of 32 or 64 bits for most modern-day machines or programs. A tree structure with n pointers will have a storage complexity of $\mathcal{O}(n \log n)$ whereas a pointer-less tree only occupies $\mathcal{O}(n)$. For pointer-less trees, to get at the elements of the structure, rank and select functions [3] are used, and that only requires simple arithmetic to find the parent's and child's positions. This is the premise that compact data structures are based on. In this work, we will use k^2-raster from Ladra et al. [20], a concept which was developed from k^2-tree, also a type of compact data structure, as well as the idea of using recursive decomposition of quadtrees. The results of k^2-raster were quite favorable for the data sets that were used. Therefore, we are extending their approach for hyperspectral images and investigate whether it would be possible to use that structure for 3D hyperspectral images. The Results section will show us that the results are quite competitive compared to other commonly-used classical compression techniques. There is a bit-rate reduction of up to 55% for the testing images. Upon more experimentation with predictive and differential preprocessing, a further bit-rate reduction of up to 64% can be attained. For that reason, we are proposing in this paper our encoder using the predictor or differential method on k^2-raster for hyperspectral images.

2.1. Quadtrees

Quadtree structures [21], which have been used in many kinds of data representations such as image processing and computer graphics, are based on the principle of recursive decomposition. As there are many variants of quadtree, we will describe the one that is pertinent to our discussion: region quadtree. Basically, a quadtree is a tree structure where each internal node has 4 children. Given a 2D square matrix, it is partitioned recursively into four equal subquadrants. If a tree is built to represent this, it will have a root node at level 0 with 4 children nodes at level 1, each child representing a node and a subquadrant. Next, if the subquadrant has a size larger than 2^2, then each of these subquadrants will be partitioned to give 4 more children and a new level 2 is added to the tree. Note that the tree nodes are traversed in a left to right order.

Considering a matrix of size $n \times n$ where n is a power of 2, it is recursively divided until each subquadrant has a size of 2^2. For example, if the size of the matrix is 8×8, after the recursive division of matrix, $(8^2)/(2^2) = 16$ subquadrants are obtained. It should be noted that the value of n in the image matrix needs to be a power of 2. Otherwise, the matrix has to be enlarged widthwise and heightwise to a value which is the next power of 2, and these additional pixels will be padded with zeros. As k^2-trees are based on quadtrees, the division and the resulting tree of a quadtree are very similar to those of a k^2-tree. Figure 2 illustrates how a quadtree's recursive partitioning works.

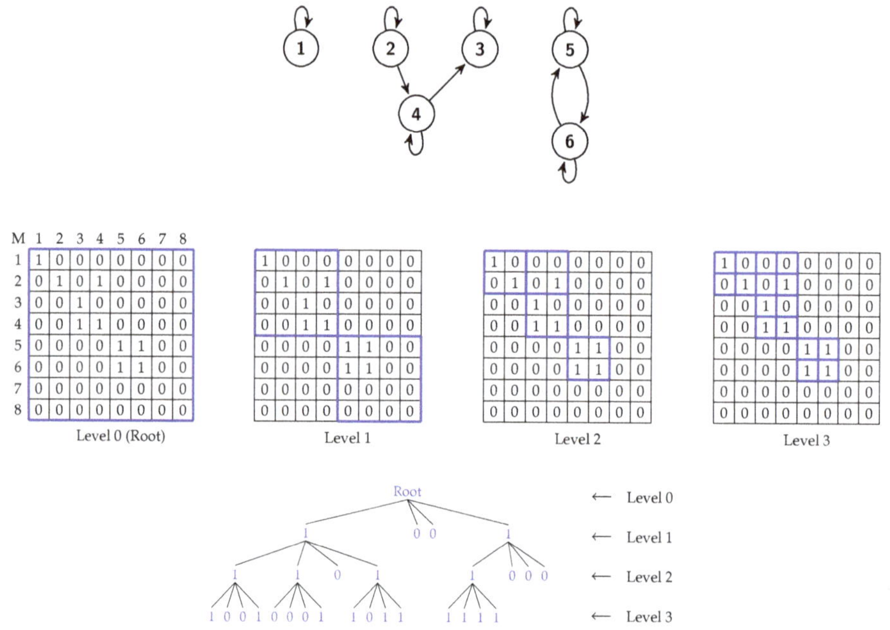

Figure 2. A graph of 6 nodes (top) with its 8×8 binary adjacency matrix at various stages of recursive partitioning. At the bottom, a k^2-trees (k=2) is constructed from the matrix.

2.2. LOUDS

k^2-tree is based on unary encoding and LOUDS, which is a compact data structure introduced by Guy Jacobson in his paper and thesis [2,3]. A bit string is formed by a breadth-first traversal (going from left to right) of an ordinal (rooted, ordered) tree structure. Each parent node is encoded with a string of '1' bits whose length indicates the number of children it has and each string ends with a '0' bit. If the parent node has no children, only a single '0' bit suffices.

The parent and child relationship can be computed by two cornerstone functions for compact data structures: rank and select. These functions give us information about the node's first-child, next-sibling(s), and parent, without the need of using pointers. They are described below:

$\mathrm{rank}_b(m)$ returns the number of bits which are set to b, left of position m (inclusive) in the bitmap where b is 0 or 1.

$\mathrm{select}_b(i)$ returns the position of the i-th b bit in the bitmap where b is 0 or 1.

By default, b is 1, i.e., $\mathrm{rank}(m) = \mathrm{rank}_1(m)$. These operations are inverses of each other. In other words, $\mathrm{rank}(\mathrm{select}(m)) = \mathrm{select}(\mathrm{rank}(m)) = m$. Since a linear scan is required to process the rank and select functions, the worst-case time complexity will be $\mathcal{O}(n)$.

To clarify how these functions work, consider the binary trees depicted in Figure 3 where the one on the left shows the values and the one on the right shows the numbering of the same tree. If the node has two children, it will be set to 1. Otherwise, it is set to 0. The values of this tree are put in a bit string shown in Figure 4. Figure 5 shows how the position of the left child, right child or parent of a certain node m is computed with the rank and select functions. An example follows:

To find the left child of node 8, we first need to compute rank(8), which is the total number of 1's from node 1 up to and including node 8 and the answer is 7. Therefore, the left child is located in 2*rank(8) = 2*7 = 14 and the right child is in 2*rank(8)+1 = 2*7+1 = 15. The parent of node 8 can be found by computing select($\lfloor 8/2 \rfloor$) or select($\lfloor 4 \rfloor$). The answer can be arrived at by counting the total number of bits starting from node 1, skipping the ones with '0' bits. When we get to node 4 which gives us a total bit count of 4, we then know that node 4 is where the parent of node 8 is.

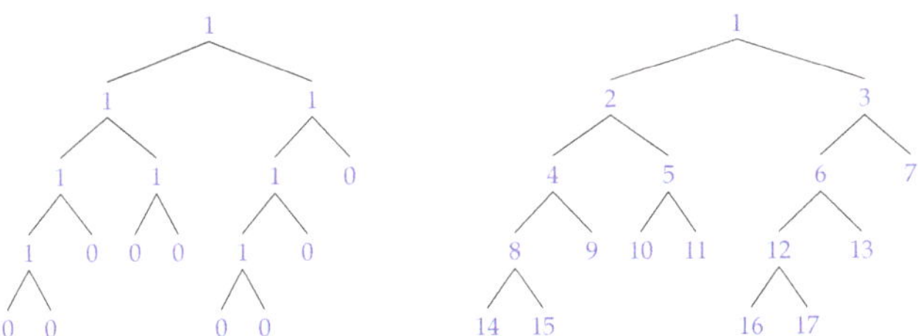

Figure 3. A binary tree example for LOUDs. The one on the left shows the values of the nodes and the one on the right shows the same tree with the numbering of the nodes in a left-to-right order. In this case the numbering starts with 1 at the root.

m	1	2	3	4	5	6	7	8	9	10	11	12	13	14	15	16	17
bit	1	1	1	1	1	1	0	1	0	0	0	1	0	0	0	0	0

Figure 4. A bit string with the values from the binary tree in Figure 3.

m	1	2	3	4	5	6	8	12
Left child(m) = 2 · rank(m)	2	4	6	8	10	12	14	16
Right child(m) = 2 · rank(m)+1	3	5	7	9	11	13	15	17
Parent(m) = select($\lfloor m/2 \rfloor$)	-	1	1	2	2	3	4	6

Figure 5. With the rank and select functions listed in the first column, we can navigate the binary tree in Figure 3 and compute the position node for the left child, right child or parent of the node.

In the next section, we will explain how the rank function can be used to determine the children's positions in a k^2-tree, thus enabling us to query the values of the cells.

2.3. k^2-Tree

Originally proposed for compressing Web graphs, k^2-tree is a LOUDS variant compact data structure [22]. The tree represents a binary adjacency matrix of a graph (see Figure 2). It is constructed by recursively partitioning the matrix into square submatrices of equal size until each submatrix reaches a size of k x k where $k \geq 2$. During the process of partitioning, if there is at least one cell in the submatrix that has a value of 1, the node in the tree will be set to 1. Otherwise, it will be set to 0 (i.e., it is a leaf and has no children) and this particular submatrix will not be partitioned any further. Figure 2 illustrates an example of a graph of 6 nodes, its 8 × 8 binary adjacency matrix at various stages of recursive partitioning, and the k^2-tree that is constructed from the matrix.

The values of k^2-trees are basically stored in two bitmaps denoted by T (tree) and L (leaves). The values are traversed in a breadth-first fashion starting with the first level. The T bitmap stores the bits at all levels except the last one where its bits will be stored in the L bitmap. Note that the bit values of T which are either 0 or 1 will be stored as a bit vector. To illustrate this with an example, we again make use of the binary matrix in Figure 2. The T bitmap contains all the bits from levels 1 and 2. Thus the T bitmap has the following bits: 1001 1101 1000 (see Figure 6). The bits from the last level, level 3, will be stored in the L bitmap with the following bits: 1001 0001 1011 1111.

Consider a set S with elements from 1 to n, to find the child's or the parent's position of a certain node m in a k^2-tree, we perform the following operations:

$$\text{first-child}(m) \leftarrow \text{rank}(m) \cdot k^2 \text{ where } 1 \leq m \leq \|S\|$$
$$\text{parent}(m) \leftarrow \text{select}(\lfloor m/k^2 \rfloor) \text{ where } 1 \leq m \leq \|S\|$$

Once again using the k^2-tree in Figure 2 as an example, with the T bitmap (Figure 6) and the rank and select functions, we can navigate the tree and obtain the positions of the first child and the parent. Figure 7 shows how the nodes of the k^2-tree are numbered.

Ex. Locate the first child of node 8:
$\text{rank}_1(8) * 4 = 6 * 4 = 24$
(There are 6 one bits in the T bitmap starting from node 0 up to and including node 8.)

Ex. Locate the parent of node 11:
$\text{select}_1(\lfloor 11/4 \rfloor) = \text{select}_1(2) = 3$
(Start counting from node 0, skipping all nodes with '0' bits, and node 3 is the first node that gives a total number of 1-bit count of 2. Therefore, node 3 is the parent.

Node	0	1	2	3	4	5	6	7	8	9	10	11
Bit	1	0	0	1	1	1	0	1	1	0	0	0

Figure 6. A T bitmap with the first node labeled as 0.

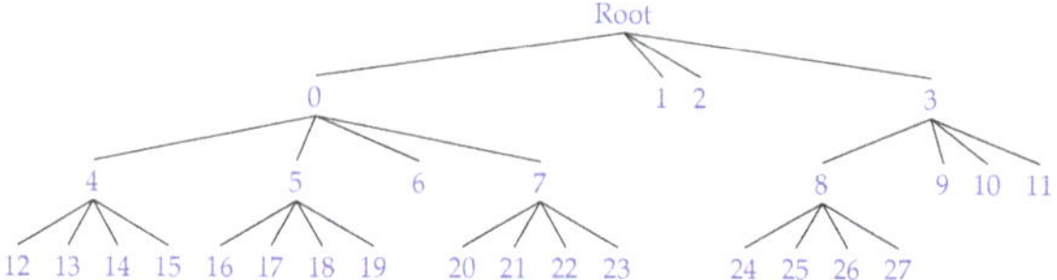

Figure 7. An example showing how the rank function is computed to obtain the children's position on a k^2-tree node (k=2) based on the tree in Figure 2. It starts with 0 on the first child of the root (first level) and the numbering traverses from left to right and from top to bottom.

It was shown that k^2-tree gave the best performance when the matrix was sparse with large clusters of 0's or 1's [20].

2.4. DACs

This section describes DACs which is used in k^2-raster to directly access variable-length codes. Based on the concept of compact data structures, DACs were proposed in the papers published by Brisaboa et al. in 2009 and 2013 [23,24] and the structure was proven to yield good compression ratios for variable-length integer sequences. By means of the rank function, it gains fast direct access to any position of the sequence in a very compact space. The original authors also asserted that it was better suited for a sequence of integers with a skewed frequency distribution toward smaller integer values.

Different types of encoding are used for DACs and the one that we are interested in for k^2-raster is called Vbyte coding. Consider a sequence of integers x. Each integer, which is represented by $\lfloor \log_2 x_i \rfloor + 1$ bits, is broken into blocks of bits of size S. Each block is stored as a chunk of $S + 1$ bits. The chunk that holds the most significant bits has the highest bit set to 0 while the other chunks have their highest bit set to 1. For example, if we have an integer 20 (10100_2) which is 5 bits long and if the block size is $S = 3$, then we can have 2 chunks denoted by the following: $\underline{0}010\ \underline{1}100$.

To show how the chunks are organized and stored, we again illustrate it with an example. If we have 3 integers of variable length 20 (10100_2), 6 (110_2), 73 (1001001_2) and each block size is 3, then the three integers have the following representations.

20   $\underline{0}010\ \underline{1}100$ ($B_{1,2}A_{1,2}\ B_{1,1}A_{1,1}$)
6    $\underline{0}110$ ($B_{2,1}A_{2,1}$)
73   $\underline{0}001\ \underline{1}001\ \underline{1}001$ ($B_{3,3}A_{3,3}\ B_{3,2}A_{3,2}\ B_{3,1}A_{3,1}$)

We will store them in three chunks of arrays A and bitmaps B. This is depicted in Figure 8. To retrieve the values in the arrays A, we make use of the corresponding bitmaps B with the rank function.

More information on DACs and the software code can be found in the papers [23,24].

C_1	A_1	100 ($A_{1,1}$)	110 ($A_{2,1}$)	001 ($A_{3,1}$)
	B_1	1 ($B_{1,1}$)	0 ($B_{2,1}$)	1 ($B_{3,1}$)
C_2	A_2	010 ($A_{1,2}$)	001 ($A_{3,2}$)	
	B_2	0 ($B_{1,2}$)	1 ($B_{3,2}$)	
C_3	A_3	001 ($A_{3,3}$)		
	B_3	0 ($B_{3,3}$)		

Figure 8. Organization of 3 Directly Addressable Codes (DACs) clusters.

2.5. k^2-Raster

k^2-raster is a compact data structure that allows us to store raster pixels in reduced space. It consists of several basic components: bitmaps, DACs and LOUDS. Similar to a k^2-tree, the image matrix is partitioned recursively until each subquadrant is of size k^2. The resulting LOUDS tree topology contains the bitmap T where the elements are accessed with the rank function. Unlike k^2-tree, at each tree level, the maximum and minimum values of each subquadrant are stored in two bitmaps which are respectively called $Vmax$ and $Vmin$. However, to compress the structure further, the maximum and minimum values of each level are compared with the corresponding values of the parent and their differences will replace the stored values in the $Vmax$ and $Vmin$ bitmaps. The rationale behind all this is to obtain smaller values for each node so as to get a better compression with DACs. An example of a simple 8×8 matrix is given to illustrate this point in Figure 9. A k^2-raster is constructed from this matrix with maximum and minimum values stored in each node in Figure 10. The structure is further modified, according to the above discussion, to form a tree with smaller maximum and minimum values and this is shown in Figure 11.

Next, with the exception of the root node at the top level, the $Vmax$ and $Vmin$ bitmaps at all levels are concatenated to form $Lmax$ and $Lmin$ bitmaps. The root's maximum ($rMax$) and minimum ($rMin$) values are integer values and will remain uncompressed.

For an image of size $n \times n$ with n bands, the time complexity to build all the k^2-rasters is $\mathcal{O}(n^3)$ [22]. To query a cell from the structure, which has a tree height of at most $\lceil \log_k n \rceil$ levels, the time complexity to extract a codeword at a single $Lmax$ level is $\mathcal{O}(\log_k n)$, and this is the worst-case time to traverse from the root node to the last level of the structure. The number of levels, $\mathcal{L}$, in $Lmax$ can be obtained from the maximum integer in the sequence and with this, we can compute the time complexity for a cell query, which is $\mathcal{O}(\log_k n \cdot \mathcal{L})$ [23,25].

To sum up, a k^2-raster structure is composed of a bitmap T, a maximum bitmap $Lmax$, a minimum bitmap $Lmin$, a root maximum $rMax$ integer value and a root minimum $rMin$ integer value.

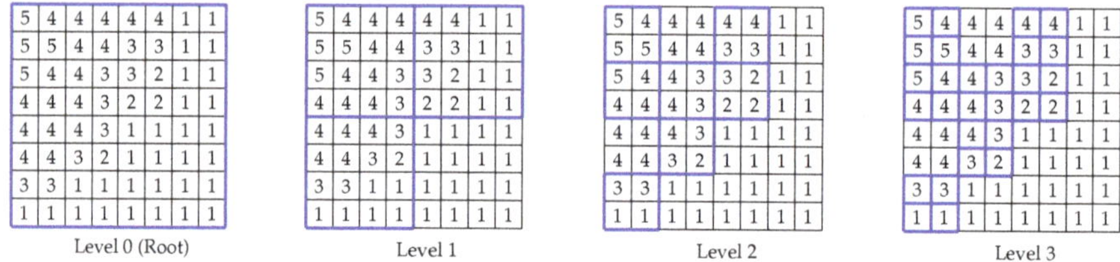

Level 0 (Root)   Level 1   Level 2   Level 3

Figure 9. An example of an 8×8 matrix for k^2-raster. The matrix is recursively partitioned into square subquadrants of equal size. During the process, unless all the cells in a subquadrant have the same value, the partitioning will continue. Otherwise the partitioning of this particular subquadrant will end at this point.

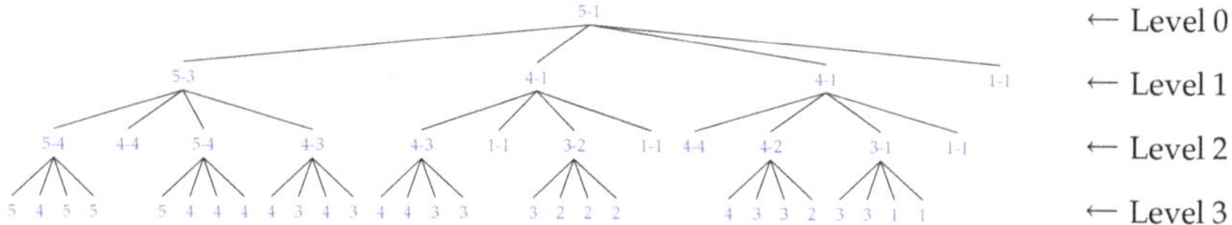

$\leftarrow$ Level 0
$\leftarrow$ Level 1
$\leftarrow$ Level 2
$\leftarrow$ Level 3

Figure 10. A k^2-raster ($k = 2$) tree storing the maximum and mininum values for each quadrant of every recursive subdivision of the matrix in Figure 9. Every node contains the maximum and minimum values of the subquadrant, separated by a dash. On the last level, only one value is shown as each subquadrant contains only one cell.

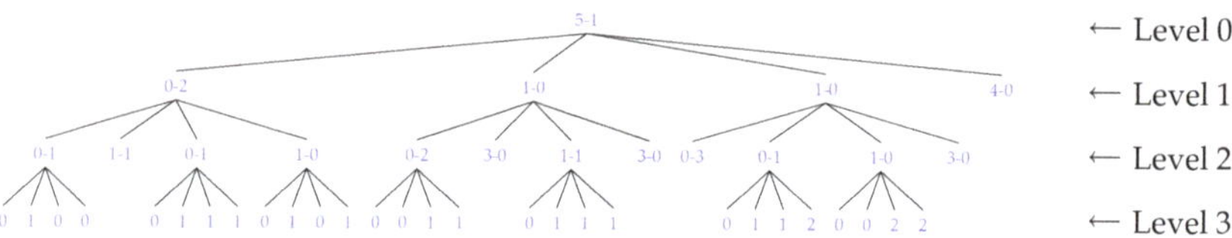

$\leftarrow$ Level 0
$\leftarrow$ Level 1
$\leftarrow$ Level 2
$\leftarrow$ Level 3

Figure 11. Based on the tree in Figure 10, the maximum value of each node is subtracted from that of its parent while the minimum value of the parent is subtracted from the node's minimum value. These differences will replace their corresponding values in the node. The maximum and minimum values of the root remain the same.

2.6. Predictive Method

As mentioned in the Introduction, an interband predictor called 3D-CALIC was proposed by Wu et al. in 2000 and another predictor called M-CALIC by Magli et al. in 2004. Our predictor is based on the idea of least squares method and the use of reference bands that were discussed in both the 3D-CALIC [14] and M-CALIC [15] papers. Consider two neighboring or close neighboring bands of the same hyperspectral image. These bands can be represented by two vectors $\mathbf{X} = (x_1, x_2, x_3, ..., x_{n-1}, x_n)$ and $\mathbf{Y} = (y_1, y_2, y_3, ..., y_{n-1}, y_n)$ where x_i and y_i are two pixels that are located at the same spatial position but in different bands, and n is the number of pixels in each band. We can then employ the close similarity between the bands to predict the pixel value in the current band $\mathbf{Y}$ using the corresponding pixel value in band $\mathbf{X}$, which we designate as the reference band.

A predictor for a particular band can be built from the linear equation:

$$\hat{\mathbf{Y}} = \alpha \mathbf{X} + \beta \tag{1}$$

so as to minimize $||\hat{Y} - Y||_2^2$ where $\hat{Y}$ is the predicted value and Y is the actual value of the current band. The optimal values for α and β should minimize the prediction error of the current pixel and can be obtained by using the least squares solution:

$$\hat{\alpha} = \frac{n \sum_{i=1}^{n} x_i y_i - \sum_{i=1}^{n} x_i \sum_{i=1}^{n} y_i}{n \sum_{i=1}^{n} x_i^2 - (\sum_{i=1}^{n} x_i)^2} , \tag{2}$$

$$\hat{\beta} = \frac{n \sum_{i=1}^{n} y_i - \hat{\alpha} \sum_{i=1}^{n} x_i}{n} \tag{3}$$

where n is the size of each band, i.e., the height multiplied by the width, $\hat{\alpha}$ the optimal value of α and $\hat{\beta}$ the optimal value of β.

The difference between the actual and predicted pixel values of a band is known as the residual value or the prediction error. When all the pixel values in the current band are calculated, these prediction residuals will be saved in a vector, which will later be used as input to a k^2-raster.

In other words, for a particular pixel in the current band and the corresponding pixel in the reference band, δ_i being the residual value, y_i the actual value of the current band, and x_i the value of the reference band, to encode, the following equation is used:

$$\delta_i = y_i - (\hat{\alpha} \cdot x_i + \hat{\beta}) . \tag{4}$$

To decode, the following equation is used:

$$y_i = \delta_i + (\hat{\alpha} \cdot x_i + \hat{\beta}) . \tag{5}$$

The distance from the reference band affects the residual values. The closer the current band is to the reference band, the smaller the residual values would tend to be. We can arrange the bands into groups. For example, the first band can be chosen as the reference and the second, third and fourth bands will have their residual values calculated with respect to the first band. And the next group starts with the fifth band as the reference band, etc.

For this coding method, the group size (stored as a 2-byte short integer) as well as the $\hat{\alpha}$ and $\hat{\beta}$ values for each band (stored as 8-byte double's) will need to be saved for use in both the encoder and the decoder. Note that the size of these extra data is insignificant - which generally comes to around 3.5 kB - compared to the overall size of the structure.

2.7. Differential Method

In the differential encoding, which is a special case of the predictor where $\alpha = 1$ and $\beta = 0$, the residual value is obtained by simply taking the difference between the reference band and the current band. For a particular pixel in the current band and the corresponding pixel in the reference band, δ_i being the residual value, y_i the actual value of the current band, and x_i the value of the reference band, to encode, the following equation is used:

$$\delta_i = y_i - x_i . \tag{6}$$

To decode, the following equation is used:

$$y_i = \delta_i + x_i . \tag{7}$$

Like the predictor, we can use the first band as the reference band and the next several bands can use this reference band to find the residual values. Again, the grouping is repeated up to the last band. For this coding method, only the group size (stored as a 2-byte short integer) needs to be saved.

2.8. Related Work

Since the publication of the proposals on k^2-tree and k^2-raster, more research has been done to extend the capabilities of the structures to 3D where the first and second dimensions represent the spatial element and the third dimension the time element.

Based on their previous research of k^2-raster, Silva-Coira et al. [26] proposed a structure called Temporal k^2-raster (T $-$ k^2raster) which represents a time-series of rasters. It takes advantage of the fact that in a time-series, the values in a matrix M_1 are very close to, if not the same as, the next matrix M_2 or even the one after that, M_3, along the timeline. The matrices can then be grouped into τ time instants where the values of the elements of the first matrix in the group is subtracted from the corresponding ones in the current matrix. The result will be smaller integer values that would help form a more compact tree as there are likely to be more zeros in the tree than before. Their experimental results bear this out. When the τ value is small ($\tau = 4$), the sizes are small. However, as would be expected, the results are not as favorable when the τ value becomes larger ($\tau = 50$). Akin to the Temporal k^2-raster, the differential encoding on k^2-raster that we are proposing in this paper also exploits the similarity between neighboring matrices or bands in a hyperspectral image to form a more compact structure.

Another study on compact representation of raster images in a time-series was proposed earlier this year by Cruces et al. in [27]. This method is based on the 3D to 2D mapping of a raster where 3D tuples $<x, y, z>$ are mapped into a 2D binary grid. That is, a raster of size $w \times h$ with values in a certain range, between 0 and v inclusive will have a binary matrix of $w \times h$ columns and v+1 rows. All the rasters will then be concatenated into a 3D cube and stored as a k^3-tree.

3. Results

In this section we describe some of the experiments that were performed to show the use of compact data structures, prediction and differential encoding for real-time processing and compression. First, we show the results with other compression algorithms and techniques that are currently in use such as gzip, bzip2, xz, M-CALIC [15] and CCSDS 123.0-B-1 [9]. Then we compare the build time and the data access time for k^2-raster with and without prediction and differential encoding. Next, we show the results of different rates in k^2-raster that are produced as different k-values are applied. Similarly, the results of different group sizes for prediction and differential encoding are shown. Finally, the predictive method and the differential method are compared.

Experiments were conducted using hyperspectral images from different sensors: Atmospheric Infrared Sounder (AIRS), Airborne Visible/Infrared Imaging Spectrometer (AVIRIS), Compact Reconnaissance Imaging Spectrometer for Mars (CRISM), Hyperion, and Infrared Atmospheric Sounding Interferometer (IASI). Except for IASI, all of them are publicly available for download (http://cwe.ccsds.org/sls/docs/sls-dc/123.0-B-Info/TestData). Table 1 gives more detailed information on these images. The table also shows the bit-rate reduction for using k^2-raster with and without prediction. Performance in terms of bit rate and entropy is evaluated for them.

For best results in k^2-raster for the testing images, we used the optimal k-value, and also in the case of the predictor and the differential encoder, the optimal group size for each image was used. The effects of using different k-values and different group sizes will be discussed and tested in two of the subsections below.

To build the structure of k^2-raster and the cell query functions, a program in C was written. The algorithms presented in the paper by Ladra et al. [20] were the basis and reference for writing the code. The DACs software that was used in conjunction with our program is available at the Universidade da Coruña's Database Laboratory (Laboratorio de Bases de Datos) website (http://lbd.udc.es/research/DACS/). The package is called "DACs, optimization with no further restrictions". As for the predictive and differential methods, another C program was written to perform the tasks needed to give us the results that we will discuss below. All the code was compiled using gcc or g++ 5.4.0 20160609 with -Ofast optimization.

Table 1. Hyperspectral images used in our experiments. It also shows the bit rate and bit rate reduction using k^2-raster with and without the predictor. x is the image width, y the image height and z the number of spectral bands. The unit bpppb stands for bits per pixel per band.

Sensor	Name	C/U*	Acronym	Original Dimensions ($x \times y \times z$)	Bit Depth (bpppb)	Optimal k-Value	k^2-Raster Bit Rate (bpppb)	k^2-Raster Bit-Rate Reduction (%)	k^2-Raster+ Predictor Bit Rate (bpppb)	k^2-Raster+ Predictor Bit-Rate Reduction (%)
AIRS	9	U	AG9	$90 \times 135 \times 1501$	12	6	9.49	21%	6.76	44%
	16	U	AG16	$90 \times 135 \times 1501$	12	6	9.12	24%	6.63	45%
	60	U	AG60	$90 \times 135 \times 1501$	12	6	9.81	18%	7.06	41%
	126	U	AG126	$90 \times 135 \times 1501$	12	6	9.61	20%	7.05	41%
	129	U	AG129	$90 \times 135 \times 1501$	12	6	8.65	28%	6.47	46%
	151	U	AG151	$90 \times 135 \times 1501$	12	6	9.53	21%	7.02	41%
	182	U	AG182	$90 \times 135 \times 1501$	12	6	9.68	19%	7.19	40%
	193	U	AG193	$90 \times 135 \times 1501$	12	6	9.44	21%	7.06	41%
AVIRIS	Yellowstone sc. 00	C	ACY00	$677 \times 512 \times 224$	16	6	9.61	40%	6.87	57%
	Yellowstone sc. 03	C	ACY03	$677 \times 512 \times 224$	16	6	9.42	41%	6.72	58%
	Yellowstone sc. 10	C	ACY10	$677 \times 512 \times 224$	16	4	7.57	53%	5.84	64%
	Yellowstone sc. 11	C	ACY11	$677 \times 512 \times 224$	16	6	8.81	45%	6.52	59%
	Yellowstone sc. 18	C	ACY18	$677 \times 512 \times 224$	16	6	9.78	39%	7.04	56%
	Yellowstone sc. 00	U	AUY00	$680 \times 512 \times 224$	16	9	11.92	25%	9.04	44%
	Yellowstone sc. 03	U	AUY03	$680 \times 512 \times 224$	16	9	11.74	27%	8.87	45%
	Yellowstone sc. 10	U	AUY10	$680 \times 512 \times 224$	16	9	9.99	38%	8.00	50%
	Yellowstone sc. 11	U	AUY11	$680 \times 512 \times 224$	16	9	11.27	30%	8.77	45%
	Yellowstone sc. 18	U	AUY18	$680 \times 512 \times 224$	16	9	12.15	24%	9.29	42%
CRISM	frt000065e6_07_sc164	U	C164	$640 \times 420 \times 545$	12	6	10.08	16%	10.02	16%
	frt00008849_07_sc165	U	C165	$640 \times 450 \times 545$	12	6	10.37	14%	10.33	14%
	frt0001077d_07_sc166	U	C166	$640 \times 480 \times 545$	12	6	11.05	8%	11.08	8%
	hrl00004f38_07_sc181	U	C181	$320 \times 420 \times 545$	12	5	9.97	17%	9.52	21%
	hrl0000648f_07_sc182	U	C182	$320 \times 450 \times 545$	12	5	10.11	16%	9.84	18%
	hrl0000ba9c_07_sc183	U	C183	$320 \times 480 \times 545$	12	5	10.65	11%	10.59	12%

Table 1. *Cont.*

Sensor	Name	C/U*	Acronym	Original Dimensions ($x \times y \times z$)	Bit Depth (bpppb)	Optimal k-Value	k^2-raster Bit Rate (bpppb)	k^2-Raster Bit-Rate Reduction (%)	k^2-Raster+ Predictor Bit Rate (bpppb)	k^2-Raster+ Predictor Bit-Rate Reduction (%)
	Agricultural 2905 [†]	C	HCA1	$256 \times 2905 \times 242$	12	8	8.20	32%	7.47	38%
	Agricultural 3129 [†]	C	HCA2	$256 \times 3129 \times 242$	12	8	8.08	33%	7.50	37%
	Coral Reef [†]	C	HCC	$256 \times 3127 \times 242$	12	8	7.38	39%	7.41	38%
	Urban [†]	C	HCU	$256 \times 2905 \times 242$	12	8	8.59	28%	7.83	35%
Hyperion	Filtered Erta Ale [†]	U	HFUEA	$256 \times 3187 \times 242$	12	8	6.84	43%	5.99	50%
	Filtered Lake Monona [†]	U	HFULM	$256 \times 3176 \times 242$	12	8	6.79	43%	6.06	49%
	Filtered Mt. St. Helena [†]	U	HFUMS	$256 \times 3242 \times 242$	12	8	6.78	43%	5.88	51%
	Erta Ale [†]	U	HUEA	$256 \times 3187 \times 242$	12	8	7.57	37%	6.99	42%
	Lake Monona [†]	U	HULM	$256 \times 3176 \times 242$	12	8	7.52	37%	7.08	41%
	Mt. St. Helena [†]	U	HUMS	$256 \times 3242 \times 242$	12	8	7.49	38%	6.93	42%
IASI	Level 0 1 [‡]	U	I01	$60 \times 1528 \times 8359$	12	4	5.93	51%	4.69	61%
	Level 0 2 [‡]	U	I02	$60 \times 1528 \times 8359$	12	4	5.90	51%	4.75	60%
	Level 0 3 [‡]	U	I03	$60 \times 1528 \times 8359$	12	4	5.42	55%	4.58	62%
	Level 0 4 [‡]	U	I04	$60 \times 1528 \times 8359$	12	4	6.23	48%	4.90	59%

*: Calibrated or Uncalibrated; †: Cropped to $256 \times 512 \times 242$; ‡: Cropped to $60 \times 256 \times 8359$.

The machine that these experiments ran on has an Intel Core 2 Duo CPU E7400 @2.80GHz with 3072KB of cache and 3GB of RAM. The operating system is Ubuntu 16.04.5 LTS with kernel 4.15.0-47-generic (64 bits).

To ensure that there was no loss of information, the image was reconstructed by reverse transformation and verified to be identical to the original image in the case of predictive and differential methods. For k^2-raster, after saving the structure to disk, we made sure that the original image could be reconstructed from the saved data.

3.1. Comparison with Other Compression Algorithms

Both k^2-raster with and without predictive and differential encoding were compared to other commonly-used compression algorithms such as gzip, bzip2, xz, and specialized algorithms such as M-CALIC and CCSDS 123.0-B-1. The results for the comparison are shown in Table 2 and depicted in Figure 12.

It can be seen that k^2-raster alone already performed better than gzip. When it was used with the predictor, it produced a bit rate that was basically on a par with and sometimes better than other compression algorithms such as xz or bzip2. However, it could not attain the bit-rate level done by CCSDS 123.0-B-1 or M-CALIC. This was to be expected as both are specialized compression techniques, and CCSDS 123.0-B-1 is considered a baseline against which all compression algorithms for hyperspectral images are measured. Nevertheless, k^2-raster provides direct access to the elements without full decompression, and this is undoubtedly the major advantage it has over all the aforementioned compression algorithms.

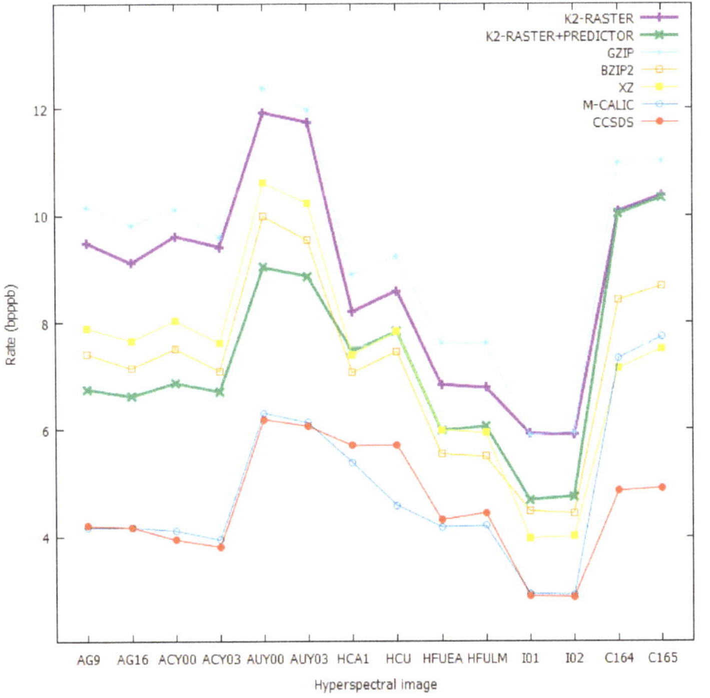

Figure 12. A rate (bpppb) comparison with other compression techniques.

Table 2. A rate (bpppb) comparison with other compression techniques. The optimal values for all compression algorithms (except for M-CALIC, CCSDS 123.0-B-1) are highlighted in red. Results for CCSDS 123.0-B-1 are from [28].

Sensor	Name	C/U *	Acronym	k^2-Raster	k^2-Raster + Predictor	k^2-Raster + Differential	gzip	bzip2	xz	M-CALIC	CCSDS 123.0-B-1
						Compression Technique (bpppb)					
AIRS	9	U	AG9	9.49	6.76	7.52	10.16	7.42	7.90	4.19	4.21
	16	U	AG16	9.12	6.63	7.29	9.82	7.15	7.66	4.19	4.18
	60	U	AG60	9.81	7.06	7.82	10.53	7.71	8.23	4.41	4.36
	126	U	AG126	9.61	7.05	7.78	10.33	7.64	8.10	4.39	4.38
	129	U	AG129	8.65	6.47	6.96	9.50	6.68	7.22	4.08	4.12
	151	U	AG151	9.53	7.02	7.74	10.31	7.43	7.97	4.39	4.41
	182	U	AG182	9.68	7.19	7.94	10.64	7.79	8.33	4.45	4.42
	193	U	AG193	9.44	7.06	7.77	10.15	7.47	7.94	4.42	4.42
AVIRIS	Yellowstone sc. 00	C	ACY00	9.61	6.87	7.79	10.12	7.51	8.04	4.12	3.95
	Yellowstone sc. 03	C	ACY03	9.42	6.72	7.65	9.59	7.10	7.62	3.95	3.82
	Yellowstone sc. 10	C	ACY10	7.57	5.84	6.26	7.41	5.30	5.73	3.31	3.36
	Yellowstone sc. 11	C	ACY11	8.81	6.52	6.85	9.04	6.65	7.07	3.71	3.63
	Yellowstone sc. 18	C	ACY18	9.78	7.04	7.53	10.00	7.45	7.95	4.09	3.90
	Yellowstone sc. 00	U	AUY00	11.92	9.04	10.04	12.39	9.99	10.61	6.32	6.20
	Yellowstone sc. 03	U	AUY03	11.74	8.87	9.91	11.98	9.54	10.23	6.14	6.07
	Yellowstone sc. 10	U	AUY10	9.99	8.00	8.57	10.17	7.71	8.40	5.53	5.58
	Yellowstone sc. 11	U	AUY11	11.27	8.77	9.21	11.49	9.08	9.66	5.91	5.84
	Yellowstone sc. 18	U	AUY18	12.15	9.29	9.92	12.29	9.90	10.58	6.33	6.21
CRISM	frt000065e6_07_sc164	U	C164	10.08	10.02	10.06	10.98	8.42	7.15	7.34	4.86
	frt00008849_07_sc165	U	C165	10.37	10.33	10.37	11.03	8.68	7.51	7.73	4.91
	frt0001077d_07_sc166	U	C166	11.05	11.08	11.14	11.20	9.04	7.64	8.44	5.44
	hrl00004f38_07_sc181	U	C181	9.97	9.52	9.52	10.77	8.28	8.20	7.09	4.27
	hrl0000648f_07_sc182	U	C182	10.11	9.84	9.86	10.90	8.53	7.90	7.28	4.49
	hrl0000ba9c_07_sc183	U	C183	10.65	10.59	10.64	10.87	8.52	7.28	7.91	4.96

Table 2. *Cont.*

Sensor	Name	C/U [*]	Acronym	Compression Technique (bpppb)							
				k^2-Raster	k^2-Raster + Predictor	k^2-Raster + Differential	gzip	bzip2	xz	M-CALIC	CCSDS 123.0-B-1
	Agricultural 2905 [†]	C	HCA1	8.20	7.47	7.47	8.90	7.07	7.40	5.39	-
	Agricultural 3129 [†]	C	HCA2	8.08	7.50	7.50	8.84	7.04	7.35	5.28	5.70
	Coral Reef [†]	C	HCC	7.38	7.41	7.41	7.45	5.74	5.90	4.59	5.42
	Urban [†]	C	HCU	8.59	7.83	7.83	9.24	7.46	7.83	5.25	5.71
Hyperion	Filtered Erta Ale [†]	U	HFUEA	6.84	5.99	6.15	7.63	5.55	6.00	4.19	4.32
	Filtered Lake Monona [†]	U	HFULM	6.79	6.06	6.18	7.61	5.50	5.94	4.21	4.45
	Filtered Mt. St. Helena [†]	U	HFUMS	6.78	5.88	6.15	7.18	5.44	5.74	4.11	4.35
	Erta Ale [†]	U	HUEA	7.57	6.99	7.06	8.69	6.41	6.73	4.87	4.32
	Lake Monona [†]	U	HULM	7.52	7.08	7.13	8.69	6.46	6.74	4.94	4.45
	Mt. St. Helena [†]	U	HUMS	7.49	6.93	7.04	8.26	6.28	6.48	4.82	4.36
IASI	Level 0 1 [‡]	U	I01	5.93	4.69	5.01	5.90	4.48	3.98	2.94	2.89
	Level 0 2 [‡]	U	I02	5.90	4.75	5.03	5.96	4.44	4.01	2.92	2.88
	Level 0 3 [‡]	U	I03	5.42	4.58	4.79	5.25	3.94	3.75	2.92	2.88
	Level 0 4 [‡]	U	I04	6.23	4.90	5.20	6.30	4.71	4.24	2.97	2.90

[*]: Calibrated or Uncalibrated; [†]: Cropped to 256 × 512 × 242 except for CCSDS 123.0; [‡]: Cropped to 60 × 256 × 8359 except for CCSDS 123.0.

3.2. Build Time

Both the time to build the k^2-raster only and the time to build k^2-raster with predictive and differential preprocessing were measured. They were then compared against the time to compress the data with gzip. The results are presented in Table 3. We can see that the build time for k^2-raster only took half as long as with gzip. Comparing the predictive and the differential methods, the time difference is small although it generally took longer to build the former than the latter due to the additional time needed to compute the values of $\hat{\alpha}$ and $\hat{\beta}$. Both, however, still took less time to build than gzip compression.

Table 3. A comparison of build time (in seconds) using k^2-raster only and k^2-raster with predictive and differential methods.

Hyperspectral Image	Build Time (s)			Gzip Compression (s)
	k^2-Raster	k^2-Raster + Predictor	k^2-Raster + Differential	
AG9	1.86	2.23	2.12	3.18
AG16	1.78	2.22	2.09	3.49
ACY00	8.32	10.11	9.49	15.01
ACY03	8.26	10.00	9.47	15.32
AUY00	5.56	7.39	6.84	12.10
AUY03	5.59	7.38	6.76	12.68
C164	17.84	21.32	21.59	27.94
C165	17.89	22.83	22.92	30.83
HCA1	1.98	2.67	2.47	5.59
HCA2	1.98	2.64	2.42	5.80
HFUEA	2.38	3.01	3.05	7.59
HFULM	2.41	3.04	2.87	7.57
HFUMS	2.33	2.95	2.76	8.26
I01	14.58	18.62	16.56	31.59
I02	14.66	17.49	16.66	29.64

3.3. Access Time

Several tests were conducted to see what the access time was like to query the cells in each image and we found that the time for a random cell access took longer for a predictor compared to just using the k^2-raster. This was expected but we should bear in mind that the bit rates are reduced when a predictor is used, thus decreasing storage size and transmission rate. Note that the last column also lists the time to decompress a gzip image file and it took at least 4 or 5 times longer than using a predictor to randomly access the data 10^5 times. Table 4 shows the results of access time in milliseconds for 100,000 iterations of random cell query done by getCell(), a function which was described in the paper from Ladra et al. [20] for accessing pixel values in a k^2-raster.

Table 4. A comparison of access time (in milliseconds) using k^2-raster only and k^2-raster with predictive and differential encoders.

Hyperspectral Image	100,000 Iterations of Random Access (ms)			Gzip Decompression (ms)
	k^2-Raster	k^2-Raster + Predictor	k^2-Raster + Differential	
AG9	90	125	92	474
AG16	85	121	85	459
ACY00	275	485	426	1949
ACY03	269	474	424	1912
AUY00	151	489	402	1941
AUY03	151	485	402	1957
C164	273	400	381	4048
C165	301	420	397	4382
HCA1	77	131	127	735
HCA2	76	121	118	737
HFUEA	93	150	129	684
HFULM	92	148	129	680
HFUMS	91	146	134	670
I01	155	222	244	2517
I02	168	236	255	2396

3.4. Use of Different k-Values

With k^2-raster, we found that different k-values used in the structure would produce different bit rates and different access time. In general, for most of our testing images the k-value is at its optimal bit-rate level when it is between 4 and 9. The reason is that as the k-value increases, the height of the constructed tree becomes smaller. Therefore, the number of nodes in the tree will decrease and so will the size of the bitmaps *Lmax* and *Lmin* that need to be stored in the structure. Table 5 shows the bit rates of some of the testing images between $k = 2$ and $k = 20$. Additionally, experiments show that as the k-value becomes higher, the access time also becomes shorter, as can be seen in Table 6. As the k-value gets larger, the tree becomes shorter, thus making it faster to traverse from the top level to a lower level when searching for a particular node in the tree. As there is a trade-off between storage size and access time, for the experiments, the k-value that produces the lowest bit rate for the image was used.

For those who would like to know which k-value would give the best or close to the best rate, we recommend them to use a value of 6 as a general rule. This can be seen from Table 5 where the difference in the rate produced by this value and the one by the optimal k-value averages out to be only about 0.19 bpppb.

3.5. Use of Different Group Sizes

Tests were performed to see how the group size affects the predictive and differential methods. The group sizes were 2, 4, 8, 12, 16, 20, 24, 28 and 32. The results in Table 7 and Figure 13 show that for most images, they are at their optimal bit rates when the size is 4 or 8. The best bit-rate values are highlighted in red. For the range of group size tested, we can also see that except for the CRISM scenes (which consist of pixels with low spatial correlation, thus leading to inaccurate prediction), the bit rates for the predictor are always lower than the ones for differential encoding, irrespective of the group size.

For users who are interested in knowing which group size is the best to apply to the predictive and differential methods, a size of 4 is recommended for general use as the difference in bit rate produced by this group size and the one by the optimal group size averages out to be about 0.06 bpppb.

For the rest of the experiments, the optimal group size for each image was used to obtain the bit rate.

Table 5. Rates (bpppb) for different k-values for some of the testing images. The k-value with the lowest rate is in red.

Hyperspectral Image	$k = 2$	3	4	5	6	7	8	9	10	11	12	13	14	15	16	17	18	19	20
AG9	13.06	10.11	10.03	10.47	9.49	9.98	10.68	9.89	10.65	-	11.23	10.33	11.29	9.53	11.57	11.72	10.78	12.52	12.13
AG16	12.72	9.78	9.66	10.11	9.12	9.57	10.32	9.51	10.29	-	10.82	9.98	10.86	9.17	11.11	11.28	10.32	12.07	11.68
ACY00	12.34	10.20	9.76	-	9.61	9.91	-	9.69	9.83	9.87	9.95	10.24	10.20	-	-	-	-	-	-
ACY03	11.81	9.87	9.56	-	9.42	9.71	-	9.50	9.65	9.70	9.76	10.01	9.98	-	-	-	-	-	-
AUY00	15.31	12.93	12.20	-	12.08	12.35	-	11.92	12.11	12.13	12.17	12.52	12.43	-	-	-	-	-	-
AUY03	15.03	12.60	12.00	-	11.90	12.20	-	11.74	11.93	11.94	12.00	12.34	12.25	-	-	-	-	-	-
C164	12.60	10.42	10.17	-	10.08	-	-	10.34	10.20	10.76	10.48	-	-	-	-	-	-	-	-
C165	12.84	10.67	10.48	-	10.37	-	-	10.54	10.51	10.79	11.03	-	-	-	-	-	-	-	-
HCA1	10.79	9.41	8.85	8.45	8.74	9.36	8.20	8.51	8.68	8.85	8.88	8.92	9.21	-	-	-	-	-	-
HCC	9.43	8.12	7.79	7.41	7.75	8.40	7.38	7.67	7.85	8.06	8.12	8.26	8.56	-	-	-	-	-	-
HFUEA	8.82	7.80	7.30	7.24	7.41	8.07	6.84	7.25	7.43	7.66	7.68	7.71	8.07	-	-	-	-	-	-
HFULM	8.69	7.70	7.20	7.13	7.33	8.02	6.79	7.21	7.40	7.64	7.66	7.68	8.05	-	-	-	-	-	-
I01	8.03	-	5.93	-	-	6.45	-	-	-	-	-	-	-	-	6.59	7.79	8.30	8.73	6.36
I02	8.02	-	5.90	-	-	6.48	-	-	-	-	-	-	-	-	6.64	7.92	8.46	8.97	6.45

Table 6. Access time (ms) for different *k*-values for some of the testing images. The best access time is in red.

Hyperspectral Image	Access Time (ms)																		
	$k=2$	3	4	5	6	7	8	9	10	11	12	13	14	15	16	17	18	19	20
AG9	345	167	130	114	91	84	83	82	82	-	60	59	56	56	55	52	60	51	47
AG16	334	152	114	108	85	79	78	80	75	-	55	54	53	53	59	51	58	48	45
ACY00	3553	1085	573	-	291	225	-	152	133	125	114	110	104	-	-	-	-	-	-
ACY03	3521	1112	572	-	277	223	-	149	131	122	112	120	102	-	-	-	-	-	-
AUY00	3569	1135	592	-	292	228	-	153	133	126	115	113	106	-	-	-	-	-	-
AUY03	3559	1123	585	-	279	221	-	152	133	124	115	109	103	-	-	-	-	-	-
C164	2924	964	606	-	272	-	-	159	161	138	131	-	-	-	-	-	-	-	-
C165	3754	1017	555	-	290	-	-	178	156	152	145	-	-	-	-	-	-	-	-
HCA1	1179	384	213	154	124	106	80	78	69	71	62	61	62	-	-	-	-	-	-
HCC	1203	406	233	172	139	123	95	94	86	85	83	79	79	-	-	-	-	-	-
HFUEA	1409	465	262	184	148	127	93	95	89	84	77	76	76	-	-	-	-	-	-
HFULM	1427	467	262	193	155	130	94	96	87	90	79	81	80	-	-	-	-	-	-
I01	999	-	779	-	-	679	-	-	-	-	-	-	-	-	610	709	715	728	450
I02	1047	-	759	-	-	698	-	-	-	-	-	-	-	-	651	746	746	730	472

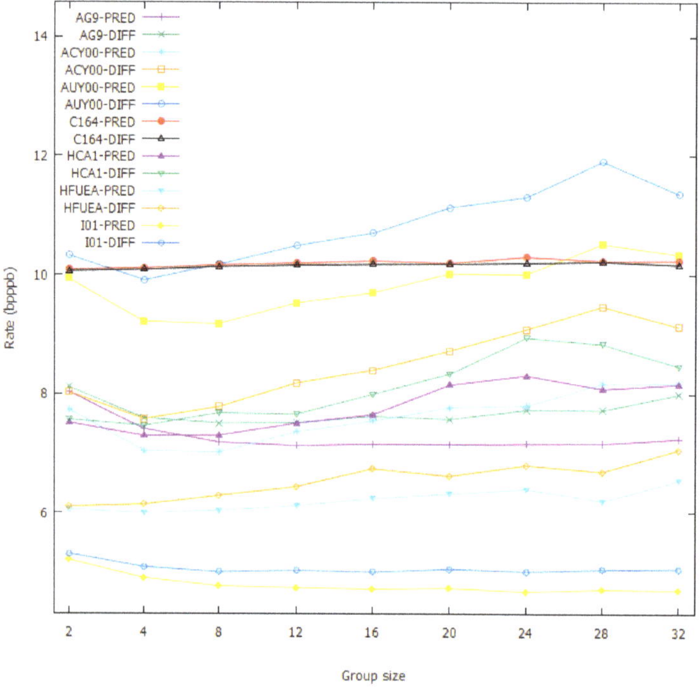

Figure 13. A rate (bpppb) comparison of different group sizes.

Table 7. A rate (bpppb) comparison of different group sizes using the predictive and the differential methods. The optimal values are highlighted in red.

Hyperspectral Image	Group Size								
	2	4	8	12	16	20	24	28	32
AG9-Pred	8.03	7.41	7.19	7.13	7.16	7.15	7.17	7.17	7.25
AG9-Diff	8.11	7.60	7.51	7.52	7.63	7.58	7.74	7.73	8.00
ACY00-Pred	7.74	7.05	7.03	7.36	7.55	7.78	7.81	8.18	8.18
ACY00-Diff	8.03	7.58	7.79	8.18	8.40	8.73	9.09	9.48	9.13
AUY00-Pred	9.94	9.22	9.18	9.53	9.70	10.03	10.02	10.53	10.34
AUY00-Diff	10.33	9.91	10.18	10.49	10.71	11.14	11.32	11.91	11.36
C164-Pred	10.08	10.11	10.17	10.20	10.23	10.20	10.31	10.24	10.24
C164-Diff	10.06	10.08	10.13	10.16	10.18	10.20	10.31	10.22	10.17
HCA1-Pred	7.51	7.30	7.31	7.51	7.65	8.15	8.31	8.09	8.16
HCA1-Diff	7.57	7.47	7.68	7.66	8.00	8.34	8.95	8.84	8.47
HFUEA-Pred	6.07	6.01	6.05	6.13	6.25	6.33	6.41	6.21	6.55
HFUEA-Diff	6.11	6.15	6.30	6.44	6.75	6.63	6.81	6.70	7.07
I01-Pred	5.22	4.91	4.78	4.75	4.73	4.74	4.69	4.73	4.71
I01-Diff	5.32	5.10	5.01	5.04	5.01	5.06	5.02	5.05	5.06

3.6. Predictive and Differential Methods

The proposed differential and predictive methods were used to transform these images into data with lower bit rates. They were then used as input to k^2-raster to further reduce their bit rates. Their performance was compared together with Reversible Haar Transform at levels 1 and 5, and the results are presented in Table 8. Figure 14 shows the entropy comparison of Yellowstone03 using differential and predictive methods while Figure 15 shows the bit rate comparison between the two methods. Both show us that the proposed algorithm has brought benefits by lowering the entropy and the bit rates. The data for reference bands are left out of the plots so that the reader can have a clearer overall picture of the bit rate comparison.

Compared to other methods, the predictive method outperforms others, with the exception of Reversible Haar Transform level 5. However, it should be noted that while the predictive and differential methods require only two pixels (reference pixel and current pixel) to perform the reverse transformation, it would be a much more involved process to decode data using Reversible Haar Transform at a higher level. The experiments show that for all the testing images, the predictive method in almost all bands perform better than the differential method. This can be explained by the fact that in predictive encoding the values of α and β in Equation (1) take into account not only the spectral correlation, but also the spatial correlation between the pixels in the bands when determining the prediction values. This is not the case with differential encoding whose values are only taken from the spectral correlation.

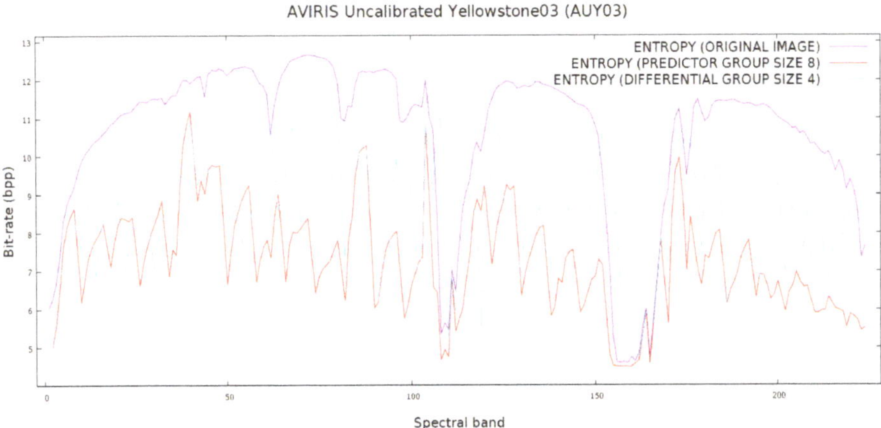

Figure 14. An entropy comparison of Yellowstone03 using differential and predictive methods. Data for reference bands are not included.

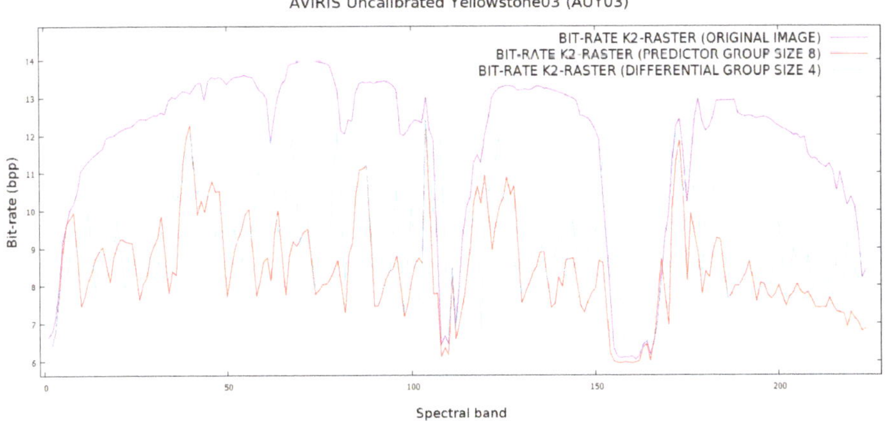

Figure 15. A bit rate comparison of Yellowstone03 using differential and predictive methods on k^2-raster. Data for reference bands are not included.

Table 8. A rate (bpppb) comparison using different transformed methods: predictor, differential, reversible Haar level 1 and reversible Haar level 5 on k^2-raster. The optimal values are highlighted in red.

Hyperspectral Image	Transformation Type				
	Without Transformation	Predictor	Differential	Reversible Haar (Level 1)	Reversible Haar (Level 5)
AG9	9.49	6.76	7.52	8.10	6.83
AG16	9.12	6.63	7.29	7.81	6.60
ACY00	9.63	6.87	7.79	8.01	7.00
ACY03	9.44	6.72	7.65	7.86	6.87
AUY00	11.92	9.04	10.04	10.33	9.35
AUY03	11.74	8.87	9.91	10.18	9.23
C164	10.08	10.02	10.06	10.01	9.83
C165	10.37	10.33	10.37	10.33	10.16
HCA1	8.20	7.47	7.47	7.37	7.05
HCC	7.38	7.50	7.50	6.71	6.54
HFUEA	6.84	5.99	6.15	7.12	6.75
HFULM	6.79	6.06	6.18	7.14	6.83
I01	5.93	4.69	5.01	5.26	4.54
I02	5.90	4.75	5.03	5.26	4.57

4. Conclusions

In this work, we have shown that using k^2-raster structure can help reduce the bit rates of a hyperspectral image. It also provides easy access to its elements without the need for initial full decompression. The predictive and differential methods can be applied to further reduce the rates. We performed experiments that showed that if the image data are first converted by either a predictive method or a differential method, we can gain more reduction in bit rates, thus making the storage capacity or the transmission volume of the data even smaller. The results of the experiments verified that the predictor indeed gives a better reduction in bit rates than the differential encoder and is preferred to be used for hyperspectral images.

For future work, we are interested in exploring the possibility of modifying the elements in a k^2-raster. This investigation is based on the dynamic structure, dk^2-tree, as discussed in the papers by de Bernardo et al. [29,30]. Additionally, we would like to improve on the variable-length encoding which is currently in use with k^2-raster, and hope to further reduce the size of the structure [23,24].

Author Contributions: Conceptualization, K.C., D.E.O.T., I.B. and J.S.-S.; methodology, K.C., D.E.O.T., I.B. and J.S.-S.; software, K.C.; validation, K.C., I.B. and J.S.-S.; formal analysis, K.C., D.E.O.T., I.B. and J.S.-S.; investigation, K.C., D.E.O.T., I.B. and J.S.-S.; resources, K.C., D.E.O.T., I.B. and J.S.-S.; data curation, K.C., I.B. and J.S.-S.; writing—original draft preparation, K.C., I.B. and J.S.-S.; writing—review and editing, K.C., I.B. and J.S.-S.; visualization, K.C., I.B. and J.S.-S.; supervision, I.B. and J.S.-S.; project administration, I.B. and J.S.-S.; funding acquisition, I.B. and J.S.-S.

Funding: This research was funded by the Spanish Ministry of Economy and Competitiveness and the European Regional Development Fund under grants RTI2018-095287-B-I00 and TIN2015-71126-R (MINECO/FEDER, UE) and BES-2016-078369 (Programa Formación de Personal Investigador), and by the Catalan Government under grant 2017SGR-463.

Acknowledgments: The authors would like to thank Magli et al. for the M-CALIC software that they provided us in order to perform some of the experiments in this research work.

Conflicts of Interest: The authors declare no conflict of interest.

Abbreviations

The following abbreviations are used in this manuscript:

AIRS	Atmospheric Infrared Sounder
AVIRIS	Airborne Visible InfraRed Imaging Spectrometer
CALIC	Context Adaptive Lossless Image Compression
CCSDS	Consultative Committee for Space Data Systems
CRISM	Compact Reconnaissance Imaging Spectrometer for Mars
DACs	Directly Addressable Codes
IASI	Infrared Atmospheric Sounding Interferometer
JPEG 2000	Joint Photographic Experts Group 2000
KLT	Karhunen–Loève Theorem
LOUDS	Level-Order Unary Degree Sequence
MDPI	Multidisciplinary Digital Publishing Institute
PCA	Principal Component Analysis
SOAP	Short Oligonucleotide Analysis Package

References

1. Navarro, G. *Compact Data Structures: A Practical Approach*; Cambridge University Press: Cambridge, UK, 2016.
2. Jacobson, G. Succinct Static Data Structures. Ph.D. Thesis, Carnegie-Mellon, Pittsburgh, PA, USA, 1988.
3. Jacobson, G. Space-efficient static trees and graphs. In Proceedings of the Annual Symposium on Foundations of Computer Science (FOCS), Research Triangle Park, NC, USA, 30 October–1 November 1989; pp. 549–554.
4. Grossi, R.; Gupta, A.; Vitter, J.S. High-order entropy-compressed text indexes. In Proceedings of the 14th Annual ACM-SIAM Symposium on Discrete Algorithms, Baltimore, MD, USA, 12–14 January 2003; Volume 72, pp. 841–850.
5. Ferragina, P.; Manzini, G. Opportunistic data structures with applications. In Proceedings of the 41st Annual Symposium on Foundations of Computer Science, Redondo Beach, CA, USA, 12–14 November 2000; p. 390.
6. Burrows, M.; Wheeler, D. *A Block Sorting Lossless Data Compression Algorithm*; Technical Report; Digital Equipment Corporation: Maynard, MA, USA, 1994.
7. Langmead, B.; Trapnell, C.; Pop, M.; Salzberg, S.L. Ultrafast and memory-efficient alignment of short DNA sequences to the human genome. *Genome Biol.* **2009**, *10*, R25. [CrossRef] [PubMed]
8. Li, R.; Yu, C.; Li, Y.; Lam, T.; Yiu, S.; Kristiansen, K.; Wang, J. SOAP2: an improved ultrafast tool for short read alignment. *Bioinformatics* **2009**, *25*, 1966–1967. [CrossRef] [PubMed]
9. Consultative Committee for Space Data Systems (CCSDS). *Image Data Compression CCSDS 123.0-B-1*; Blue Book; CCSDS: Washington, DC, USA, 2012.
10. Jolliffe, I.T. *Principal Component Analysis*; Springer: Berlin, Germany, 2002; p. 487.
11. Taubman, D.S.; Marcellin, M.W. *JPEG 2000: Image Compression Fundamentals, Standards and Practice*; Kluwer Academic Publishers: Boston, MA, USA, 2001.
12. Wu, X.; Memon, N. CALIC—A context based adaptive lossless image CODEC. In Proceedings of the 1996 IEEE International Conference on Acoustics, Speech, and Signal Processing Conference Proceedings, Atlanta, GA, USA, 9 May 1996.
13. Wu, X.; Memon, N. Context-based adaptive, lossless image coding. *IEEE Trans. Commun.* **1997**, *45*, 437–444. [CrossRef]
14. Wu, X.; Memon, N. Context-based lossless interband compression—Extending CALIC. *IEEE Trans. Image Process.* **2000**, *9*, 994–1001. [PubMed]
15. Magli, E.; Olmo, G.; Quacchio, E. Optimized onboard lossless and near-lossless compression of hyperspectral data using CALIC. *IEEE Geosci. Remote Sens. Lett.* **2004**, *1*, 21–25. [CrossRef]
16. Kiely, A.; Klimesh, M.; Blanes, I.; Ligo, J.; Magli, E.; Aranki, N.; Burl, M.; Camarero, R.; Cheng, M.; Dolinar, S.; et al. The new CCSDS standard for low-complexity lossless and near-lossless multispectral and hyperspectral image compression. In Proceedings of the 6th International WS on On-Board Payload Data Compression (OBPDC), ESA/CNES, Matera, Italy, 20–21 September 2018.

17. Fjeldtvedt, J.; Orlandić, M.; Johansen, T.A. An efficient real-time FPGA implementation of the CCSDS-123 compression standard for hyperspectral images. *IEEE J. Sel. Top. Appl. Earth Obs. Remote Sens.* **2018**, *11*, 3841–3852. [CrossRef]

18. Báscones, D.; González, C.; Mozos, D. Hyperspectral image compression using vector quantization, PCA and JPEG2000. *Remote Sens.* **2018**, *10*, 907. [CrossRef]

19. Guerra, R.; Barrios, Y.; Díaz, M.; Santos, L.; López, S.; Sarmiento, R. A new algorithm for the on-board compression of hyperspectral images. *Remote Sens.* **2018**, *10*, 428. [CrossRef]

20. Ladra, S.; Paramá, J.R.; Silva-Coira, F. Scalable and queryable compressed storage structure for raster data. *Inf. Syst.* **2017**, *72*, 179–204. [CrossRef]

21. Samet, H. The Quadtree and related hierarchical data structures. *ACM Comput. Surv. (CSUR)* **1984**, *16*, 187–260. [CrossRef]

22. Brisaboa, N.R.; Ladra, S.; Navarro, G. k^2-trees for compact web graph representation. In *International Symposium on String Processing and Information Retrieval*; Springer: Berlin/Heidelberg, Germany, 2009; pp. 18–30.

23. Brisaboa, N.R.; Ladra, S.; Navarro, G. DACs: Bringing direct access to variable-length codes. *Inf. Process Manag.* **2013**, *49*, 392–404. [CrossRef]

24. Brisaboa, N.R.; Ladra, S.; Navarro, G. Directly addressable variable-length codes. In *International Symposium on String Processing and Information Retrieval*; Springer: Berlin/Heidelberg, Germany, 2009; pp. 122–130.

25. Silva-Coira, F. Compact Data Structures for Large and Complex Datasets. Ph.D. Thesis, Universidade da Coruña, A Coruña, Spain, 2017.

26. Cerdeira-Pena, A.; de Bernardo, G.; Fariña, A.; Paramá, J.R.; Silva-Coira, F. Towards a compact representation of temporal rasters. In *String Processing and Information Retrieval*; SPIRE 2018; Lecture Notes in Computer Science; Springer: Cham, Switzerland, 2018; Volume 11147.

27. Cruces, N.; Seco, D.; Gutiérrez, G. A compact representation of raster time series. In Proceedings of the Data Compression Conference (DCC) 2019, Snowbird, UT, USA, 26–29 March 2019; pp. 103–111.

28. Álvarez Cortés, S.; Serra-Sagristà, J.; Bartrina-Rapesta, J.; Marcellin, M. Regression Wavelet Analysis for Near-Lossless Remote Sensing Data Compression. *IEEE Trans. Geosci. Remote Sens.* **2019**. [CrossRef]

29. De Bernardo, G.; Álvarez García, S.; Brisaboa, N.R.; Navarro, G.; Pedreira, O. Compact querieable representations of raster data. In *International Symposium on String Processing and Information Retrieval*; Springer: Cham, Switzerland, 2013; pp. 96–108.

30. Brisaboa, N.R.; De Bernardo, G.; Navarro, G. Compressed dynamic binary relations. In Proceedings of the 2012 Data Compression Conference, Snowbird, UT, USA, 10–12 April 2012; pp. 52–61.

Article

Compression of Hyperspectral Scenes through Integer-to-Integer Spectral Graph Transforms [†]

Dion Eustathios Olivier Tzamarias *, Kevin Chow, Ian Blanes and Joan Serra-Sagristà

Department of Information and Communications Engineering, Universitat Autònoma de Barcelona, Campus UAB, 08193 Cerdanyola del Vallès, Barcelona, Spain; kevin.chow@uab.cat (K.C.); ian.blanes@uab.cat (I.B.); joan.serra@uab.cat (J.S.-S.)

* Correspondence: dion.tzamarias@uab.cat
† This paper is an extended version of our paper published in in the Proceedings of the 6th ESA/CNES International Workshop on On-Board Payload Data Compression (OBPDC), Matera, Italy, 20–21 September 2018, titled *Compression Of hyperspectral images with graph wavelets*.

Received: 30 August 2019; Accepted: 26 September 2019; Published: 30 September 2019

Abstract: Hyperspectral images are depictions of scenes represented across many bands of the electromagnetic spectrum. The large size of these images as well as their unique structure requires the need for specialized data compression algorithms. The redundancies found between consecutive spectral components and within components themselves favor algorithms that exploit their particular structure. One novel technique with applications to hyperspectral compression is the use of spectral graph filterbanks such as the GraphBior transform, that leads to competitive results. Such existing graph based filterbank transforms do not yield integer coefficients, making them appropriate only for lossy image compression schemes. We propose here two integer-to-integer transforms that are used in the biorthogonal graph filterbanks for the purpose of the lossless compression of hyperspectral scenes. Firstly, by applying a Triangular Elementary Rectangular Matrix decomposition on GraphBior filters and secondly by adding rounding operations to the spectral graph lifting filters. We examine the merit of our contribution by testing its performance as a spatial transform on a corpus of hyperspectral images; and share our findings through a report and analysis of our results.

Keywords: hyperspectral image coding; graph filterbanks; integer-to-integer transforms; graph signal processing

---

1. Introduction

Hyperspectral images are representations of scenes across many bands of the electromagnetic spectrum (otherwise called spectral components). These images have been used to classify areas of a landscape [1], identify clouds [2] or seeds located on the Earth's surface [3] and even forecast geological events such as landslides [4]. Due to their large size, these images require the use of efficient compression techniques in the form of specialized image transforms. The above mentioned transforms can often be computationally costly but one could strive towards real-time compression of remotely-sensed hyperspectral images, when processed accordingly through on-ground, high-performance parallel computing facilities.

Transform-based lossy compression techniques used on hyperspectral images are commonly related to one of two families. The first group of algorithms are based on the Discrete Wavelet transform (DWT) [5–8]. On one hand the advantage of these transforms manifests through their low computational complexity, though they are not adaptive to the input signal. On the other hand, algorithms such as those in References [9–12] are based on the Karhunen-Loeve transform (KLT), which is adaptive at the expense of a high computational complexity. Meanwhile the techniques used for the lossless compression of hyperspectral images are generally based on a predictive coding model [13,14] though lossy predictive techniques [15] have the benefit of a lower computational complexity. Some lossless predictive techniques [16,17] are implemented in the CCSDS-123.0-B-2 [18] standard that also supports near-lossless compression and whose parameter tuning is explored in Reference [19].

This paper exploits a recently emerged family of wavelet transforms which are based on graph signal processing, a field in which data structures are represented as signals lying on the nodes of graphs while weighted graph edges denote the degree of similarity between nodes [20,21]. Graph wavelet transforms are derived from the graph structures, which in turn, one can also construct from images. Applying such transforms on hyperspectral images has been shown to lead to competitive compression results [22].

The strength of graph wavelet transforms in the field of image compression stems from the Graph Fourier Transform (GFT), an equivalent to the classical Fourier transform that is formed by the eigenvectors of the graph Laplacian, a matrix constructed from the values of the weighted edges of the graph. Just like the KLT, the GFT is an adaptive transform but instead of following a statistical approach, it attempts to encode image structures that are embedded in the graph edges. Thus, an important attribute of the GFT is related to its flexibility, since one can decide on the degree of accuracy with which image structures are represented on the graph [23]. It has also been shown [24] that not only does the GFT approximate the KLT for a piece-wise first-order autoregressive process but also that the GFT optimally decorrelates images following a Gauss-Markov random field model [25]. Consequently, graph wavelet transforms combine the GFT with a multiresolution analysis, resulting in a powerful tool for image compression.

A brief review of graph wavelet transforms follows. One can organize these transforms into two general groups—vertex domain and spectral domain designs. The former are based on spatial features of the graph such as the degree of connectivity between nodes. These vertex domain designs, though, lack spectral localization. While the energy of the resulting transformed signal is not concentrated around central graph frequencies, the vertex domain designs are described by a perfect localization on the vertex domain, meaning that one can define the number of nodes that will update the value of a vertex after the transform. One specific vertex domain design is the lifting-based graph wavelets [26] that use distinct groups of nodes to compute the update and detail wavelet coefficients.

Transforms that belong in the latter category of spectral designs exploit characteristics of the graph spectrum (not to be confused with the spectrum of hyperspectral images) in the form of the eigenvectors and the eigenvalues of the graph. One key feature of these designs is good spectral as well as vertex domain localization. One of the first spectral graph wavelet transforms are the diffusion wavelets [27] as well as the spectral graph wavelets [28]. While these spectral graph transforms are over-complete (the number of wavelet coefficients surpass the number of signal samples), this problem is solved by the two-channel graph wavelet filterbanks [29] that uses quadrature mirror filters (QMF) on bipartite graphs. These filters are not compactly supported and produce transforms that are not well localized on the vertex domain. The next iteration of such transforms are the biorthogonal graph wavelet filterbanks (introducing the GraphBior transform) [30] which are compactly supported as well as critically sampled. Since then several improvements and variations of the biorthogonal graph filterbanks have been proposed by including spectral sampling [31] or by introducing the M-channel graph wavelet filterbanks that can be implemented on large sparse graphs [32]. One of these variations uses polyphase transform matrices [33], proposing graph lifting structures [34] in the spectral domain for biorthogonal graph filterbanks.

One major issue of the filters developed for biorthogonal graph filterbanks, regarding image compression, is that they are only suited for lossy compression schemes. This drawback originates from the fact that the graph wavelet coefficients arising from these transforms are not integer. Thus a necessary quantization step is required rendering such compression schemes lossy. Furthermore, the use of spectral graph transforms for the compression of hyperspectral images spawns further difficulties—a huge amount of side information is required, due to the necessity to transmit the graph structure to the decoder.

In this paper, we provide transforms that allow us to construct a lossy-to-lossless compression scheme for hyperspectral images, using biorthogonal graph filterbanks. We developed two integer graph-transforms, both suitable for the biorthogonal graph filterbanks.

Our first approach is to modify the filtering process of GraphBior so that the resulting analysis wavelet coefficients are integer. We solve this problem by computing the Triangular Elementary Rectangular Matrix (TERM) [35] decomposition of the spectral GraphBior filter. At first glance this solutions seems promising but due to the high complexity of the TERM factorization process and the large size of the GraphBior filters, one could argue otherwise. Thus we partition each GraphBior filter in tiles and compute the TERM factorization of each tile in parallel. This process dramatically decreases the time complexity of the TERM factorization of GraphBior.

Our second integer spectral graph transform is achieved by introducing rounding operations within the spectral graph lifting structures proposed in Reference [34]. Hence, we transform it into an integer-to-integer graph wavelet transform.

Using our proposed transforms mentioned above, we design a lossy-to-lossless extension to the scheme developed for the lossy compression of hyperspectral images through GraphBior in Reference [22]. The compression scheme published in Reference [22] tackles the issue of transmitting the graph structure to the decoder by assembling consecutive hyperspectral components into packets called band groups.

In this paper we evaluate and analyze the performance of our lossy-to-lossless compression scheme on a variety of hyperspectral images. We then explore the use of our graph transforms for hyperspectral images. Additional, we explore the effect that different parameters have on our compression scheme. Such parameters are the size of the tiles as well as the size of band groups. Experimental results are provided comparing the performance of the proposed techniques for several images from the CCSDS MHDC corpus of hyperspectral images [36].

This paper is structured as follows. First, in Section 2, we present a brief overview of graph signal processing and biorthogonal graph filterbanks. Then, the derivation of the TERM decomposition of the GraphBior transform, the introduction of the integer spectral graph lifting transform, as well as the adopted coding strategy are discussed in Section 3. In Section 4 we detail the setting of our experiments and showcase our results. Our conclusions are stated in Section 5.

2. Graphs and Biorthogonal Graph Wavelets

This section provides a brief introduction on graph signal processing and GraphBior. A graph $G = (V, E)$ is composed by a set of nodes V that are linked to each other by a set of edges E. There is a large variety of graphs, each with its own special properties. One relevant example is the bipartite graph, whose nodes can be arranged in 2 subsets such that there are no edges connecting nodes of the same subset. In other words, one needs only two colors in order to color the nodes of the graph in a way that no two nodes of the same color are connected through an edge. Two examples of bipartite graphs can be seen in Figure 1b,c.

The edges of a graph can be weighted, in a way that nodes can be connected strongly or weakly by a non discrete measure. The adjacency matrix $A \in \mathbb{R}^{|V| \times |V|}$, of a graph, is the symmetric matrix that describes the strength of all the possible connections between all nodes of a graph. We establish matrix A in such a way that any of its elements located in row i and column j is a real number $a_{ij} \in [0, 1]$ representing the weight of the edge between node i and j. The stronger the connection between two nodes, the higher the value of the weight, with the value 1 describing an edge of the highest strength. Conversely, weaker connections are represented by lower adjacency values such that a weight value equal to 0 represents a non existing edge between nodes. Using the Adjacency matrix one can compute the normalized Laplacian matrix that is defined as $\widetilde{L} = I - D^{-1/2} A D^{-1/2}$, where D is the diagonal degree matrix whose ith element is equal to the sum of the elements of A situated on its ith row. In graph signal processing the normalized Laplacian matrix is of vital importance since it embeds within its structure the spectral information of the graph. Due to its symmetric nature, the eigenvectors of $\widetilde{L}$ create a complete orthogonal basis that can describe any vector in $\mathbb{R}^{|V|}$ as a linear combination of its basis. Additionally the eigenvalues of $\widetilde{L}$ are known as the spectrum of the graph and portray a notion of frequency.

We can also assign a real value to each node of the graph such that node i has a value of f_i. A graph signal $f \in \mathbb{R}^{|V|}$ is expressed mathematically as a vector and represents the collective values of the nodes. A graph signal of low frequency is expected to vary slowly, meaning that strongly connected nodes will support very similar graph signal values, while a high frequency signal is expected to assign very dissimilar values to strongly connected nodes. In other words a low frequency signal reflects the natural connectivity characteristics of the nodes of the graph whereas a high frequency signal goes against the connectivities imposed by the values of the graph edges.

The GraphBior transform, just as classical wavelet filterbanks, filters in the analysis step a signal into low pass and high pass signals that are later downsampled in order to decrease the number of graph wavelet coefficients. In order to guarantee reversibility of the transform, GraphBior exploits a particular characteristic of the spectrum of bipartite graphs (called spectral folding) [30]. Thus an arbitrary graph first needs to be decomposed into a series of bipartite subgraphs. These subgraphs share the same set of nodes as the original graph but their sets of edges do not intersect. The GraphBior filterbanks then makes use of these graphs by creating low and high pass filters.

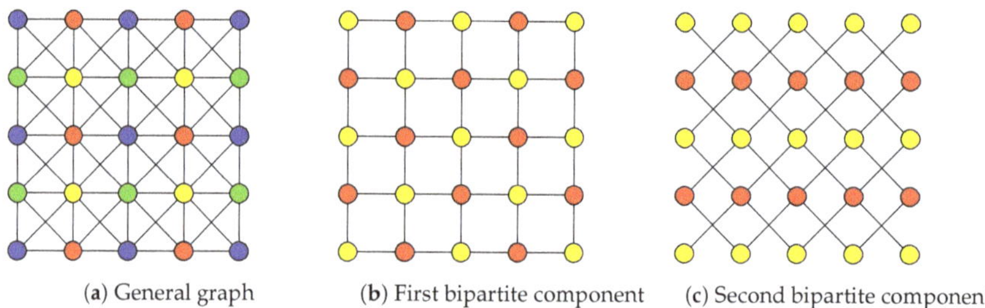

(**a**) General graph   (**b**) First bipartite component   (**c**) Second bipartite component

Figure 1. A graph (**a**) is decomposed into bipartite components (**b**) and (**c**). Node colors are the result of the graph coloring process, required for the graph bipartition.

3. Lossy-to-Lossless Graph Wavelet Filterbanks

Our work in this paper strives towards a lossy-to-lossless extension of Reference [22] for the compression of hyperspectral images using the GraphBior filterbanks. In this following section we go through the compression scheme that is employed, which follows the one introduced in Reference [22]. Then we propose an integer-to-integer version of the GraphBior filters, by applying a TERM decomposition, as well as tiling to decrease the time complexity of the TERM factorization. Additionally, we propose a second integer-to-integer graph transform suited for the biorthogonal graph filterbanks by modifying the spectral graph lifting structures in Reference [34].

3.1. Compression Scheme

Our compression scheme consists of 3 modules—one whose purpose is to calculate the biorthogonal graph filterbank transform, an encoder and a decoder.

The first module is where a single hyperspectral component is used for the construction of a biorthogonal graph filterbank transform and follows the scheme introduced in Reference [22]. The component is first mapped into a graph, then decomposed into a series of bipartite graphs which will finally be used to create the graph transform. We design a graph G by representing the pixels of a component as graph nodes and connect neighboring nodes with vertical, horizontal and diagonal edges resulting in a 8-regular graph just as the one shown in Figure 1a. Edges between two connected nodes are strong if the luminance values of these two pixels are similar and low when they are not.

Specifically a graph edge a_{ij} is calculated through the Gaussian kernel $a_{ij} = e^{-\frac{(f_i - f_j)^2}{\sigma}}$ where f_i and f_j are the values of pixels i and j and σ is a scaling factor. Once the graph $G = (V, E)$ has been constructed, it is decomposed into a series of n bipartite subgraphs $B_i = (V_i, E_i)$ with $i = 1, \ldots, n$. The bipartition should be done in a way where each bipartite subgraph has the same vertex set as the original graph, whereas the edges of E are distributed among the subgraphs B_i in such a way that no single edge of the original graph is present in more than one bipartite subgraphs. Thus, the union of all the sets of edges from all the bipartite graphs is equal to the set of edges of the original graph. In other words, $V_i = V$, $\cup_i E_i = E$ and $E_i \cap E_j = \emptyset$ for $i \neq j$. To decompose a graph into a series of bipartite subgraphs, we utilize Harary's decomposition [37]. Other decomposition techniques are found in References [38–40]. In the case of the 8-regular graph, shown in Figure 1a, the bipartition resulting from Harary's algorithm is very intuitive and is not computationally intensive. By discarding all diagonal edges from the graph in Figure 1a, we construct the first bipartite graph, shown in Figure 1b and by discarding all horizontal and vertical edges from Figure 1a, we construct the second bipartite graph, shown in Figure 1c. Hence, an 8-regular graph G, constructed out of a single component, is decomposed into two bipartite graphs that are utilized for the computation of a biorthogonal graph filterbank transform, just as shown in Figure 2.

It is necessary, for the reversibility of the transform, that the encoder and the decoder have both access to the same graph structure. This ensures that both modules construct the analysis and synthesis filters out of the same graph. The first module is designed in a way that circumvents the transmission of the graph structure to the decoder without the need of side information. This is done by only using decoded components for the calculation of the graph wavelet transforms. Specifically, components are bundled into packets of consecutive components called band groups. Each band group should preferably consist of the same number, ω, of components. Once the decoder has decompressed a band group we extract the last of its components. Using that component we create the graph G and thus the graph wavelet transform. This process can be executed in the decoder and the encoder simultaneously. Once the graph wavelet transform has been learned, it is then applied to the next band group, in the analysis step within the encoder as well as in the synthesis step at the decoder. A schematic representing the computation and usage of the graph filterbank transform is depicted in Figure 3a.

The next module is the encoder, whose first operation upon receiving an image is to partition all but the first component into band groups. The first component is then compressed in a lossless manner by a spatial DWT and transmitted to the decoder. Thus the first component can be used to create the biorthogonal graph filterbank transform for the first band group. Remaining components are partitioned in band groups and encoded sequentially. The next step involves an optional RKLT or DWT transform across the spectral direction of the band group. This step is followed by the application of a spatial graph wavelet transform and the resulting graph wavelet coefficients are then quantized and coded by a plain entropy encoder. Accordingly, the module of the decoder is composed by an entropy decoder, a dequantizer, the synthesis process of the biorthogonal graph filterbank and the optional inverse spectral transform. The encoder and decoder modules are depicted in Figure 3b. By introducing our integer-to-integer filters for the biorthogonal graph filterbank we obtain a lossy-to-lossless compression scheme using graph wavelets.

It is important to note that this manuscripts aims to study graph wavelet transforms when applied spatially to hyperspectral images. For that reason, we have followed a minimalist approach to designing our overall compression scheme. By avoiding complex quantization or entropy encoding stages, with complex overall interactions, we focus our efforts to the proposed graph transforms and to measure their performance with less interference. While a more complex integration of post-transform encoder parts would be possible, due to the adaptive nature of the biorthogonal graph filterbanks, it is expected that these transforms perform competitively, regardless of the probabilistic characteristics of the pre-transform data. For theoretical result on transform performance, readers can see Reference [25], where it is proven that the GFT optimally decorrelates images following a Gauss-Markov random field model.

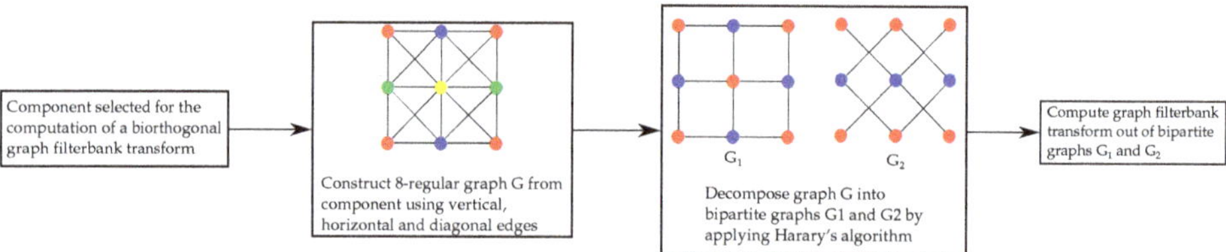

Figure 2. Construction of the bipartite graphs out of a selected component leading to the calculation of a graph filterbank transform.

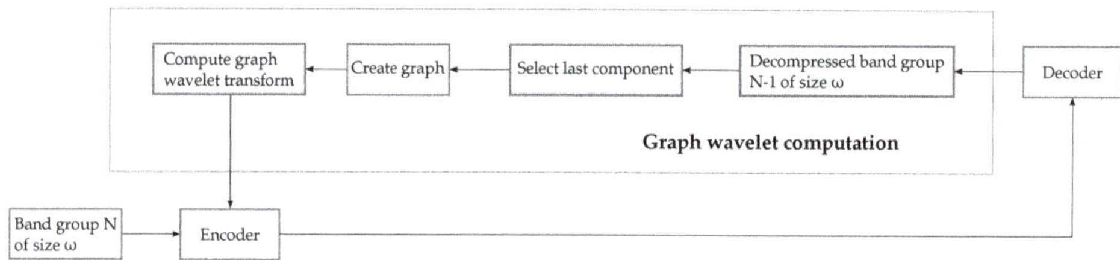

(**a**) Computation of graph wavelet filterbank transform

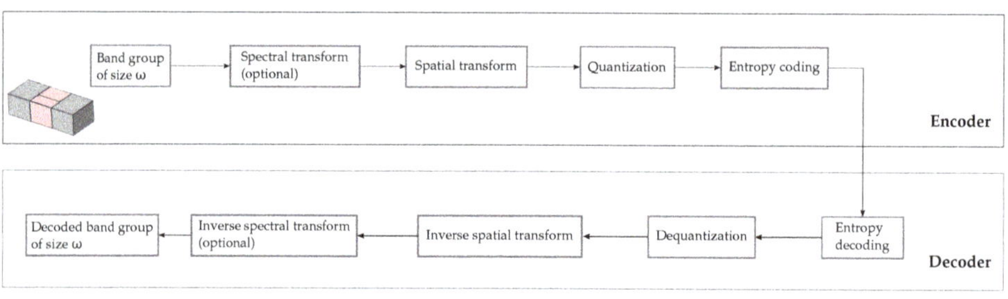

(**b**) Compression scheme

Figure 3. Schematics of the compression scheme used.

3.2. TERM factorization of GraphBior

In order for the proposed graph compression scheme, based on the biorthogonal graph filterbanks, to develop a lossless behavior we require an appropriate transform. Specifically, an integer-to-integer spectral graph wavelet transform. One of our solutions is to apply the TERM factorization process and the computational scheme introduced in Reference [35] on the GraphBior transform matrix.

However, one cannot decompose any arbitrary square transform matrix through TERM. The necessary condition in order to apply such a factorization dictates that the determinant of the transform should be strictly equal to 1. By design the GraphBior transform is reversible, which means that the determinant of the GraphBior analysis filter $T_a \in \mathbb{R}^{|V| \times |V|}$ is equal to an arbitrary non zero real number d.

We scale the elements of T_a by $d^{-1/|V|}$, without the loss of graph spectral information, to force its determinant to 1. Therefore we create a new matrix $T'_a = T_a \cdot d^{-1/|V|}$ whose determinant is equal to 1 and can follow the decomposition and computational scheme of Reference [35]. Applying the TERM factorization to the transform T'_a and by following the computation scheme introduced in Reference [35] we manage to create an integer-to-integer variation of the GraphBior transform. We shall abbreviate our proposed TERM GraphBior transform as IGB (integer GraphBior).

This factorization procedure raises the concern for certain drawbacks, specifically if we consider the large size of the GraphBior transform matrix. These square nonsingular matrices have as many rows (or columns) as the total number of pixels in each hyperspectral component. Due to its cubical computational complexity, the factorization of a GraphBior transform for an entire component, through TERM, is computationally intensive. As a first measure we use low order polynomials to compute the GraphBior filters. For that reason we use the GraphBior(1,1) rather than the better performing GraphBior(5,5) transform, since the former produces a sparser transform matrix. Additionally we split each component that is used to create the transform into smaller square tiles in an effort to accelerate the factorization process.

3.3. Tiling

In order to alleviate the time complexity from the TERM factorization, we developed a divide and conquer method. We initially split the component that is used to derive the graph wavelet transform into multiple small square tiles. From each tile, we compute its corresponding GraphBior transform to which we apply the TERM factorization. For the selection of a suitable tile size, in this subsection we study the effect on the performance of our compression scheme with respect to the tile size. Therefore we have experimented with IGB on our compression scheme, with no spectral transforms and using 3 different tile sizes (8 by 8, 16 by 16 and 32 by 32 square tile sizes). We tested the performance of IGB on the hyperspectral images, listed in Table 1, that contain several hyperspectral scenes available at the CCSDS website [36]. In Table 2 we display the entropy, in bits per pixels per component (bpppc), at which IGB becomes lossless, while using $\omega = 2$. We observe that the entropy where the IGB transform becomes lossless decreases as the tile size increases. This first observation encourages us to use a larger tiles for the lossless case. The results from Figure 4 lead to the same conclusions when testing the lossy performance of IGB. Specifically, in Figure 4 we display the performances of IGB when using tiles of 8 by 8 and 16 by 16 pixels, relative to using larger tiles of 32 by 32 pixels. Larger tile sizes are not included in our experiments as we are limited by the high computational complexity of the TERM decomposition. In Figure 5 we have repeated the same experiment on other hyperspectral images. Regarding the two AVIRIS images, due to their much larger spectral size, we have not experimented with the largest tile option as we have done with the Landsat image. The experimental conclusions also agree with our intuitive interpretation of the graph biorthogonal filterbanks since when using larger tiles, the graph can exploit spatial redundancies on a bigger area of each component. This is because the graph edges that connect adjacent nodes that belong to different tiles are not utilized. Since the number of discarded edges rises as the tiles becomes smaller so does the performance of IGB deteriorate. Therefore it is beneficial to keep the sizes of the tiles as large as possible but small enough to allow a fast TERM factorization. Luckily this solution can be parallelized since we could process every tile of every component of the same group at the same time.

Table 1. The hyperspectral images with their dimensions used in our experiments.

Name	Instrument	Calibrated	Across-Track	Along-Track	Spectral Dimension
Yellowstone sc. 0 cal.	AVIRIS	Yes	512	512	224
Yellowstone sc. 0 raw	AVIRIS	No	512	512	224
Lake Monona	Hyperion	Yes	512	256	242
Mt. St. Helens	Hyperion	Yes	512	256	242
Agriculture	Landsat	No	512	512	6

Table 2. Rates at which integer GraphBior (IGB) achieves a lossless compression using $\omega = 2$ for various tile sizes and multiple images. Units are in bpppc.

Image	IGB		
	Tiles of 8 by 8	**Tiles of 16 by 16**	**Tiles of 32 by 32**
Yellowstone sc. 0 cal.	7.14	6.95	-
Yellowstone sc. 0 raw	9.15	9.02	-
Lake Monona	6.40	6.31	6.26
Mt. St. Helens	6.78	6.68	6.63
Agriculture	4.39	4.27	4.21

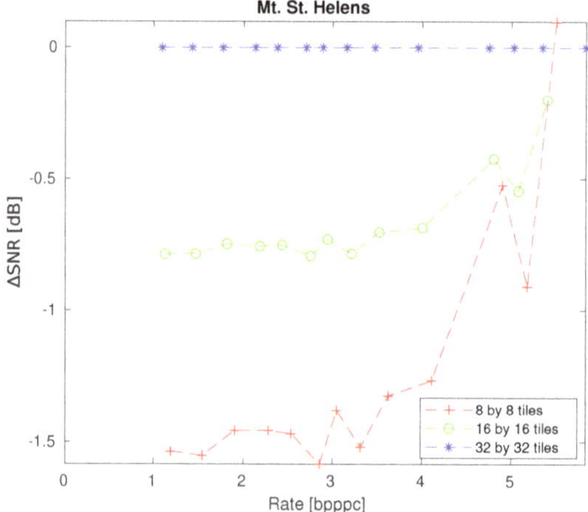

Figure 4. Relative rate-distortion plots for IGB, using $\omega = 2$, when varying the tile size. Results are relative to those of the largest tile size.

3.4. Integer-to-Integer Spectral Graph Lifting

The main disadvantage of the IGB arises from the computational complexity of the TERM factorization step. As we have seen in the previous section although the tiling strategy reduces the time complexity for the TERM factorization, the performance of our compression scheme decreases in relation to the size of the tiles. For that reason we introduce a second integer-to-integer graph filterbank transform using the spectral graph lifting filters [34].

These filters are arranged into Type 1 and 2 polyphase transform matrices (PTM) which have an upper and lower triangular structure respectively. The final transform matrix is calculated by multiplying alternatively Type 1 PTMs with Type 2 PTMs. One can notice that both types of PTMs are unit triangular rectangular matrices, which means that by following the TERM filtering process we can modify the spectral graph lifting filters into integer-to-integer transforms. Due to their particular structure, by introducing rounding operations after applying each PTM matrix multiplication, the transform becomes integer and reversible. We experiment with the integer-to-integer spectral graph lifting using the quad kernel (ISGL$_Q$) and the integer-to-integer spectral graph lifting using the dual kernel (ISGL$_D$) designs [41].

It should be noted that the computational complexity of this transform is comparable to the one of the GraphBior transform and can be easily computed from an entire component without the need of tiling operations. Specifically the computational complexity of the costlier ISGL$_Q$ is comparable to the one of GraphBior(5,5) since the former requires a computation of a 10th and 13th degree matrix polynomial, whereas the latter one of 10th and one of 11th. This means that by implementing the filtering process with Chebyshev polynomials [28] one can use the same order M to approximate similarly well the GraphBior(5,5) as well as the ISGL$_Q$ filters resulting in a computational complexity of $O(M|E|)$ per filter, where E is the set of edges of the graph. By performing the appropriate modifications, our proposed ISGL$_Q$ compresses and decompresses the calibrated AVIRIS image from Table 1 in 4.5 min, whereas the IGB required more than 1 h to successfully perform the same action on a computer using an Intel Core i7-7700HQ CPU and 16GB of RAM.

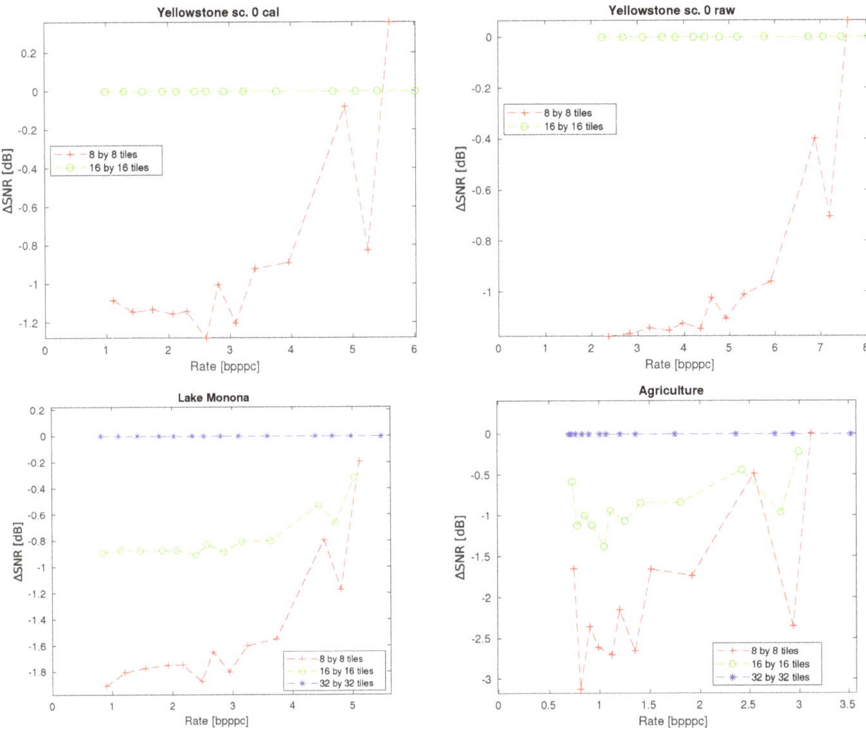

Figure 5. Relative rate-distortion plots for IGB, using $\omega = 2$, when varying the tile size, for multiple images. Results are relative to those of the largest tile.

4. Experimental Results

In this section we describe the setting of our experiments and analyze our results. We provide several comparisons testing the performance of our compression scheme using our proposed transforms, in both progressive lossy-to-lossless and lossless settings. We should note that regarding the rate computations, we compute the average of the entropies of each compressed component instead of using any particular entropy encoder to discard any bias introduced by an entropy encoder fine-tuned for a specific transform. All experiments have been performed using a prototype encoder implementation in Matlab R2017b. Initially we experiment solely on spatial transforms. We compare our proposed transforms against the GraphBior transform, with and without dividing the components into tiles as well as the reversible 5/3 integer wavelet transform (DWT). All GraphBior transforms are using the maximally flat GraphBior(1,1) filters. The spectral graph lifting designs are using the maximum flatness dual (with the $B_{7,3}$ Parametric-Bernstein-Polynomial for ISGL$_D$) and quad (with the $B_{3,1}$ Parametric-Bernstein-Polynomial for ISGL$_Q$) kernel designs without normalization parameters [41]. We provide further comparisons between the spatial transforms by including spectral transforms such as the Reversible Karhunen Loeve Transform (RKLT) [42] and the DWT.

We have repeated our experiments on crops of the Aviris Yellowstone scene 00, the Hyperion images Lake Monona and Mt. St. Helens and the Landsat Agriculture image. The dimensions of the cropped images that have been used are displayed in Table 1. More precisely, we have discarded the last rows and columns of each component for both Aviris as well as the Landsat scenes. Regarding the Hyperion images we have discarded the last columns and the first 1350 and 760 rows for Lake Monona and Mt. St. Helens respectively.

Next we explore the version of the compression scheme where we include a spectral transform on each band group before applying the spatial transform. The graph transforms are always computed from the luminance values of components and not the spectrally transformed results. We experiment with the Reversible Karhunen Loeve Transform (RKLT) [42] and DWT as spectral transforms. The spatial transforms compared in this experiment are the proposed IGB, the tiled GraphBior, the DWT, the ISGL$_Q$ and ISGL$_D$. In this final experiment, we also include a comparison of our overall compression scheme with the CCSDS-123.0-B-2 standard, tuned according to Reference [19].

4.1. Parameter Variations

We first study the effect that different values of ω have on IGB, without taking into account spectral transforms. To search for the most beneficial choices of ω to IGB we use tiles of large size.

In Figure 6 the relative rate-distortion plot displays the decrease of performance when larger values of ω are compared against the smallest $\omega = 2$. One can clearly conclude that it is beneficial to choose smaller values of the parameter ω when no spectral transform is used. We repeat the experiment with several other hyperspectral images and display the results in Figure 7. Specifically, it is for $\omega = 2$ that tends to lead to optimal results in all cases. This result agrees with our expectations since components closer in the spectral dimension tend to have high correlation. Thus, the component that has been used to create the IGB transform will capture sometimes more accurately the spatial redundancies of its neighboring component rather than of one located further away in the spectral dimension. Though this can lead to detrimental results for low ω values, when high quantization occurs. As we can see from Figures 6 and 7c,d at low rates, larger parameters ω outperform lower ones.

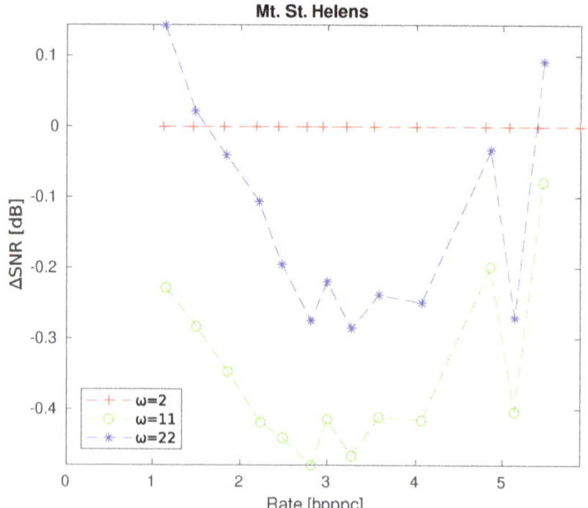

Figure 6. Relative rate-distortion plot for IGB using tiles of 16 by 16 when varying ω. The reference curve coincides with tuning the parameter $\omega = 2$.

This could be explained by high quantization, when the surviving features preserved in a decompressed component are general enough to construct a graph that represents sufficiently well a larger number of components. Furthermore, the smaller the parameter ω, the higher the probability of outlier components (ones that do not share high correlation with neighboring components such as dead bands) to be used for the graph creation, which may result in poor reconstruction results of components of an entire band group.

From Table 3 we also observe that the entropy where the compression scheme using the IGB transform becomes lossless, increases with ω. This explains why at high rates, higher ω values can be seen to sometimes perform better. This is also noticeable in classical reversible integer wavelet transforms. Their rate distortion curves tend to plateau close to the bitrate at which they achieve a reversible compression. As a result, sometimes, they are surpassed by the distortion curves corresponding to methods that become reversible at higher bitrates (since they still continue to grow with a higher gradient).

Table 3. Rates at which IGB achieves a lossless compression for different values of ω. The tiles size is set to 16 by 16. The spectral size of the hyperspectral image should be a multiple of ω, resulting in empty cells. Units are in bpppc.

Image	IGB						
	$\omega = 2$	$\omega = 3$	$\omega = 8$	$\omega = 11$	$\omega = 14$	$\omega = 16$	$\omega = 22$
Yellowstone sc. 0 cal.	6.95		7.00		7.06	7.14	
Yellowstone sc. 0 raw	9.02		9.06		9.14	9.15	
Lake Monona	6.31			6.36			6.37
Mt. St. Helens	6.68			6.75			6.75
Agriculture	4.27	4.39					

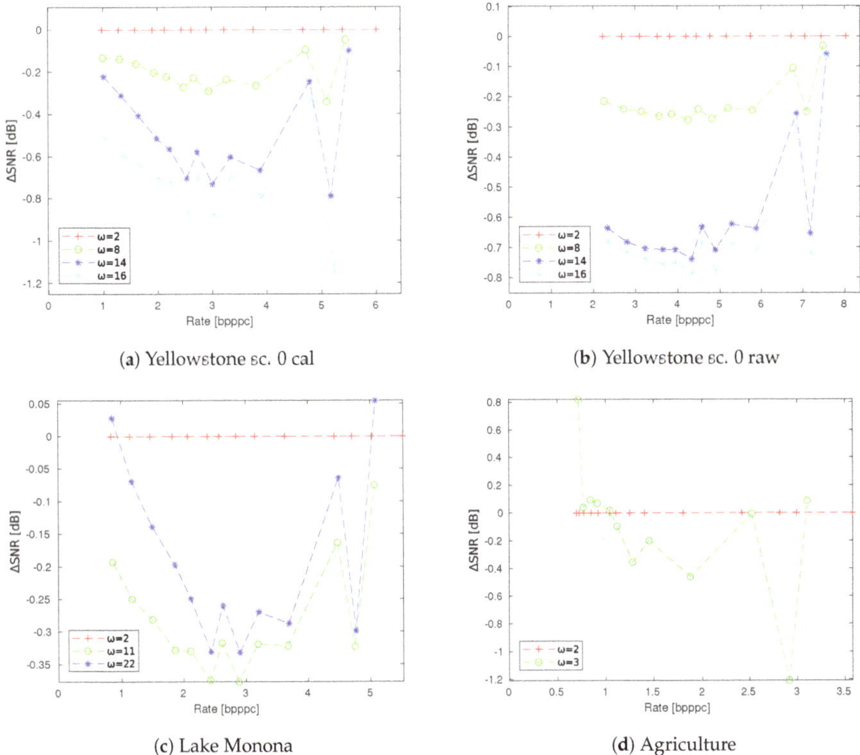

(a) Yellowstone sc. 0 cal

(b) Yellowstone sc. 0 raw

(c) Lake Monona

(d) Agriculture

Figure 7. Relative rate-distortion plots for IGB using tiles of 16 by 16 when varying the parameter ω for multiple images. Results are relative to those of smallest value of ω.

4.2. Spatial Transformations

Adopting our previous findings regarding the parameter ω as well as the size of the tiles, we compare our proposed reversible transforms against the DWT and the GraphBior with and without using tiles. All transforms are applied spatially whereas spectral transforms are not included in the subsequent experiments. It should be noted that all spatial transforms are tested using the compression scheme introduced in Section 3.

For this experiment, the rate-distortion plot, as well as the relative rate-distortion plot using the DWT transform as reference can be observed in Figure 8a,b respectively. From the rate-distortion plots of Figure 8 we observe that the $ISGL_Q$ performs the best for low and medium rates, whereas GraphBior provides the best results for high rates. It is important to mention, though, that in most cases IGB slightly outperforms the DWT as well as the $ISGL_D$ at low rates. Moreover, although in general, tiled GraphBior outperforms IGB, at low rates, both perform similarly. This same behavior is also observed between classical discrete wavelet transforms and their reversible counterparts. Given the previous observation and since the results of IGB improve as the tile size increases, one could speculate that a IGB that does not use any tiles would perform similarly to GraphBior for low rates. This experiment has been repeated on different hyperspectral scenes and the results can be observed in Figure 9.

In Table 4 we can see the rates at which the IGB, DWT, $ISGL_Q$ and $ISGL_D$ achieve a lossless compression. We also observe that in most cases the entropy where IGB achieves a lossless compression is the lowest.

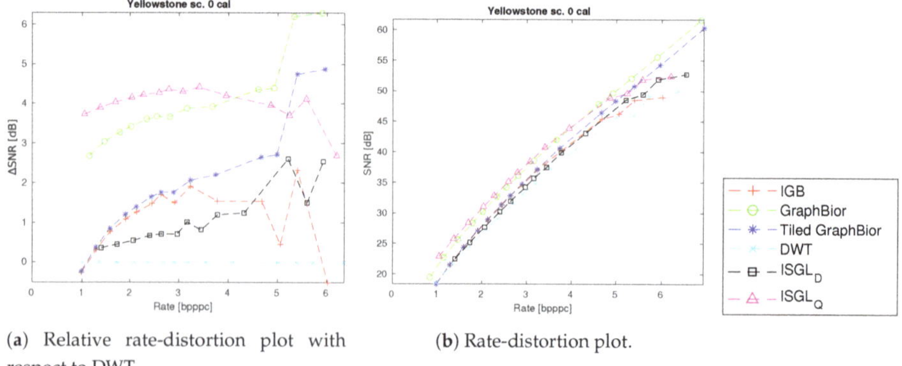

(**a**) Relative rate-distortion plot with respect to DWT. (**b**) Rate-distortion plot.

Figure 8. Comparison of spatial transforms. Tiles of 16 by 16 and $\omega = 2$ were used.

Table 4. Rates at which each integer transform achieves a lossless compression. No spectral transform is used. The parameter ω is set to 2. The IGB transform uses tiles of 16 by 16 for the Yellowstone images and 32 by 32 for Lake Monona, Mt. St. Helens and Agriculture. Units are in bpppc.

Image	Transform			
	IGB	**DWT**	**$ISGL_D$**	**$ISGL_Q$**
Yellowstone sc. 0 cal.	6.95	7.29	7.44	7.11
Yellowstone sc. 0 raw	9.02	9.32	9.47	9.19
Lake Monona	6.26	6.23	6.50	6.28
Mt. St. Helens	6.63	6.57	6.93	6.67
Agriculture	4.21	4.37	4.67	4.37

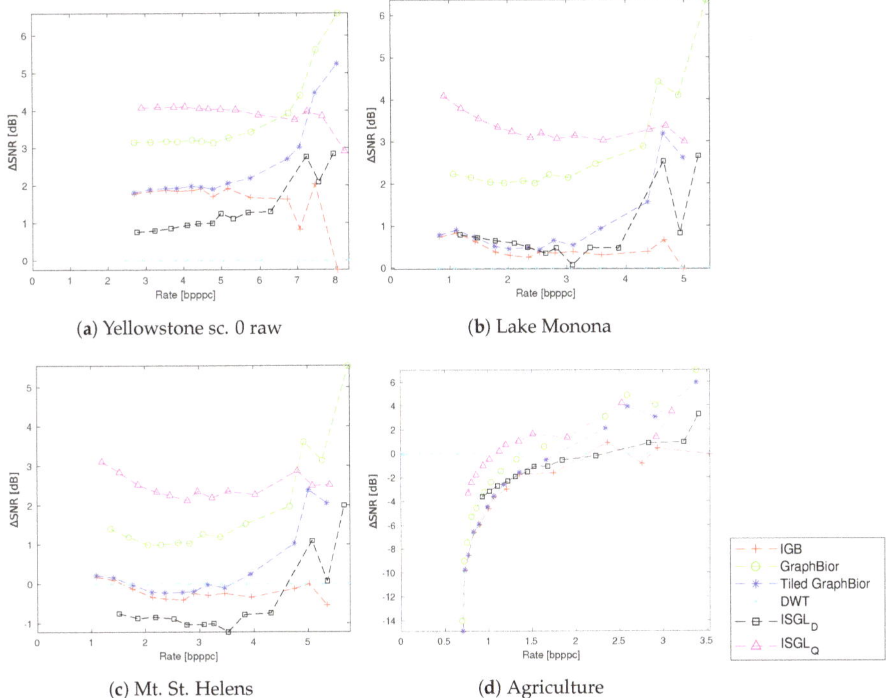

(**a**) Yellowstone sc. 0 raw

(**b**) Lake Monona

(**c**) Mt. St. Helens

(**d**) Agriculture

Figure 9. Relative rate-distortion plots comparing spatial transforms using $\omega = 2$ for multiple images. Results are relative to DWT. The tiles sizes are 16 by 16 for the Yellowstone sc. 0 raw image and 32 by 32 for the rest.

4.3. Spectral and Spatial Transformations

Our comparisons on strictly spatial transforms are ensued by experiments that include the DWT and the RKLT as spectral transforms, in the compression scheme mentioned in Section 3. Succeeding the spectral transform, we then compare our proposed reversible transforms against the DWT and the tiled GraphBior applied in the spatial dimension of the hyperspectral images. Our overall compression scheme including classical spectral and graph spatial transforms is also compared against the CCSDS-123.0-B-2 standard.

In Figure 10a,b we present the rate-distortion plot as well as the relative rate-distortion plot. The results are relative to applying a spectral RKLT along the entire hyperspectral image, followed by a spatial DWT.

Regarding the comparison between the transforms used in our compression scheme, we observe that a spectral RKLT followed by the proposed ISGL$_Q$ or ISGL$_D$ transforms perform the best. Several additional comparisons are done on multiple hyperspectral images and from our results in Figure 11 we observe that the spectral RKLT and the spatial ISGL$_Q$ systematically outperforms the spectral RKLT and spatial DWT, mostly at medium to high rates. The spectral RKLT and spatial ISGL$_D$, also, usually surpasses the reference method but provides less important results when compared to using the spatial ISGL$_Q$ instead. On the other hand the spectral RKLT and the spatial IGB is almost always providing worse results when compared to the spectral RKLT and the spatial DWT.

Our conclusions from the comparison of the spatial transforms done in Section 4.2 are also evident in our current experiment. Regardless of the spectral transform, the spatial IGB coincides with the spatial tiled GraphBior for low entropy values. When we apply DWT as a spectral transform, out of the spatial transforms, the ISGL$_Q$ performs the best followed mostly by the DWT and the ISGL$_D$.

From Table 5, we observe that mostly the spectral RKLT and the spatial DWT achieve a lossless compression at lower rates. Only in the case of the Hyperion images (Lake Monona and Mt. St. Helens) the spectral RKLT followed by the spatial $ISGL_Q$ performs better.

When comparing the compression efficiency of the latest CCSDS-123.0-B-2 standard against our overall proposed compression scheme, we observe the usual pattern where transform-based methods tend to perform better at lower rates while the CCSDS standard yields better results at mid to high rates. In Figures 10 and 11, a spectral RKLT followed by a spatial DWT often outperforms CCSDS for low rates. Furthermore, CCSDS results are also improved by the proposed compression scheme using a spectral RKLT and a spatial $ISGL_Q$ for low-rate results in Figure 11a,b. In the mid to high rate regions, the CCSDS standard clearly outperforms all transform-based methods. In addition, for pure lossless compression, the CCSDS standard consistently achieves best results (Table 5). It should be noted though that the CCSDS standard is a highly refined method, whereas the proposed compression scheme here presented is mainly designed to compare the graph transforms. Thus, it lacks many of the techniques that the CCSDS standard incorporates, such as quantization enhancements, noise tolerance or entropy encoding with adaptive statistics.

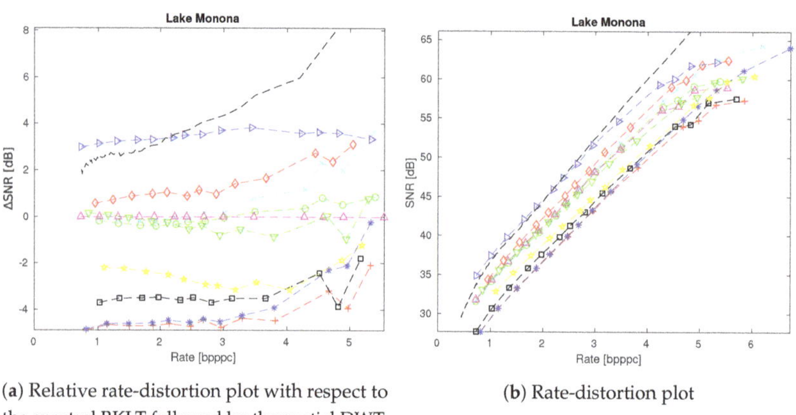

(**a**) Relative rate-distortion plot with respect to the spectral RKLT followed by the spatial DWT.

(**b**) Rate-distortion plot

Figure 10. Comparison of spectral + spatial transforms. Tiles of 16 by 16 and $\omega = 11$ was used.

Table 5. Rates at which each of the integer transforms achieves a lossless compression. RKLT and DWT are used as spectral transforms. For the methods that use tiles, their size has been set to 16 by 16. The Yellowstone images are evaluated at $\omega = 8$, the Lake Monona and Mt. St. Helens are evaluated at $\omega = 11$ and the Agriculture image is evaluated at $\omega = 3$. For each column, the spectral transform is mentioned first followed by the spatial transform (spectral transform + spatial transform). Units are in bpppc.

Image	Transforms								CCSDS
	DWT + IGB	RKLT + IGB	DWT + DWT	DWT + ISGL$_D$	DWT + ISGL$_Q$	RKLT + DWT	RKLT + ISGL$_D$	RKLT + ISGL$_Q$	
Yellowstone sc. 0 cal.	5.03	4.41	4.73	5.36	5.05	3.74	4.65	4.36	4.04
Yellowstone sc. 0 raw	7.17	6.47	6.97	7.54	7.25	5.93	6.72	6.44	6.19
Lake Monona	6.69	6.19	6.56	6.89	6.66	6.35	6.34	6.13	6.10
Mt. St. Helens	7.06	6.52	6.90	7.32	7.05	6.58	6.71	6.47	6.37
Agriculture	4.21	3.96	3.96	4.63	4.34	3.68	4.37	4.08	3.62

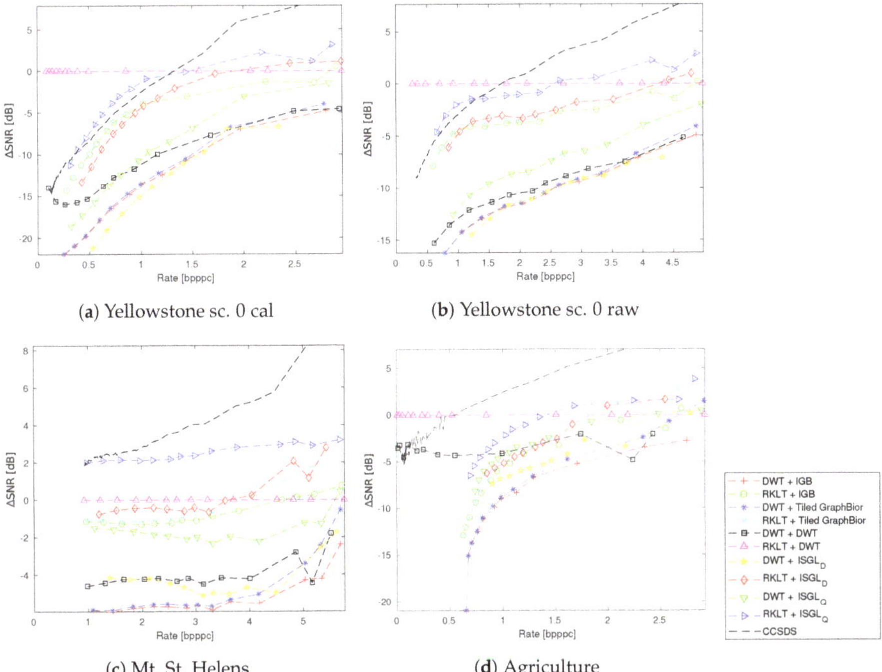

(**a**) Yellowstone sc. 0 cal   (**b**) Yellowstone sc. 0 raw

(**c**) Mt. St. Helens   (**d**) Agriculture

Figure 11. Relative rate-distortion plots comparing spectral + spatial transforms for multiple images. Results are relative to the spectral RKLT followed by the spatial DWT. Tiles of 16 by 16 were used. The parameter ω is set to 8 for Yellowstone sc. 0 cal., to 11 for Mt. St. Helens and to 3 for Agriculture.

5. Conclusions

In this paper we study the suitability of graph transforms for lossy-to-lossless hyperspectral compression. The adopted compression scheme organizes the components into packets, called band groups and transforms each one of them through graph wavelet filterbanks. We introduce two spatial, integer-to-integer, biorthogonal graph filterbank transforms. The first transform is calculated by applying a TERM factorization on the GraphBior filterbank, whereas the second one is designed by modifying the spectral graph lifting transform [34]. The high computational complexity of the TERM decomposition is addressed and is solved through processing the GraphBior transform in tiles. Our theoretical interpretations as well as our experimental results show that it is advantageous to use larger tiles and give us valuable insight about the choice of sizes of band groups. Our experiments without spectral transformations suggest that it is preferable to use small band groups, while our integer-to-integer transforms outperform the DWT in the lossy regime. Further-more we show that one of our integer transforms performs similarly to the DWT for the case of lossless compression. Additional experimental results including spectral transforms on each bandgroup show that our proposition improves on the results obtained by using the spectral RKLT along all the spectral dimension of the hyperspectral image followed by the spatial DWT in the lossy setting, as well as, in some cases, at the lossless one.

Author Contributions: Conceptualization, D.E.O.T., K.C., I.B. and J.S.-S.; methodology, D.E.O.T., K.C., I.B. and J.S.-S.; software, D.E.O.T.; validation, D.E.O.T., I.B. and J.S.-S.; formal analysis, D.E.O.T., K.C., I.B. and J.S.-S. ; investigation, D.E.O.T., K.C., I.B. and J.S.-S.; resources, D.E.O.T., K.C., I.B. and J.S.-S.; data curation, D.E.O.T., I.B. and J.S.-S.; writing–original draft preparation, D.E.O.T.; writing–review and editing, D.E.O.T., I.B. and J.S.-S.; visualization, D.E.O.T., I.B. and J.S.-S.; supervision, I.B. and J.S.-S.; project administration, I.B. and J.S.-S.; funding acquisition, I.B. and J.S.-S.

Funding: This research was funded by the Spanish Ministry of Economy and Competitiveness and the European Regional Development Fund under grants RTI2018-095287-B-I00 and TIN2015-71126-R (MINECO/FEDER, UE) and BES-2016-078369 (Programa Formación de Personal Investigador), and by the Catalan Government under grant 2017SGR-463.

Conflicts of Interest: The authors declare no conflict of interest.

References

1. Liang, H.; Li, Q. Hyperspectral imagery classification using sparse representations of convolutional neural network features. *Remote Sens.* **2016**, *8*, 99. [CrossRef]
2. Thompson, D.R.; Green, R.O.; Keymeulen, D.; Lundeen, S.K.; Mouradi, Y.; Nunes, D.C.; Castaño, R.; Chien, S.A. Rapid spectral cloud screening onboard aircraft and spacecraft. *IEEE Trans. Geosci. Remote Sens.* **2014**, *52*, 6779–6792. [CrossRef]
3. Zhang, X.; Liu, F.; He, Y.; Li, X. Application of hyperspectral imaging and chemometric calibrations for variety discrimination of maize seeds. *Sensors* **2012**, *12*, 17234–17246. [CrossRef] [PubMed]
4. Wang, X.; Niu, R. Spatial forecast of landslides in three gorges based on spatial data mining. *Sensors* **2009**, *9*, 2035–2061. [CrossRef]
5. Tang, X.; Pearlman, W.A. Three-dimensional wavelet-based compression of hyperspectral images. In *Hyperspectral Data Compression*; Springer: Boston, MA, USA, 2006; pp. 273–308.
6. Fowler, J.E.; Rucker, J.T. Three-dimensional wavelet-based compression of hyperspectral imagery. In *Hyperspectral Data Exploitation: Theory and Applications*; John Wiley & Sons, Inc.: Hoboken, NJ, USA, 2007; pp. 379–407.
7. Karami, A.; Yazdi, M.; Mercier, G. Compression of hyperspectral images using discerete wavelet transform and tucker decomposition. *IEEE J. Sel. Top. Appl. Earth Obs. Remote Sens.* **2012**, *5*, 444–450. [CrossRef]
8. Tang, X.; Pearlman, W.A.; Modestino, J.W. Hyperspectral image compression using three-dimensional wavelet coding. In *Image and Video Communications and Processing 2003*; International Society for Optics and Photonics: Bellingham, WA, USA, 2003; Volume 5022, pp. 1037–1047.
9. Du, Q.; Fowler, J.E. Hyperspectral image compression using JPEG2000 and principal component analysis. *IEEE Geosci. Remote Sens. Lett.* **2007**, *4*, 201–205. [CrossRef]
10. Penna, B.; Tillo, T.; Magli, E.; Olmo, G. Transform coding techniques for lossy hyperspectral data compression. *IEEE Geosci. Remote Sens. Lett.* **2007**, *45*, 1408–1421. [CrossRef]
11. Penna, B.; Tillo, T.; Magli, E.; Olmo, G. A new low complexity KLT for lossy hyperspectral data compression. In Proceedings of the 2006 IEEE International Symposium on Geoscience and Remote Sensing, Denver, CO, USA, 31 July–4 August 2006; pp. 3525–3528.
12. Galli, L.; Salzo, S. Lossless hyperspectral compression using KLT. In Proceedings of the 2004 IEEE International Geoscience and Remote Sensing Symposium (IGARSS 2004), Anchorage, AK, USA, 20–24 September 2004; Volume 1.
13. Rizzo, F.; Carpentieri, B.; Motta, G.; Storer, J.A. Low-complexity lossless compression of hyperspectral imagery via linear prediction. *IEEE Signal Process. Lett.* **2005**, *12*, 138–141. [CrossRef]
14. Pizzolante, R.; Carpentieri, B. Multiband and lossless compression of hyperspectral images. *Algorithms* **2016**, *9*, 16. [CrossRef]

15. Abrardo, A.; Barni, M.; Magli, E. Low-complexity predictive lossy compression of hyperspectral and ultraspectral images. In Proceedings of the 2011 IEEE International Conference on Acoustics, Speech and Signal Processing (ICASSP), Prague, Czech Republic, 22–27 May 2011; pp. 797–800.

16. Klimesh, M. Low-complexity adaptive lossless compression of hyperspectral imagery. In Proceedings of the Satellite Data Compression, Communications, and Archiving II, San Diego, CA, USA, 13–17 August 2006; Volume 6300, p. 63000N.

17. Klimesh, M.; Kiely, A.; Yeh, P. Fast lossless compression of multispectral and hyperspectral imagery. In Proceedings of the 2nd International Workshop on On-Board Payload Data Compression (OBPDC), Toulouse, France, 28–29 October 2010; Volume 8, pp. 28–29.

18. Consultative Committee for Space Data Systems. *Lossless and Near-Lossless Multispectral and Hyperspectral Image Compression. Recommendation for Space Data System Standards, CCSDS 123.0-B-2*; CCSDS: Washington, DC, USA, 2019.

19. Blanes, I.; Kiely, A.; Hernández-Cabronero, M.; Serra-Sagristà, J. Performance Impact of Parameter Tuning on the CCSDS-123.0-B-2 Low-Complexity Lossless and Near-Lossless Multispectral and Hyperspectral Image Compression Standard. *Remote Sens.* **2019**, *11*, 1390. [CrossRef]

20. Shuman, D.I.; Narang, S.K.; Frossard, P.; Ortega, A.; Vandergheynst, P. The emerging field of signal processing on graphs: Extending high-dimensional data analysis to networks and other irregular domains. *IEEE Signal Process. Mag.* **2013**, *30*, 83–98. [CrossRef]

21. Ortega, A.; Frossard, P.; Kovačević, J.; Moura, J.M.; Vandergheynst, P. Graph signal processing: Overview, challenges, and applications. *Proc. IEEE* **2018**, *106*, 808–828. [CrossRef]

22. Zeng, J.; Cheung, G.; Chao, Y.H.; Blanes, I.; Serra-Sagristà, J.; Ortega, A. Hyperspectral image coding using graph wavelets. In Proceedings of the IEEE International Conference on Image Processing (ICIP), Beijing, China, 17–20 September 2017.

23. Cheung, G.; Magli, E.; Tanaka, Y.; Ng, M.K. Graph spectral image processing. *Proc. IEEE* **2018**, *106*, 907–930. [CrossRef]

24. Hu, W.; Cheung, G.; Ortega, A.; Au, O.C. Multiresolution graph fourier transform for compression of piecewise smooth images. *IEEE Trans. Image Process.* **2014**, *24*, 419–433. [CrossRef] [PubMed]

25. Zhang, C.; Florêncio, D. Analyzing the optimality of predictive transform coding using graph-based models. *IEEE Signal Process. Lett.* **2012**, *20*, 106–109. [CrossRef]

26. Narang, S.K.; Ortega, A. Lifting based wavelet transforms on graphs. In Proceedings of the APSIPA ASC 2009: Asia-Pacific Signal and Information Processing Association, 2009 Annual Summit and Conference, Sapporo, Japan, 4–7 October 2009; pp. 441–444.

27. Coifman, R.R.; Maggioni, M. Diffusion wavelets. *Appl. Comput. Harmonic Anal.* **2006**, *21*, 53–94. [CrossRef]

28. Hammond, D.K.; Vandergheynst, P.; Gribonval, R. Wavelets on graphs via spectral graph theory. *Appl. Comput. Harmonic Anal.* **2011**, *30*, 129–150. [CrossRef]

29. Narang, S.K.; Ortega, A. Perfect reconstruction two-channel wavelet filter banks for graph structured data. *IEEE Trans. Signal Process.* **2012**, *60*, 2786–2799. [CrossRef]

30. Narang, S.K.; Ortega, A. Compact support biorthogonal wavelet filterbanks for arbitrary undirected graphs. *IEEE Trans. Signal Process.* **2013**, *61*, 4673–4685. [CrossRef]

31. Sakiyama, A.; Watanabe, K.; Tanaka, Y.; Ortega, A. Two-Channel Critically Sampled Graph Filter Banks with Spectral Domain Sampling. *IEEE Trans. Signal Process.* **2019**, *67*, 1447–1460. [CrossRef]

32. Li, S.; Jin, Y.; Shuman, D.I. Scalable M-Channel Critically Sampled Filter Banks for Graph Signals. *IEEE Trans. Signal Process.* **2019**, *67*, 3954–3969. [CrossRef]

33. Tay, D.B.; Ortega, A. Bipartite graph filter banks: Polyphase analysis and generalization. *IEEE Trans. Signal Process.* **2017**, *65*, 4833–4846. [CrossRef]

34. Tay, D.B.; Ortega, A.; Anis, A. Cascade and Lifting Structures in the Spectral Domain for Bipartite Graph Filter Banks. In Proceedings of the 2018 Asia-Pacific Signal and Information Processing Association Annual Summit and Conference (APSIPA ASC), Honolulu, HI, USA, 12–15 November 2018; pp. 1141–1147.

35. Hao, P.; Shi, Q. Matrix factorizations for reversible integer mapping. *IEEE Trans. Signal Process.* **2001**, *49*, 2314–2324.

36. Consultative Committee for Space Data Systems, Test Data. Available online: http://cwe.ccsds.org/sls/docs/sls-dc/123.0-B-Info/TestData (accessed on 3 March 2019).

37. Harary, F.; Hsu, D.; Miller, Z. The biparticity of a graph. *J. Graph Theory* **1977**, *1*, 131–133. [CrossRef]

38. Narang, S.K.; Ortega, A. Multi-dimensional separable critically sampled wavelet filterbanks on arbitrary graphs. In Proceedings of the 2012 IEEE International Conference on Acoustics, Speech and Signal Processing (ICASSP), Kyoto, Japan, 25–30 March 2012; pp. 3501–3504.

39. Nguyen, H.Q.; Do, M.N. Downsampling of signals on graphs via maximum spanning trees. *IEEE Trans Signal Process.* **2014**, *63*, 182–191. [CrossRef]

40. Zeng, J.; Cheung, G.; Ortega, A. Bipartite approximation for graph wavelet signal decomposition. *IEEE Trans Signal Process.* **2017**, *65*, 5466–5480. [CrossRef]

41. Tay, D.B.; Zhang, J. Techniques for constructing biorthogonal bipartite graph filter banks. *IEEE Trans Signal Process.* **2015**, *63*, 5772–5783. [CrossRef]

42. Hao, P.; Shi, Q. Reversible integer KLT for progressive-to-lossless compression of multiple component images. In Proceedings of the 2003 International Conference on Image Processing (Cat. No. 03CH37429), Barcelona, Spain, 14–17 September 2003; Volume 1, pp. 1–633.

Article

Performance Impact of Parameter Tuning on the CCSDS-123.0-B-2 Low-Complexity Lossless and Near-Lossless Multispectral and Hyperspectral Image Compression Standard [†]

Ian Blanes [1,*], Aaron Kiely [2], Miguel Hernández-Cabronero [1] and Joan Serra-Sagristà [1]

[1] Department of Information and Communications Engineering, Universitat Autònoma de Barcelona, Campus UAB, 08193 Cerdanyola del Vallès, Spain; mhernandez@deic.uab.cat (M.H.-C.); joan.serra@uab.cat (J.S.-S.)

[2] NASA Jet Propulsion Laboratory (JPL), California Institute of Technology, 4800 Oak Grove Drive, Pasadena, CA 91109, USA; Aaron.B.Kiely@jpl.nasa.gov

* Correspondence: ian.blanes@uab.cat

† This paper is an extended version of our paper published in The 6th ESA/CNES International Workshop on On-Board Payload Data Compression.

Received: 11 April 2019; Accepted: 6 June 2019; Published: 11 June 2019

Abstract: This article studies the performance impact related to different parameter choices for the new CCSDS-123.0-B-2 *Low-Complexity Lossless and Near-Lossless Multispectral and Hyperspectral Image Compression* standard. This standard supersedes CCSDS-123.0-B-1 and extends it by incorporating a new near-lossless compression capability, as well as other new features. This article studies the coding performance impact of different choices for the principal parameters of the new extensions, in addition to reviewing related parameter choices for existing features. Experimental results include data from 16 different instruments with varying detector types, image dimensions, number of spectral bands, bit depth, level of noise, level of calibration, and other image characteristics. Guidelines are provided on how to adjust the parameters in relation to their coding performance impact.

Keywords: on-board data compression; CCSDS 123.0-B-2; near-lossless hyperspectral image compression

---

1. Introduction

It is well known that space-borne remote-sensing instruments are often the source of large volumes of data and that, due to constraints on the down-link channel, these data need to be compressed [1–3]. In this regard, the Consultative Committee for Space Data Systems (CCSDS) has standardized several data-compression techniques [4–6].

Very recently, the CCSDS has superseded Issue 1 of the *Lossless Multispectral & Hyperspectral Image Compression* standard [7] with Issue 2 titled *Low-Complexity Lossless and Near-Lossless Multispectral and Hyperspectral Image Compression* (CCSDS-123.0-B-2) [8]. The original issue of the standard employed the fast lossless compression algorithm [9,10] to achieve state-of-the-art compression performance, whilst being implemented in resource-constrained hardware available for space operation [11–19].

The new Issue 2 extends the previous issue, primarily by incorporating support for near-lossless compression, while retaining lossless compression capabilities (and all other features) of Issue 1.

The cornerstone behind Issue 2 of the standard is the newly-available option for near-lossless compression provided by an in-loop quantizer embedded in the prediction stage of the compressor. In addition, Issue 2 incorporates several other features including: *prediction representatives, weight exponent offsets, narrow prediction modes,* and a new *hybrid entropy coder.* The authors assume that readers are familiar with its contents (an overview of the new standard is provided in [20]).

As with the previous Issue 1, several tunable parameters are available to implementers and end-users. Employing different settings for these parameters may allow an implementer to achieve different trade-offs between implementation complexity and compression efficiency or may allow an end-user to fine-tune compression performance for particular datasets. This paper studies these parameters and provides some guidelines on how to adjust them to achieve high coding efficiency based on a representative corpus of multi- and hyper-spectral images.

The current study includes revisiting our previous assessment of parameter settings under lossless compression [21,22] while considering the newly-available coding options, in addition to providing new guidelines for both old and new coding options for near-lossless compression.

This paper is organized as follows. Section 2 describes the experimental methodology. Section 3 reports the principal experimental results and provides usage guidelines. Finally, some conclusions close this document.

2. Experimental Approach and Default Settings

The new CCSDS-123.0-B-2 provides more than twenty configuration parameters, with most of these parameters having a direct impact on the compression performance of an implementation. Moreover, the performance impact of one parameter may depend on the values set for all the remaining ones. Given these interactions and the large number of parameters, any exhaustive exploration of the configuration space will rapidly hit a complexity wall. Hence, any reasonable experimental study requires a pragmatic approach to the exploration of the configuration space.

It is expected that different encoding configurations yield not only different coding performances, i.e., smaller compressed files, but also different implementation considerations, such as FPGA area utilization, required memory buffers, etc. However, these are strongly dependent on the implementation strategy and technology employed. This paper does not try to address these implementation considerations and focuses exclusively on coding performance, which does not depend on the implementation approach.

The approach followed in the parameter study presented in this paper is as follows. First, an initial configuration for a CCSDS-123.0-B-2 encoder has been drawn from the authors' prior experiences and previous parameter assessments [22] (Annex C). Then, the effects of each of the parameters have been studied (one or a few at a time), and the configuration has been adjusted based on the outcomes of those analyses. Several iterations have been performed until a reasonable final configuration reached a local performance maximum.

This article reports the studies for parameters with the most significant performance impacts, where one or more parameters are studied independently, while the rest are set to default values, as reported in Tables 1 and 2. Defaults are based on the final configuration described in the previous paragraph.

In order to obtain relevant experimental results, a curated corpus of images is employed, encompassing images acquired by 16 different instruments, of varying processing levels, bit depths, and dimensions. The corpus is a superset of the publicly-available corpus used in the course of developing the CCSDS-123.0-B-1 recommendation [23]. Table 3 summarizes the corpus images and their properties. The number of images available for the each instrument is indicated through the use of $\times$ and $\{\cdot\}$ in the width and height columns of the table.

The default values for three of the parameters presented in Table 1 are adjusted to account for the specific characteristics of each instrument whenever experimental data presented in Section 3 suggest that the same parameter value is not adequate for all instruments. These adjustments are as follows. For the *local sum type* parameter, a value of *wide column-oriented* is employed for the CRISM, Hyperion, M3, and MODIS day and night images, while a value of *wide neighbor-oriented* is employed otherwise. A *full prediction mode* is employed for all images, except for the AIRS, AVIRIS 12-bit, CASI, CRISM, Hyperion, IASI, M3, and MODIS day and night images. The value of the *sample representative damping* parameter is set to five for the AVIRIS NG, HICO, and SFSI images; to three for the AIRS, AVIRIS 12-bit raw, CASI, CRISM, Hyperion, and M3 target images; and to zero for the rest of the images.

Table 1. Default predictor settings employed in the experimental results, unless otherwise indicated.

Parameter Name	Symbol	Default Value	Description
Sample Encoding Order		Band interleaved by pixel (BIP)	The order in which mapped quantizer indexes are encoded by the entropy coder
Entropy Coder Type		Hybrid	Indicates which entropy coding option is employed
Quantizer Fidelity Control Method		*Absolute* error limits or *lossless*	Enables or disables absolute and relative error limits
Number of Prediction Bands	P	3	Number of previous bands used to perform prediction
Register Size	R	64	Size of the register used in prediction calculation
Local Sum Type		(Image dependent)	Identifies neighborhood used to calculate local sums
Prediction Mode		(Image dependent)	Indicates whether directional local differences are used in the prediction calculation
Weight Component Resolution	Ω	19	Determines the number of bits used to represent each weight vector component
Weight Initialization Method		*Default*	Determines initial values of weight vector components
Weight Initialization Table	$\{\Lambda_z\}$	(Unused)	Defines the initial weight components under *custom* weight initialization
Weight Initialization Resolution	Q	(Unused)	Determines the precision of the initial weight components under *custom* initialization
Weight Update Scaling Exponent Initial Parameter	$\nu_{\min}$	-1	Determines initial rate at which the predictor adapts the weight vector to input
Weight Update Scaling Exponent Final Parameter	$\nu_{\max}$	$7 - \phi_z$	Determines the final rate at which the predictor adapts the weight vector to input
Weight Update Scaling Exponent Change Interval	t_{inc}	2^6	Determines the interval between increments to the weight update scaling exponent
Weight Update Scaling Exponent Offsets	$\{\zeta_z^{(i)}\},$ $\{\zeta_z^*\}$	All are set to 0	Offsets the exponent of each predictor weight update calculation
Periodic Error Limit Updating		Disabled	Enables or disables periodic updates of error limits
Sample Representative Resolution	Θ	3	Determines the precision of sample representative calculations
Sample Representative Offset[1]	$\{\psi_z\}$	7	Controls the offset from the center of the quantizer bin in the sample representative calculation.
Band Varying Offset		Disabled (all ψ_z are equal)	Enables the use of different values for each element in $\{\psi_z\}$
Sample Representative Damping	$\{\phi_z\}$	(Image dependent)	Controls the tradeoff between the predicted sample value and offset bin center in the sample representative calculation
Band Varying Damping		Disabled (all ϕ_z are equal)	Enables the use of different values for each element in $\{\phi_z\}$

[1] When the *quantizer fidelity control method* is set to *lossless*, the *sample representative offset* parameter does not influence compression performance. However, its actual value is set to 0, as mandated in the standard.

Table 2. Default entropy encoder settings employed in the experimental results, unless otherwise indicated.

Parameter Name	Symbol	Default Value	Description
Unary Length Limit	$U_{\max}$	18	Limits the maximum length of any encoded sample
Rescaling Counter Size	γ^*	6	Determines the interval between rescaling of the counter and accumulator
Initial Count Exponent	γ_0	1	Sets the initial counter value
Initial High-resolution Accumulator Value	$\{\tilde{\Sigma}_z(0)\}$	0	Sets an initial accumulator value for each band

Table 3. Summary of the corpus images and their properties.

Instrument	Image Type	Bit Depth	Number of Bands	Width	Height
AIRS	raw	14	1501	90	135×10
AVIRIS	raw	12	224	{614, 680}	512
AVIRIS	calibrated	16	224	677	512×5
AVIRIS	raw	16	224	680	512×5
AVIRIS-NG	radiance	14	432	598	512×4
AVIRIS-NG	raw	14	432	640	512×4
CASI	raw	12	72	405	2852
CASI	raw	12	72	406	1225
CRISM	FRT, raw	12	545	640	$\{420 \times 4, 450, 480 \times 2, 510 \times 2\}$
CRISM	HRL, raw	12	545	320	$\{420, 450 \times 2, 480\}$
CRISM	MSP, raw	12	74	64	2700×7
HICO	calibrated	14	128	512	2000×2
Hyperion	raw	12	242	256	{1024, 3176, 3187, 3242}
IASI	calibrated	12	8461	66	60×4
Landsat	raw	8	6	1024	1024×3
M3	global, radiance	12	85	304	512×4
M3	global, raw	12	86	320	512×2
M3	target, raw	12	260	640	512×3
MODIS	night, raw	12	17	2030	1354×5
MODIS	day, raw	12	14	2030	1354×5
MODIS	500 m, raw	12	5	4060	2708×5
MODIS	250 m, raw	12	2	8120	5416×5
MSG	calibrated	10	11	3712	3712×3
Pleiades	HR, simulated	12	4	224	{2456, 3928}
Pleiades	misregistered, sim.	12	4	296	2448×4
SFSI	raw	12	240	496	140×1
SPOT-5	HRG , processed	8	3	1024	1024×3
Vegetation	raw	10	4	1728	$\{10{,}080 \times 2, 10193 \times 2\}$

3. Experimental Results

This section analyzes how the principal parameters for the new Issue 2 affect compression performance. The experimental results are organized into four subsections, each focusing on one of the key new features introduced in Issue 2, examining both lossless and near-lossless compression.

While experiments have been performed for a large number of images, due to space constraints, this document only reports results covering the most significant behaviors for some of the experiments. In addition, averaged results are reported when multiple images are available for an instrument type for the same reason (averages are weighted by image size).

Other experimental restrictions, intended to obtain a reasonably-sized set of results, are as follows. When compression is not lossless, reconstructed image fidelity is controlled using *band-independent absolute error limits* via the value of the integer *absolute error limit constant* parameter A^*. Note that setting $A^* = 0$ yields lossless compression and is equivalent to selecting the fidelity control method to be *lossless*, apart from some minor differences in the compressed image header. Relative error limits are not used, and periodic error limit updating is not used, i.e., a fixed value of A^* is used for the entire image (see [24,25] for alternative error limit adjustment strategies). Per-band parameter adjustment and custom weight initialization are purposely left out of the article as well. In addition, the studies related to the following less relevant parameters are omitted for conciseness: weight component resolution, register size, sample-adaptive and block-adaptive encoder initialization.

Even with all the aforementioned constraints, the experimental results presented in this section are the outcome of more than two million individual compression experiments where an image is compressed with an implementation fully compliant with CCSDS-123.0-B-2. In terms of the volume of data processed, more than 100 TB of image data have been compressed to produce these results.

3.1. Local Sum Type, Prediction Mode, and Number of Prediction Bands

The choices of *local sum type*, *prediction mode*, and *number of prediction bands* (P) determine the prediction neighborhood, i.e., the neighboring samples that directly influence the prediction of a given image sample.

Different choices for these three parameters strongly influence the implementation complexity due to the varying data dependencies of each mode. In particular, to facilitate pipelining in a hardware implementation, Issue 2 introduces a new option to use *narrow* local sums as an alternative to the existing local sums, which are now called *wide*. However, note that the use of narrow local sums and full prediction mode might be an unlikely combination in practice since full prediction mode has data dependencies that may quash any advantage provided by narrow local sums over their wide counterparts.

Figure 1 reports results for all choices of the local sum type and prediction mode under a varying number of prediction bands and an absolute error limit. As in Issue 1, the choice of both local sum type and prediction mode can have a huge impact on compression performance, and the optimum choices strongly depend on the image type. However, results suggest that the best choices remain constant regardless of the number of prediction bands employed and that the same recommendations available for Issue 1 [22] are applicable to Issue 2 when in lossless mode; i.e., the use of column-oriented local sums and reduced mode is still recommended for images that exhibit significant streaking artifacts parallel to the y direction (e.g., AVIRIS-NG raw images), whereas for images without such artifacts, neighbor oriented local sums provide the best performance.

Varying the absolute error limit constant appears to generally have little impact on which choice of local sum type and prediction mode is optimum. Results may vary at very high (in relation to image bit depth) quantization levels, as indicated for the Landsat images.

The use of narrow local sums nearly always results in some performance penalty. The exception here is the use of neighbor-oriented local sums on AVIRIS-NG images. However, in this case, the lower complexity column-oriented local sums perform substantially better anyway. Narrow and wide column-oriented local sums differ only for the first image frame (i.e., at $y = 0$), and thus tend to yield only a small performance difference as long as image height is not small. For neighbor-oriented local sums, the penalty for using narrow instead of wide sums becomes significant (10 percent or more in several cases) at larger values of absolute error.

Regarding the selection of the number of prediction bands, as shown in Figures 1 and 2, using a very large value of P provides no appreciable improvement for most images. In addition, a large P value may even slightly decrease the performance for some images, which could be explained by the slower adaptation caused by a larger weight vector. A default value of $P = 3$ appears reasonable for both lossless and near-lossless compression.

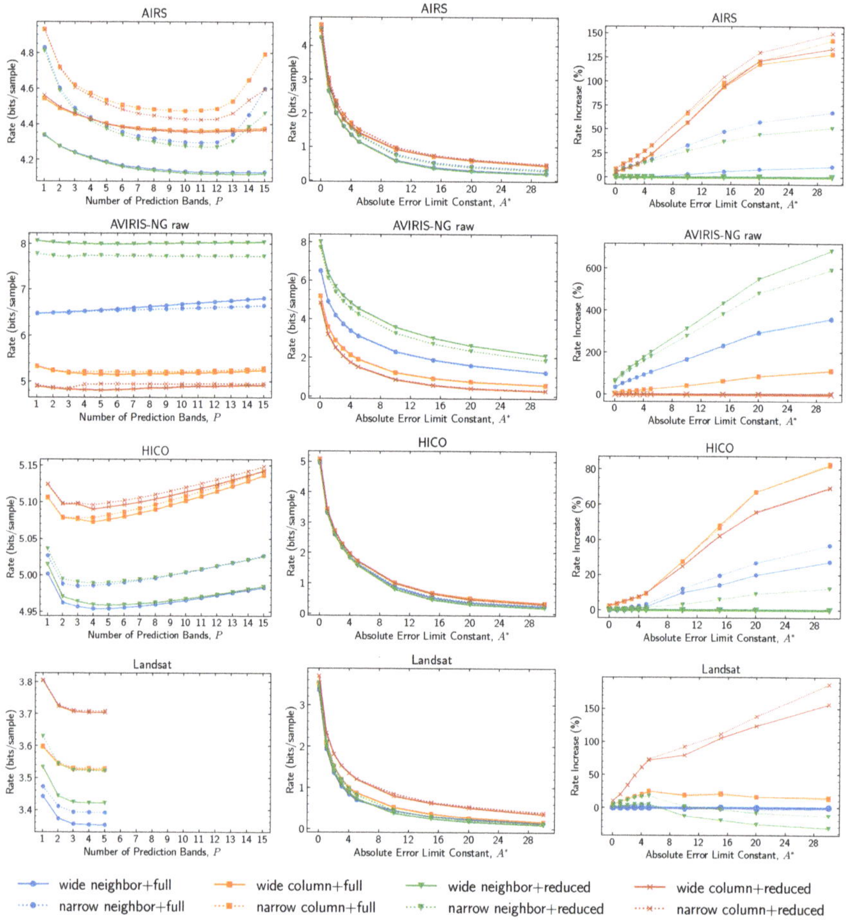

Figure 1. Compressed bit rate for different choices of prediction mode and local sum type. Plots in the left column are for lossless compression ($A^* = 0$). Plots in the right column are relative to the adjusted combination of the local sum type and prediction mode described in Section 2.

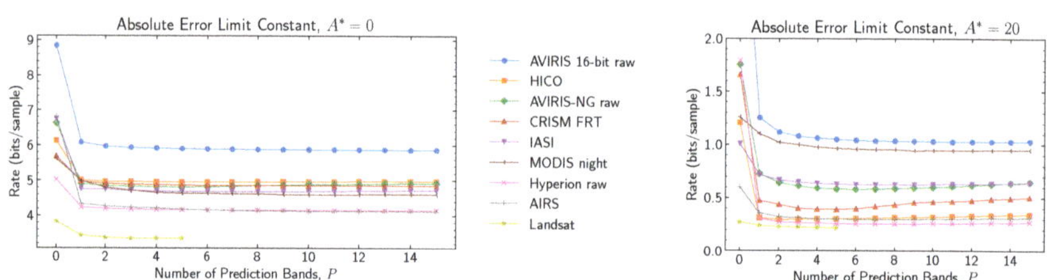

Figure 2. Average compressed bit rate performance as a function of P.

3.2. Adaptation Rate and Sample Representatives

The rate at which the predictor adapts to varying image characteristics is controlled by $\nu_{\min}$ and t_{inc} initially and by $\nu_{\max}$ once a steady state is reached. Figure 3 shows the performance of different steady-state learning rates, $\nu_{\max}$, under varying quantization levels. The results indicate that adequate values for $v_{\max}$ are generally invariant to the quantization level, though a more pronounced decrease in performance as $v_{\max}$ becomes smaller is evident.

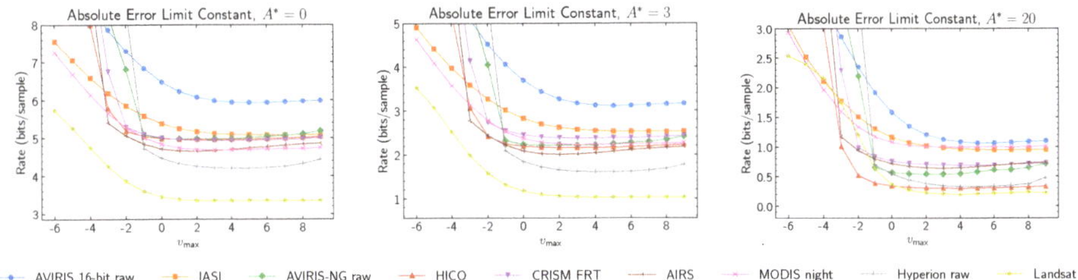

Figure 3. Average compressed data rate as a function of $v_{\max}$ when $v_{\min} = -6$ and $t_{\mathrm{inc}} = 2^7$.

Figure 4 reports on the performance impact of extreme choices for $v_{\min}$ and t_{inc}. In images with few samples per band, parameters affecting the initial learning rate control predictor learning rate for most (or all) samples in each band, thus, have a significant impact on overall prediction performance. This effect diminishes as images grow in spatial size.

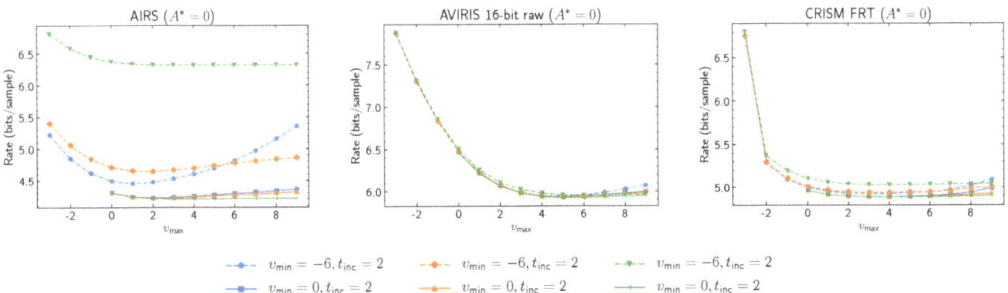

Figure 4. Average compressed data rate for different choices of parameters that affect the adaptation of the predictor to the image.

The newly-introduced sample representatives and the associated *damping* parameter ϕ_z interact significantly with the predictor learning rate. The damping parameter interacts with the predictor feedback loop in a manner that may prevent image noise from influencing the training process, and thus improve coding performance for noisy images. Figure 5 shows coding performances obtained for varying damping values in relation to $v_{\max}$. Substantial benefits may be obtained by properly-selected damping values. In our experiments, using sample representative *resolution* $\Theta = 3$, the best results have been obtained by employing a strong damping value of five for AVIRIS-NG, HICO, and SFSI; by employing a medium value of three for AIRS, AVIRIS 12-bit raw, CASI, CRISM, Hyperion, and M3 target; and no damping ($\phi_z = 0$) for the remaining instruments. Visual inspection of image noise levels seem to corroborate that noise levels are a determining factor in the selection of the damping value. However, these results do not exclude the possibility that other factors might also be relevant to the selection of damping values, such as instrument resolution or other forms of signal distortion different than noise.

When damping values are examined in relation to $v_{\min}$ and $v_{\max}$ in Figures 5 and 6, larger damping values have been found to be related to a decreased performance at higher $v_{\max}$ values. Setting $v_{\max} = 7 - 2^{-\Theta+3}\phi_z$ has been found to be a good choice in the experimental results. Similarly, higher damping values seem to produce decreased performances when used with smaller $v_{\min}$ values. Thus, higher damping values seem to narrow the desirable predictor learning rates.

Figure 7 reports the relation between damping value ϕ_z and the quantity of samples necessary to reach a steady predictor state, as controlled by t_{inc}. Curiously, for the $v_{\min}$ and $v_{\max}$ values set as described above, varying ϕ_z does not seem to influence the choice of t_{inc}.

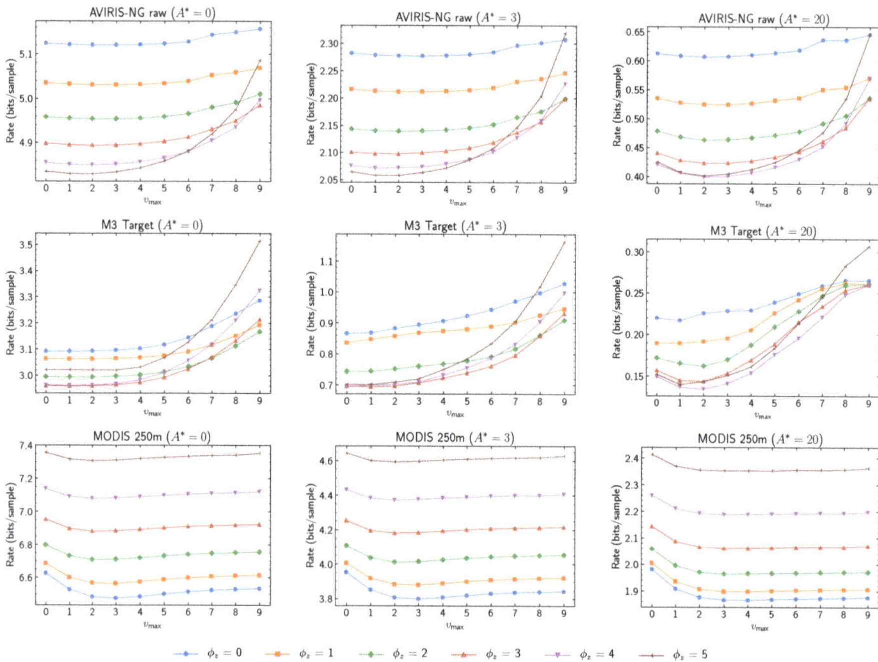

Figure 5. Average compressed data rate as a function of $\nu_{\max}$ for different values of the sample representative damping value, ϕ_z, and absolute error limit constant A^*, when $\Theta = 3$.

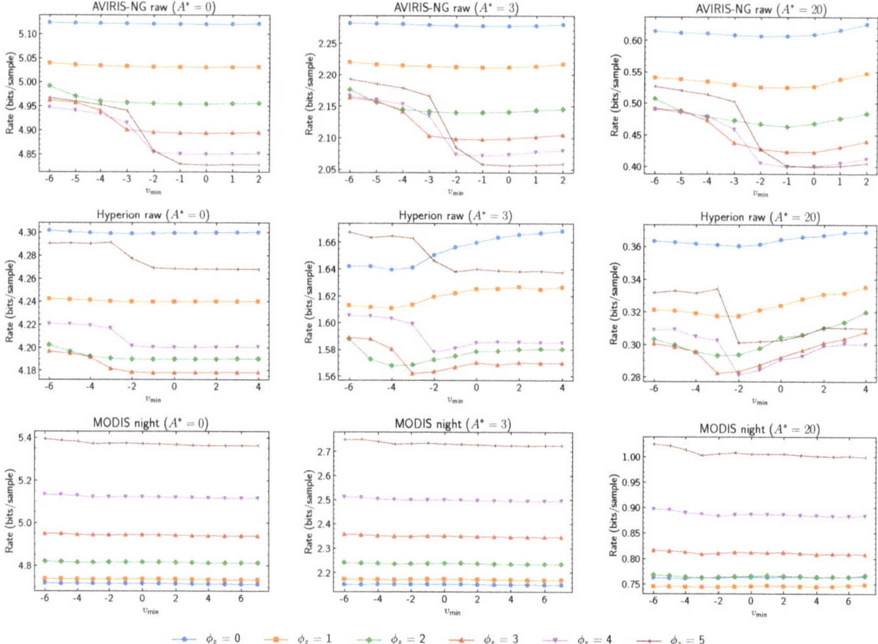

Figure 6. Average compressed data rate as a function of $\nu_{\min}$ for various values of the sample representative damping value, ϕ_z, and absolute error limit constant, A^*, when $\Theta = 3$.

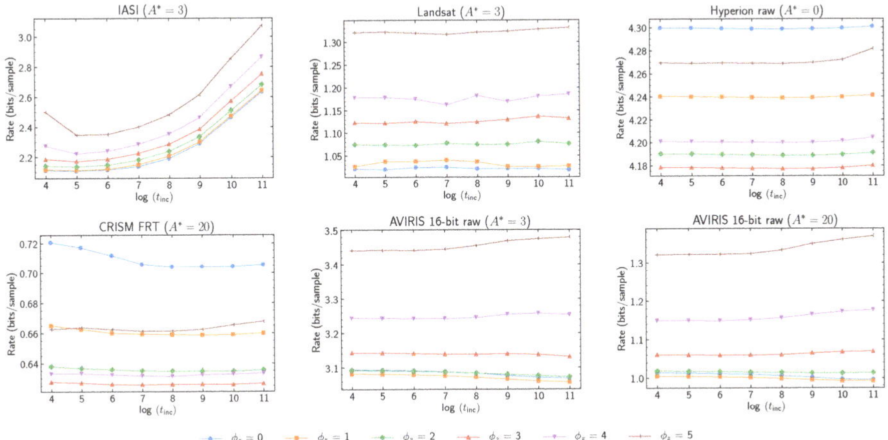

Figure 7. Average compressed data rate as a function of t_{inc} for various values of the sample representative damping value, ϕ_z, and absolute error limit constant, A^*, when $\Theta = 3$.

The *offset* parameter, ψ_z, also affects sample representative values, though this value has no effect when compression is lossless. For properly-selected values of ϕ_z, experimental results suggest setting $\psi_z = 2^\Theta - 1$ is a reasonable default choice (Figure 8).

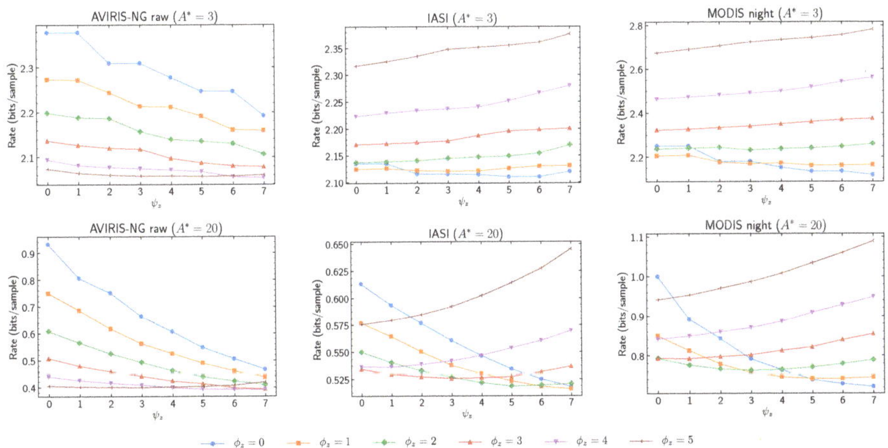

Figure 8. Average compressed data rate as a function of ψ_z for various values of sample representative damping value ϕ_z and absolute error limit constant A^*, when $\Theta = 3$.

3.3. Weight Component Resolution and Register Size

The precision with which the predictor stores its weight vector (i.e., its internal state) is controlled by the *weight component resolution* parameter, Ω. Together with the *register size* parameter, R, both parameters regulate the bit depths required for the multipliers employed in the predictor calculation. For each multiplier, the value of Ω controls the depth of one of its inputs (the other is controlled by image bit depth), while the value of R enables the multipliers to provide results modulo R, thus limiting its output bit depth.

The effects of varying weight resolutions are reported in Figures 9 and 10. For lossless compression, Figure 9 shows the relation between weight resolution and rate, both in absolute and relative terms. It can be observed that as Ω is decreased, prediction accuracy is negatively impacted by less accurate weight vector coefficients, and thus, the rate is increased. For the images tested, weight resolution can be decreased down to 11 bits with a rate increase smaller than 5%. For near-lossless compression, Figure 10 shows the relation between weight resolution and rate increase over results for $\Omega = 19$. It can be observed that the small perturbations in the curves for $A^* = 0$ are significantly amplified as A^* increases. While the predictor is approximately linear, lower values of Ω seem to magnify the non-linear interactions that finite-precision operations have on least significant bits.

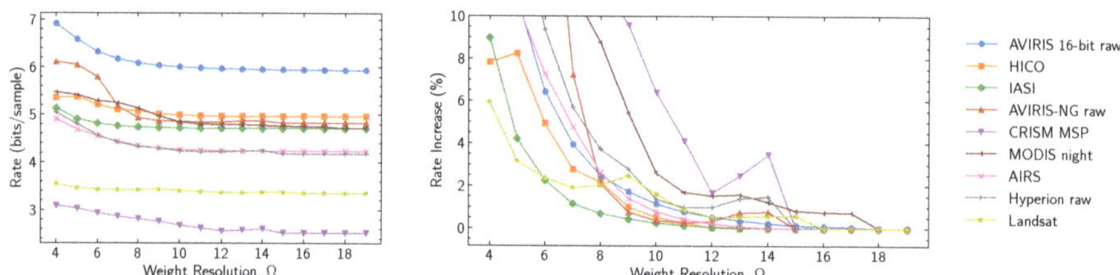

Figure 9. Average bit rate as a function of weight resolution Ω for lossless compression ($A^* = 0$). Curves in the right plot are relative to $\Omega = 19$.

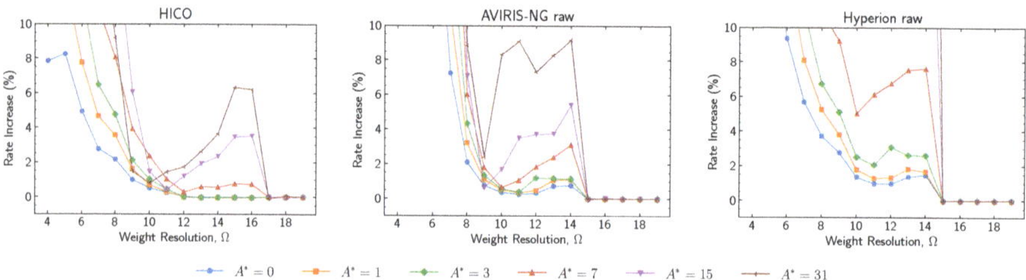

Figure 10. Change in the compressed data rate as a function of weight resolution Ω (and relative to $\Omega = 19$) for multiple absolute error limit constants A^*.

The effects of varying the register size parameter, R, are shown in Figure 11. For values of R over a certain threshold (shaded in gray in the figure plots), prediction results can be proven invariant [22] (pp. 4–7).While decreasing register size by one or two bits under this threshold does not increase the rate significantly, decreasing a few more bits rapidly yields very large increases in the rate.

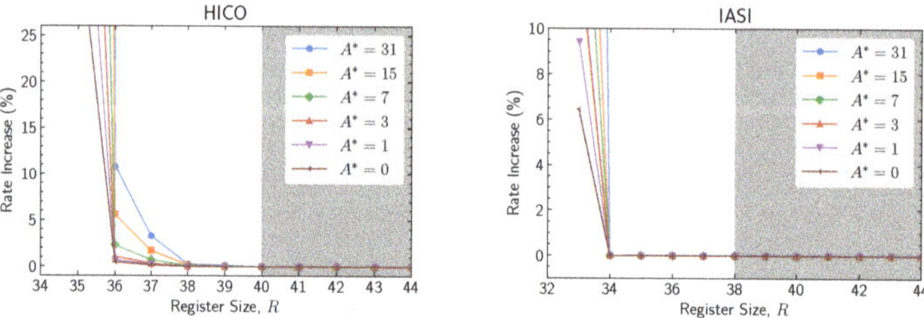

Figure 11. Change in compressed data rate as a function of register size R for multiple absolute error limit constants A^*. Results are relative to $R = 64$.

3.4. Weight Exponent Offsets

The use of weight exponent offsets $\{\zeta_z^{(i)}, \zeta^*\}$ allows fine tuning of the predictor learning rate for each band, and within a band for each predictor input as defined by the prediction mode and number of prediction bands. Hence, weight exponent offsets can be seen as an extension to the learning rate control provided through $\nu_{\min}$ and $\nu_{\max}$.

A brute force approach has been followed to understand the potential gains that could yield well-informed choices of weight exponent offsets for some of the corpus images. The procedure employed is as follows. First, optimal values for v_{min} and v_{max} are found through exhaustive search. Then, all choices of weight exponent offsets are tested for the first image band. The best choice is kept for that band, and the same procedure is repeated for each remaining band, one by one, until all weight exponent offsets are set.

Results are reported in Table 4. When a fixed damping value is used for an entire image, as studied in this article, the potential coding performance gains obtained by weight exponent offsets are scant.

Table 4. Data reductions obtained for weight exponent offsets set through exhaustive search.

Image	*Data Reduction (%)*
AVIRIS Yellowstone Scene 0 raw	0.3%
Landsat Agriculture	0.2%
MODIS 500 m A2001123.0000	0.5%
Pleiades Montpellier Misreg0	0.1%
Vegetation 1 1b	0.3%

Nonetheless, given that good choices for v_{max} vary from instrument to instrument, it is foreseen that an end-user may want to adjust the predictor learning rate for each image band. In particular, the end-user may want to do so whenever different per-band noise profiles recommend the use of different per-band sample representative damping values, or for images with large signal energy variation among bands. The authors have observed gains of up to 7% in a synthetic scenario where an instrument with large energy variations is simulated. For this purpose, a synthetic image is produced from a high SNR image by simulating 32 consecutive very dark bands followed by 32 consecutive very bright bands. Gains are observed when the synthetic image is encoded with band-dependent absolute error limits so that both dark and bright bands are encoded at roughly 1.5 bits per sample. However, setting per-band damping values requires a careful instrument modeling, which is out of the scope of this article.

3.5. Entropy Encoder

Issue 2 incorporates a new hybrid entropy encoder in addition to the sample-adaptive and block-adaptive encoders available in Issue 1. At high rates, the hybrid encoder encodes most samples using a family of codes that are equivalent to those used by the sample-adaptive coder, and thus has nearly identical performance [22]. However, the hybrid encoder has substantially better performance than the Issue 1 entropy encoders at low bit rates (Figure 12).

The sample-adaptive encoder is unable to reach rates lower than 1 bit per sample (due to design constraints), whereas the block-adaptive encoder is able to do so, but with non-negligible rate overhead over the hybrid encoder. In addition, the block-adaptive encoder may have poor performance when encoding in band interleaved by pixel (BIP) order, because samples from different bands (which may have different statistical behavior) are jointly encoded in the same block. Setting encoding order to band sequential (BSQ) encoding order somewhat mitigates these issues. However, the use of BSQ encoding order may not be suitable for all instruments due to large buffering requirements when acquisition order is not BSQ. In this case, similar performance may be often obtained by band interleaved by line (BIL) encoding order (Figure 13).

Code selection statistics used by the hybrid entropy coder are periodically rescaled at an interval determined by parameter γ^*; this is the parameter that has the most significant impact on hybrid entropy coder performance. Within each band, each time rescaling occurs, an additional bit is output to the compressed bitstream so that the decompressor can reconstruct the code selection statistics. At lower bit rates, this overhead can become significant, and larger values of γ^* are required to diminish its impact. This can be seen in Figure 14, where γ^* is varied. Results are reported in relation to the obtained data rate by adjusting the absolute error limit constant A^* (via trial-and-error) to achieve bit rates of approximately 0.1, 0.5, 1, 2, 3, and 4 bits per sample for the optimal γ^* value.

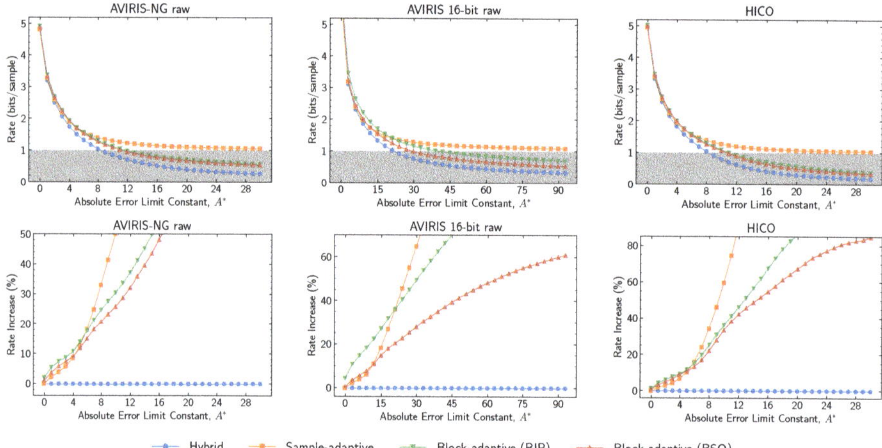

Figure 12. Compressed data rate for each entropy encoder as a function of absolute error limit constant A^* as the absolute value (top) and relative increases over the hybrid entropy encoder (bottom). Parameters for the sample-adaptive encoder are $K = 0$, $\gamma_0 = 1$, and $\gamma^* = 6$. Parameters for the block-adaptive encoder are $J = 64$ and $r = 4096$.

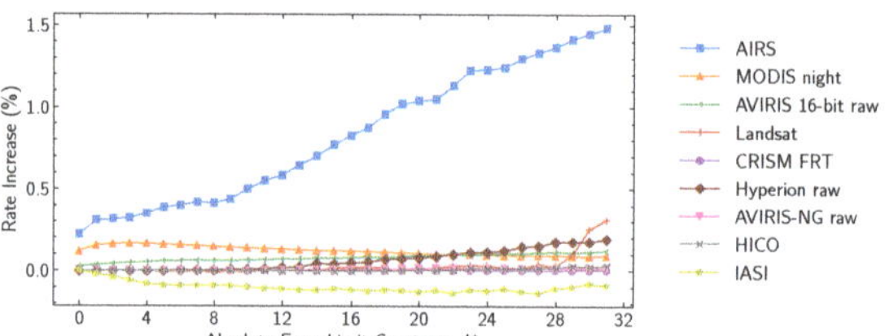

Figure 13. Change in compressed data rate for the block-adaptive entropy encoder when encoding in BIL order relative to when encoding in BSQ order. Parameters for the block-adaptive encoder are $J = 64$ and $r = 4096$.

The results suggest the use of $\gamma^* \approx 6$ for a high rate and to transition to larger γ^* values at rates below 1 bit per sample. At high compressed data rates, these results are comparable to those for the sample-adaptive encoder due to their similar operation. Inflection points are due to the logarithmic nature of the parameter, where a one-unit increase doubles the size of the renormalization interval and makes local statistics less important to the entropy encoder. For instance, this explains the CRISM FRT plot, where the results tend to stabilize as γ^* grows.

Results regarding the operation of the block-adaptive encoder under near-lossless mode for high compressed data rates are now reported. The encoder operation is controlled by the *reference sample interval* parameter, r, and the *block size* parameter, J, with results reported respectively in Figures 15 and 16. The reference sample interval enables efficient encoding of "zero-block" runs (sequences of blocks of zero values). However, the practical effects of this parameter are negligible except for combinations of very small values of r with high values of A^*. Regarding the block size values, for small absolute error limit constant values, where it makes sense to use the block-adaptive encoder, employing a block size of 64 yields better performance, even if this trend is reversed for larger A^* values.

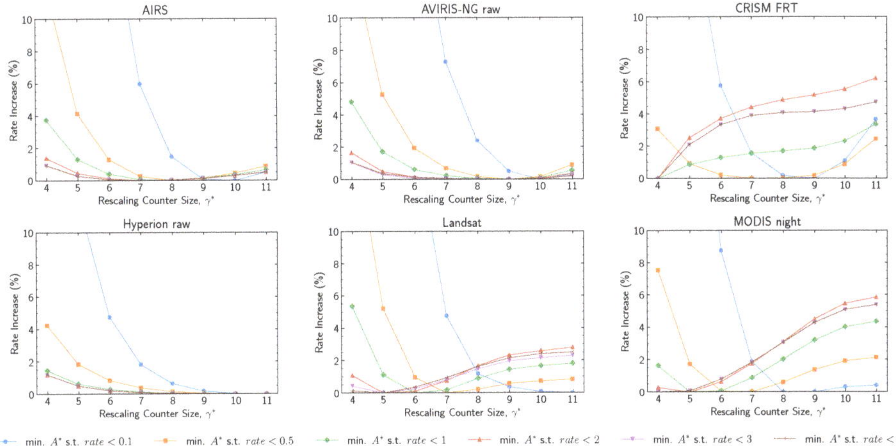

Figure 14. Change in compressed data rate (i.e., relative to an optimal selection of γ^*) as a function of rescaling counter size γ^*.

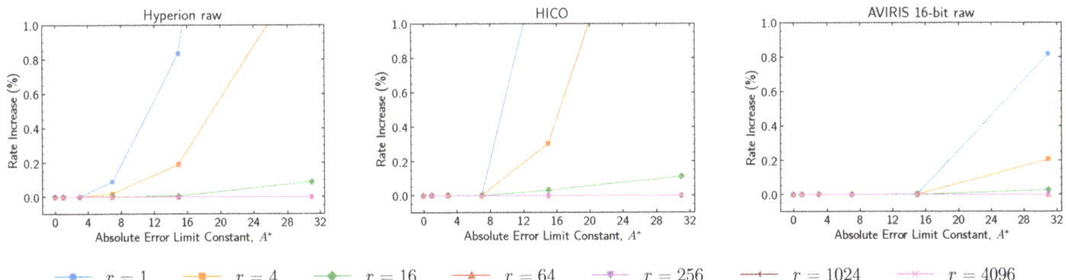

Figure 15. Average compressed data rate when employing the block-adaptive entropy encoder as a function of absolute error limit constant A^*. Results are relative to $r = 4096$. Block size J is set to 64, and images are encoded in BIL order.

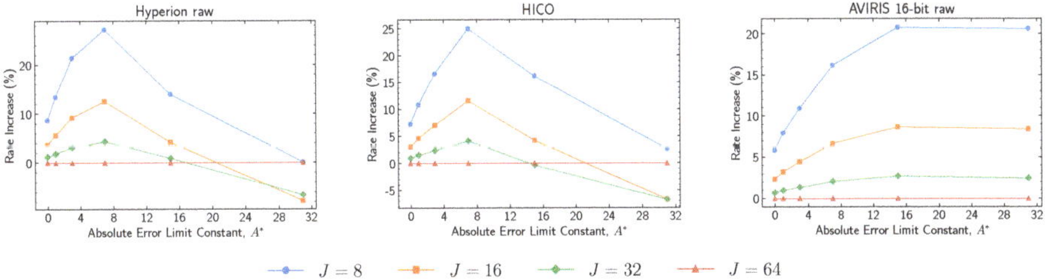

Figure 16. Change in compressed data rate when employing the block-adaptive entropy encoder for multiple block sizes J as a function of absolute error limit constant A^*. Results are relative to $J = 64$. Reference sample interval r is set to 4096, and images are encoded in BIL order.

4. Conclusions

This article examined performance trade-offs among key parameters and options in the new Issue 2 of "Low-Complexity Lossless and Near-Lossless Multispectral and Hyperspectral Image Compression."

The behavior of the new near-lossless capability was studied along with the effects of four key new features: narrow local sums, sample representatives, weight exponent offsets, and the new hybrid entropy encoder.

Regarding narrow local sums, their use tends to incur a minimal to moderate coding performance penalty. In particular, this penalty tends to be fairly small for column-oriented local sums when image height is large.

In the study of sample representatives, for many images, well-chosen nonzero values of parameters ϕ_z and ψ_z were shown to provide a noticeable improvement in compression performance compared to the obvious alternative of setting the sample representative equal to the center of the quantizer bin (achieved by setting $\phi_z = \psi_z = 0$). However, at larger values of ϕ_z, performance becomes more sensitive to prediction adaptation rates as controlled though $v_{\min}$ and $v_{\max}$. It is recommended to decrease $v_{\max}$ as ϕ_z increases.

Concerning the use of weight exponent offsets, for the images tested in this article and in combination with band-independent sample representative parameters, the use of nonzero weight exponent offsets has shown scant performance increments. The significant band-to-band variation in signal energy needed to motivate the use of nonzero weight exponent offsets may be uncommon in practice, based on the images in the corpus.

Regarding the variables controlling data widths in the predictor, the recommended value for Ω is 19 bits. For very small absolute error limits, this value can be lowered down to 11 bits with a rate increase of less than 5% for the images tested. Parameter R can be set one or two bits below the proven invariance threshold at insignificant rate variations. Setting values larger than necessary yields no benefits.

Finally, under near-lossless compression, the hybrid entropy encoder has shown substantial coding gains compared to the legacy coding options, particularly as the maximum allowed error increases. Experimental results suggest that higher values of γ^* improve performance at lower bit rates. Regarding the block-adaptive entropy encoder, larger block sizes have been shown to yield improved coding performances when the maximum allowed error was small.

Author Contributions: Conceptualization, I.B., A.K., and J.S.-S.; methodology, I.B.; software, I.B. and M.H.-C.; validation, I.B., A.K., M.H.-C., and J.S.-S.; formal analysis, I.B. and A.K.; investigation, I.B., A.K., and J.S.-S.; resources, I.B. and M.H.-C.; data curation, I.B.; writing, original draft preparation, I.B., A.K., and J.S.-S.; writing, review and editing, I.B., A.K., and J.S.-S.; visualization, I.B.; supervision, I.B., A.K., and J.S.-S.; project administration, I.B., A.K., and J.S.-S.; funding acquisition, I.B., A.K., and J.S.-S.

Funding: This work was supported in part by the Centre National d'Études Spatiales, by the Spanish Ministry of Economy and Competitiveness and the European Regional Development Fund under Grants TIN2015-71126-R and RTI2018-095287-B-I00 (MINECO/FEDER, UE), and by the Catalan Government under Grant 2017SGR-463. This project has received funding from the European Union's Horizon 2020 research and innovation program under Grant Agreement No. 776151. The research conducted by A. Kiely was carried out at the Jet Propulsion Laboratory, California Institute of Technology, under a contract with the National Aeronautics and Space Administration.

Conflicts of Interest: The funders had no role in the design of the study; in the collection, analyses, or interpretation of the data; in the writing of the manuscript; nor in the decision to publish the results.

References

1. Motta, G.; Rizzo, F.; Storer, J.A. *Hyperspectral Data Compression*; Springer Science & Business Media: New York, NY, USA, 2006.

2. Blanes, I.; Serra-Sagrista, J.; Marcellin, M.W.; Bartrina-Rapesta, J. Divide-and-conquer strategies for hyperspectral image processing: A review of their benefits and advantages. *IEEE Signal Process. Mag.* **2012**, *29*, 71–81. [CrossRef]

3. Báscones, D.; González, C.; Mozos, D. Hyperspectral Image Compression Using Vector Quantization, PCA and JPEG2000. *Remote Sens.* **2018**, *10*, 907. [CrossRef]

4. Consultative Committee for Space Data Systems. *Lossless Data Compression. Recommendation for Space Data System Standards, CCSDS 121.0-B-2*; CCSDS: Washington, DC, USA, 2012.

5. Consultative Committee for Space Data Systems. *Image Data Compression. Recommendation for Space Data System Standards, CCSDS 122.0-B-2*; CCSDS: Washington, DC, USA, 2017.

6. Consultative Committee for Space Data Systems. *Spectral Preprocessing Transform for Multispectral and Hyperspectral Image Compression. Recommendation for Space Data System Standards, CCSDS 122.1-B-1*; CCSDS: Washington, DC, USA, 2017.

7. Consultative Committee for Space Data Systems. *Lossless Multispectral & Hyperspectral Image Compression. Recommendation for Space Data System Standards, CCSDS 123.0-B-1*; CCSDS: Washington, DC, USA, 2012.

8. Consultative Committee for Space Data Systems. *Lossless and Near-Lossless Multispectral and Hyperspectral Image Compression. Recommendation for Space Data System Standards, CCSDS 123.0-B-2*; CCSDS: Washington, DC, USA, 2019.

9. Klimesh, M. Low-complexity adaptive lossless compression of hyperspectral imagery. *Proc. SPIE* **2006**, *6300*, 63000N-1–63000N-9.

10. Klimesh, M.; Kiely, A.; Yeh, P. Fast lossless compression of multispectral and hyperspectral imagery. In Proceedings of the 2nd International WS on On-Board Payload Data Compression (OBPDC), ESA/CNES, Tolouse, France, 28–29 October 2010.

11. Keymeulen, D.; Aranki, N.; Bakhshi, A.; Luong, H.; Sarture, C.; Dolman, D. Airborne demonstration of FPGA implementation of Fast Lossless hyperspectral data compression system. In Proceedings of the 2014 IEEE NASA/ESA Conference on Adaptive Hardware and Systems (AHS), Leicester, UK, 14–18 July 2014; pp. 278–284.

12. Santos, L.; Berrojo, L.; Moreno, J.; López, J.F.; Sarmiento, R. Multispectral and hyperspectral lossless compressor for space applications (HyLoC): A low-complexity FPGA implementation of the CCSDS 123 standard. *IEEE J. Sel. Top. Appl. Earth Obs. Remote Sens.* **2016**, *9*, 757–770. [CrossRef]

13. Theodorou, G.; Kranitis, N.; Tsigkanos, A.; Paschalis, A. High performance CCSDS 123.0-B-1 multispectral & hyperspectral image compression implementation on a space-grade SRAM FPGA. In Proceedings of the 5th International Workshop On-Board Payload Data Compression, Frascati, Italy, 28–29 September 2016; pp. 28–29.

14. Báscones, D.; González, C.; Mozos, D. FPGA implementation of the CCSDS 1.2. 3 standard for real-time hyperspectral lossless compression. *IEEE J. Sel. Top. Appl. Earth Obs. Remote Sens.* **2018**, *11*, 1158–1165. [CrossRef]

15. Báscones, D.; González, C.; Mozos, D. Parallel Implementation of the CCSDS 1.2. 3 Standard for Hyperspectral Lossless Compression. *Remote Sens.* **2017**, *9*, 973. [CrossRef]

16. Tsigkanos, A.; Kranitis, N.; Theodorou, G.A.; Paschalis, A. A 3.3 Gbps CCSDS 123.0-B-1 multispectral & Hyperspectral image compression hardware accelerator on a space-grade SRAM FPGA. *IEEE Trans. Emerg. Top. Comput.* **2018**. [CrossRef]

17. Fjeldtvedt, J.; Orlandić, M.; Johansen, T.A. An efficient real-time FPGA implementation of the CCSDS-123 compression standard for hyperspectral images. *IEEE J. Sel. Top. Appl. Earth Obs. Remote Sens.* **2018**, *11*, 3841–3852. [CrossRef]

18. Orlandić, M.; Fjeldtvedt, J.; Johansen, T.A. A Parallel FPGA Implementation of the CCSDS-123 Compression Algorithm. *Remote Sens.* **2019**, *11*, 673. [CrossRef]

19. Rodríguez, A.; Santos, L.; Sarmiento, R.; Torre, E.D.L. Scalable Hardware-Based On-Board Processing for Run-Time Adaptive Lossless Hyperspectral Compression. *IEEE Access* **2019**, *7*, 10644–10652. [CrossRef]

20. Kiely, A.; Klimesh, M.; Blanes, I.; Ligo, J.; Magli, E.; Aranki, N.; Burl, M.; Camarero, R.; Cheng, M.; Dolinar, S.; et al. The new CCSDS standard for low-complexity lossless and near-lossless multispectral and hyperspectral image compression. In Proceedings of the 6th International WS on On-Board Payload Data Compression (OBPDC), ESA/CNES, Matera, Italy, 20–21 September 2018.

21. Augé, E.; Sánchez, J.E.; Kiely, A.; Blanes, I.; Serra-Sagristà, J. Performance impact of parameter tuning on the CCSDS-123 lossless multi- and hyperspectral image compression standard. *J. Appl. Remote Sens.* **2013**, *7*. [CrossRef]

22. Consultative Committee for Space Data Systems. *Lossless Multispectral & Hyperspectral Image Compression. Report Concerning Space Data System Standards, CCSDS 120.2-G-1*; CCSDS: Washington, DC, USA, 2012.

23. Consultative Committee for Space Data Systems. Testing and Evaluation Image Corpus of CCSDS-123.0-B-1. Available online: https://cwe.ccsds.org/sls/docs/SLS-DC/123.0-B-Info/TestData/ (accessed on 10 June 2019).

24. Valsesia, D.; Magli, E. A Novel Rate Control Algorithm for Onboard Predictive Coding of Multispectral and Hyperspectral Images. *IEEE Trans. Geosci. Remote Sens.* **2014**, *52*, 6341–6355. [CrossRef]
25. Conoscenti, M.; Coppola, R.; Magli, E. Constant SNR, Rate Control, and Entropy Coding for Predictive Lossy Hyperspectral Image Compression. *IEEE Trans. Geosci. Remote Sens.* **2016**, *54*, 7431–7441. [CrossRef]

 *remote sensing*

Technical Note

A Task-Driven Invertible Projection Matrix Learning Algorithm for Hyperspectral Compressed Sensing

Shaofei Dai [1,2,*], Wenbo Liu [1,2], Zhengyi Wang [1,2] and Kaiyu Li [1,2]

1. College of Automation Engineering, Nanjing University of Aeronautics and Astronautics, Nanjing 211006, China; wenboliu@nuaa.edu.cn (W.L.); 13685111061@nuaa.edu.cn (Z.W.); LKY_401@nuaa.edu.cn (K.L.)
2. Non-Destructive Testing and Monitoring Technology for High-Speed Transport Facilities Key Laboratory of Ministry of Industry and Information Technology, Nanjing 211006, China
* Correspondence: daishaofei@nuaa.edu.cn

Abstract: The high complexity of the reconstruction algorithm is the main bottleneck of the hyperspectral image (HSI) compression technology based on compressed sensing. Compressed sensing technology is an important tool for retrieving the maximum number of HSI scenes on the ground. However, the complexity of the compressed sensing algorithm is limited by the energy and hardware of spaceborne equipment. Aiming at the high complexity of compressed sensing reconstruction algorithm and low reconstruction accuracy, an equivalent model of the invertible transformation is theoretically derived by us in the paper, which can convert the complex invertible projection training model into the coupled dictionary training model. Besides, aiming at the invertible projection training model, the most competitive task-driven invertible projection matrix learning algorithm (TIPML) is proposed. In TIPML, we don't need to directly train the complex invertible projection model, but indirectly train the invertible projection model through the training of the coupled dictionary. In order to improve the accuracy of reconstructed data, in the paper, the singular value transformation is proposed. It has been verified that the concentration of the dictionary is increased and that the expressive ability of the dictionary has not been reduced by the transformation. Besides, two-loop iterative training is established to improve the accuracy of data reconstruction. Experiments show that, compared with the traditional compressed sensing algorithm, the compressed sensing algorithm based on TIPML has higher reconstruction accuracy, and the reconstruction time is shortened by more than a hundred times. It is foreseeable that the TIPML algorithm will have a huge application prospect in the field of HSI compression.

Keywords: compressed sensing; hyperspectral image; invertible projection; coupled dictionary; singular value; task-driven learning

Citation: Dai, S.; Liu, W.; Wang, Z.; Li, K. A Task-Driven Invertible Projection Matrix Learning Algorithm for Hyperspectral Compressed Sensing. *Remote Sens.* **2021**, *13*, 295. https://doi.org/10.3390/rs13020295

Received: 8 December 2020
Accepted: 14 January 2021
Published: 15 January 2021

Publisher's Note: MDPI stays neutral with regard to jurisdictional claims in published maps and institutional affiliations.

1. Introduction

Spectral images with a spectral resolution in the 10-2 order of magnitude are called hyperspectral images (HSI). HSI is one of the main components of modern remote sensing [1]. In recent decades, the popularity of hyperspectral technology has continued to rise. One of the main reasons for the higher visibility of hyperspectral imaging is the richness of the spectral information collected by this sensor. This function has positioned the hyperspectral analysis technology as the mainstream solution for land area analysis and the identification and differentiation of visually similar surface materials. Therefore, hyperspectral technology has become more and more important and is widely used in various applications, such as precision agriculture, environmental monitoring, geology, urban surveillance and homeland security, food quality inspection, etc. However, hyperspectral image processing is accompanied by a large amount of data management, which affects real-time performance on the one hand, and, on the other hand, the demand for on-board storage resources. In addition, the latest technological advances are introducing hyperspectral cameras with a higher spectrum and spatial resolution to the market. From the perspective of onboard processing, communication, and storage, all of these make efficient data processing more challenging [2–5].

In order to solve the storage and transmission problems that need to be faced in processing multi-dimensional hyperspectral image data, data compression technology is usually selected. The original digital signal is refined and expressed, and its storage space and transmission bandwidth requirements are reduced by data compression technology. However, in traditional data compression technology, a large amount of redundant data is collected, and then compression technology is used to remove these redundant data. Although the redundant information of data is reduced, huge resources are wasted in this process [6].

Compressed Sensing (CS) theory was proposed in 2006 [7], and was widely used in wireless networks, imaging technology, target positioning, the direction of arrival estimation, and other fields [8,9].

As shown in Figure 1, in the compressed sensing technology, data compression is set up at the signal acquisition end, the information characteristics of the data are directly collected, and the original signal is reconstructed with high precision at the reconstruction end.

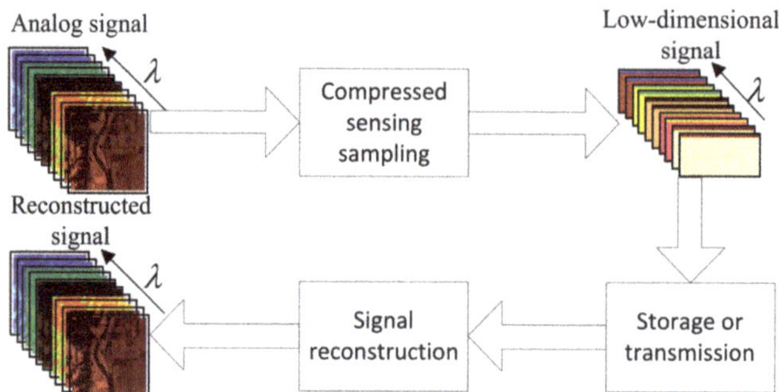

Figure 1. Block diagram of compressed sensing.

As shown in Equation (1), compressed sensing theory shows [7] that an over-complete dictionary $D \in R^{m \times s}$ is used, so that the original signal x is sparse enough under dictionary D, and its sparse coefficient is expressed as $\alpha \in R^s$. Equation (2) is the mathematical expression of compressed sensing. The original signal is observed by the observation matrix $\Phi \in R^{d \times m}$ to obtain the observation signal $y \in R^d$ ($d \ll m$). In this process, signal compression is realized. When the dimension d of the observation matrix is higher than a certain lower limit, the original signal x can be uniquely reconstructed from the observation signal y [10].

$$x = D\alpha \tag{1}$$

$$y = \Phi x \tag{2}$$

In CS theory, the compressibility, sparsity, and incoherence of the signal are fused, and in addition, compression and sampling are combined. The measurement signal is projected into the observation space containing all the effective information of the signal so that the sparse signal of limited dimensions can be sampled at a sampling frequency far less than the Nyquist theorem requires, making it possible to sample less than twice the original signal bandwidth [11].

The compression rate and reconstruction accuracy of the CS system mainly depend on the following three aspects:

(1)   Sparse expression

To express the signal in a refined manner, the signal is usually transformed into a new basis or framework, which is also called a dictionary. The more sparsely the data is represented by the dictionary, the larger the compression ratio can be obtained under the condition of ensuring the image reconstruction error. Therefore, dictionary learning plays an important role in data compression based on sparse decomposition. At present, the commonly used dictionaries in compressed sensing include the discrete cosine transform (DCT) dictionary [12,13], the discrete wavelet transform (DWT) dictionary [9,14], and the K-SVD dictionary [15–17], Gabor dictionary [15], CDL dictionary [18], etc.

(2)   Establishment of measurement matrix

When the original signal is compressed by the observation matrix, it must obey the Restricted Isometry Property (RIP). Under RIP conditions, the dimension d of the observation matrix of the CS system must be higher than a certain limit value. At the same time, the measurement matrix Φ and the sparse basis D meet incoherence, and the observation signal y can be uniquely reconstructed [7]. At present, in compressed sensing, the commonly used measurement matrices include random Gaussian matrix, random Bernoulli matrix, random Fourier matrix, and random Hadamard matrix [19,20], etc.

(3)   Sparse reconstruction

The signal reconstruction problem of compressed sensing is to solve the underdetermined equations $y = \Phi x$ to obtain the original signal x based on the known measurement value y and measurement matrix Φ. In order to ensure that the original signal is efficiently and stably reconstructed, many excellent sparse signals have been proposed. Commonly used sparse reconstruction algorithms include Orthogonal Matching Pursuit (OMP), Stagewise Orthogonal Matching Pursuit (StOMP), Compressive sampling matching pursuit (CoSaMP), Regularized Orthogonal Matching Pursuit (ROMP), Generalized Regularized Orthogonal Matching Pursuit (GROMP), Smoothed Projected Landweber (SPL) [21–25], etc.

Although compressed sensing technology has been proven to have great research significance in data compression, however, the features of HSI are complex. In order to improve the sparsity of the signal, the redundant dictionary with high redundancy is used as its sparse domain. When HSI is reconstructed, a large number of iterative calculations and inversion operations are required, and the traversal process of finding the optimal atom in each iteration is very time-consuming. A lot of time will be wasted in the sparse reconstruction process, which greatly limits the application of compressed sensing technology in the field of HSI compression.

Aiming at the time-consuming and low reconstruction accuracy of the reconstruction algorithm of compressed sensing technology, in the paper, we proposed a task-driven invertible projection matrix learning algorithm. The problems of long reconstruction time and insufficient reconstruction accuracy of HSI compression algorithm based on solving compressed sensing are solved by our proposed algorithm.

In the algorithm proposed in this paper, prior knowledge of data is used to train an invertible projection matrix $U \in R^{d \times m}$. The projection matrix can project the original signal x into the low-dimensional observation signal y, and the projection formula can be expressed as $y = Ux$. Different from the traditional observation matrix Φ, the process in which the original signal x is projected by the projection matrix U to the low-dimensional observation signal y is invertible, and the inverse process can be expressed as $x' = U^T y$, where U^T represents the transposition of U, and x' is the reconstruction signal. Compared with the sparse reconstruction of traditional compressed sensing, when the algorithm proposed in this paper is used to train the projection matrix U as the observation matrix of the compressed sensing algorithm. In the signal reconstruction process, there is no need to perform tedious iteration and inversion operations, which results in a lot of time being saved and increases the real-time performance of the compressed sensing algorithm.

On the basis of this algorithm, we study a hyperspectral compressed sensing algorithm based on a task-driven invertible projection matrix learning algorithm. In order to prove the effectiveness of the algorithm proposed in this paper, this algorithm was compared with the most competitive compressed sensing algorithm based on DCT dictionary, CDL dictionary, and K-SVD dictionary. Besides, in order to verify the real-time performance of the signal reconstruction process of the compressed sensing algorithm based on the algorithm proposed in this paper, in the experiment, this algorithm was compared with the current best compressed sensing reconstruction algorithm, such as OMP, StOMP, CoSaMP, GOMP, GROMP, and SPL, etc. [21–25].

2. Principles and Methods

2.1. Constraints on Invertible Projection Transformation

Assuming that the projection matrix $U \in R^{d \times m}$, the original signal $x \in R^{m \times 1}$ can be accurately reconstructed from the low-dimensional signal y by the projection transposed matrix U^T, and the solution to the projection matrix U can be described as

$$U = \underset{U}{\operatorname{argmin}} \left\| x - U^T U x \right\|_F^2 \tag{3}$$

Model (3) can be equivalent to

$$
\begin{aligned}
U &= \underset{U}{\operatorname{argmin}} \left\| x - U^T U x \right\|_F^2 \\
&= \underset{U}{\operatorname{argmin}} \left\| x^T x - x^T U^T U x \right\|_F^2 \\
&= \underset{U}{\operatorname{argmin}} \left\| x^T x - y^T y \right\|_F^2
\end{aligned}
\tag{4}
$$

where x is the high-dimensional original signal, and $y = Ux$, y is the low-dimensional signal of the high-dimensional original signal x.

The high-dimensional signal x and its low-dimensional signal y have the same or similar sparse representation coefficients $\alpha \in R^{s \times 1}$ [26]. It can be described as

$$
\begin{cases}
x = D\alpha \\
y = P\alpha
\end{cases}
\tag{5}
$$

where $D \in R^{m \times s}$ is the high-dimensional dictionary of high-dimensional signal x, $P \in R^{d \times s}(d << m)$ is the low-dimensional dictionary of low-dimensional signal y, and dictionary P is constructed by D through linear projection mapping U. The projection formula is $P = UD$ [26].

Therefore, model (3) can be described as

$$
\begin{aligned}
U &= \underset{U}{\operatorname{argmin}} \left\| x - U^T U x \right\|_F^2 \\
&= \underset{U}{\operatorname{argmin}} \left\| \alpha^T D^T D\alpha - \alpha^T P^T P\alpha \right\|_F^2 \\
&= \underset{U}{\operatorname{argmin}} \left\| D^T D - P^T P \right\|_F^2 \\
&= \underset{U}{\operatorname{argmin}} \left\| D^T D - :UD)^T UD \right\|_F^2
\end{aligned}
\tag{6}
$$

From the derivation process of model (6) we can know that the optimal projection matrix U under model (3) is equivalent to the optimal value under model (6). Since the high-dimensional dictionary D can be regarded as fixed when solving model (6), the optimal projection matrix U in the solution model (6) can be regarded as the optimal $P = UD$, so that model (6) is obtained the minimum Frobenius norm. The problem of calculating the projection matrix U can be converted to calculating the low-dimensional matrix P indirectly, which can be described as

$$P = \underset{P}{\arg\min} \left\| D^T D - P^T P \right\|_F^2 \tag{7}$$

In model (7), P can be considered as a principal components analysis (PCA) dimensionality reduction of dictionary D. Although PCA has the best performance in the sense of mean square error, the PCA algorithm requires the covariance matrix of the signal to be calculated in advance, and many covariance calculations are required. Different from the idea of PCA, the low-dimensional dictionary $P(P = UD)$ is trained by model (7) to make model (7) obtain the optimal value. From the derivation process of model (6), we can know that this projection matrix U is also the optimal value of model (3). Therefore, Equations (3) and (7) have the same extreme points. In the process of training model (7), in order to ensure sufficient reconstruction accuracy of the signal, we only need to ensure that model (7) obtains the optimal value, then model (3) will also be optimal. Model (3), in order to solve the invertible transformation problem, is converted to the problem of solving model (7), that is, extracting principal components from dictionary D. In order to ensure that more principal components of high-dimensional dictionary D are retained in low-dimensional dictionary P, dictionary D is required to concentrate energy. We will give a solution in Section 2.2.

2.2. Singularity Transformation

For a general dictionary after dimensionality reduction, more principal components cannot be retained [16], so the singular value decomposition of dictionary D can be described as

$$D = M \Lambda V^T = \begin{bmatrix} M_h & M_l \end{bmatrix} \begin{bmatrix} \Lambda_h & 0 \\ 0 & \Lambda_l \end{bmatrix} \begin{bmatrix} V_h & V_l \end{bmatrix}^T \tag{8}$$

where M is the left singular matrix of matrix D, Λ is the singular value of matrix D, V is the right singular matrix of matrix D, and Λ_h is the first h rows and h columns of Λ.

Take $U = M_h{}^T$, then

$$D^T D = V \Lambda^T \Lambda V^T = V(\Lambda_h{}^T \Lambda_h + \Lambda_l{}^T \Lambda_l) V^T \tag{9}$$

$$P^T P = (UD)^T UD = V_h \Lambda_h^T \Lambda_h V_h^T \tag{10}$$

Therefore, the first h singular values Λ_h of the matrix D are larger (the last l singular values Λ_l are smaller and not 0), and more principal components of the high-dimensional dictionary D are retained in the low-dimensional dictionary P, in other words, $\| D^T D - P^T P \|_F^2$ is smaller.

In order to make the first h dimension of dictionary D gather more energy, it should satisfy:

(1) The expression coefficient of signal x under dictionary D is sparse enough;
(2) the first h singular values of dictionary D are large enough, and the last l singular values are small enough, and the value is not 0 [16].

The singular values often correspond to the important information implicit in the matrix, and the importance is positively related to the magnitude of the singular values. The more "singular" the matrix is, the less singular value contains more matrix information, and the smaller the information entropy of the matrix, and the more relevant its row (or column) vectors are [27]. Aiming at the characteristics of matrix singular values, in the paper, matrix singular transformation is proposed to increase the redundancy of the dictionary without reducing the expressive ability of the dictionary and the rank of the matrix.

The singular value transformation matrix Θ is defined as

$$\Theta(t,r) = Q(1,0)\ldots \underbrace{Q(n-t+1,r)\ldots Q(n,r)}_{t} = \begin{bmatrix} 1 & 0 & \cdots & r & \cdots & r \\ 0 & 1 & \cdots & 0 & \cdots & 0 \\ 0 & 0 & \cdots & 0 & \cdots & 0 \\ 0 & 0 & \cdots & 1-r & \cdots & 0 \\ \cdots & \cdots & \cdots & \cdots & \cdots & \cdots \\ 0 & 0 & \cdots & 0 & \cdots & 1-r \end{bmatrix}_{n\times n} \tag{11}$$

where, the variables t and r are transformation adjustment parameters, and $Q(i,r)$ represents the values of the first row and the i-th row of the i-th column vector of the unit square $E_{n\times n}$ are r and $1-r$, respectively.

Let the dictionary $D = [d_1\ d_2\ d_3\ \ldots\ d_n]$ and the original data set be $X = [x_1, x_2, \ldots, x_n]$ ($x_i \in R^m$). Assuming data x_i under dictionary D indicates that the sparse coefficient has four non-zero terms, expressed as $\alpha_i = [\ \ldots\ a_o\ \ldots\ a_m\ \ldots\ a_p\ \ldots\ a_q\ \ldots\ ]^T$ ($\alpha_i \in R^s$). The regulating parameter of singular value transformation $t = n - 1$.

The result of dictionary D undergoing singular value Θ transformation is:

$$\hat{D} = D\Theta(n-1,r) = \begin{bmatrix} d_1 & dw_2 & \cdots & dw_n \end{bmatrix} \tag{12}$$

and

$$dw_k = rd_1 + (1-r)d_k \tag{13}$$

Assuming that the data x_i under the dictionary $\hat{D}$ represents the sparsity coefficient $\beta = [b_1\ b_2\ b_3\ \ldots\ b_n]^T$, then:

$$D\alpha_i = a_o d_o + a_m d_m + a_p d_p + a_q d_q \tag{14}$$

$$\hat{D}\beta = b_1 d_1 + b_2 dw_2 + \ldots + b_n dw_n] \tag{15}$$

$$x_i = D\alpha_i = \hat{D}\beta \tag{16}$$

According to Equations (14)–(16), it can be determined that the coefficient of data x_i is still sparse under dictionary $\hat{D}$, and the sparse coefficient is:

$$\begin{cases} b_1 = \frac{r(a_o + a_m + a_p + a_q)}{1-r} \\ b_o = \frac{a_o}{(1-r)} \\ b_m = \frac{a_m}{(1-r)} \\ b_p = \frac{a_p}{(1-r)} \\ b_q = \frac{a_q}{(1-r)} \\ b_i = 0 \\ \scriptstyle i\neq 1,o,m,p,q \end{cases} \tag{17}$$

Therefore, not only the redundancy of dictionary D can be increased by the singular value Θ transformation, which makes the energy of dictionary D more concentrated, but it also hardly affects the expression ability of dictionary D.

2.3. Task-Driven Invertible Projection Matrix Learning

In order to obtain a reversible projection transformation, we have to use prior knowledge to solve the invertible projection matrix. However, as we all know, it is very difficult to solve the low-dimensional invertible projection matrix U directly. Fortunately, it has been proven in Section 2.1 that the low-dimensional invertible projection matrix U can be solved indirectly by solving model (7). Model (7) is the problem of extracting principal components from dictionary D. Based on the theory in Section 2.2, a task-driven invertible projection matrix learning algorithm (TIPML) is proposed. In this algorithm, dictionary learning and singular value Θ transformation are used as the core of the algorithm, and a dual-loop iterative training mechanism is established based on the core.

As shown in Figure 2, in order to ensure that more principal components of high-dimensional dictionary D are retained in the low-dimensional dictionary P. In TIPML, the singular value Θ transformation is first used to make the energy of the dictionary D more concentrated. Next, the sparse representation coefficient A of the training data under the low-dimensional dictionary P is used to train the high-dimensional dictionary D. In this step, the low-dimensional dictionary P is used to train the high-dimensional dictionary D, which further increases the coupling between the two dictionaries and lets the principal components of the high-dimensional dictionary D be more retained in the low-dimensional dictionary P.

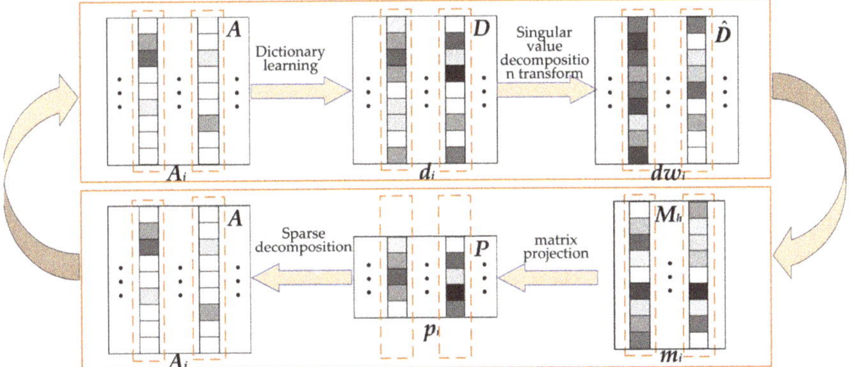

Figure 2. Block diagram of TIPML (task-driven invertible projection matrix learning algorithm).

The specific implementation steps of the TIPML algorithm are shown in Algorithm 1.

Algorithm 1: TIPML

Input: Training data set X, number of iterations T, the singular value Θ transformation parameter t,r, low-dimensional dictionary P dimension h, Signal sparsity K, and dimension of the data column N.
Output: Projection matrix U.

1: Initialization: Split the data set X into data columns x_i ($N \times 1$), $I = 1,2,\ldots,i$ is the index of the data column x_i. Assuming $X = [x_1, x_2, \ldots, x_n]$ ($x_i \in R^N$). The initial dictionary D is set as a DCT dictionary, $D = [d_1, d_2, \ldots, d_n]$, ($d_i \in R^N$).

2: Repeat
3:    Do singular value Θ transformation to dictionary D: $D = D\Theta(t,r)$.
4:    Singular value decomposition: $D = M\Lambda V^T$.
5:    Calculate low-dimensional dictionary: $P = M_h{}^T D$.
6:    Based on the low-dimensional dictionary P, the OMP algorithm is used to sparse the data set X to obtain the sparse coefficient $A = [a_1, a_2, \ldots, a_n]$, update the index $j = 1$ of the dictionary atom.
7:    Repeat
8:      The error is calculated: $E_j = X - \sum_{l \neq j} d_l a_l{}^T$.
9:      The error is decomposed by SVD (rank-1 decomposition) into: $E_j \approx u\lambda v^T$.
10:    Update dictionary: $d_j = u$, Update sparsity coefficient: $a_j = \lambda v$.
11:    j = j + 1
12:    Until $j > n$
13: Until $\frac{\sum_i \Lambda_h}{\sum_i \Lambda}$ is big enough, or reach the maximum number of iterations T.
14: Calculate the projection matrix: $U = M_h{}^T$.

---

3. Simulation Experiment and Results

In order to verify the effectiveness of the algorithm proposed in this paper, the AVIRIS hyperspectral data image Indian Pine dataset [28] was selected as the experimental dataset. These hyperspectral data include 220 bands, and each pixel is stored in a 16-bit integer format. In the experiment, the most competitive sparse expression dictionary and reconstruction algorithm in compressed sensing were selected and compared with the algorithm proposed in this paper. Sparse expression dictionaries included DCT dictionary, KSVD dictionary, and CDL dictionary, and reconstruction algorithms included StOMP, OMP, GOMP, GROMP, CoSaMP, and SPL. Besides, the data compression algorithm PCA [29] has also been selected for comparison with the algorithm proposed in this paper. The average peak signal-to-noise ratio (PSNR), Spectral Angle Mapper (SAM), and reconstruction time T were selected as the evaluation indicators of the experiment, and the PSNR and reconstruction time T of the reconstructed image was obtained by setting different sampling rates (SR). The software and hardware environment of the test experiment are CPU: Intel(R) Core (TM) i7-9750H 2.6GHz, 8GB memory, and Windows 10 and MatLab 2019a.

The PSNR calculation formula is

$$\text{PSNR} = 10\log_{10}\left(\frac{MAX^2}{\frac{1}{m\bullet n}\sum_{i=1,j=1}^{m,n}[x(i,j)-x'(i,j)]^2}\right) \tag{18}$$

where x is the original image, x' is the reconstructed image, m and n respectively represent the length and width of the image, and MAX is the maximum value of image pixels.

The sampling rate calculation formula is

$$SR = \frac{M_\lambda}{N_\lambda} \tag{19}$$

where N_λ is the dimension of the original data, and M_λ is the dimension of compressed sensing sampling.

The SAM calculation formula is

$$SAM = \cos^{-1}\frac{x^T x'}{(x^T x)^{\frac{1}{2}}(x'^T x')^{\frac{1}{2}}} \tag{20}$$

where x is the original image, x' is the reconstructed image, x^T and x'^T are the transpose of matrix x and x', respectively, and $\cos^{-1}$ is the arc cosine function.

As shown in Figure 3, the simulation experiment mainly included three stages, namely offline learning, data sampling (encoding), and data reconstruction (decoding). In the experiment, the data of the Indian Pine data set was divided into two parts: The training data set and the test data set. The transformation adjustment parameters of the TIPML algorithm were set to 't = 255' and 'r = 0.3'. In addition, the data set was processed in blocks with a block size of 16 × 16, and each small block was arranged into a column vector with a size of 256 × 1. First, the training set Z was used as the TIPML training sample to obtain the projection matrix U. Then, the projection matrix U was used as a reduced-dimensional sampling matrix of the test data X to obtain the observation data Y. Finally, the transposed U^T of the projection matrix was used to back-project the low-dimensional sampling data Y to obtain the reconstructed data.

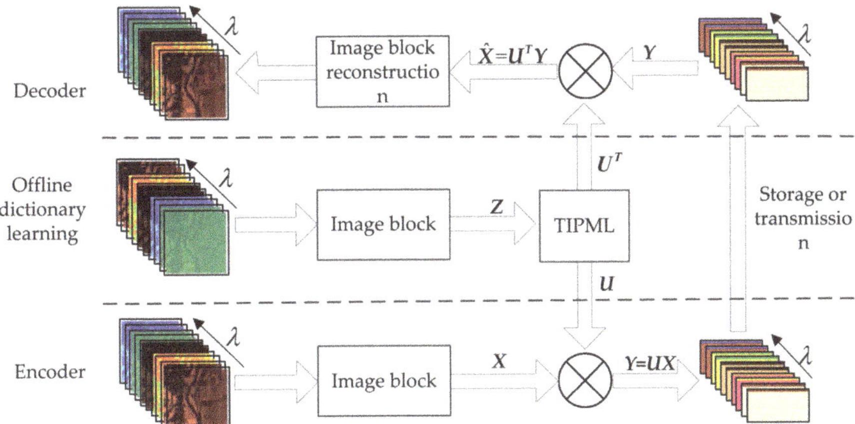

Figure 3. Flow chart of simulation experiment.

As shown in Figures 4–6, the PSNR of the reconstructed image of the algorithm proposed in this paper, CDL-OMP algorithm, and KSVD-StOMP algorithm under different sampling rates (SR) are respectively given. By comparing the reconstructed images, it can be known that the KSVD-StOMP algorithm has the lowest image reconstruction accuracy at the same sampling rate. The reconstruction accuracy of the algorithm proposed in this paper is visually similar to that of the CDL-OMP algorithm. At low sampling rates (SR = 0.1), there is still a good reconstruction effect.

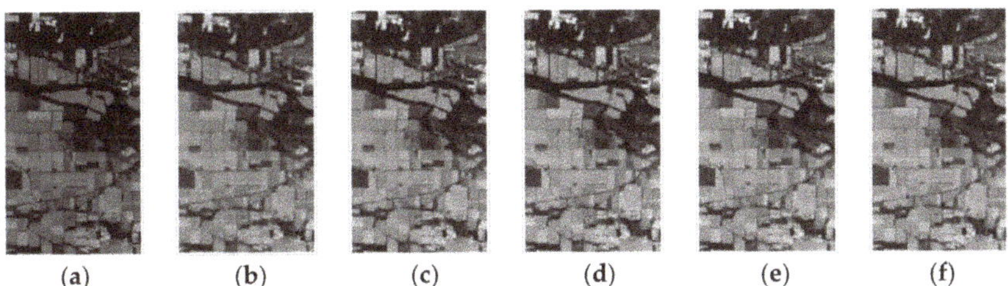

Figure 4. The original image and reconstructed image with band number 180, the compressed sensing algorithm is based on the TIPML algorithm, (**a**)The original images; (**b**) The sampling rate of the reconstructed image is SR = 0.1; (**c**) The sampling rate of the reconstructed image is SR = 0.2; (**d**) The sampling rate of the reconstructed image is SR = 0.3; (**e**) The sampling rate of the reconstructed image is SR = 0.4; (**f**) The sampling rate of the reconstructed image is SR = 0.5.

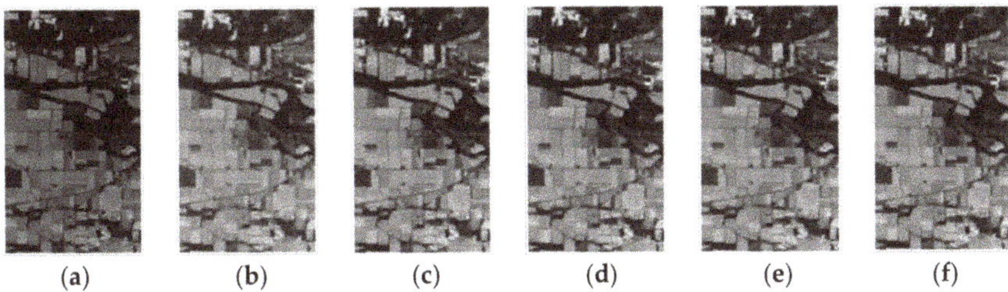

Figure 5. The Original image and reconstructed image with band number 180, the compressed sensing algorithm is based on the CDL-OMP algorithm, (**a**) The original images; (**b**) The sampling rate of the reconstructed image is SR = 0.1; (**c**) The sampling rate of the reconstructed image is SR = 0.2; (**d**) The sampling rate of the reconstructed image is SR = 0.3; (**e**) The sampling rate of the reconstructed image is SR = 0.4; (**f**) The sampling rate of the reconstructed image is SR = 0.5.

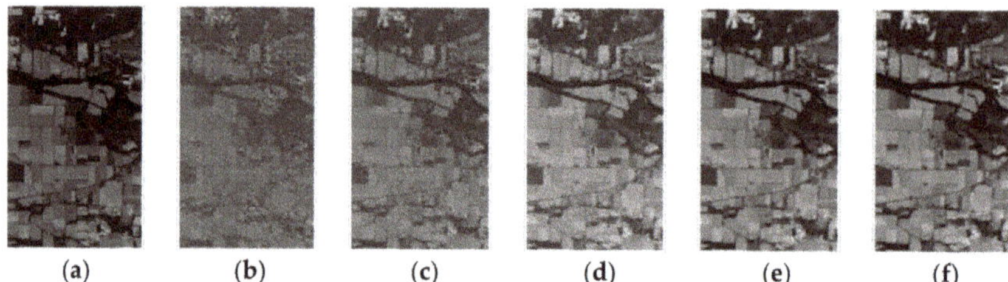

(a)   (b)   (c)   (d)   (e)   (f)

Figure 6. The Original image and reconstructed image with band number 180, the compressed sensing algorithm is based on KSVD-StOMP algorithm, (**a**) The original image; (**b**) The sampling rate of the reconstructed image is SR = 0.1; (**c**) The sampling rate of the reconstructed image is SR = 0.2; (**d**) The sampling rate of the reconstructed image is SR = 0.3; (**e**) The sampling rate of the reconstructed image is SR = 0.4; (**f**) The sampling rate of the reconstructed image is SR = 0.5.

As shown in Figure 7, at different sampling rates, the original spectral line and the reconstructed spectral line at the (256, 1) pixel point, the experimental comparison methods are the algorithm proposed in this paper, CDL-OMP algorithm, and KSVD-StOMP algorithm. Comparing the experimental results, it can be seen that when the sampling rate SR = 0.2 in Figure 7b, the reconstructed spectrum line of the algorithm proposed in this paper almost coincides with the original spectrum line. When the sampling rate is low, the reconstructed spectrum of the proposed algorithm and the original spectrum have some errors, but the reconstruction effect is still slightly better than the CDL-OMP algorithm, and far better than the KSVD-StOMP algorithm. This is because when SR = 0.1 in Figure 7a, the dimension of the projection matrix $M_\lambda = 26$, which is much smaller than the original signal dimension $N_\lambda = 256$. When the original signal is projected into a low-dimensional space, too much information is lost, resulting in reduced reconstruction accuracy.

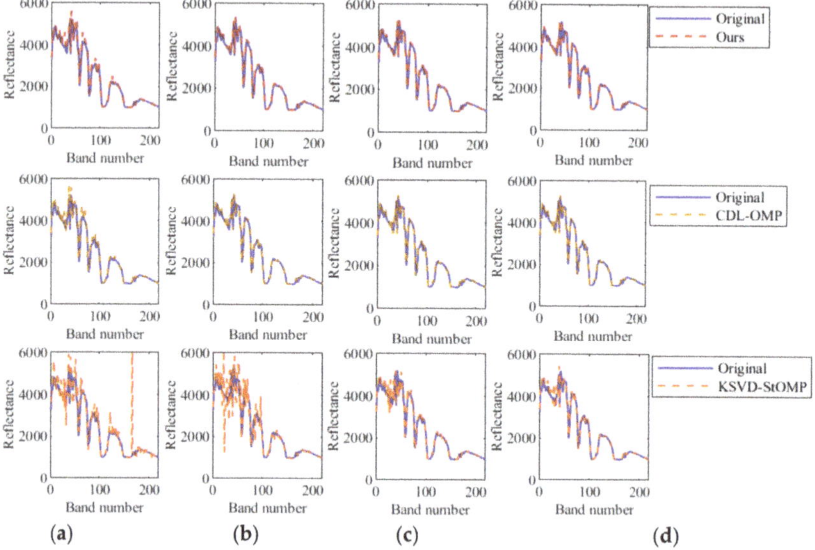

(a)   (b)   (c)   (d)

Figure 7. The original and reconstructed spectral lines of the pixel at coordinates (256, 1). From top to bottom, they are TIPML algorithm, CDL-OMP, and KSVD-StOMP. From the first column to the fourth column (**a**) The sampling rate SR = 0.1; (**b**) The sampling rate SR = 0.2; (**c**) The sampling rate SR = 0.3; (**d**) The sampling rate SR = 0.4.

In order to further verify the superiority of the algorithm proposed in this paper, more compressed sensing algorithms and PCA algorithm were selected. As shown in Table 1, the PSNR under the same sampling rate in the table is the average of all PSNRs of 220-band images. The results in the comparison table show that the reconstructed image accuracy of the compressed sensing method based on the algorithm proposed in this paper is better than other algorithms, and it still has a higher reconstruction accuracy at low sampling rates.

Table 1. The reconstructed average peak signal-to-noise ratio (PSNR) (dB) of hyperspectral images in all bands under different sampling rates.

Methods	Sampling Rate				
	0.1	0.2	0.3	0.4	0.5
Ours	61.20	63.39	64.17	64.86	65.63
PCA	54.09	56.15	57.84	59.46	61.10
KSVD-StOMP	51.06	54.10	59.21	60.87	61.51
KSVD-OMP	31.37	28.30	26.93	27.08	27.20
KSVD-GOMP	33.85	30.31	29.02	28.61	28.74
KSVD-GROMP	34.39	33.59	33.57	35.04	35.72
KSVD-CoSaMP	48.70	49.61	50.30	50.18	50.26
DCT-SPL	38.27	46.83	54.10	58.81	60.70
DCT-StOMP	30.44	30.97	31.52	32.17	32.96
CDL-StOMP	60.67	63.06	63.33	63.33	63.33
CDL-OMP	60.67	63.06	63.33	63.33	63.33
CDL-CoSaMP	48.79	49.93	49.94	49.94	49.94
CDL-GROMP	14.83	22.90	28.39	33.27	37.21
CDL-GOMP	9.19	13.73	17.85	21.60	24.74

As shown in Table 2, the SAM under the same sampling rate in the table is the average of all SAMs of 220-band images. For the convenience of comparison, the SAM data in the table is displayed in the form of scientific notation. The results in the comparison table show that the reconstructed image SAM still has a lower SAM at low sampling rates. In other words, the reconstructed spectrum is more similar to the original spectrum.

Table 2. The reconstructed average Spectral Angle Mapper (SAM) (Scientific notation) of hyperspectral images in all bands under different sampling rates.

Methods	Sampling Rate				
	0.1	0.2	0.3	0.4	0.5
Ours	1.17×10^{-8}	1.16×10^{-8}	1.18×10^{-8}	1.18×10^{-8}	1.14×10^{-8}
PCA	1.20×10^{-8}	1.19×10^{-8}	1.22×10^{-8}	1.21×10^{-8}	1.18×10^{-8}
KSVD-StOMP	8.39×10^{-4}	1.16×10^{-8}	1.19×10^{-8}	1.18×10^{-8}	1.15×10^{-8}
KSVD-OMP	1.45×10^{-1}	6.06×10^{-3}	5.03×10^{-4}	1.18×10^{-8}	1.32×10^{-3}
KSVD-GOMP	5.34×10^{-2}	1.15×10^{-8}	4.82×10^{-3}	1.18×10^{-8}	1.97×10^{-3}
KSVD-GROMP	1.22×10^{-3}	1.16×10^{-8}	3.19×10^{-2}	1.17×10^{-8}	1.07×10^{-2}
KSVD-CoSaMP	1.16×10^{-8}	1.16×10^{-8}	1.19×10^{-8}	1.16×10^{-8}	1.15×10^{-8}
DCT-SPL	1.14×10^{-8}	1.18×10^{-8}	1.15×10^{-8}	1.17×10^{-8}	1.15×10^{-8}
CDL-StOMP	1.17×10^{-8}	1.16×10^{-8}	1.19×10^{-8}	1.17×10^{-8}	1.15×10^{-8}
CDL-OMP	1.17×10^{-8}	1.16×10^{-8}	1.19×10^{-8}	1.17×10^{-8}	1.15×10^{-8}
CDL-CoSaMP	1.65×10^{-3}	7.16×10^{-1}	1.18×10^{-8}	1.75×10^{-2}	8.31×10^{-2}
CDL-GROMP	4.55×10^{-4}	1.20×10^{-4}	1.32×10^{-3}	9.83×10^{-4}	2.88×10^{-4}
CDL-GOMP	4.55×10^{-4}	1.20×10^{-4}	1.32×10^{-3}	9.83×10^{-4}	2.88×10^{-4}

For compressed sensing algorithms, the running time T of the reconstruction algorithm is a very important evaluation index. As shown in Table 3, the running time T under the same sampling rate in the table is the mean value of the running time of 220-band image reconstruction, and the time unit recorded in the table is milliseconds. By comparing the experimental results, it can be seen that the running time T of the algorithm proposed in this paper is much lower than that of the traditional compressed sensing algorithm, and the running time is shortened by at least a hundred times. This is because the projection matrix

U trained by the algorithm proposed in this paper is approximately invertible. In the reconstruction process, only the low-dimensional sampling matrix Y is left multiplied by U^T. However, traditional compression sparse reconstruction algorithms require multiple matrix inversion and iterative operations. These operations are very time-consuming. Therefore, the real-time performance of the algorithm proposed in this paper is much higher than other algorithms. Besides, compared to data compression algorithm PCA, when we know the covariance matrix of the source in advance and find the eigenvalues, the time consumption of the TIPML algorithm is similar to that of the PCA algorithm. However, in practical applications, the covariance and eigenvalue of the source are unknown. We need to calculate the covariance and eigenvalue of the source, which will cause a lot of time to be consumed. Therefore, compared to traditional compressed sensing algorithms and traditional PCA data compression algorithms, the TIPML algorithm is more competitive.

Table 3. The reconstructed average time (millisecond) of hyperspectral images in all bands under different sampling rates.

Methods	Sampling Rate				
	0.1	0.2	0.3	0.4	0.5
Ours	0.3	0.3	0.4	0.4	0.4
PCA	0.3	0.4	0.4	0.4	0.4
KSVD-StOMP	210.8	443.7	667.1	934.6	1235.0
KSVD-OMP	264.9	401.8	545.3	594.0	613.5
KSVD-GOMP	101.7	170.4	206.2	246.8	252.3
KSVD-GROMP	84.6	131.7	181.0	186.7	189.3
KSVD-CoSaMP	302.7	527.7	597.3	829.7	877.4
DCT-SPL	354.8	355.1	354.8	357.0	359.1
DCT-StOMP	199.4	437.6	732.2	1085.2	1517.4
CDL-StOMP	228.2	503.3	763.0	1054.0	1384.2
CDL-OMP	209.5	462.7	696.1	955.6	1261.4
CDL-CoSaMP	341.3	553.0	635.0	895.7	943.9
CDL-GROMP	75.5	111.8	150.4	153.0	157.3
CDL-GOMP	167.7	253.3	317.4	365.8	375.5

4. Conclusions

HSI is the main tool for remote sensing and earth observation. The amount of valuable industrial and scientific data retrieved on the ground can be greatly increased by data compression technology. The resources of spaceborne equipment are very precious, and compression is placed on the sampling end by compressed sensing technology, which can save a lot of time and resources for hyperspectral imaging technology. Therefore, compressed sensing technology has huge application prospects in the field of hyperspectral compression. However, compressed sensing technology needs to solve underdetermined equations. In traditional algorithms, a lot of time and storage resources are spent in the sparse reconstruction process. Therefore, the high complexity of data reconstruction is also the biggest drawback of compressed sensing technology.

Aiming at the high computational complexity of compressed sensing technology and insufficient reconstruction accuracy, a task-driven invertible projection matrix learning algorithm was proposed by us. Our main contribution:

(1) Derived the equivalent model of the invertible projection model theoretically, which converts the complex invertible projection training model into a coupled dictionary training model;

(2) proposed a task-driven invertible projection matrix learning algorithm for invertible projection model training;

(3) based on a task-driven reversible projection matrix learning algorithm, established a compressed sensing algorithm with strong real-time performance and high reconstruction accuracy.

Experimental verification has shown that the compressed sensing technology based on the algorithm proposed in this paper not only has higher reconstruction accuracy than traditional compressed sensing technology, but also has improved real-time performance by more than a hundred times. It is foreseeable that the algorithm proposed in this paper will have great application prospects in the field of hyperspectral image compression. In addition, the algorithm proposed in this paper can't only be used in the fields of one-dimensional signal compression, two-dimensional image compression, and data denoising, but it will also have greater research value on hardware platforms such as FPGA and embedded devices due to its extremely low the complexity.

Author Contributions: Conceptualization, S.D.; methodology, S.D.; software, S.D.; validation, S.D.; formal analysis, S.D.; investigation, S.D.; resources, Z.W.; data curation, S.D.; writing—original draft preparation, S.D.; writing—review and editing, S.D.; visualization, S.D.; supervision, W.L and K.L.; project administration, W.L. and K.L.; funding acquisition, W.L. and K.L. All authors have read and agreed to the published version of the manuscript.

Funding: This research was funded by National Key Research and Development Program of China grant number 2018YFB2003304, National Natural Science Foundation of China grant number 61871218, Fundamental Research Funds for the Central Universities grant number NJ2019007 and NJ2020014, National Key Research and Development Program of China grant number 2017YFF0107304, 2017YFF0209700, 2016YFB1100205, and 2016YFF0103702.

Institutional Review Board Statement: Not applicable.

Informed Consent Statement: Not applicable.

Data Availability Statement: All data included in this study are available upon request by contact with the corresponding author.

Conflicts of Interest: The authors declare no conflict of interest.

References

1. Parente, M.; Kerekes, J.; Heylen, R. A Special Issue on Hyperspectral Imaging [From the Guest Editors]. *IEEE Geosci. Remote Sens. Mag.* **2019**, *7*, 6–7. [CrossRef]
2. Vohland, M.; Jung, A. Hyperspectral Imaging for Fine to Medium Scale Applications in Environmental Sciences. *Remote Sens.* **2020**, *12*, 2962. [CrossRef]
3. Saari, H.; Aallos, V.-V.; Akujärvi, A.; Antila, T.; Holmlund, C.; Kantojärvi, U.; Mäkynen, J.; Ollila, J. Novel Miniaturized Hyperspectral Sensor for UAV and Space Applications. In Proceedings of the SPIE 7474, Sensors, Systems, and Next-Generation Satellites XIII, 74741M, SPIE Remote Sensing, Berlin, Germany, 31 August–3 September 2009. [CrossRef]
4. Renhorn, I.G.E.; Axelsson, L. High spatial resolution hyperspectral camera based on exponentially variable filter. *Opt. Eng.* **2019**, *58*, 103106. [CrossRef]
5. Pu, H.; Lin, L.; Sun, D.-W. Principles of Hyperspectral Microscope Imaging Techniques and Their Applications in Food Quality and Safety Detection: A Review. *Compr. Rev. Food Sci. Food Saf.* **2019**, *18*, 853–866. [CrossRef] [PubMed]
6. Wang, L.; Lu, K.; Liu, P. Compressed Sensing of a Remote Sensing Image Based on the Priors of the Reference Image. *IEEE Geosci. Remote Sens. Lett.* **2015**, *12*, 736–740. [CrossRef]
7. Donoho, D.L. Compressed sensing. *IEEE Trans. Inf. Theory* **2006**, *52*, 1289–1306. [CrossRef]
8. Hsu, C.; Lin, C.; Kao, C.; Lin, Y. DCSN: Deep Compressed Sensing Network for Efficient Hyperspectral Data Transmission of Miniaturized Satellite. *IEEE Trans. Geosci. Remote Sens.* **2020**. [CrossRef]
9. Biondi, F. *In Compressed Sensing Radar-New Concepts of Incoherent Continuous Wave Transmissions*; IEEE: Piscataway, NJ, USA, 2015; pp. 204–208.
10. Candes, E.J.; Romberg, J.; Tao, T. Robust uncertainty principles: Exact signal reconstruction from highly incomplete frequency information. *IEEE Trans. Inf. Theory* **2004**, *52*, 489–509. [CrossRef]
11. Candes, E.J.; Tao, T. Near-Optimal Signal Recovery from Random Projections: Universal Encoding Strategies. *Inf. Theory IEEE Trans.* **2006**, *52*, 5406–5425. [CrossRef]
12. Xue, H.; Sun, L.; Ou, G. Speech reconstruction based on compressed sensing theory using smoothed L0 algorithm. In Proceedings of the 2016 8th International Conference on Wireless Communications & Signal Processing (WCSP), Yangzhou, China, 13–15 October 2016; pp. 1–4. [CrossRef]
13. Luo, H.; Zhang, N.; Wang, Y. Modified Smoothed Projected Landweber Algorithm for Adaptive Block Compressed Sensing Image Reconstruction. In Proceedings of the 2018 International Conference on Audio, Language and Image Processing (ICALIP), Shanghai, China, 16–17 July 2018; pp. 430–434. [CrossRef]

14. Matin, A.; Dai, B.; Huang, Y.; Wang, X. Ultrafast Imaging with Optical Encoding and Compressive Sensing. *J. Light. Technol.* **2019**, 761–768. [CrossRef]
15. Dao, P.; Li, X.; Griffin, A. Quantitative Comparison of EEG Compressed Sensing using Gabor and K-SVD Dictionaries. In Proceedings of the 2018 IEEE 23rd International Conference on Digital Signal Processing (DSP), Shanghai, China, 19–21 November 2018; pp. 1–5. [CrossRef]
16. Sana, F.; Katterbauer, K.; Al-Naffouri, T. Enhanced recovery of subsurface geological structures using compressed sensing and the Ensemble Kalman filter. In Proceedings of the 2015 IEEE International Geoscience and Remote Sensing Symposium (IGARSS), Milan, Italy, 26–31 July 2015; pp. 3107–3110. [CrossRef]
17. Dehkordi, R.; Khosravi, H.; Ahmadyfard, A. Single image super-resolution based on sparse representation using dictionaries trained with input image patches. *IET Image Process.* **2020**, *14*, 1587–1593. [CrossRef]
18. Silong, Z.; Yuanxiang, L.; Xian, W. Nonlinear dimensionality reduction method based on dictionary learning. *Acta Autom. Sin.* **2016**, *42*, 1065–1076.
19. Vehkaperä, M.; Kabashima, Y.; Chatterjee, S. Analysis of Regularized LS Reconstruction and Random Matrix Ensembles in Compressed Sensing. *IEEE Trans. Inf. Theory* **2016**, *62*, 2100–2124. [CrossRef]
20. Ziran, W.; Huachuang, W.; Jianlin, Z. Structural optimization of measurement matrix in image reconstruction based on compressed sensing. In Proceedings of the 2017 7th IEEE International Conference on Electronics Information and Emergency Communication (ICEIEC), Macau, China, 21–23 July 2017; pp. 223–227. [CrossRef]
21. Zhang, W.; Tan, C.; Xu, Y. Electrical Resistance Tomography Image Reconstruction Based on Modified OMP Algorithm. *IEEE Sens. J.* **2019**, *19*, 5723–5731. [CrossRef]
22. Qian, Y.; Zhu, D.; Yu, X. SAR high-resolution imaging from missing raw data using StOMP. *J. Eng.* **2019**, *2019*, 7347–7351. [CrossRef]
23. Davenport, M.; Needell, D.; Wakin, M. Signal Space CoSaMP for Sparse Recovery with Redundant Dictionaries. *IEEE Trans. Inf. Theory* **2013**, *59*, 6820–6829. [CrossRef]
24. Chen, S.; Feng, C.; Xu, X. Micro-motion feature extraction of narrow-band radar target based on ROMP. *J. Eng.* **2019**, *2019*, 7860–7863. [CrossRef]
25. Trevisi, M.; Akbari, A.; Trocan, M. Compressive Imaging Using RIP-Compliant CMOS Imager Architecture and Landweber Reconstruction. *IEEE Trans. Circuits Syst. Video Technol.* **2020**, *30*, 387–399. [CrossRef]
26. Wei, X. Reconstructible Nonlinear Dimensionality Reduction via Joint Dictionary Learning. *IEEE Trans. Neural Netw. Learn. Syst.* **2019**, *30*, 175–189. [CrossRef] [PubMed]
27. Shufang, X. *Theory and Method of Matrix Calculation*; Peking University Press: Beijing, China, 1995; p. 7.
28. Baumgardner, M.F.; Biehl, L.L.; Landgrebe, D.A. 220 Band AVIRIS Hyperspectral Image Data Set: 12 June 1992 Indian Pine Test Site 3. In *Purdue University Research Repository*; Purdue University: West Lafayette, IN, USA, 2015. [CrossRef]
29. Lee, C.; Youn, S.; Jeong, T.; Lee, E.; Serra-Sagristà, J. Hybrid Compression of Hyperspectral Images Based on PCA with Pre-Encoding Discriminant Information. *IEEE Geosci. Remote Sens. Lett.* **2015**, *12*, 1491–1495. [CrossRef]

Technical Note

High-Performance Lossless Compression of Hyperspectral Remote Sensing Scenes Based on Spectral Decorrelation

Miguel Hernández-Cabronero [1,*], Jordi Portell [2,3], Ian Blanes [1,4] and Joan Serra-Sagristà [1,4]

[1] Group on Interactive Coding of Images (GICI), Universitat Autònoma de Barcelona, Campus UAB, 08193 Cerdanyola del Vallès, Spain; ian.blanes@uab.cat (I.B.); joan.serra@uab.cat (J.S.-S.)
[2] Department of FQA, Institut de Ciències del Cosmos (ICCUB), Universitat de Barcelona (IEEC-UB), Martí i Franquès 1, 08028 Barcelona, Spain; jportell@fqa.ub.edu
[3] DAPCOM Data Services S.L., Vilabella 5-7, 08500 Vic, Spain
[4] Institut d'Estudis Espacials de Catalunya (IEEC/CERES), Gran Capità 2-4, 08034 Barcelona, Spain
* Correspondence: miguel.hernandez@uab.cat

Received: 30 July 2020; Accepted: 10 September 2020; Published: 11 September 2020

Abstract: The capacity of the downlink channel is a major bottleneck for applications based on remote sensing hyperspectral imagery (HSI). Data compression is an essential tool to maximize the amount of HSI scenes that can be retrieved on the ground. At the same time, energy and hardware constraints of spaceborne devices impose limitations on the complexity of practical compression algorithms. To avoid any distortion in the analysis of the HSI data, only lossless compression is considered in this study. This work aims at finding the most advantageous compression–complexity trade-off within the state of the art in HSI compression. To do so, a novel comparison of the most competitive spectral decorrelation approaches combined with the best performing low-complexity compressors of the state is presented. Compression performance and execution time results are obtained for a set of 47 HSI scenes produced by 14 different sensors in real remote sensing missions. Assuming only a limited amount of energy is available, obtained data suggest that the FAPEC algorithm yields the best trade-off. When compared to the CCSDS 123.0-B-2 standard, FAPEC is 5.0 times faster and its compressed data rates are on average within 16% of the CCSDS standard. In scenarios where energy constraints can be relaxed, CCSDS 123.0-B-2 yields the best average compression results of all evaluated methods.

Keywords: multispectral; hyperspectral; CCSDS; FAPEC; data compression; transform

---

1. Introduction

Hyperspectral imaging (HSI) is one of the main components of modern remote sensing [1]. In recent years, new public and private organizations have joined the remote sensing ecosystem [2]. As a result, the number of deployed HSI sensors has grown significantly, especially for Earth Observation tasks, and this trend is predicted to continue in the future [3]. Parallel to this increment of active sensors, substantial efforts are being devoted to improving the analysis of scenes obtained with HSI sensors [4]. Classification, unmixing, and segmentation are among the problems currently more actively researched [4–15]. Thanks to the increased availability of HSI data and the accuracy of their interpretation, many commercial and scientific applications are becoming feasible, and attainable for a larger base of users. Some of the most widespread applications include resource management, e.g., in agriculture, mining. and forestry; geological structure observation; and disaster monitoring and response coordination [2,16–18].

One of the main bottlenecks when retrieving HSI information from spaceborne sensors is the downlink channel capacity [19]. This bottleneck is emphasized by the large size of HSI data.

Retrieved scenes typically contain from tens to hundreds of spectral bands. For comparison, traditional color images contain only three bands. The inclusion of more than three bands is needed for a more detailed observation of the electromagnetic spectrum, but they multiply the number of samples produced by the sensor. Furthermore, newly deployed sensors also tend to exhibit larger spatial resolutions, i.e., a higher number of samples per spectral band. As a result, the latest sensors—such as NASA's HyspIRI—can produce up to 5 Terabytes of data per day [20], i.e., significantly more than can be transmitted through existing space links [19].

In order to maximize the effective capacity of downlink channels, compression is routinely employed to reduce the volumes of data to be transmitted [19]. In particular, lossless compression allows data volume reduction, while at the same time enabling perfect reconstruction of the scenes. Thus, decompressed scenes will produce the exact same result as the original scenes when subject to any deterministic analysis algorithm. This includes most classification, unmixing, and segmentation algorithms [4].

This paper investigates two aspects of lossless compression specific to HSI remote sensing—spectral band redundancy—and the performance of low-complexity algorithms. Redundancy among spectral bands in HSI is typically higher than in traditional color images. Thus, to attain competitive coding efficiency, it is paramount to effectively exploit this redundancy. In turn, low complexity in the compression pipeline, including the spectral decorrelation stage, is needed due to the limitations in spendable energy and available hardware in spaceborne sensors [19,21]. These two aspects of compression must be jointly considered so that obtained conclusions are relevant in practical remote sensing HSI scenarios.

Hereafter, a selection of the best performing, reversible spectral decorrelation transforms is studied, including the integer wavelet transform (IWT [22]), the pairwise-orthogonal transform (POT [23]), and the regression wavelet analysis (RWA [24]) transform. These transforms are evaluated in combination with some of the most competitive low-complexity compression algorithms including the Fully Adaptive Prediction Error Coder (FAPEC [25]), the CCSDS 122.1-B-1 standard [26], and the well-known JPEG-LS [27] algorithm. For completeness, the aforementioned transform-compressor combinations are also compared to CCSDS 123.0-B-2 [28] and M-CALIC [29], which exploit redundancy across bands by means of prediction. This work extends upon our previous work [30] by considering a wider selection of spectral decorrelation transforms, compression algorithms, and test scenes. This provides a more complete analysis of the state-of-the-art and increases the significance of the obtained results Note that, even though hardware implementations have been described for some of the aforementioned algorithms (see, e.g., in [31–33]), this study focuses on software implementations. This approach is consistent with the current trend in Earth Observation, in which constellations of small satellites and cubesats (mainly from the private sector) are being deployed [2]. These constellations often employ commercial-off-the-shelf hardware [3], as opposed to the custom hardware implementations traditionally used in larger satellites launched by space agencies. By analyzing the efficiency of software implementations, results become more relevant to the foreseeable future of Earth Observation.

The rest of this paper is structured as follows. Section 2 contains a summary of the spectral decorrelation transforms and compression methods considered for this study. Section 3 presents and analyzes experimental results based on a comprehensive corpus of real hyperspectral scenes. Conclusions are drawn in Section 4.

2. Methods and Materials

State-of-the-art data compression algorithms often consist of two conceptually separated parts: data decorrelation and entropy coding. Data decorrelation exploits the fact that neighboring pixels—either within a band or across several of them—tend to have similar values. Decorrelated samples normally exhibit less variability, i.e., lower entropy rates. The entropy coding stage then compacts decorrelated samples in a reversible way. The output compressed data size depends directly on the entropy of those samples, hence the importance of effective decorrelation.

This work focuses on compression pipelines in which decorrelation is divided into two consecutive stages: spectral and spatial decorrelation. Typically, scenes are first processed along the spectral axis, although some compressors invert this order or skip some of the stages. The resulting transformed bands are often processed separately, i.e., they are spatially decorrelated and entropy coded. The compressed data obtained for each band is then packed together to form the final compressed bitstream, which can also be output to a file. A diagram of this general design is provided in Figure 1. This approach allows combining and comparing different methods for each stage in order to get the best compression results. At the same time, it helps to obtain an overall low-complexity algorithm. As discussed in the introduction, both aspects are critical for HSI remote sensing.

The remainder of this section provides some background on the spectral decorrelation methods considered in this study in Section 2.1. The spatial decorrelation and entropy coding algorithms with which they are combined are described in Section 2.2. In order to obtain representative results, a comprehensive corpus of hyperspectral scenes has been selected. This corpus is presented in Section 2.3.

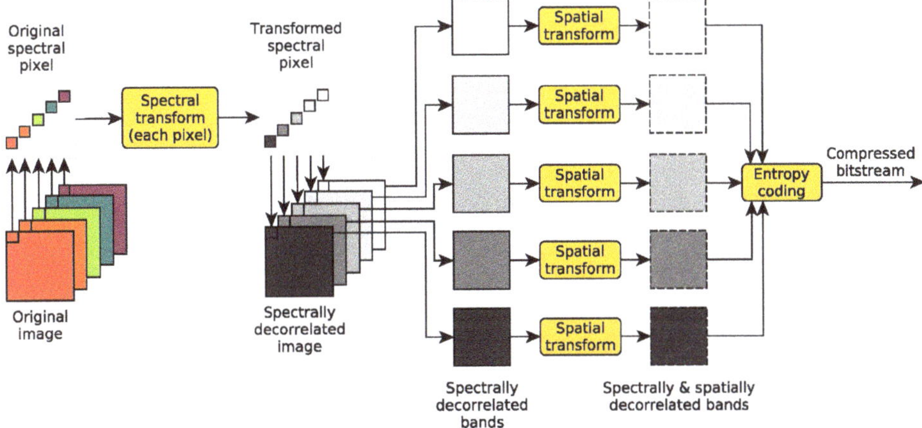

Figure 1. Main stages of transform-based hyperspectral scene compression.

2.1. Spectral Decorrelation Methods

Spectral decorrelation transforms are applied to co-located samples across different spectral bands. This process is repeated until all spatial positions are decorrelated. Three spectral transforms have been selected for this work due to their competitive decorrelation-complexity trade-offs: the integer wavelet transform (IWT), the pairwise-orthogonal transform (POT), and the regression wavelet analysis (RWA) transform. These are briefly described next in chronological order of publication.

The IWT [22] is a well-known discrete wavelet transform (DWT). It is a linear, multilevel filter that produces integer coefficients. This allows reversible transformation, which is required to attain lossless compression. In each level, the IWT filter separates the input signal into so-called low-frequency and high-frequency sub-bands. After that, the low frequencies of one level are the input of the next level. In this work, five levels of the LeGall 5/3 integer wavelet decomposition are employed except for the FAPEC compressor described below, which applies three levels of 9/7-M integer wavelet decomposition. One important property of the IWT is that it can be implemented very efficiently using so-called lifting steps [34]. This approach significantly reduces the number of arithmetic operations needed per input sample, thus enhancing throughput.

The POT [23] is a low-complexity, reversible approximation of the Karhunen-Loève Transform (KLT). The KLT provides perfect decorrelation, i.e., zero covariance between different spectral bands, which normally results in very high compression performance. However, the KLT is generally considered too computationally intensive for low-complexity scenarios. In turn, the POT offers much lower computational cost, while trying to reach the compression performance gains of the KLT. To achieve this goal, the POT applies a divide-and-conquer strategy that decorrelates band pairs in a multilevel structure. This is opposed to the KLT, which decorrelates all spectral bands simultaneously.

The RWA [24] is a reversible transform that operates in two steps. First, a reversible wavelet transform is applied along the spectral axis. The Haar wavelet is used for this purpose due to its low computational cost and reasonable decorrelating power. In the second step, pyramidal nonlinear estimators are used to predict coefficients in high-frequency sub-bands using coefficients from low-frequency sub-bands.

2.2. Low-Complexity Compression Methods

In this study, three compression algorithms are applied after each of the spectral decorrelation transforms described in Section 2.1: CCSDS 122.1-B-1, FAPEC, and JPEG-LS. Two additional compressors are considered in the study, CCSDS 123.0-B-2 and M-CALIC, which do not exactly follow the structure depicted in Figure 1. Instead, they apply their own spectral decorrelation methods, so they are not combined with any of the aforementioned spectral transforms. FAPEC also follows a slightly different approach, as described later in this section.

The CCSDS 122.1-B-1 [26] standard (hereafter CCSDS 122.1) was proposed by the Consultative Committee for Space Data Systems (CCSDS) as an extension to the CCSDS 122.0-B-2 standard [28] (All CCSDS standards are available free of charge at https://public.ccsds.org/Publications/default. aspx). The CCSDS agglutinates the most important space agencies in the world, as well as other important space industry actors, with the goal of creating state-of-the-art standards for space communications. The CCSDS 122 standard follows closely the structure shown in Figure 1. Both the IWT and the POT are available in the spectral preprocessing stage, as well as a no-transform option that allows combination with other transforms including the RWA. In the spatial decorrelation stage, CCSDS 122.1 employs a 5-level 2D 5/3 IWT, which is then followed by a bitplane encoder. Note that the advantages of bitplane encoding are most evident for lossy compression, although this coding regime is not explored in this study.

FAPEC [25] contains a highly efficient entropy coder that works in a block-wise manner. In each block, the probability density function of its samples is computed, and used to determine which of the available coding modes to use. High-entropy blocks are output without further processing to reduce computational complexity and avoid data expansion. Samples in medium-entropy blocks are encoded using variable-length codes. Several code tables are available depending on the statistics of each block. Low-entropy blocks, where most samples are in $\{-1, 0, +1\}$, are divided in groups of six samples and compressed using fixed-length codes. Finally, runs of 5 or more zeros are encoded using run-length coding. The block-wise code selection of FAPEC is highly robust to outliers and other noise sources in the data, while being suitable for the complexity restrictions of space hardware. This entropy coder is applied to the samples after decorrelation. When no spectral transform is selected, decorrelation consists in 3 levels of 2D 9/7-M IWT. When the POT or the RWA are selected, they are applied before the same 2D IWT. For the IASI instrument, a unit-delay predictor is used instead of the 2D 9/7-M IWT, due to the better performance observed empirically. Note that, in its current version, FAPEC does not directly support the application of only spatial 2D IWT to scenes with more than one band. Therefore, to apply the 2D IWT in the aforementioned cases, scenes are compressed as a single band of width w and height $h \cdot z$, where w, h and z are the width, height and number of bands of the original scene, respectively. This reorganization is performed after the spectral transform, when one is applied. Finally, when the IWT is selected, two different approaches are used based on the number of spectral bands. If the scene contains 22 bands or fewer, three spatial 2D 9/7-M

IWT decomposition levels are applied on each image band, and then the unit-delay differentiator is applied along the spectral axis. If the scene contains 23 or more bands, the unit-delay differentiator is substituted by three levels of 9/7-M IWT, also along the spectral axis. Hence, regardless of the number of bands, FAPEC does not exactly follow the diagram shown in Figure 1 when the IWT is selected. Details and justification of this design decision are provided in [25,35]. Note that this approach is in contrast to CCSDS 122.1 with the IWT, which always applies spectral decorrelation before the spatial 2D IWT. In all cases, FAPEC is invoked with parameters -chunk 128M -bl 128 -mt 1 -ow -noatt -be, adding -dwt when the spectral IWT is selected. This sets the block size to 128 samples, and forces a single thread of execution. In addition, parameter -signed is used when appropriate.

The JPEG-LS [27] is a well-known compression standard that offers lossless compression, not to be confused with JPEG. JPEG-LS uses a simple predictor to spatially decorrelate input samples. This predictor uses only three of the nearest neighboring pixels, namely, those at the North, West, and Northwest positions with respect to the pixel being coded. Entropy coding is performed using Golomb-Rice codes, in combination with context modeling to further improve compression performance. Context modeling uses one additional neighbor, situated at the Northeast position relative to the current pixel. An additional run-length coding mode is also available for low entropy distributions. In this work, the reference implementation (Available at https://github.com/thorfdbg/libjpeg) is employed, using parameters -ls 0 -m 0. As JPEG-LS does not support compression of more than three spectral bands, scenes are compressed as a single band of with w and height $h \cdot z$, as described above for CCSDS 123.0-B-2.

The CCSDS 123.0-B-2 [28] standard is the latest compression algorithm published by the CCSDS. Unlike FAPEC, CCSDS 122.1, and JPEG-LS, CCSDS 123.0-B-2 combines spectral and spatial decorrelation in a single prediction stage (Note that FAPEC provides a prediction-based multiband algorithm that combines spatial and spectral decorrelation, as described in [25]. Notwithstanding, this paper focuses on transform-based compression, so this alternative mode is not considered). This prediction stage uses so-called local sums and local differences, computed from neighboring pixels from the same and previously coded bands. These are combined to obtain a prediction of the current pixel based on a weighted average. Weights employed for prediction are automatically updated based on the encoded values. CCSDS 123.0-B-2 allows three different entropy coders to compress the prediction errors of the previous stage: block-adaptive, sample-adaptive, and hybrid. The hybrid is the latest entropy coder published by the CCSDS, and is designed to yield competitive performance for a wide range of data entropies. This coder updates two variables—the accumulator and the counter—whenever a sample is coded. These variables provide an adaptive estimation of the data source's entropy. This estimation is used to determine what coding mode is to be used for the current sample. In addition to Golomb power-of-two coding, 16 variable-to-variable codes are defined for low-entropy scenarios. In this work, the default parameters described in [36] are used for CCSDS 123.0-B-2.

M-CALIC [29] is the last compressor considered in this work. Spectral decorrelation is performed by linear prediction using co-located samples in previous bands. Prediction errors are then compressed using an adaptive arithmetic coder with 256 defined contexts. These contexts implicitly perform spatial decorrelation by using neighboring samples at the same spectral band to determine the coding context to be used.

2.3. Test Corpus

In order to obtain a representative corpus of test scenes, 47 scenes have been selected from those available in the "CCSDS MHDC" image corpus, which has been assembled over the years by the *Multispectral Hyperspectral Data Compression Working Group* (MHDC WG) of the CCSDS to provide a suitable testing framework for the compression of satellite images (See https://cwe.ccsds.org/sls/docs/SLS-DC/123.0-B-Info/TestData/README.txt for futher details.). These images were obtained by 14 different instruments deployed in real missions. Both hyperspectral and multispectral scenes are used so that results are illustrative of currently deployed sensors. The corpus includes mainly raw (unprocessed) scenes and some calibrated scenes. Dynamic ranges vary from 8 to 15 bits per sample, although most scenes employ at least 12 bits per sample. Note that all scenes are stored in files using 2 bytes per sample, regardless of the actual dynamic range. A summary of the selected scenes and their properties is provided in Table 1. For illustrative purposes, crops of two sample scene bands are shown in Figure 2. Moreover, the zero-order entropies of each spectral band of three different scenes are plotted in Figure 3. All scenes are publicly available (Download links for the test scenes can be found at at http://cwe.ccsds.org/sls/docs/sls-dc/123.0-B-Info/TestData), except for those produced by the IASI and MSG instruments due to licensing restrictions.

(a)   (b)

Figure 2. Crops (512 × 512) of two scenes from the test corpus. (**a**) Band 0 of the *vgt_1b* scene from the Vegetation sensor; (**b**) band 5 of the *agriculture* scene from the Landsat sensor (over Northern Mexico). Scene brightness has been adjusted for display purposes.

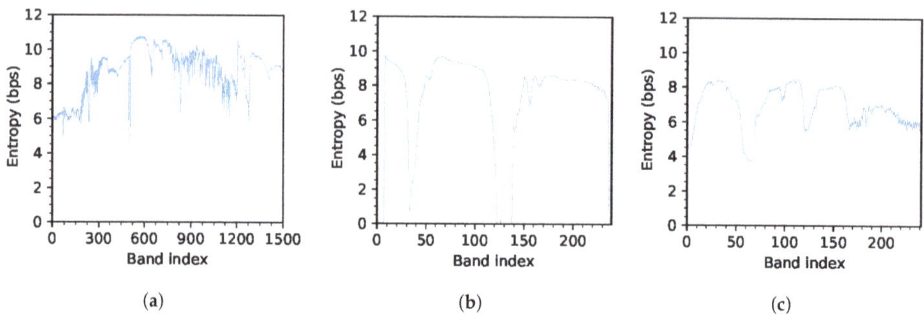

(a)   (b)   (c)

Figure 3. Zero-order entropies in bits per sample for each band. (**a**) *gran9* scene, AIRS instrument (Pacific Ocean, Daytime); (**b**) *mantar_Rad_rmnoise* scene, SFSI instrument (human-made targets); (**c**) *Geo_Sample* scene, Hyperion instrument (over Goldfield/Cuprite, Nevada).

Table 1. Summary of employed test corpus scenes and some of their properties.

Instrument	Scene Type	Dynamic range (bits)	#Bands	Width	Height	#Scenes
AIRS	raw	12	1501	90	135	1
AVIRIS	raw	15	224	680	512	1
	raw	10	224	614	512	1
	calibrated	13	224	677	512	1
CASI	raw	12, 13, 15	72	406	1225	3
CRISM	raw	11	107	640	510	2
	raw	12, 13	438	640	510	2
	raw	12, 13	545	640	510	2
	raw	13	545	320	450	2
	calibrated	11	74	64	2700	2
Hyperion	raw	12	242	256	1024	3
IASI L1C	calibrated	15	8461	60	1530	1
Landsat	raw	8	6	1024	1024	3
M3	raw	12	260	640	512	2
	raw	11, 12	86	320	512	2
MODIS	raw	12	17	1354	2030	2
	raw	12, 13	14	1354	2030	2
	raw	12, 13	5	2708	4060	2
	raw	12	2	5416	8120	2
MSG	calibrated	10	11	3712	3712	1
PLEIADES	calibrated	12	4	224	2465	1
	calibrated	12	4	224	2448	3
SFSI	calibrated	15	240	452	140	1
	raw	9, 11	240	496	140	2
SPOT5	calibrated	8	3	1024	1024	1
Vegetation	raw	10	4	1728	10,080	2

3. Experimental Results

Experimental results for the aforementioned spectral decorrelation methods and compression algorithms are presented in this section. The POT implementation of CCSDS 122.1 is also applied to FAPEC and JPEG-LS. The IWT implementation of CCSDS 122.1 is also applied to JPEG-LS, while FAPEC uses its own. All implementations are in C/C++, except for the RWA transform, which is available as Matlab code (The RWA can be downloaded from http://gici.uab.cat/GiciWebPage/downloads.php). All time measurements are obtained as the average over 10 repetitions on a dedicated Intel(R) Core(TM) i7-8700 CPU @ 3.20 GHz machine, using a single thread for the experiments. The user and system CPU times are considered here (as opposed to real time) to remove measurement dependencies on the data storage hardware.

3.1. Spectral Transforms

Two important properties of the spectral decorrelation transform described in Section 2.1 are analyzed next: transformed entropy and dynamic range expansion.

Entropy rates before and after spectral decorrelation provide valuable information about how much redundancy is removed. Entropy is defined as $-\sum_i p_i \cdot \log_2(p_i)$, where p_i is the probability of the i-th sample value. This probability is computed empirically based on each symbol's observed frequency. The distribution of zero-order entropy for the original and transformed scenes are shown in Figure 4a. The vertical axis of the figure indicates the fraction of the images whose entropy falls within the range indicated by the width of each vertical bar. Note that a relative frequency of 1 is equivalent to 100% of the images. No attempt is made to measure spatial redundancy present within the scene bands. Note that some of the analyzed spectral transforms produce side information, namely, the POT and the RWA. For these, the size of the side information is added to the zero-order entropy. Results are expressed in bits per sample (bps), where each pixel of each spectral band is considered a different sample. As expected, all spectral transforms reduce entropy rates. After the IWT, average entropy for all 47 scenes is 1.1 bps lower with respect to the average entropy of the original scenes. In turn, the POT and the RWA transform reduce entropy rates by 2.3 bps and 2.6 bps, respectively.

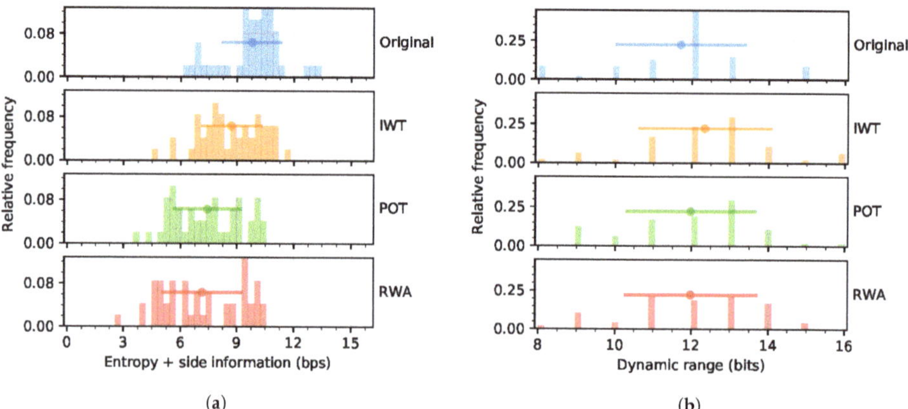

Figure 4. Property histograms for original and spectrally decorrelated scenes. (**a**) Entropy including side information; (**b**) dynamic range. Solid dots represent average values and horizontal lines span two standard deviations.

The dynamic range of an scene is hereinafter defined as the maximum number of bits required to store any of its samples. Transformed scenes typically exhibit larger dynamic ranges. This expansion is of interest because larger samples may slow down compression, e.g., when a bitplane encoder is employed or when additional bytes are needed to represent each sample. The distribution of dynamic ranges for the original and transformed scenes are shown in Figure 4b. The vertical axis indicates relative frequency as defined for Figure 4a above, considering dynamic range instead of entropy. All transforms introduce relatively small dynamic range expansions for the test corpus. The IWT introduces the largest expansion of all tested transforms, i.e., 0.60 bps. The POT and RWA transform both increase it by 0.23 bps. It is worth noting that all transformed scenes have dynamic ranges of 16 bits or less. Thus, compressors applied after these transforms need not support sample sizes beyond 2 bytes.

3.2. Lossless Compression Rates

Compressed data rates for the spectral transforms and compressors described in Section 2 have been obtained for all scenes of the test corpus. Table 2 shows average results for each instrument expressed in bps. Average results for all scenes are also shown for ease of comparison. Note that these averages are obtained as the arithmetic mean of the bps for each scene, and this mean is not weighted based on the number of samples of each scene. Consistent with the previous section, bits per sample are obtained assuming that each pixel of each spectral band is a different sample. For ease of comparison, Figure 5 displays the compression efficiency of all tested compressors, relative to that of CCSDS 123.0-B-2, averaged for each instrument. This relative efficiency is defined as the compressed data size of CCSDS 123.0-B-2 divided by the compressed data size of each algorithm, including side information when applicable.

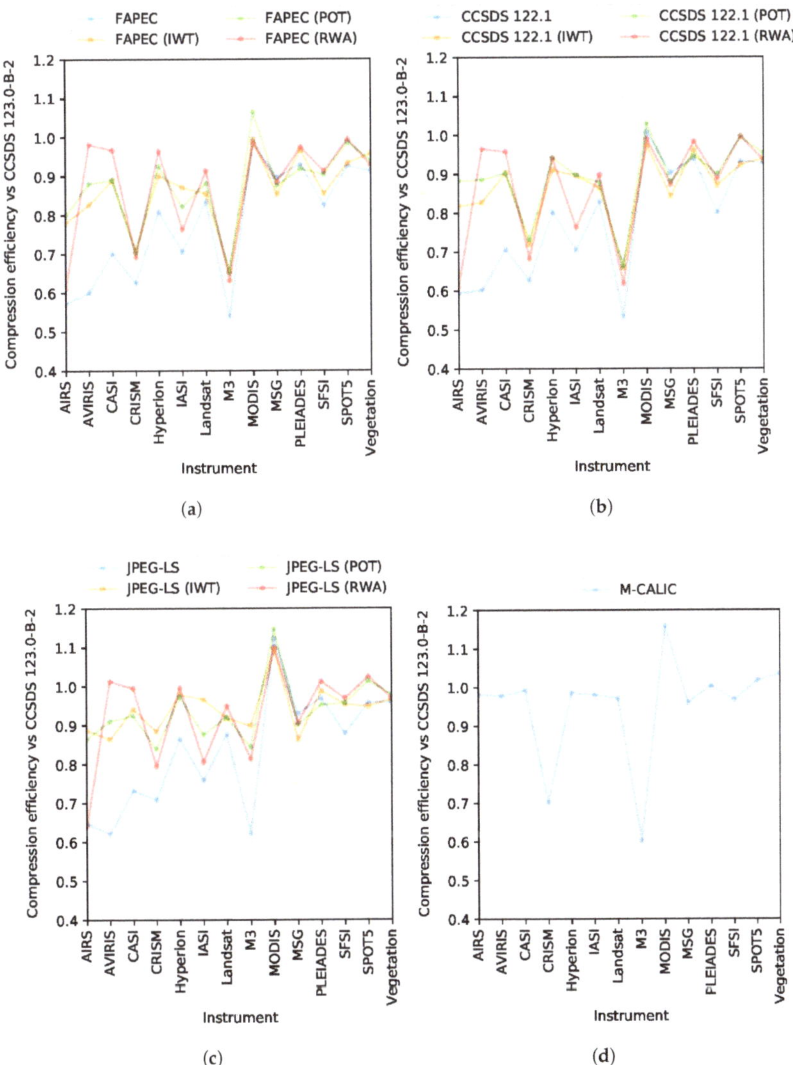

Figure 5. Average compression efficiency relative to CCSDS 123.0-B-2 for each instrument. Values larger than 1 indicate better performance than CCSDS 123.0-B-2. (**a**) FAPEC compressors; (**b**) CCSDS 122.1 compressors; (**c**) JPEG-LS compressors; (**d**) M-CALIC compressor.

The lowest average compressed data rates, i.e., those with smallest compressed files, are obtained with the CCSDS 123.0-B-2 standard. This can be explained by the fact that CCSDS 123.0-B-2 uses the most complex decorrelation approach and entropy coding mechanism of all tested compressors. Both decorrelation and entropy coding are adaptive at the sample level, which improves compressed data rates. As discussed in the next section, this is also at the cost of slower execution times. When combined with the most advantageous spectral decorrelation strategy (when applicable), compressors sorted by compressed data rates averaged for all test images are CCSDS 123.0-B-2 (4.84 bps), JPEG-LS (5.01 bps, IWT), M-CALIC (5.27 bps), CCSDS 122.1 (5.46 bps, POT), and FAPEC

(5.54 bps, POT). Thus, on average, the best configuration of all codecs are within 16% of CCSDS 123.0-B-2.

Table 2. Average lossless compression rate in bits per sample. Column *No* indicates compression is applied to the original scenes. Results include side information overhead when applicable. The best average results per instrument are highlighted in bold face.

| | FAPEC | | | | CCSDS 122.1 | | | | JPEG-LS | | | | CCSDS 123.0-B-2 | M-CALIC |
Corpus	No	IWT	POT	RWA	No	IWT	POT	RWA	No	IWT	POT	RWA	No	No
AIRS	7.18	5.27	5.15	6.75	6.92	5.03	4.65	6.74	6.35	4.63	4.76	6.43	**4.11**	4.19
AVIRIS	7.12	5.20	4.88	4.40	7.08	5.18	4.84	4.46	6.89	4.98	4.73	**4.27**	4.36	4.43
CASI	8.53	6.76	6.75	6.24	8.46	6.66	6.67	6.29	8.17	6.40	6.51	6.07	**6.06**	6.10
CRISM	6.19	5.48	5.57	5.69	6.17	5.41	5.33	5.76	5.44	4.36	4.60	4.92	**3.89**	5.69
Hyperion	5.25	4.72	4.59	4.41	5.28	4.66	4.51	4.52	4.91	4.34	4.37	4.27	**4.25**	4.31
IASI	9.56	7.73	8.20	8.83	9.57	7.52	7.50	8.83	8.88	6.97	7.69	8.36	**6.74**	6.87
Landsat	4.02	3.92	3.80	3.66	4.04	3.88	3.81	3.73	3.83	3.66	3.64	3.53	**3.36**	3.47
M3	5.04	4.05	4.10	4.23	5.07	4.04	4.03	4.31	4.38	2.93	3.12	3.24	**2.65**	4.51
MODIS	6.77	6.71	6.25	6.77	6.67	6.83	6.46	6.73	6.03	6.06	**5.77**	6.04	6.66	5.90
MSG	3.93	4.13	4.03	3.98	3.90	4.18	4.00	4.04	3.79	4.08	3.91	3.89	**3.52**	3.67
PLEIADES	7.74	7.45	7.83	7.39	7.66	7.48	7.60	7.32	7.43	7.28	7.56	**7.11**	7.20	7.17
SFSI	4.88	4.70	4.43	4.39	5.02	4.59	4.45	4.52	4.56	4.21	4.21	4.17	**4.02**	4.13
SPOT5	5.70	5.65	5.35	5.31	5.66	5.72	5.30	5.30	5.51	5.56	5.21	**5.15**	5.28	5.19
Vegetation	5.60	5.33	5.44	5.52	5.52	5.44	5.36	5.48	5.33	5.28	5.22	5.27	5.11	**4.93**
All scenes	6.26	5.61	5.54	5.59	6.23	5.60	5.46	5.62	5.75	5.01	5.05	5.11	**4.84**	5.27

For some of the instruments considered separately, CCSDS 123.0-B-2 yields larger differences when compared to the most advantageous transform for each compressor. For example, rates 53% better than FAPEC and CCSDS 122.1 are observed for M3 scenes, 41% better than FAPEC and 37% better than CCSDS 122.1 for CRISM data, and 25% better than FAPEC for the AIRS instrument. On the other hand, some combinations outperform CCSDS 123.0-B-2 for other instruments, e.g., JPEG-LS with the POT is 13% better for MODIS and with the RWA transform is 2% better for AVIRIS.

Regarding the optimum transforms for each compressor, FAPEC typically performs best with the POT or the RWA transform. On average, the POT provides a compression gain of 0.72 bps as compared to not using any transform, 0.07 bps better than using the spectral IWT. In turn, the RWA transform yields average bitrates 0.05 bps higher than using the POT transform, although the RWA provides the best results for 8 of the 14 tested instruments. CCSDS 122.1 achieves the best rates when using POT for two-thirds of the instruments, with RWA leading on the rest. Finally, JPEG-LS benefits specially from RWA on half of the tested instruments, with the IWT on less than a third, and with the POT on just two instruments.

Consistent with the entropy reduction results shown in Section 3.1, all tested transforms typically improve average compressed data rates. There are only a few exceptions on specific instruments and compressors, such as MSG (where the use of a spectral transform always leads to worse results) or MODIS with several combinations (CCSDS 122.1 with IWT or RWA, and JPEG-LS with IWT or RWA). Moreover, PLEIADES and SPOT5 exhibit a few exceptions. Results suggest that instruments with a small number of bands or a high spatial resolution are prone to get worse lossless compression rates when prepending some of these spectral decorrelation transforms.

In general, the best transform for spectral decorrelation is compressor-dependent. This is due to the different assumptions made in the design of the different algorithms on the distribution of the samples. Furthermore, each transform spreads the noise present in the least significant bits of the original samples differently, e.g., due to sensor shot noise. Notwithstanding these differences, average results for all transforms are within 0.16 bps and 0.10 bps for CCSDS 122.1 and JPEG-LS, respectively. Note that results for CCSDS 123.0-B-2 and M-CALIC after spectral decorrelation are not presented, because both compressors are designed assuming pixel distributions typical for the original domain.

Table 3. Average compression time in seconds. Column *No* indicates compression is applied to the original scenes. Results include spectral transformation times when applicable. Best results for each instrument are highlighted in bold font.

	FAPEC				CCSDS 122.1				JPEG-LS				CCSDS 123.0-B-2	M-CALIC
Corpus	No	IWT	POT	RWA	No	IWT	POT	RWA	No	IWT	POT	RWA	No	No
AIRS	0.81	**0.25**	2.67	2.26	14.28	11.66	11.39	11.32	2.48	4.59	5.17	4.61	1.55	15.46
AVIRIS	2.44	**1.16**	7.20	6.71	55.32	45.17	46.56	42.93	9.97	11.43	13.76	13.08	7.32	61.33
CASI	1.08	**0.56**	3.48	2.88	30.31	25.77	27.60	25.87	4.97	5.69	7.00	6.30	3.53	30.55
CRISM	3.25	**1.30**	9.28	9.57	66.32	63.61	67.18	69.52	11.51	14.21	17.18	17.47	7.11	77.90
Hyperion	2.15	**0.87**	6.97	6.05	39.14	36.16	38.14	38.33	7.63	9.76	12.11	11.22	6.36	52.33
IASI	**9.87**	14.73	120.43	208.01	723.38	615.70	623.41	751.97	117.99	175.99	217.58	301.01	82.63	724.83
Landsat	0.16	**0.09**	0.49	0.38	3.43	3.34	3.57	3.38	0.71	0.87	1.03	0.91	0.65	4.39
M3	1.54	**0.71**	4.79	4.62	31.53	29.18	31.34	32.12	5.67	7.16	8.79	8.74	4.00	40.89
MODIS	1.82	**1.63**	4.48	4.00	42.26	42.50	42.85	42.53	6.99	8.58	9.66	9.16	5.69	39.55
MSG	4.85	**4.38**	12.78	11.74	81.68	86.69	91.35	87.04	16.38	21.41	24.84	23.96	15.33	114.45
PLEIADES	0.07	**0.05**	0.23	0.13	2.22	2.19	2.30	2.20	0.38	0.46	0.53	0.43	0.29	2.05
SFSI	0.44	**0.21**	1.53	1.37	9.94	8.99	9.66	9.48	1.79	2.30	2.87	2.69	1.62	13.18
SPOT5	0.07	**0.05**	0.22	0.16	2.12	2.17	2.16	2.10	0.39	0.48	0.53	0.47	0.32	2.04
Vegetation	2.23	**2.02**	5.48	4.46	47.40	46.51	49.74	47.38	8.66	10.58	11.95	10.98	7.06	47.86
All scenes	1.97	**1.30**	7.55	9.15	52.49	48.28	50.08	52.76	9.04	11.81	14.23	15.73	6.51	57.55

Table 4. Average decompression time in seconds. Column *No* indicates decompression is applied to the original scenes. Results include inverse spectral transformation times when applicable. Best results for each instrument are highlighted in bold font.

	FAPEC				CCSDS 122.1				JPEG-LS				CCSDS 123.0-B-2	M-CALIC
Corpus	No	IWT	POT	RWA	No	IWT	POT	RWA	No	IWT	POT	RWA	No	No
AIRS	0.74	**0.27**	2.38	1.14	10.05	8.07	7.58	7.19	2.43	4.34	4.73	3.40	1.86	13.62
AVIRIS	2.35	**1.11**	6.35	3.81	38.27	30.76	31.11	27.77	9.91	11.04	12.62	9.82	8.56	55.63
CASI	1.02	**0.52**	3.00	1.70	21.04	17.65	18.49	17.10	5.01	5.60	6.51	5.06	4.10	27.99
CRISM	3.18	**1.28**	8.11	5.01	44.05	42.32	43.79	44.78	11.30	13.76	15.59	12.68	8.51	71.52
Hyperion	2.20	**0.91**	6.22	3.45	26.63	24.96	25.62	24.95	7.30	9.39	10.89	8.15	7.49	46.73
IASI	**12.43**	18.24	106.99	176.51	522.67	445.48	429.82	568.51	118.92	168.05	200.70	265.14	95.24	682.36
Landsat	0.16	**0.09**	0.45	0.27	2.19	2.21	2.29	2.11	0.67	0.89	0.94	0.76	0.72	4.05
M3	1.56	**0.71**	4.20	2.54	19.34	18.12	19.13	19.48	5.41	6.93	7.96	6.40	4.81	36.87
MODIS	1.72	**1.53**	3.97	2.81	28.74	29.96	29.05	28.80	6.95	9.21	9.09	7.93	6.49	37.32
MSG	4.80	**4.26**	11.40	8.21	52.00	57.06	58.33	53.58	15.76	21.24	22.80	19.54	17.64	102.83
PLEIADES	0.06	**0.04**	0.19	0.10	1.52	1.55	1.58	1.51	0.38	0.48	0.51	0.40	0.31	1.92
SFSI	0.45	**0.22**	1.34	0.78	6.38	6.15	6.35	6.04	1.72	2.21	2.59	1.99	1.85	11.78
SPOT5	0.07	**0.04**	0.19	0.13	1.42	1.53	1.42	1.41	0.37	0.52	0.48	0.42	0.34	1.87
Vegetation	2.11	**1.96**	4.94	3.41	31.12	31.53	31.65	31.08	8.31	10.99	11.08	9.62	8.04	44.01
All scenes	1.98	**1.35**	6.67	6.51	35.87	33.23	33.30	35.93	8.93	11.58	13.06	12.82	7.59	53.11

3.3. Throughput Results

The throughput yielded by the compressors described in the previous section is analyzed next. Average compression times for each instrument and for all scenes is provided in Table 3. Execution time is measured as the total user and system time reported by the operating system (Ubuntu 18.04 LTS). This is to avoid dependencies with the employed storage medium, and to provide a better surrogate for energy consumption in the comparison. It should be highlighted that all tested compressors are software implementations. Some differences in the level of optimization of these implementations were observed, also between different compression modes of the same algorithm.

The fastest of all tested compressors is FAPEC. When executed without any spectral decorrelation transform, it is 3.30 times faster than the next fastest non-FAPEC compressor, i.e., CCSDS 123.0-B-2.

When the IWT is applied both spatially and spectrally with FAPEC, it is 5.0 times faster than CCSDS 123.0-B-2. The faster execution speed with the IWT, as compared to not using any spectral decorrelation, is due to the better decorrelation power of the IWT, and its low computational complexity. After applying the IWT, the entropy coder has to deal with a more favorable distribution of input symbols. This results in the emission of fewer bits per sample, and contributes to requiring shorter execution times. Thanks to the low complexity of the IWT, the execution time improvements due to its decorrelation power often outweigh the cost of applying this transform.

When compared to the fastest configuration of CCSDS 122.1 and JPEG-LS, FAPEC with the IWT is 37.1 times and 7.0 times faster, respectively. CCSDS 122.1 achieves average execution times significantly slower than FAPEC, JPEG-LS, and CCSDS 123.0-B-2. This is explained by the use of a bitplane encoder which imposes several passes on each sample when implemented on software. In turn, M-CALIC is the slowest of all tested compressors, 9.1% slower than the slowest configuration of CCSDS 122.1. These results can be explained by the use of an arithmetic coder in the entropy coding stage of M-CALIC. It is also interesting to analyze average compression time as a function of the employed spectral transform. As can be observed, compression using the IWT is faster than compression using the POT or the RWA transform. This is due to the lower complexity of the IWT. For the CCSDS 122.1 and JPEG-LS compressors, coding with the IWT takes respectively 3.7% and 17.0% less than with the fastest of POT and RWA. Significantly larger differences are observed for FAPEC and the IWT. As mentioned before, this can be explained by the use of a more efficient implementation of the IWT, more tightly integrated in the compression pipeline. For most instruments, the RWA attains higher throughput than the POT. Notwithstanding, the throughput of RWA for the IASI instrument is significantly lower due to the very high number of bands, which makes the throughput of POT better on average. For completeness, average decompression times—including inverse spectral transformation, when applicable—are presented in Table 4. Decompression times are generally lower than compression times. An exception to this is CCSDS 123.0-B-2, whose entropy decoder processes data in inverse order [28]. Another exception is FAPEC with the spectral IWT. This is due to an asymmetry between the optimization degree of the compressor (meant to be executed onboard spacecraft hardware) and the decompressor (meant to be executed on the ground).

Additional analysis is devoted to jointly compare compression performance and execution times. Figure 6a provides average compression speed as a function of the obtained compressed data rates for a selection of spectral transforms and compressors. For each compressor, the spectral decorrelation method that yields the best average compressed data rate is included. FAPEC with the IWT is also included for completeness. Figure 6b shows average decompression speed as a function of the obtained compressed data rates for the same selection as Figure 6a. For medium and low compression throughput, i.e., below 12×10^6 samples/s, CCSDS 123.0-B-2 dominates over JPEG-LS with spectral IWT, CCSDS 122.1 with spectral POT, and M-CALIC and FAPEC with the POT. It can also be observed that FAPEC with the spectral IWT yields significantly higher compression speeds, reaching 60×10^6 samples/s on average. As discussed above, the additional speed is traded for about 16% higher compressed data rates. Decompression throughput follows a pattern similar to compression, with the difference that most compressors achieve higher speeds. As execution speed is directly related to energy consumption, and assuming energy is a limited resource in the system where compression is to be performed, results suggest that FAPEC yields the most desirable trade-off when energy and compression performance are jointly considered.

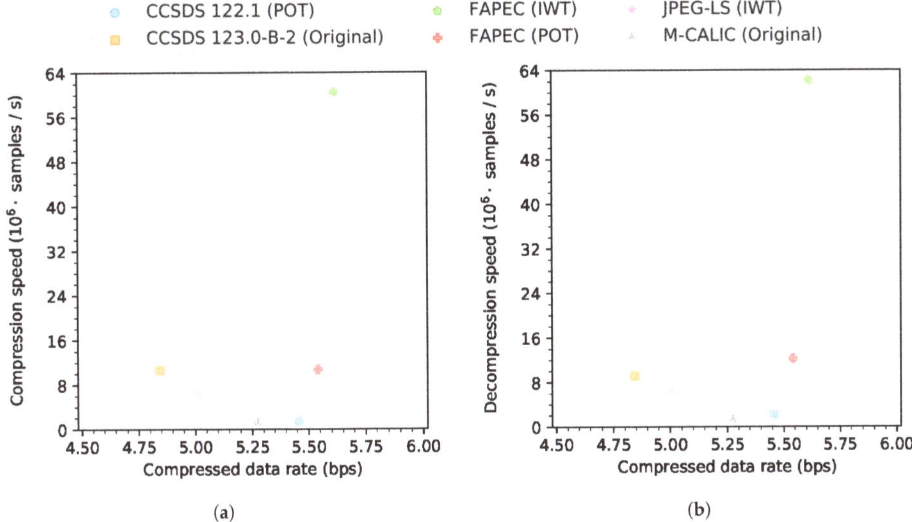

Figure 6. Average (**a**) compression and (**b**) decompression speed as a function of the compressed data rate for all scenes in the corpus. The spectral decorrelation transform applied to the data is shown between brackets, and *Original* indicates no transform. When applicable, speeds include the spectral decorrelation stage times, and compressed data rates include side information overhead.

4. Conclusions

Hyperspectral imaging (HSI) is a staple tool of remote sensing and Earth observation. Data compression can significantly increase the amount of valuable industrial and scientific data retrieved on the ground. In particular, scene compression based on spectral decorrelation transforms enables important bitrate reductions at a reasonable computational complexity. This work presents a thorough analysis of the IWT, the POT, and the RWA transform as spectral decorrelators, in combination with FAPEC, CCSDS 122.1, and JPEG-LS coding techniques. Results for CCSDS 123.0-B-2 and M-CALIC are also shown for completeness. Compression rates and speeds on a large variety of scenes have been analyzed here, considering only lossless regimes. This analysis was based on software implementations, which enable more flexibility than their hardware equivalents (when available), especially on small satellites and cubesats. To the best of our knowledge, such a thorough comparison is not previously available in the literature. Experiments suggest that the use of a spectral decorrelation transform improves compression rates for most HSI instruments. The most advantageous transform depends on each kind of instrument and on the compressor to be used. Both the POT and the RWA transform can provide significantly better compression rates than IWT in specific cases, yet at a higher computing cost. Results indicate that the best trade-off is attained by the IWT followed by FAPEC. On average, this combination produces compressed data rates within 16% of CCSDS 123.0-B-2, which typically yields the smallest compressed data volumes. At the same time, FAPEC with its IWT is 5.0 times faster than CCSDS 123.0-B-2.

Author Contributions: Conceptualization, J.S.-S., J.P., M.H.-C., and I.B.; methodology, J.P., J.S.-S., M.H.-C., and I.B.; software, J.P., M.H.-C., and I.B.; validation, J.P. and J.S.-S.; formal analysis, M.H.C.; investigation, J.P., J.S.-S., and M.H.-C.; resources, I.B.; data curation, M.H.-C. and I.B.; writing—original draft preparation, M.H.C.; writing—review and editing, J.P., J.S.-S., M.H.-C., and I.B.; visualization, M.H.-C.; supervision, J.S.-S. and J.P.; project administration, J.S.-S. and J.P.; funding acquisition, J.P., J.S.-S., M.H.-C., and I.B. All authors have read and agreed to the published version of the manuscript.

Remote Sens. **2020**, *12*, 2955

Funding: This research was funded by the Spanish Ministry of Economy and Competitiveness and the European Regional Development Fund under grants RTI2018-095287-B-I00, TIN2015-71126-R, RTI2018-095076-B-C21 (MINECO/FEDER, UE); by the ESA Business Incubation Programme and Barcelona Activa; by the Catalan Government under grant 2017SGR-463; by the postdoctoral fellowship programme Beatriu de Pinós, reference 2018-BP-00008, funded by the Secretary of Universities and Research (Government of Catalonia); and by the Horizon 2020 programme of research and innovation of the European Union under the Marie Skłodowska-Curie grant agreement #801370.

Conflicts of Interest: The authors declare the following conflicts of interest. J.P. is CTO of DAPCOM, original developer of the FAPEC algorithm. M.H-C., I.B., and J.S. have participated in the standardization of several CCSDS standards, including CCSDS 122.1-B-1 and CCSDS 123.0-B-2. The funders had no role in the design of the study; in the collection, analyses, or interpretation of data; in the writing of the manuscript; or in the decision to publish the results.

Abbreviations

The following abbreviations are used in this manuscript.

bps	Bits per sample
CALIC	Context-based, Adaptive, Lossless Image Coder
CCSDS	Consultative Committee for Space Data Systems
DWT	Discrete Wavelet Transform
FAPEC	Fully Adaptive Prediction Error Coder
HSI	HyperSpectral Imagery
IWT	Integer Wavelet Transform
POT	Pair-Orthogonal Transform
RWA	Regression Wavelet Analysis

References

1. Parente, M.; Kerekes, J.; Heylen, R. A Special Issue on Hyperspectral Imaging [From the Guest Editors]. *IEEE Geosci. Remote Sens. Mag.* **2019**, *7*, 6–7. [CrossRef]
2. Malyy, M.; Tekic, Z.; Golkar, A. What Drives Technology Innovation in New Space? A Preliminary Analysis of Venture Capital Investments in Earth Observation Start-Ups. *IEEE Geosci. Remote Sens. Mag.* **2019**, *7*, 59–73.
3. Denis, G.; Claverie, A.; Pasco, X.; Darnis, J.P.; de Maupeou, B.; Lafaye, M.; Morel, E. Towards disruptions in Earth observation? New Earth Observation systems and markets evolution: Possible scenarios and impacts. *Acta Astronaut.* **2017**, *137*, 415–433.
4. Sun, W.; Du, Q. Hyperspectral Band Selection: A Review. *IEEE Geosci. Remote Sens. Mag.* **2019**, *7*, 118–139. [CrossRef]
5. Li, S.; Song, W.; Fang, L.; Chen, Y.; Ghamisi, P.; Benediktsson, J.A. Deep Learning for Hyperspectral Image Classification: An Overview. *IEEE Trans. Geosci. Remote Sens.* **2019**, *57*, 6690–6709. [CrossRef]
6. Duan, P.; Kang, X.; Li, S.; Ghamisi, P.; Benediktsson, J.A. Fusion of Multiple Edge-Preserving Operations for Hyperspectral Image Classification. *IEEE Trans. Geosci. Remote Sens.* **2019**, *57*, 10336–10349. [CrossRef]
7. Su, Y.; Li, J.; Plaza, A.; Marinoni, A.; Gamba, P.; Chakravortty, S. DAEN: Deep Autoencoder Networks for Hyperspectral Unmixing. *IEEE Trans. Geosci. Remote Sens.* **2019**, *57*, 4309–4321. [CrossRef]
8. Chen, Y.; Zhu, K.; Zhu, L.; He, X.; Ghamisi, P.; Benediktsson, J.A. Automatic Design of Convolutional Neural Network for Hyperspectral Image Classification. *IEEE Trans. Geosci. Remote Sens.* **2019**, *57*, 7048–7066. [CrossRef]
9. Haut, J.M.; Gallardo, J.A.; Paoletti, M.E.; Cavallaro, G.; Plaza, J.; Plaza, A.; Riedel, M. Cloud Deep Networks for Hyperspectral Image Analysis. *IEEE Trans. Geosci. Remote Sens.* **2019**, *57*, 9832–9848. [CrossRef]
10. Tu, B.; Zhang, X.; Kang, X.; Wang, J.; Benediktsson, J.A. Spatial Density Peak Clustering for Hyperspectral Image Classification With Noisy Labels. *IEEE Trans. Geosci. Remote Sens.* **2019**, *57*, 5085–5097. [CrossRef]
11. Bhardwaj, K.; Patra, S.; Bruzzone, L. Threshold-Free Attribute Profile for Classification of Hyperspectral Images. *IEEE Trans. Geosci. Remote Sens.* **2019**, *57*, 7731–7742. [CrossRef]
12. Lu, X.; Dong, L.; Yuan, Y. Subspace Clustering Constrained Sparse NMF for Hyperspectral Unmixing. *IEEE Trans. Geosci. Remote Sens.* **2020**, *58*, 3007–3019. [CrossRef]

13. Della Porta, C.J.; Bekit, A.A.; Lampe, B.H.; Chang, C. Hyperspectral Image Classification via Compressive Sensing. *IEEE Trans. Geosci. Remote Sens.* **2019**, *57*, 8290–8303. [CrossRef]

14. Nalepa, J.; Myller, M.; Kawulok, M. Validating Hyperspectral Image Segmentation. *IEEE Geosci. Remote Sens. Lett.* **2019**, *16*, 1264–1268. [CrossRef]

15. Hong, D.; Wu, X.; Ghamisi, P.; Chanussot, J.; Yokoya, N.; Zhu, X.X. Invariant Attribute Profiles: A Spatial-Frequency Joint Feature Extractor for Hyperspectral Image Classification. *IEEE Trans. Geosci. Remote Sens.* **2020**, *58*, 3791–3808. [CrossRef]

16. Theiler, J.; Ziemann, A.; Matteoli, S.; Diani, M. Spectral Variability of Remotely Sensed Target Materials: Causes, Models, and Strategies for Mitigation and Robust Exploitation. *IEEE Geosci. Remote Sens. Mag.* **2019**, *7*, 8–30. [CrossRef]

17. Zhong, Y.; Wang, X.; Xu, Y.; Wang, S.; Jia, T.; Hu, X.; Zhao, J.; Wei, L.; Zhang, L. Mini-UAV-Borne Hyperspectral Remote Sensing: From Observation and Processing to Applications. *IEEE Geosci. Remote Sens. Mag.* **2018**, *6*, 46–62. [CrossRef]

18. Khan, M.J.; Khan, H.S.; Yousaf, A.; Khurshid, K.; Abbas, A. Modern Trends in Hyperspectral Image Analysis: A Review. *IEEE Access* **2018**, *6*, 14118–14129. [CrossRef]

19. Qian, S.E. *Optical Satellite Data Compression and Implementation*; SPIE: Bellingham, WA, USA, 2013.

20. Turpie, K.; Veraverbeke, S.; Wright, R.; Anderson, M.; Quattrochi, D. *NASA 2014 The Hyperspectral Infrared Imager (HyspIRI)—Science Impact of Deploying Instruments on Separate Platforms*; Techreport JPL-Publ-14-13; Jet Propulsion Lab: Pasadena, CA, USA, 2014.

21. Qian, S.E. *Optical Satellite Signal Processing and Enhancement*; SPIE: Bellingham, WA, USA, 2013.

22. Calderbank, A.R.; Daubechies, I.; Sweldens, W.; Yeo, B.-L. Lossless image compression using integer to integer wavelet transforms. In Proceedings of the International Conference on Image Processing, Santa Barbara, CA, USA, 26–29 October 1997; Volume 1, pp. 596–599.

23. Blanes, I.; Serra-Sagristà, J. Pairwise Orthogonal Transform for Spectral Image Coding. *IEEE Trans. Geosci. Remote Sens.* **2011**, *49*, 961–972. [CrossRef]

24. Amrani, N.; Serra-Sagristà, J.; Laparra, V.; Marcellin, M.W.; Malo, J. Regression Wavelet Analysis for Lossless Coding of Remote-Sensing Data. *IEEE Trans. Geosci. Remote Sens.* **2016**, *54*, 5616–5627. [CrossRef]

25. Portell, J.; Iudica, R.; García-Berro, E.; Villafranca, A.; Artigues, G. FAPEC, a versatile and efficient data compressor for space missions. *Int. J. Remote Sens.* **2018**, *39*, 2022–2042. [CrossRef]

26. Consultative Committee for Space Data Systems (CCSDS). Spectral Preprocessing Transform for Multispectral and Hyperspectral Image Compression. In *Blue Book*; Issue 1, Number CCSDS 122.1-B-1; CCSDS: Washington, DC, USA, 2017.

27. Weinberger, M.J.; Seroussi, G.; Sapiro, G. The LOCO-I lossless image compression algorithm: Principles and standardization into JPEG-LS. *IEEE Trans. Image Process.* **2000**, *9*, 1309–1324. [CrossRef] [PubMed]

28. Consultative Committee for Space Data Systems (CCSDS). Low-Complexity Lossless and Near-Lossless Multispectral and Hyperspectral Image Compression. In *Blue Book*; Issue 2, Number CCSDS 123.0-B-2; CCSDS: Washington, DC, USA, 2019.

29. Magli, E.; Olmo, G.; Quacchio, E. Optimized onboard lossless and near-lossless compression of hyperspectral data using CALIC. *IEEE Geosci. Remote Sens. Lett.* **2004**, *1*, 21–25. [CrossRef]

30. Portell, J.; Blanes, I.; Hernández-Cabronero, M.; Serra-Sagristà, J.; Iudica, R.; Villafranca, A.G. Prepending spectral decorrelating transforms to FAPEC: A competitive high-performance approach for remote sensing data compression. In Proceedings of the On-Board Payload Data Compression (OBPDC), Matera, Italy, 20–21 September 2018.

31. Tsigkanos, A.; Kranitis, N.; Theodorou, G.A.; Paschalis, A. A 3.3 Gbps CCSDS 123.0-B-1 Multispectral Hyperspectral Image Compression Hardware Accelerator on a Space-Grade SRAM FPGA. *IEEE Trans. Emerg. Top. Comput.* **2018**, *1*. [CrossRef]

32. Keymeulen, D.; Shin, S.; Riddley, J.; Klimesh, M.; Kiely, A.; Liggett, E.; Sullivan, P.; Bernas, M.; Ghossemi, H.; Flesch, G.; et al. High Performance Space Computing with System-on-Chip Instrument Avionics for Space-based Next Generation Imaging Spectrometers (NGIS). In Proceedings of the 2018 NASA/ESA Conference on Adaptive Hardware and Systems (AHS), Edinburgh, UK, 6–9 August 2018; pp. 33–36.

33. Santos, L.; Gomez, A.; Sarmiento, R. Implementation of CCSDS Standards for Lossless Multispectral and Hyperspectral Satellite Image Compression. *IEEE Trans. Aerosp. Electron. Syst.* **2019**, *56*, 1120–1138. [CrossRef]

34. Daubechies, I.; Sweldens, W. Factoring wavelet transforms into lifting steps. *J. Fourier Anal. Appl.* **1998**, *4*, 247–269. [CrossRef]

35. Artigues, G.; Portell, J.; Villafranca, A.G.; Ahmadloo, H.; García-Berro, E. Discrete wavelet transform fully adaptive prediction error coder: Image data compression based on CCSDS 122.0 and fully adaptive prediction error coder. *J. Appl. Remote Sens.* **2013**, *7*, 074592.

36. Blanes, I.; Kiely, A.; Hernández-Cabronero, M.; Serra-Sagristà, J. Performance Impact of Parameter Tuning on the CCSDS-123.0-B-2 Low-Complexity Lossless and Near-Lossless Multispectral and Hyperspectral Image Compression Standard. *Remote Sens.* **2019**, *11*, 1390. [CrossRef]

MDPI
St. Alban-Anlage 66
4052 Basel
Switzerland
Tel. +41 61 683 77 34
Fax +41 61 302 89 18
www.mdpi.com

Remote Sensing Editorial Office
E-mail: remotesensing@mdpi.com
www.mdpi.com/journal/remotesensing